Organic Chemistry

With a focus on organic chemistry students at all levels, problems are incorporated into the body of the text in an effort to engage students more directly in chemistry. Arrowless mechanisms seek to help students develop their electron-pushing skills and intuition through repeated practice. By design this volume is more actively engaging than a traditional textbook. In addition, the historical development of ideas is presented to help frame and centre these concepts for the reader. Primary and summative sources are given for all topics covered. The sources provide definitive information for the reader and ensure that all information is supported by peer-reviewed, experimental sources.

Features:

■ The development of key ideas is presented in their historical context.

■ All information presented is supported through citations to chemical literature.

■ Problems are incorporated into the body of the text, including arrowless mechanisms which encourage students to engage more actively and to develop their electron-pushing skills and intuition.

■ International Union of Pure and Applied Chemistry style and technical guidelines are followed throughout the text.

■ The problems, text, and presentation are based on years of classroom refinement of teaching pedagogy.

Dr William Tucker – his passion for chemistry was inspired by his high school teacher, Gary Osborn. He left Maine to pursue Chemistry at Middlebury College, and after graduating in 2010 he decided to pursue a PhD in organic chemistry at the University of Wisconsin-Madison. At the University of Wisconsin-Madison, he worked in the laboratory of Dr Sandro Mecozzi, where he developed semifluorinated triphilic surfactants for hydrophobic drug delivery. After earning his PhD in 2015, he took a fellowship at Boston University as a Postdoctoral Faculty Fellow. There he co-taught organic chemistry while working in the laboratory of Dr John Caradonna. In the Caradonna laboratory, he worked on developing a surface-immobilized iron-oxidation catalyst for the oxidation of C–H bonds using dioxygen from the air as the terminal oxidant. Throughout all of this work, his passion has always been for teaching and working with students both in and out of the classroom. He has been lucky for the past six years to work at Concord Academy, where his students have, through their questions, pushed him to think deeper and more critically about chemistry. Their curiosity inspires him, and their inquisitiveness inspired his writing.

Organic Chemistry

Structure, Function, and Practice

William B. Tucker

CRC Press
Taylor & Francis Group
Boca Raton London New York

CRC Press is an imprint of the
Taylor & Francis Group, an **informa** business

Open-access funding provided by the Concord Academy Class of 1972 Green Seed Fund.

First edition published 2024
by CRC Press
2385 NW Executive Center Drive, Suite 320, Boca Raton FL 33431

and by CRC Press
4 Park Square, Milton Park, Abingdon, Oxon, OX14 4RN

CRC Press is an imprint of Taylor & Francis Group, LLC

Library of Congress Cataloging–in–Publication Data
Names: Tucker, William (Chemist), author.
Title: Organic chemistry: structure, function, and practice / William B. Tucker.
Description: First edition. | Boca Raton, FL: CRC Press, 2024. | Includes bibliographical references and index. |
Identifiers: LCCN 2023057349 (print) |
LCCN 2023057350 (ebook) | ISBN 9781032766362 (hbk) | ISBN 9781032766058 (pbk) |
ISBN 9781003479352 (ebk)
Subjects: LCSH: Chemistry, Organic.
Classification: LCC QD251.3 .T83 2024 (print) | LCC QD251.3 (ebook) |
DDC 547–dc23/eng/20240326
LC record available at https://lccn.loc.gov/2023057349
LC ebook record available at https://lccn.loc.gov/2023057350

ISBN: 9781032766362 (hbk)
ISBN: 9781032766058 (pbk)
ISBN: 9781003479352 (ebk)

DOI: 10.1201/9781003479352

Typeset in Palatino
by Deanta Global Publishing Services, Chennai, India

Access the Support Material: www.routledge.com/9781032766362

This book is dedicated to
Gary Osborn
Beloved Winslow High School (Winslow, ME) chemistry teacher
and inspiration of my love of and interest in chemistry.

Contents

This book was the creation of a personal project to combine years of teaching and learning about organic chemistry into a single source. To that end, I endeavoured to accomplish several things to make this book the most useful for the broadest population. First, this book is written to be accessible to any student – in high school or in college – who has had a full year of introductory general chemistry. Second, this book is designed to be more engaging than a typical textbook with mechanisms presented without arrows and problems embedded in the text. Arrowless mechanisms encourage the reader to review the steps presented and to practice the skill of showing electron movement by adding in the appropriate arrows. The embedded problems come from discussion worksheets, homework problems, and exam problems that have been used by the author. Together, these active pieces of the book constitute about half of the book's total page count. Third, to provide the most definitive information for the reader, all the original references (or summative references where an original reference could not be found) for the chemistry and ideas presented are cited with footnotes. Fourth, the traditional organic chemistry content is supplemented with a discussion of diversity, equity, inclusion, and belonging (DEIB) and the incorporation of both computational chemistry and green chemistry. In addition, in situations where a reaction or mechanism is presented and no single set of conditions or single mechanism is presented in other textbooks, every effort has been made to find references which support the conditions presented or the mechanism presented here. Some mechanisms have been omitted entirely because of the mechanism's difficulty or when a single, referenceable mechanism was not available. These references are also meant to provide an interested reader with a point of entry into understanding the origins of these ideas, though I should caution that a working knowledge of German, French, and English is necessary to read all the references presented. Finally, organic chemistry is a broad field and a living one, as such someone looking for mastery should be prepared to utilize many other books and teachers than just this one book. I hope you find this supportive and helpful in your efforts to learn organic chemistry.

Will Tucker, PhD
Science Teacher
Concord Academy
Concord, MA

Attribution

All chemical structures were drawn using ChemDraw 22.2 (RRID:SCR_016768). Registered Trademark of PerkinElmer Informatics. https://perkinelmerinformatics.com/products/research/chemdraw (accessed 30 April 2023).

All computationally derived models and energy values were created and calculated using the platform WebMO: Schmidt, J.R.; Polik, W.F. *WebMO*, version 20.0; WebMO LLC: Holland, MI, US, 20; https://www.webmo.net (accessed 30 April 2023).

Calculations at the B3LYP/6-31G(d) level of theory and basis set were performed using Gaussian 09: Gaussian 09, Revision A.02, M. J. Frisch, G. W. Trucks, H. B. Schlegel, G. E. Scuseria, M. A. Robb, J. R. Cheeseman, G. Scalmani, V. Barone, G. A. Petersson, H. Nakatsuji, X. Li, M. Caricato, A. Marenich, J. Bloino, B. G. Janesko, R. Gomperts, B. Mennucci, H. P. Hratchian, J. V. Ortiz, A. F. Izmaylov, J. L. Sonnenberg, D. Williams-Young, F. Ding, F. Lipparini, F. Egidi, J. Goings, B. Peng, A. Petrone, T. Henderson, D. Ranasinghe, V. G. Zakrzewski, J. Gao, N. Rega, G. Zheng, W. Liang, M. Hada, M. Ehara, K. Toyota, R. Fukuda, J. Hasegawa, M. Ishida, T. Nakajima, Y. Honda, O. Kitao, H. Nakai, T. Vreven, K. Throssell, J. A. Montgomery, Jr., J. E. Peralta, F. Ogliaro, M. Bearpark, J. J. Heyd, E. Brothers, K. N. Kudin, V. N. Staroverov, T. Keith, R. Kobayashi, J. Normand, K. Raghavachari, A. Rendell, J. C. Burant, S. S. Iyengar, J. Tomasi, M. Cossi, J. M. Millam, M. Klene, C. Adamo, R. Cammi, J. W. Ochterski, R. L. Martin, K. Morokuma, O. Farkas, J. B. Foresman, and D. J. Fox, Gaussian, Inc., Wallingford CT, 2016.

Visualizations of orbitals and electrostatic potential maps were produced using NBO 7: E. D. Glendening, K. Badenhoop, A. E. Reed, J. E. Carpenter, J. A. Bohmann, C. M. Morales, P. Karafiloglou, C. R. Landis, and F. Weinhold, Theoretical Chemistry Institute, University of Wisconsin, Madison (2018).

Acknowledgements

I want to thank the following people for their thoughtful review of this work and their feedback and criticism that has helped to make this work better.

Grace Delgado, Editor
Jake Klineman, *Concord Academy Class of 2025*, Editor
Dr John Snyder, *Boston University Professor of Chemistry*
Max Hall, *Concord Academy Science Teacher*
Edoardo Takacs, *Concord Academy Class of 2023*
Ava Driggers, *Concord Academy Class of 2024*
Isabella Ginsburg, *Concord Academy Class of 2023*
Aidan Quealy, *Concord Academy Class of 2024*

1 Introduction

Chemistry is a branch of science that focuses on the study of matter and change. Organic chemistry focuses specifically on carbon-containing matter and the chemical reactions of carbon-containing compounds. This book has two aims. The first is to help readers develop a sense of the structure of carbon-containing compounds and the underlying physical chemistry properties that help us understand their structure. The second aim of this book is to help develop an understanding of the relationship between chemical structure and reactivity. The majority of the chapters in this book are case studies that focus on a specific structural motif, what chemists call a functional group, and how and why that functional group reacts the way it does.

An important point to address is why you, the reader, are reading this book and studying organic chemistry. It may be because it is a required course for your major or because you are animated by the intricacies of the subject. In either case, an important point to hold onto is that organic chemistry is not a set of facts to memorize nor a prescriptive programme that one should know for its own sake. Rather, organic chemistry provides one with the understanding of carbon-chemical architecture, tools to help one design and then build these edifices, and knowledge to understand the chemistry that occurs in cells. This work has given rise to molecules like dyes that make clothes, paints, and materials more vibrant; pharmaceuticals that treat disease and improve the quality of life; and materials that help make human life safer, more energy efficient, and more comfortable. Therefore, while the traditional definition is the study of carbon-containing compounds and the reactions of those compounds, a more compelling definition is "the science of the transformation of matter".[1]

INCLUSION

Organic chemistry, as a discipline, has an emphasis on names and history. Throughout this book, you will see the last names of the chemists who discovered a given reaction or principle, for example: a Walden inversion, the Diels-Alder reaction, and the Hammond-Leffler postulate. Further, to ensure that the information presented here is fully supported by the scientific literature, original and summative references are presented for the reactions, principles, and concepts presented. If you were to research the people named by organic chemistry and those cited in this text, you would be able to identify a stark trend: They are almost all white, they are almost all European or American, they are almost all men, almost all came from wealthy backgrounds, and all were, so far as the author knows, cisgender and heterosexual.

The artefacts of organic chemistry's homogeneous history, a human endeavour that was developed by only a small subsection of humanity, are seen through to the present day: racism and sexism in organic chemistry and science broadly, the underrepresentation of Black, Pacific Islander, and Latine chemists in degree recipients and in organic chemistry jobs (researchers, professors, and teachers), and the lower likelihood of non-white, nonmale applicants receiving funding from grant agencies. What created this homogeneous history of organic chemistry? Organic chemistry rose to significance in scientific laboratories in Europe during the Industrial Revolution, where it first provided a means of producing dyestuffs and later pharmaceuticals from coal tar. In the 18th, 19th, and early 20th centuries, only people who were wealthy, white, and male were afforded the privilege of pursuing science in the centres of learning and research in Europe and later America. And so, the system has historically been built for and by these people. With the rise of funding from the federal government, part of the Cold War, the study of organic chemistry started to become accessible to those who were from lower socioeconomic classes. With the work of countless activists in the Civil Rights, Women's Liberation, and LGBTQIA Liberation movements, the study of organic chemistry started to become accessible to those who were non-white, nonmale, non-cisgender, and non-heterosexual. With these changes, organic chemistry has become more diverse, but it is still harder for some people than it is for others to progress through the system. What is important as you read this text is to consider the folks – non-white, nonmale, non-cisgender, and non-heterosexual – who are not cited or recognized in chemical history. These people have always been a part of organic chemistry and they have done significant research in the field of organic chemistry. As organic chemistry slowly becomes more inclusive, it is incumbent upon us to both recognize those folks omitted by organic chemistry's named past and to help expand and make space for those folks that can and should be part of organic chemistry's present and future,[2] because to not do so is actively harmful.[3]

DOI: 10.1201/9781003479352-1

SUSTAINABILITY AND GREEN CHEMISTRY

Industrial organic chemistry has its origins in the use and manipulation of coal tar. And when the author took his first organic chemistry course, the description was, "this is the study of petroleum and everything we can do with it". While organic chemistry is still a materially intensive field that is heavily connected to the petroleum industry, it is important to highlight chemical sustainability and green chemistry. To think about sustainability in organic chemistry, let us first consider organic chemistry in its fullest context. Organic chemistry uses materials and energy, both commonly from fossil fuel feedstocks, to produce new materials, all of which ultimately needs to be disposed of and, ideally, recycled. This chemical synthesis and material consumption are in connection to the Earth's land, air, and water. In addition, chemical synthesis and material consumption are in service of human society, which is sustained by and is responsible for the stewardship of Earth. Sustainability is an emergent property of these synthesis-consumption, biosphere, and human systems. So, making organic chemistry sustainable should not be thought of as a single change to a single process, but rather sustainable organic chemistry is a systemic and holistic approach to the work and processes in organic chemistry, in society, and in our interactions with the planet.[4]

While sustainability is a systemic concern, green chemistry is a focus on making individual aspects of chemical processes more sustainable. A guiding framework for helping to analyze individual processes is the 12 Principles of Green Chemistry.[5] The 12 principles aim to reduce or eliminate the use or generation of hazardous substances and reduce the material cost of the process.

1. Prevention: Unwanted byproducts and waste should be minimized or, ideally, not produced.

2. Atom economy: In a reaction, what atoms are incorporated into the final material and what atoms are wasted?

3. Less hazardous chemical syntheses: If possible, and practical, the (by)products of chemical reactions should be minimally toxic.

4. Designing safer chemicals: Chemicals should be developed that accomplish desired goals and minimize any hazards.

5. Safer solvents and auxiliaries: If they cannot be avoided altogether, then chemicals with no or minimal hazard should be used for running reactions and purifying products.

6. Design for energy efficiency: When practicable, reactions should be carried out without heat or added pressure, and if either heat or pressure is needed the energy should be considered in terms of its environmental and economic impacts.

7. Use of renewable feedstocks: If possible, and practical, starting materials should come from renewable sources.

8. Reduce derivatives: Nonproductive synthetic steps, including protecting groups, should be avoided to reduce waste.

9. Catalysis: Chemicals that increase the rate of reaction without modifying the overall change in reaction energy.

10. Design for degradation: Products should be designed so that when disposed of they do not persist in the environment.

11. Real-time analysis for pollution prevention: Tools should be developed and used to monitor processes so that the formation of hazardous or toxic substances can be prevented.

12. Inherently safer chemistry for accident prevention: Improvements in safety through the reduction of hazards for researchers, technicians, and workers should be prioritized.

Through to the present day, organic chemistry is a materially intensive science, which has often overlooked its own impact on the Earth. This has led to environmental degradation and pollution in the form of contamination of soils and waterways by perfluorinated chemicals (so-called forever chemicals), lead as a component of gasoline, depletion of the ozone layer with chlorofluorocarbons (CFCs), and plastic packaging and waste. While organic chemistry has created these problems, organic chemistry is also the solution to its own problems and those of human society writ large.

In this book, green chemistry will be highlighted when it intersects with the material of a chapter. In addition, you are encouraged to keep the 12 Principles of Green Chemistry in mind as you read through the book. Advancements in green chemistry and improvements to organic chemistry's sustainability will require not only applying existing knowledge to current practices, but it will also require new innovations. As learners of chemistry, you are uniquely positioned to offer outside perspectives to those who "know" what can and cannot be done.

CLARITY

Clarity, by which we mean unambiguousness, is key in most areas of life; however, it is of paramount importance in science. Whether discussing feet or metres, 0.1 g or 0.01 g, or weight or mass it is vital that the speaker (or writer) can convey the correct information to the audience.[6] Organic chemistry has the same emphasis on quantitative measurement, significant figures, and units as any other branch of chemistry; unlike other subdisciplines of chemistry, though, organic chemistry has a unique need for clarity in terms of nomenclature. Nomenclature provides a means of converting back and forth between chemical structures and words. The core structure of the International Union of Pure and Applied Chemistry (IUPAC) nomenclature will be presented in Chapter 3. As functional groups are introduced in later chapters, the appropriate suffixes, e.g. -ol for alcohols, will be introduced at the start of each chapter.

WRITING

In this introduction, it is also essential to underscore the importance of writing. Good analytical writing necessitates several things. First, analytical, or argumentative, writing should always include three pieces: claim, evidence, and reasoning (CER). A claim is any inference that you draw based on your analysis of the information or data at hand. The evidence is the information from the problem or data from empirical observation or an experiment that you are basing that claim on (you need to clearly highlight this evidence to the audience because they may not see or notice the same evidence). Finally, the audience needs to see your reasoning, which is the logical connection that you have made to move from the evidence to the claim you are stating.

Second, good writing, of any kind, needs to follow the rules of prescriptive English grammar. While prescriptive English grammar is not how people usually talk, it is an established convention for ensuring that the grammatical construction of your writing does not stand in the way of your audience's understanding.[7] Some of the more salient points of prescriptive English grammar are:[8]

- Sentences end in periods, and punctuation is appropriately used to avoid run-on sentences, comma splices, or sentence fragments.
- Verbs end in -ed and -s when necessary and appropriate helping verbs are employed.
- Verb tense shifts are avoided.
- The correct pronoun case (I/me/my or who/whom/whose, for example) is used.
- The pronoun agrees with the antecedent. The antecedent is the word the pronoun refers to.
- The plural endings -s/-es and possessive ending 's are used appropriately.
- Quotation marks are only employed when citing others' work or ideas. In scientific writing, we rarely use quotation marks to cite other's ideas. But to ensure appropriate credit to the other, ideas are adequately referenced and noted with footnotes.
- Material taken from another source is always referenced following a standard format, for example: APA, MLA, Chicago, Turabian, or ACS. ACS reference formatting is used in this book.
- Typographical errors – including misspellings – are avoided.
- There is proper sentence sense. That is no words are omitted and ideas are not scrambled nor incomprehensible.

As a reader of this first chapter and the thoughts above on writing, you may have a question that is often received by the author. It is not a new question from folks studying chemistry and a famous example of this question comes from Allen Hoos, an American high school student, who wrote to R.B. Woodward, one of the most famous organic chemists of the 20th century:[9]

I am a sophomore at Fairview High School and I have been studying English since about the third grade. I plan to become a chemist such as you are and I wondered why I have to take so much English. Is English that important in the field of chemistry? It seems as though I would be better off taking subjects that would have closer relationships to chemistry. I would appreciate your opinions about English and its usefulness, if any, in chemistry.

Woodward's response to Allen Hoos' query is a great summary of the importance of writing and knowledge of language to chemistry:[10]

It is a valuable asset to a chemist to be able to formulate [their] ideas, describe [their] experiments, and express [their] conclusions in clear, forceful English. Further, since thought necessarily involves the use of words, thinking is more powerful, and its conclusions are more valid, in the degree to which the thinker has a command of the language.

NOTES

1. Dr Stephen Matlin

2. i) Reisman, S.E.; Sarpong, R.; Sigman, M.S.; Yoon, T.P. Organic Chemistry: A Call to Action for Diversity and Inclusion. *J. Org. Chem.*, **2020**, *85* (16), 10287–10292. DOI: 10.1021/acs.joc.0c01607. ii) Sanford, M.S. Equity and Inclusion in the Chemical Sciences Requires Actions Not Just Words. *J. Am. Chem. Soc.*, **2020**, *142* (26), 11317–11318. DOI: 10.1021/jacs.0c06482. iii) Ruck, R.T. and Faul, M.M. Update to Editorial "Gender Diversity in Process Chemistry". *Org. Process Res. Dev.*, **2021**, *25* (3), 349–353. DOI: 10.1021/acs.oprd.0c00471

3. Dunn, A.L.; Decker, D.M.; Cartaya-Marin, C.P.; Cooley, J.; Finster, D.C.; Hunter, K.P.; Jacques, D.R.N.; Kimble-Hill, A.; Maclachlan, J.L.; Redden, P.; Sigmann, S.B.; Situma, C. Reducing Risk: Strategies to Advanced Laboratory Safety through Diversity, Equity, Inclusion, and Respect. *J. Am. Chem. Soc.*, **2023**, *145*, 21, 11468–11471. DOI: 10.1021/jacs.3c03627

4. Matlin, S.; Mehta, G.; Cornell, S.E.; Krief, A.; Hopf, H. Chemistry and pathways to net zero for sustainability. *RSC Sustain.*, **2023**, *1*, 1704–1721. DOI: 10.1039/D3SU00125C

5. Anastas, P.T. and Warner, J.C. *Green Chemistry: Theory and Practice*. Oxford University Press: Oxford, **1998**.

6. For a clear example of what can happen when ambiguity comes into play, you should consider the debacle of the 1999 Mars Climate Orbiter.

7. https://osuwritingcenter.okstate.edu/blog/2020/10/30/prescriptive-and-descriptive-grammar. (Date accessed 24 August 2021)

8. https://www.ben.edu/college-of-liberal-arts/writing-program/upload/Policy-for-Use-of -Edited-Standard-Written-English.pdf (Date accessed 24 August 2021).

9. A. Hoos, letter to R. B. Woodward, Boulder, CO, 23 October 1961 (Harvard University. Records of Robert B. Woodward. HUGFP 68.8 Early subject files, 1931–1960 (bulk), 1931–1979 (inclusive), Box 8, in folder Correspondence – Requests for autographs, etc. [1960–1971]. Harvard University Archives). From: Seeman, J.I. Woodward's words: elegant and commanding. *Angew. Chem. Int. Ed.*, **2016**, *55* (41), 12898–12912. DOI: 10.1002/anie.201600811

10. R. B. Woodward, letter to A. Hoos, Boulder, CO, 1 December 1961 (Harvard University. Records of Robert B. Woodward. HUGFP 68.8 Early subject files, 1931–1960 (bulk), 1931–1979 (inclusive), Box 8, in folder Correspondence – Requests for autographs, etc. [1960–1971]. Harvard University Archives). From: Seeman, J.I. Woodward's words: elegant and commanding. *Angew. Chem. Int. Ed.*, **2016**, *55* (41), 12898–12912. DOI: 10.1002/anie.201600811

2 Electrons, Bonding, and Structure[1]

The energy and movement of electrons along with the making and breaking of bonds is at the core of what we think of as chemistry. This is especially true for organic chemistry. In this chapter, we will focus on the electronic structure of atoms and on bonding with an emphasis on bonds involving the elements in periods 1 and 2, those elements of greatest concern to organic chemists.

ELECTRONS

This chapter begins by focusing on electrons and how we came to understand atomic electron configurations before we move on to bonding. Consider the atomic emission spectrum (the light produced by atoms excited with energy, here electricity) of hydrogen (Figure 2.1). This pattern of lines is unique to hydrogen, and every element has its own unique pattern of lines in its atomic emissions spectrum.[2]

Despite discovering the atomic emission spectra nearly 200 years ago, the physical basis for the origin of these spectra was mysterious; however, scientists identified mathematical formulas that could model these lines. The most general formula developed was that proposed by Rydberg, which stated that the wavelength (λ) of a line (in metres) could be calculated using Equation 2.1.[3]

$$\frac{1}{\lambda} = R_H \left(\frac{1}{n_i^2} - \frac{1}{n_f^2} \right) \qquad \text{(Equation 2.1)}$$

λ is the wavelength (m)
R_H is the Rydberg constant for hydrogen (1.096 775 83x10^7 1/m)
n_i is the initial state (any natural number, 1, 2, 3, 4, 5, 6, 7,...)
n_f is the final state (any natural number, 1, 2, 3, 4, 5, 6, 7,...)

Using this equation, if we were to choose n_i equal to 3 and n_f equal to 2, then we would calculate the following:

$$\frac{1}{\lambda} = 1.09677583 \times 10^7 \, \frac{1}{m} \left(\frac{1}{3^2} - \frac{1}{2^2} \right) = -1523299.76 \frac{1}{m}$$

Taking the reciprocal of this value gives –6.5647x10^{-7} m or –656.47 nm (the value is negative because light is being emitted), which is in close agreement with the 656 nm line in Figure 2.1.

Problem 2.1 What are n_i and n_f for the other three lines in Figure 2.1?

Problem 2.2 Calculate the wavelength in nm for n_i equals 2 and n_f equals 1. Can humans see this light (humans see light with wavelengths roughly between 400 nm to 800 nm)?

Shortly after Rydberg proposed his formula for modelling the lines of the hydrogen spectrum, J.J. Thomson conducted unrelated experiments on cathode rays,[4] which would end up providing insight into the atomic emission spectrum problem. Thomson found that cathode rays were composed of negatively charged particles that were less massive than the lightest atom, hydrogen. This meant that these "negatively charged corpuscles" – we now call them electrons – had to be a new type of subatomic particle. The fact that atoms contained electrons – and were not infinitesimally small, unitary spheres – was revolutionary. However, the existence of electrons raised a bigger question: how were electrons arranged in an atom?

Niels Bohr was the individual who was able to take and unify these two pieces of information – electrons exist within atoms and the atomic emission spectrum is unique to each atom – into a

Figure 2.1 The atomic emission spectrum of hydrogen showing the spectral lines in the visible range and their corresponding wavelengths.

DOI: 10.1201/9781003479352-2

coherent framework and our first reasonable insight into electronic structure.[5] Bohr proposed that i) the electrons in an atom occupy specific, discrete energy levels and ii) the lines are the result of electrons moving between these discrete energy levels (here the emission of light is the result of electrons going from a higher energy level to a lower energy level). It is the fact that there are few discrete lines that means the energy levels are specific and discrete (if an electron could have any energy, we would expect a continuous rainbow of colour). Therefore, we say that the energy of electrons is quantized, which is a more concise way of saying that it exists at specific, discrete energy levels.

While Bohr's theory was part of the origin of a new model (quantum mechanics), it proved to be incorrect. Bohr still considered the electrons to be purely particles. Subsequent experimentation showed that electrons, and (sub)atomic particles generally, also behave like waves. Given that electrons are both particles and waves, the shortcomings of the Bohr model become clear. Bohr assumed that electrons were purely particles, for which we could know their precise pathway (their speed and their position). Given that electrons also exhibit wave behaviour, a new approach was necessary. Two different quantum mechanical approaches were independently developed. The first used matrices to model observable phenomena such as atomic emission spectra.[6] The second used wave functions to model the particle waves of electrons.[7] Both approaches were shown to be mathematically equivalent.[8] The study and explanation of matrix mechanics or wave functions is beyond the scope of this book. What is relevant to know is that the specific energy level (known as an orbital) of an electron can be described by three quantum numbers: the principal quantum number (n), the angular momentum quantum number (ℓ), and the magnetic quantum number (m_ℓ).[9] The energy of an electron is described by n, ℓ, and m_ℓ along with the spin quantum number (m_s).

The principal quantum number (n) is any natural number (1, 2, 3, 4, 5, 6, 7, …). This number corresponds to the shell number and describes an electron's distance from the nucleus (a smaller n value means closer to the nucleus) and its orbital energy level (a smaller n value means lower energy level). In some fields of chemistry, particularly X-ray spectroscopy, you may see that the shells are given letter designations: the first shell is the K shell, the second shell is the L shell, the third shell is the M shell, and so on.

The angular momentum quantum number (ℓ) is a whole number (0, 1, 2, 3, 4, 5, 6, …) and for a given n value ℓ is all of the values 0 to $n - 1$, for example: if $n = 1$ then $\ell = 0$; if $n = 2$ then $\ell = 0, 1$; if $n = 3$ then $\ell = 0, 1, 2$; if $n = 4$ then $\ell = 0, 1, 2, 3$. The angular moment quantum number corresponds to the subshell ($\ell = 0$ is the s subshell, $\ell = 1$ is the p subshell, $\ell = 2$ is the d subshell, and $\ell = 3$ is the f subshell) and describes the shape of the orbital.

The magnetic quantum number (m_ℓ) is an integer (0, ±1, ±2, ±3, ±4, ±5, ±6, …) and for a given ℓ value, m_ℓ is all the integers from $-\ell$ to $+\ell$, for example: if $\ell = 0$ then $m_\ell = 0$; $\ell = 1$ then $m_\ell = -1, 0, +1$; $\ell = 2$ then $m_\ell = -2, -1, 0, +1, +2$; $\ell = 3$ then $m_\ell = -3, -2, -1, 0, +1, +2, +3$. The magnetic quantum number identifies the direction of the orbital and indicates how many orbitals there are per subshell (e.g. $\ell = 0$ then $m_\ell = 0$ corresponds to the s subshell and there is one s-orbital per s-subshell; $\ell = 1$ then $m_\ell = -1, 0, +1$ corresponds to the p subshell and there are three p-orbitals per p-subshell.

The spin quantum number (m_s) describes the spin (an intrinsic property of the electron not related to our macroscopic idea of spin) of the electron and has values of $-\frac{1}{2}$ and $+\frac{1}{2}$.

Putting this together into a coherent energy diagram gives Figure 2.2, which is only a partial diagram as it shows only up to $n = 4$, $\ell = 1$.

While directly tied to and showing the quantum numbers of each level, the diagram in Figure 2.2 loses some utility and functionality as a model because of the depth of information presented. As such, Figure 2.3 is a more common representation where the quantum number information has been encoded in the shell or orbital designations that are more commonly used in chemistry in general.

Electrons occupy these atomic orbitals according to several rules and principles. According to the Aufbau principle, orbitals are filled starting from the lowest energy, typically in order of the $n + \ell$ value.[10] The Pauli exclusion principle states that no electron can have the same four quantum numbers.[11] As can be seen in Figure 2.3, this means that there are only two electrons allowed in any one orbital. To show this in Figure 2.3 or Figure 2.4 we use arrows pointed up (↑) and down (↓) to show the $+\frac{1}{2}$ and $-\frac{1}{2}$ spin values. If there are two electrons in a single orbital, they must be spin-paired (one up and one down: ↑↓). Finally, Hund's rule states that degenerate, equal energy, orbitals (p-, d-, or f-orbitals) must each be filled singly and spin-aligned (all up or all down) before two electrons can be paired in the same orbital.[12]

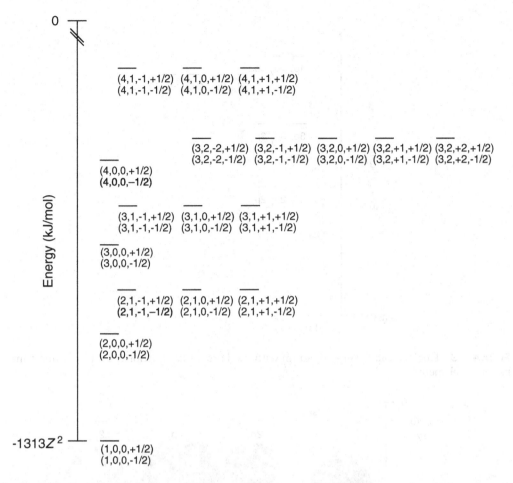

Figure 2.2 Partial orbital energy diagram with corresponding quantum number designations (Z = the atomic number of the element). The energy ordering shown here comes from the orbital energy ordering for fluorine, actual orbital energy ordering and filling may vary. Note that $-1313Z^2$ is the absolute lowest possible value for the $n = 1$ and $\ell = 1$ energy level, which is only valid for one-electron atoms and ions. With more electrons, each energy level shifts upwards due to electron-electron repulsion. Kramida, A., Ralchenko, Yu., Reader, J., and NIST ASD Team (2022). NIST Atomic Spectra Database (ver. 5.10), [Online]. Available: https://physics.nist.gov/asd [7 July 2023]. National Institute of Standards and Technology, Gaithersburg, MD. DOI: 10.18434/T4W30F

Following all the rules and principles above will give the ground-state electron configuration of an atom. With only one electron, the hydrogen electron configuration is: $1s^1$. To fully distribute the six electrons of carbon they would occupy the 1s-, 2s-, and 2p-orbitals: $1s^2 2s^2 2p^2$. An electron configuration that does not follow the Aufbau principle or Hund's rule is what we term an excited state (high energy states that can decay to a ground state by emitting light like that seen in Figure 2.1). In contrast, the Pauli exclusion principle is inviolable and any stated electron configuration that violates the Pauli exclusion principle is impossible, for example: $1s^2 2s^3 2p^1$ is an impossible configuration for carbon.

Problem 2.3 For each electron configuration identify it as ground state, excited state, or impossible. Identify what neutral atom each configuration (that is not impossible) corresponds to.

a. $1s^2 2s^2 2p^6 3s^2 3p^1$

b. $1s^2 2s^2 2p_x^2 2p_y^1 2p_z^0$

c. $1s^2 2s^2 2p^6 3s^2 3p^6 4s^2 3d^{12}$

d. $1s^1 2s^2 2p^6$

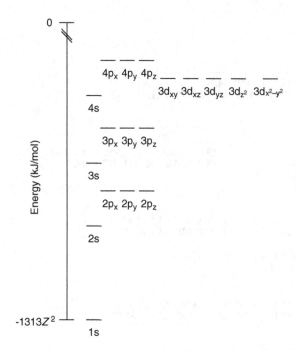

Figure 2.3 Partial orbital energy diagram with shell orbital designations (Z = the atomic number of the element).

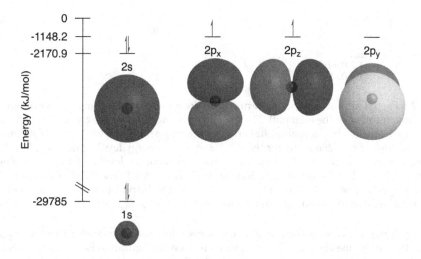

Figure 2.4 Electron configuration of a carbon atom with images of the s- and p-orbitals. Orbital images and energy values calculated at the CCSD(T)/cc-pVTZ level of theory and basis set.

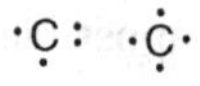

$$\cdot \ddot{C} : \quad \cdot \dot{\underset{\cdot}{C}} \cdot$$

Figure 2.5 The Lewis dot symbol for carbon showing the valence configuration (*left*) and for a hybridized (see below) carbon atom (*right*) that highlights the tendency of carbon to make four bonds.

For our discussion of bonding and structure in organic chemistry, it is important to have a sense of what these orbitals look like. Figure 2.4 shows the electron configuration of carbon including the relative energy levels and shapes of the 1s-, 2s-, and 2p-orbitals (note: s- and p- orbitals are the only ones with which organic chemists generally need to concern themselves). The s-orbitals have a spherical shape and are radially symmetric (that is every point at a defined distance r from the nucleus is the same). The 2s-orbital is larger than the 1s-orbital corresponding to its higher energy. The 2p-orbitals have a nodal plane centred on the atom and have two symmetric lobes along the x, y, or z axes. It is important to keep in mind that at any atom, any p-orbitals (or orbitals hybridized with p-orbitals) adopt perpendicular orientations.

Looking at Figure 2.4, there are two different types of electrons. Four of carbon's electrons (the $2s^2 2p^2$) are designated as valence electrons. Valence electrons are the highest-energy, outermost electrons, which are responsible for the reactions and bonding of an atom. The other two electrons (core electrons) are in a much lower-energy closed (filled) shell (1s) and not significantly involved in chemistry or bonding.

Problem 2.4 Write the ground-state electron configuration of each atom or ion.

Li$^+$:

F$^-$:

B:

N:

Problem 2.5 Define the following terms.

valence shell:

valence electron:

Problem 2.6 How many electrons are in the valence shell for each of the following?

C:

Cl:

Al:

LEWIS DOT SYMBOLS

The principles, theories, and rules discussed above for the electron configuration of an atom give us a string of alphanumeric characters that represent how the electrons are distributed in an atom. For example, the electron configuration of carbon is $1s^2 2s^2 2p^2$. This can be abbreviated using noble gas notation to highlight only the valence electrons of carbon: $[He]2s^2 2p^2$. Each of these configurations, while rigorously sound in terms of theory and experiment, lacks a certain fluidity and facility in its presentation. It is for this reason that Lewis dot symbols, developed by G.N. Lewis in 1916,[13] are still so widely employed by chemists. Consider the Lewis dot symbol for carbon (Figure 2.5).

The fact that carbon has four valence electrons is more readily apparent to both a practising and non-practising chemist with this symbology. If we consider a periodic table of Lewis dot symbols (Figure 2.6), the patterns and group trends in electron configuration become much more apparent.

Looking at the Lewis dot symbols (Figure 2.6), recall that atoms are at their most stable (lowest energy) when they have filled valence shells. Main group elements, aside from hydrogen and helium, typically follow the octet rule,[14] which states that a stable atom has eight valence electrons. Specifically, those eight valence electrons correspond to filled s and p subshells. For hydrogen and helium, there is no 1p subshell and so they only need a duet of electrons to fill the 1s subshell and to be stable. Looking at the main group elements in Figure 2.6, we can see that only group 18 elements have an octet of electrons, which is why the noble gas elements naturally exist as monatomic gases. All other elements can only achieve stability through forming di- or polyatomic molecules or compounds.

BONDING: Electronegativity (χ) and Bonding Type

Atoms in molecules and compounds are held together by bonds. Bonds can be thought of as the energy, or force, holding atoms together. More specifically, when ions and atoms are brought together, forming ionic bonds or covalent bonds, respectively, they release energy (Figure 2.7). To

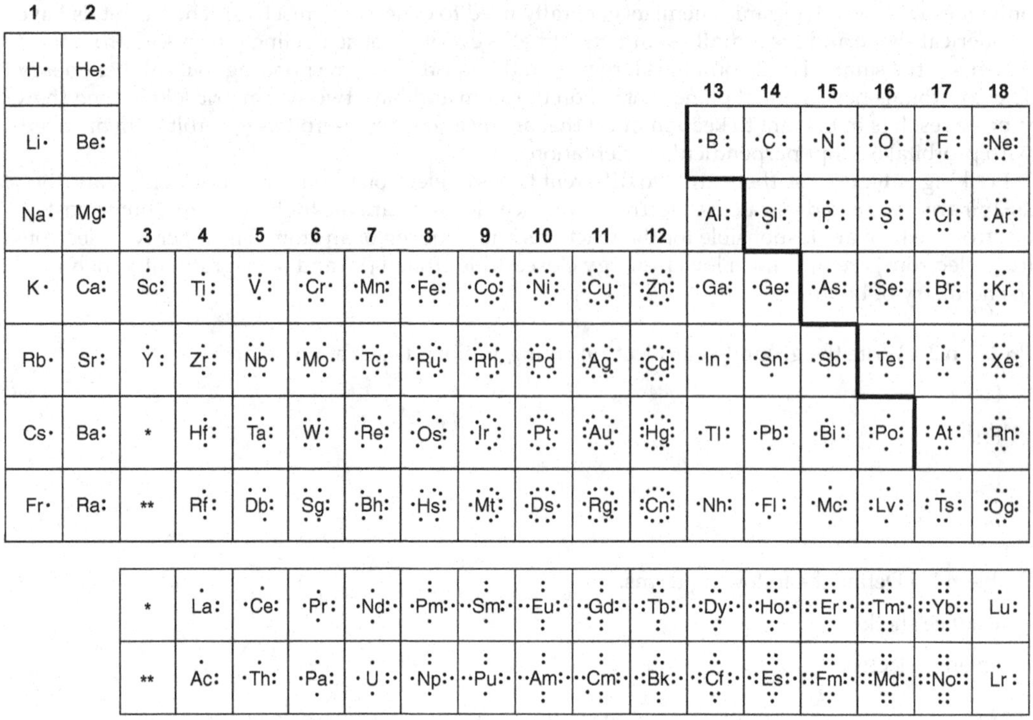

1 2 13 14 15 16 17 18
3 4 5 6 7 8 9 10 11 12
H He
Li Be B C N O F Ne
Na Mg Al Si P S Cl Ar
K Ca Sc Ti V Cr Mn Fe Co Ni Cu Zn Ga Ge As Se Br Kr
Rb Sr Y Zr Nb Mo Tc Ru Rh Pd Ag Cd In Sn Sb Te I Xe
Cs Ba * Hf Ta W Re Os Ir Pt Au Hg Tl Pb Bi Po At Rn
Fr Ra ** Rf Db Sg Bh Hs Mt Ds Rg Cn Nh Fl Mc Lv Ts Og
* La Ce Pr Nd Pm Sm Eu Gd Tb Dy Ho Er Tm Yb Lu
** Ac Th Pa U Np Pu Am Cm Bk Cf Es Fm Md No Lr

Figure 2.6 Lewis dot symbols for the elements.

Energy (kJ/mol)
800
700
600
500
400
300
200
100
0
-100
-200
-300
-400
-500
20 40 60 80 100 120 140 160 180 200 210 220 240 260 280 300 320
Distance (pm)
Ed
H H
H H

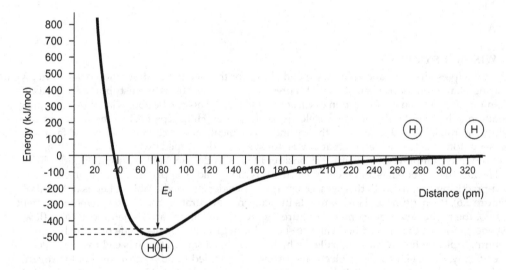

Figure 2.7 Morse potential – which shows the change in energy between two atoms as a function of internuclear distance – for dihydrogen. When the distance between the two nuclei is 74 pm, the energy decreases to a minimum. The internuclear distance (74 pm) is called the bond length; the energy it would take to pull the two hydrogen atoms apart – 436 kJ/mol – is called the dissociation energy (E_d). The sharp increase in potential energy as the atoms get closer together is due to the coulombic repulsion of the positively charged nuclei. Note that the dissociation energy (E_d) does not measure from the bottom of the potential energy curve but from the minimum energy that the molecule can have due to vibrational motion. The difference between the electronic potential minimum (black line) and the minimum energy the molecule can have is called the zero-point energy (ZPE). Reference for the Morse Potential: Morse, P.M. Diatomic Molecules According to the Wave Mechanics. II. Vibrational Levels. ***Phys. Rev.*, 1929, *34*** (1), 57–64. DOI: 10.1103/PhysRev.34.57

Table 2.1 Pauling Electronegativity (χ_P) Differences and Bond Polarity

Electronegativity difference ($\Delta\chi_P$)	Bond type
<1.7	Covalent
>1.7	Ionic

break ions and atoms in a compound apart, energy (dissociation energy, E_d) must be added to break the bond.

Atoms bond to lower their energy (increase their stability), but not all bonding is the same. Bonding exists along a continuum from purely covalent to purely ionic. We can understand why different types of bonds form if we consider the atoms' electronegativity values. Electronegativity is the ability of an atom to draw electrons toward itself through a chemical bond. The difference in electronegativity between two bonded atoms determines how electrons are shared in that bond. There is no universally agreed upon electronegativity scale, but a modified version of the one suggested by Linus Pauling in 1932 is the most common and will be used here in Figure 2.8.[15]

In a case where two atoms have the same electronegativity, the electron density is equally shared between the two atoms giving a covalent bond. In the other extreme, there is no sharing of electrons but oppositely charged ions are held together through coulombic attraction, which is called an ionic bond. An intermediate difference in electronegativity between two atoms results in an unequal sharing of electrons (polar covalent bond). A scale is provided in Table 2.1 that shows how the difference in electronegativity leads to different types of bonds. The 1.7 value is a guide for the cutoff between different bond types, but many other factors can influence and dictate the actual bond type.

The electronegativity values of these highlighted elements will be the most utilized while assessing the bonds of organic molecules. It is important to note that carbon has a moderate electronegativity of 2.55, which means that most organic molecules are made of covalent and polar covalent bonds.

Problem 2.7. Which atom in each set is more electronegative and what type of bond is likely to form?

a. carbon or nitrogen   b. chlorine or bromine   c. iron or fluorine

IONIC COMPOUNDS

Ionic compounds are formed in cases of large electronegativity differences (Table 2.1). The electronegativity difference is so large between the atoms (or groups of atoms) in these compounds that electrons are not shared between the atoms. As a result, the interacting particles are ions – charged particles – and their attraction is described as coulombic attraction. This attraction is expressed in Coulomb's law (Equation 2.3).[16]

$$F = k_e \frac{Q_1 Q_2}{r^2} \qquad \text{(Equation 2.3)}$$

F is an attractive (if the value is negative) or repulsion (if the value is positive) force (N)

k_e is Coulomb's constant ($8.988 \times 10^9 \ \frac{N\,m^2}{C^2}$)
Q_1 is the charge of ion one (C)
Q_2 is the charge of ion two (C)
r is the internuclear distance (m)

These ions form crystal lattices (Figure 2.9) in the solid phase and ion pairs or solvated ions when dissolved in polar, protic solvents (see Chapter 3: "Arrow Pushing and Solvent"). They tend to have high melting points and be solids at normal temperature and pressure. Most importantly, since crystal lattices exist as repeating unit cells of ions and not as molecular compounds – which exist as discrete sets of bonded atoms in solid, liquid, and gas phases – they are not best depicted

1	2	3	4	5	6	7	8	9	10	11	12	13	14	15	16	17	18
1 H 2.20																	2 He
3 Li 0.98	4 Be 1.57											5 B 2.04	6 C 2.55	7 N 3.04	8 O 3.44	9 F 3.98	10 Ne
11 Na 0.93	12 Mg 1.31											13 Al 1.61	14 Si 1.90	15 P 2.19	16 S 2.58	17 Cl 3.16	18 Ar
19 K 0.82	20 Ca 1.00	21 Sc 1.36	22 Ti 1.54	23 V 1.63	24 Cr 1.66	25 Mn 1.55	26 Fe 1.83	27 Co 1.88	28 Ni 1.91	29 Cu 1.90	30 Zn 1.65	31 Ga 1.81	32 Ge 2.01	33 As 2.18	34 Se 2.55	35 Br 2.96	36 Kr 3.0
37 Rb 0.82	38 Sr 0.95	39 Y 1.22	40 Zr 1.33	41 Nb 1.6	42 Mo 2.16	43 Tc 1.9	44 Ru 2.2	45 Rh 2.28	46 Pd 2.20	47 Ag 1.93	48 Cd 1.69	49 In 1.78	50 Sn 1.96	51 Sb 2.05	52 Te 2.30	53 I 2.66	54 Xe 2.6
55 Cs 0.79	56 Ba 0.89	*	72 Hf 1.3	73 Ta 1.5	74 W 2.36	75 Re 1.9	76 Os 2.0	77 Ir 2.20	78 Pt 2.28	79 Au 2.54	80 Hg 2.00	81 Tl 2.04	82 Pb 2.33	83 Bi 2.02	84 Po 2.0	85 At 2.2	86 Rn
87 Fr 0.7	88 Ra 0.9	**	104 Rf	105 Db	106 Sg	107 Bh	108 Hs	109 Mt	110 Ds	111 Rg	112 Cn	113 Nh	114 Fl	115 Mc	116 Lv	117 Ts	118 Og

*	57 La 1.1	58 Ce 1.12	59 Pr 1.13	60 Nd 1.14	61 Pm 1.2	62 Sm 1.17	63 Eu 1.1	64 Gd 1.20	65 Tb 1.2	66 Dy 1.22	67 Ho 1.23	68 Er 1.24	69 Tm 1.25	70 Yb 1.1	71 Lu 1.27
**	89 Ac 1.1	90 Th 1.0	91 Pa 1.5	92 U 1.38	93 Np 1.36	94 Pu 1.28	95 Am 1.30	96 Cm	97 Bk	98 Cf	99 Es	100 Fm	101 Md	102 No	103 Lr

Figure 2.8 Pauling electronegativity (χ_P) values.

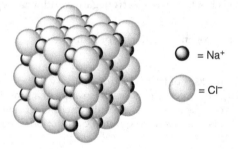

= Na⁺

= Cl⁻

Figure 2.9 An example crystal lattice of sodium chloride. Based on Figure 8 from: Barlow, W. XXVI. Geometrische Untersuchung über eine mechanische Ursache der Homogenität der Structur und der Symmetrie; mit besonderer Anwendung auf Krystallisation und chemische Verbindung. *Z. Kristallogr. Cryst. Mater.*, **1898**, *29* (1–6), 433–588. DOI: 10.1524/zkri.1898.29.1.433

by Lewis structures. Note, therefore, that ionic compounds depicted as Lewis structures are particularly misleading as to the true structure of the compound in question.

MOLECULAR COMPOUNDS

The discussion of electron configurations looked at ground and excited states, it did not touch on the idea of stable electron configurations. The most stable configurations are those of Group 18 (the noble gases), which have their s and p subshells (helium is an exception because it only has an s subshell filled). This leads to our concept of the octet rule: atoms are stable when they have eight valence electrons (or two for hydrogen and helium). This is generally true for main group elements (transition metals, lanthanides, and actinides behave somewhat differently). From a valence bond perspective, then, bonding occurs so that all atoms achieve an octet. We use Lewis structures[17] to depict this bonding in molecular compounds and polyatomic ions by connecting each atomic symbol with a line (representing a shared pair of electrons, or bond) and placing a dot on each atom to represent any unshared electrons. A good Lewis structure always: 1) displays all bonds between appropriate atoms, 2) displays all radical and lone-pair electrons, and 3) places all nonzero formal charges on the appropriate atoms. Good Lewis structures tend to: 1) place an octet on non-hydrogen atoms and 2) represent the bonding present in the actual molecule as well as possible. In the following examples, we will review how to go from a formula to a Lewis structure.

LEWIS STRUCTURE EXAMPLE 1: Methane (CH_4)

1. Identify the elements that make up the molecule (carbon and hydrogen) and the total number of valence electrons (eight total, four for the carbon atom and one for each hydrogen atom).

2. Determine the number of electrons needed to satisfy each atom's valence shell: carbon needs eight and each hydrogen needs two (eight total). For methane, then 16 electrons are needed to satisfy the valence shell of each atom. The difference between the number of electrons needed to fill the valence shell (16) and the number of valence electrons in the molecule (eight) is the number of electrons shared (eight) electrons. These are shared in bonds and since there are two electrons per bond, that means methane has four total bonds.

3. The difference between the number shared and the total valence is the number of unshared electrons. Here all eight valence electrons are shared and so that means there are no unshared (lone-pair or radical) electrons.

4. We then place the least electronegative (non-hydrogen) atom (typically carbon) in the middle of the molecule and draw the appropriate number of bonds:

$$
\begin{array}{c}
\text{H} \\
| \\
\text{H}-\text{C}-\text{H} \\
| \\
\text{H}
\end{array}
$$

5. Check to ensure that each atom has a full valence shell. (Each atom has the full number of valence electrons so this is a reasonable Lewis structure.)

LEWIS STRUCTURE EXAMPLE 2: Formaldehyde (CH_2O)

1. Identify the elements that make up a molecule (carbon, oxygen, and hydrogen) and the total number of valence electrons (12 total, four from the carbon atom, one from each hydrogen atom, and six from the oxygen atom).

2. Determine the number of electrons needed to satisfy each atom's valence shell: carbon and oxygen need eight and each hydrogen needs two (20 total). For formaldehyde, then 20 electrons are needed to satisfy the valence shell of each atom. The difference between the number of electrons needed to fill the valence shell (20) and the number of valence electrons in the molecule (12) is the number of electrons shared (eight shared electrons). These are shared in bonds and since there are two electrons per bond, that means phosgene has four total bonds.

3. The difference between the number shared (eight) and the total valence (12) is the number of unshared electrons (four or two lone pairs).

4. We then place the least electronegative (non-hydrogen) atom (typically carbon) in the middle of the molecule and draw the appropriate number of bonds. Then lone pairs are drawn to satisfy the valence shell of each atom:

$$
\begin{array}{c}
\ddot{\text{O}}\!: \\
\| \\
\text{H}-\text{C}-\text{H}
\end{array}
$$

5. Check to ensure that each atom has a full valence shell. (Each atom has the full number of valence electrons so this is a reasonable Lewis structure.)

LEWIS STRUCTURE EXAMPLE 3: Hydrogen Cyanide (HCN)

1. Identify the elements that make up a molecule (hydrogen, carbon, and nitrogen) and the total number of valence electrons (ten total, one for the hydrogen atom, four for the carbon atom, and five for the nitrogen atom).

2. Determine the number of electrons needed to satisfy each atom's valence shell: carbon and nitrogen need eight (16 total) and hydrogen needs two. For hydrogen cyanide, then 18 electrons are needed to satisfy the valence shell of each atom. The difference between the number of electrons needed to fill the valence shell (18) and the number of valence electrons in the molecule (10) is the number of electrons shared (eight shared electrons). These are shared in bonds and since there are two electrons per bond, that means hydrogen cyanide has four total bonds.

3. The difference between the number shared (eight) and the total valence (10) is the number of unshared electrons (two or one lone pair).

4. We then place the least electronegative (non-hydrogen) atom (typically carbon) in the middle of the molecule and draw the appropriate number of bonds. Then lone pairs are drawn to satisfy the valence shell of each atom.

$$H-C\equiv N:$$

5. Check to ensure that each atom has a full valence shell. (Each atom has the full number of valence electrons so this is a reasonable Lewis structure.)

FORMAL CHARGE[18]

When drawing a Lewis structure, it is critical to keep track of the number of electrons around each atom and specifically account for the gain or loss of electrons for each atom. In organic chemistry, the most common method for bookkeeping electrons is formal charge. There is a formulaic method for calculating formal charges. The formal charge is the number of valence electrons for a particular atom minus the number of bonds drawn to that atom minus the number of lone pairs drawn to that atom.

The formal charge concept assumes that all bonds are perfectly covalent. Most bonds involve some difference in electronegativity and hence some inequity in the distribution of electron density. Correct formal charges are one means of accounting for the electron count in a Lewis structure. Formal charges also often offer insights into a molecule's chemical reactivity or other physical properties.

Problem 2.8 Draw reasonable Lewis structures for each of the following polyatomic ions/molecules. Make sure to show all bonds, lone-pair electrons, and formal charges.

Sodium bicarbonate ($NaHCO_3$)

Sulfuric acid (H_2SO_4)

Acetic acid (CH_3CO_2H)

OXIDATION NUMBER[19]

Another common means of bookkeeping electrons is oxidation numbers (or oxidation states, these terms are used interchangeably). While formal charge makes the simplifying assumption that all bonds are perfectly covalent, oxidation numbers are based on the simplifying assumption that all bonds are perfectly ionic. To assign oxidation numbers, one follows a prescribed list of rules:

1. **For any atom in its elemental form, the oxidation number is always zero.**

   For example, H in H_2 has an oxidation number of 0. Each P atom in P_4 has an oxidation number of 0. Every Cu atom in copper metal has an oxidation number of 0.

2. **For any monatomic ion, the oxidation number always equals the charge.**

   For example, in the compound K_2S, the ions are K^+ and S^{2-}. The oxidation number of potassium is +1 and the oxidation number of sulfide is –2.

3. **For any polyatomic ion or molecular compound, assign oxidation numbers in the following order.**
    a. **Fluorine always has an oxidation number of –1.**
    b. **Oxygen usually has an oxidation number of –2.**

    The exceptions are when oxygen is bonded to fluorine or when oxygen is part of peroxide (oxygen has an oxidation number of –1 in peroxides).

    c. **The oxidation number of hydrogen is +1 when combined with non-metals and –1 when combined with metals.**
    In NaH sodium has an oxidation number of +1 and hydrogen has an oxidation number of –1.
    In hydrogen fluoride (HF) hydrogen has an oxidation number of +1 and fluorine has an oxidation number of –1.

    d. **Halogen atoms Cl, Br, and I usually have an oxidation number of –1, except when bonded to another element of higher electronegativity.**
    In HCl hydrogen has an oxidation number of +1 and chlorine has an oxidation number of –1.
    In ClF chlorine has an oxidation number of +1 and fluorine has an oxidation number of –1.

4. **The sum of all oxidation numbers must equal 0 for a neutral molecule or they must equal the charge of a polyatomic ion.**

    In ClF_3 the fluorine atoms each have an oxidation number of –1 and the chlorine atom has an oxidation number of +3: $1Cl + 3(–1) = 0$, $Cl = +3$.

    For CO_3^{2-} oxygen atoms each have an oxidation number of –2 and the carbon atom has an oxidation number of +4: $1C + 3(–2) = –2$, $C = +4$.

Assigning oxidation numbers to a Lewis structure is more straightforward. Each bonding pair of electrons is assigned to the more electronegative atom (unless electronegativity values are equal, in which case the electrons are split evenly). The oxidation number is then the number of valence electrons minus the number of electrons around that atom. Consider ethyl fluoride (Figure 2.10) once bonding pairs are appropriately divvied up.

The blue boxes have assigned electrons to the more electronegative atoms (bifurcating the C–C bond electrons evenly). If we then assign oxidation numbers, we have +1 for the hydrogen atoms (one valence less zero assigned). For carbon one, there are four valence electrons less seven assigned giving an oxidation number of –3. For carbon two, there are four valence electrons less five assigned giving an oxidation number of –1. For fluorine, there are seven valence electrons less eight assigned giving an oxidation number of –1.

ACTUAL CHARGE

In both formal charge (assumption: all bonds are perfectly covalent) and oxidation numbers (assumption: all bonds are ionic), there is a simplifying assumption to allow us to calculate a "charge" on each atom. However, the actual charge distribution within a molecule is heavily influenced by the electronegativity differences between atoms, which is not accounted for by formal charge, and the nature of the covalent bonding, which is not accounted for by oxidation numbers. Ideally, we would want to know the actual, empirical charges on each atom. There is, sadly, no definitive way to measure or to calculate the real charge, but we can approximate the real charge with computational analysis. One method (the details are beyond the scope of this book) is Natural Population Analysis (NPA),[20] which is a result of a Natural Bond Orbital (NBO) calculation.[21] Figure 2.11 shows the formal charges (0 for hydrogen and +1 for nitrogen), oxidation

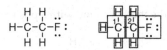

Figure 2.10   Ethyl fluoride (left) and ethyl fluoride with bonding electrons apportioned to the more electronegative atom (right).

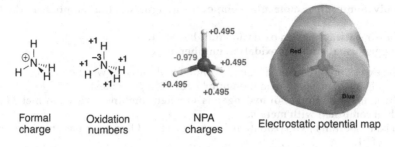

| Formal charge | Oxidation numbers | NPA charges | Electrostatic potential map |

Figure 2.11 Formal charge, oxidation numbers, NPA charges, and electrostatic potential for ammonium.

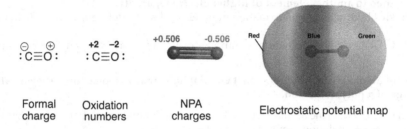

| Formal charge | Oxidation numbers | NPA charges | Electrostatic potential map |

Figure 2.12 Formal charge, oxidation numbers, NPA charges, and electrostatic potential for carbon monoxide.

numbers (+1 for hydrogen and –3 for nitrogen), and NPA results (+0.495 for hydrogen and –0.979 for nitrogen) for the ammonium cation. What is most apparent is that formal charge has no relationship to the actual distribution of electron density. Nitrogen is more electronegative than hydrogen so it should be (and is) more electron-rich. This is apparent in the oxidation numbers, charges, and electrostatic potential map (red is electron-rich and blue is electron-poor). Oxidation numbers provide a better sense of distribution but overestimate the magnitude of charge.

Figure 2.12 shows the formal charges (1– for carbon and 1+ for oxygen), oxidation numbers (+2 for carbon and –2 for oxygen), and NPA results (+0.506 for carbon and –0.506 for oxygen) for carbon monoxide. Again, formal charge does not give a good sense of the actual distribution of charge, but oxidation numbers do (again, overestimating the magnitude).

Although formal charge does not indicate actual charge, organic chemists use it almost exclusively because it has two useful features: 1) it is the easiest to calculate and 2) formal charges provide clues as to how a molecule is most likely to react. Let's consider examples.

If we consider the methyl cation (CH_3^+), the carbon has a positive formal charge (Figure 2.13). What is most important to notice is that the carbon has less than a full octet. So, the positive formal charge (and less than full octet) indicates that the carbon atom is Lewis acidic (an acceptor of electrons). In organic chemistry parlance, we term this electrophilic (Greek for "electron loving").

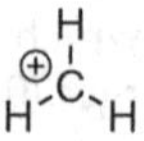

$$\overset{H}{\underset{H}{\overset{|}{\underset{\diagdown}{\overset{\oplus}{C}}}}}\diagup H$$

Figure 2.13 Lewis structure of the methyl cation with a positive formal charge.

$$\overset{\oplus}{\underset{H}{O}}\diagup H \\ \overset{\|}{\underset{H}{C}}\diagdown H$$

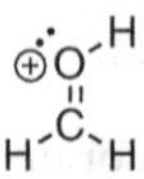

Figure 2.14 Lewis structure of protonated formaldehyde with a positive formal charge.

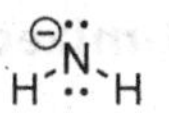

Figure 2.15 Lewis structure of the amide anion with a negative formal charge.

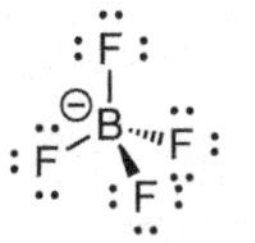

Figure 2.16 Lewis structure of tetrafluoroborate anion with a negative formal charge.

If we consider the cation of protonated formaldehyde (CH_3O^+), the oxygen has a positive formal charge (Figure 2.14). What is most important to notice is that the oxygen has a full octet. Here, the positive formal charge indicates that the atoms attached are electrophilic.

If we consider the amide anion (NH_2^-), the nitrogen has a negative formal charge (Figure 2.15). What is most important to notice is that the nitrogen has two lone pairs. So, the negative formal charge (and lone pair) indicates that the nitrogen atom is Lewis basic (a donor of electrons). In organic chemistry parlance, we term this *nucleophilic* (Greek for "nucleus loving").

If we consider the tetrafluoroborate anion (BF_4^-), the boron has a negative formal charge (Figure 2.16). What is most important to notice is that the boron has no lone pair. Here, the negative formal charge indicates that the atoms attached are nucleophilic.

Altogether, then, a positive formal charge is indicative of a site that is Lewis acidic, or electrophilic, and a negative formal charge is indicative of a site that is Lewis basic, or nucleophilic. A key challenge in organic chemistry is identifying the Lewis base (nucleophile) and the Lewis acid (electrophile).

Problem 2.9 Assign correct formal charges and oxidation numbers to each atom. Where would you predict the molecule to be most Lewis acidic and where would you predict it to be most Lewis basic?

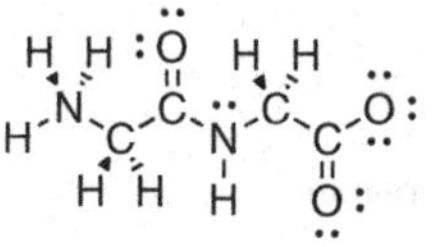

BOND LENGTH

While Lewis structures are incredibly useful, all atoms and bonds are drawn at roughly the same size in Lewis structures. As shown in Table 2.2, the length of the H–X bond varies substantially from HF to HI due to the increase in atomic radius of the halogen atoms from fluorine to iodine. The length of a bond is correlated with several properties, one of which is dissociation energy (E_d). Dissociation energy is the amount of energy it takes to pull apart two atoms that are bonded together. As can be seen in Table 2.2, shorter bonds tend to be stronger bonds and longer bonds tend to be weaker bonds (*note:* electronegativity also plays a role in the E_d values in this series). There is no standardized method for showing differences in bond length in a Lewis structure.

MOLECULAR STRUCTURE

Lewis structures were not developed to depict molecular geometry. It is, thus, completely acceptable when drawing a Lewis structure to depict a carbon atom with four bonds connected to it forming a square planar shape as shown for methane above. To have a better sense of geometry we frequently employ VSEPR theory (Table 2.3).[22] VSEPR is based upon the assumption that only the repulsion of electrons in bonding and non-bonding orbitals around an atom is responsible for the molecular geometry of that atom. It ignores the impact of important factors such as bond polarization, conjugation, orbital mixing, H-bonding, and ring strain among other factors. It is worth noting that VSEPR predictions for oxygen and other atoms with multiple lone pairs strongly diverge from those results from quantum mechanics. As such, the VSEPR predictions should be treated as predictions and not as facts.

Review Table 2.3, you can see the structures of methane, phosgene, and hydrogen cyanide that have been redrawn (if necessary) to appropriately show the geometry of each molecule. Note that

Table 2.2 HX Lewis Structures, Optimized Structures, Bond Lengths, and Dissociation Energies

H–X Lewis structure	Computed structure	H–X bond length (pm)	E_d (kJ/mol)
H–F̈:		92	569
H–C̈l:		127	431
H–B̈r:		141	364
H–Ï:		161	297

Table 2.3 VSEPR Reference for 2–4 Areas of Electron Density

Areas of electron density around the central atom (attached atoms or lone pair)	Electron geometry shape	Generic model (Ideal angles are shown, but lone-pair electrons cause a compression of those idealized angles)	Examples
2	Linear	180° A–X–A	:C≡O: H–C≡N: :O=C=O:
3	Trigonal planar	A X A 120°	F B F O C Cl Cl O S O
4	Tetrahedral	109.5° A X A A A	H C H H H N H H H O H H

a good Lewis structure does not strictly have to show molecular geometry, but it is best practice to always show the correct molecular geometry in a Lewis structure.

As the shape of a molecule is considered, review the atomic orbitals of carbon (Figure 2.5). If we consider the valence orbitals (those involved in boding), the s-orbital is spherically symmetric, and

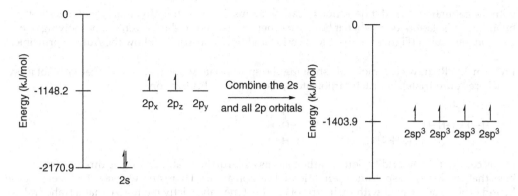

Figure 2.17 Diagram of sp³ hybridization in tetrahedral carbon atoms.

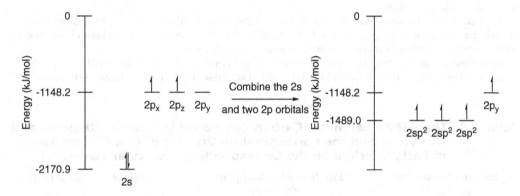

Figure 2.18 Diagram of sp² hybridization in trigonal planar carbon atoms.

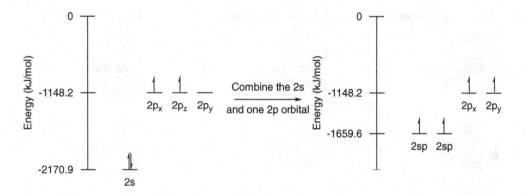

Figure 2.19 Diagram of sp hybridization in line carbon atoms.

the p-orbitals point along the x, y, and z axes. How then can we understand the linear, trigonal planar, and tetrahedral shapes carbon adopts given these orbitals? Chemists have developed the concept of hybridized orbitals, which involves the mixing of two, or more, different atomic orbitals to produce a new set of atomic hybrid orbitals. Figure 2.17, Figure 2.18, and Figure 2.19 shows the hybridization of carbon's orbitals. Note that hybridized carbon (whether sp³, sp², or sp) is shown to have four singly occupied valence orbitals (therefore it can make four bonds), which is consistent with the common Lewis dot depiction of carbon atoms. It is worth noting that carbon does not hybridize in isolation (only when it bonds to other atoms) and so the convention of showing one electron per orbital (Figure 2.17) is a convention and does not strictly follow the Aufbau principle.

Problem 2.10 Redraw each molecule showing the appropriate shape (using a wedge-dash notation where appropriate). For each carbon atom, identify its hybridization.

a. H–N–C–N–H (with O double bonded to C, lone pairs shown, H on each N)

b. H–N=C=O:

c. H–C–C–H (each C with two H, all single bonds)

Let us consider the hybridization of carbon atoms a bit further. Table 2.4 shows three hydrocarbons that contain sp^3-, sp^2-, and sp-hybridized carbon atoms. There are two ways to remember the hybrid orbitals associated with each type of carbon. One can strictly memorize that tetrahedral is sp^3, trigonal planar is sp^2, and linear is sp; however, it is more efficient to have a tool one can draw upon. The best way to determine the hybridization of a carbon atom is to use the mnemonic sp^{SN-1}, where SN is the steric number or the number of areas of electron density (bonded atoms and/or lone pair) around an atom.

A sp^3-hybridized carbon atom can make four single bonds (with the orbitals lying along the bonding axis) and will adopt a tetrahedral geometry: the four C-atom hybrid orbitals in ethane are all sp^3 orbitals creating four tetrahedrally-oriented single bonds.

An sp^2-hybridized carbon atom can make two single bonds and one double bond. The single bonds use the hybridized orbitals while the double bond uses both the hybridized orbital and the

Table 2.4 The VSEPR Geometry Carbon Can Adopt in Ethane, Ethylene, and Acetylene and the Corresponding Orbitals of One Carbon Atom in Each, Overlaid on the Corresponding Molecular Structures

Ethane – tetrahedral geometry	Ethylene – trigonal planar geometry	Acetylene – linear geometry
(ethane structure)	(ethylene structure)	H–C≡C–H
Ethane – sp^3 C-atom	Ethylene – sp^2 C-atom	Acetylene – sp C-atom

	sp^3		p		p
	sp^3		sp^2		p
	sp^3		sp^2		sp
	sp^3		sp^2		sp

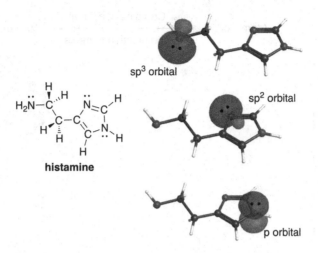

Figure 2.20 Nitrogen atom lone pairs (p, sp², sp³) in histamine.

unhybridized p-orbital. The molecule will adopt a trigonal planar geometry with the second bond of the double bond existing above and below the molecular plane.

An sp-hybridized carbon atom can make two double bonds (or one single and one triple bond) and will adopt a linear geometry with the bonds using the p-orbitals perpendicular to each other and the molecular axis. The two hybrid orbitals in acetylene are arranged linearly and 180° apart.

LONE PAIR AND ORBITALS

Finally, it is worth noting that Lewis structures show all lone-pair electrons as roughly equivalent. However, lone-pair electrons can occupy hybridized (and unhybridized) orbitals and the specific orbital (energy level) will give rise to different behaviour of each lone pair in a molecule. Consider the structure of histamine (Figure 2.20).

There is no clear way of identifying that the three different lone pairs are energetically different. If we consider the hybridization of each atom (using the sp^{SN-1} mnemonic), we can see that there is a difference in hybridization and so there will be a difference in the type of orbital each lone pair occupies. The fact that one lone pair is a p-orbital lone pair can only be inferred from delocalization, see the section "Electron Delocalization (Conjugation)" below.

Problem 2.11. What is the hybridization of each carbon in the following molecule? Predict the bond angles and shape (tetrahedral, trigonal planar, linear) for each carbon atom.

$$\begin{array}{c} \text{H} \\ | \\ \text{H--C--H} \qquad :\!\ddot{\text{O}} \quad \text{H} \\ | \qquad\qquad \| \quad | \\ :\!\text{N--C}\!\equiv\!\text{C--C--C--H} \\ | \qquad\qquad\quad | \\ \text{H--C--H} \qquad\qquad \text{H} \\ | \\ \text{H} \end{array}$$

QUANTUM THEORIES OF BONDING: Valence Bond (VB) and Molecular Orbital (MO) Theory

Lewis structures provide a way to visualize the bonding in covalent compounds. As we have seen, however, electrons in atoms are best described through a quantum mechanical approach. Lewis structures do not convey how the configuration of electrons changes as atoms combine into compounds nor do they explain the change in energy (Figure 2.7) that is observed when atoms form bonds. With the Lewis structure model as our guide, we will see in the following sections how quantum mechanics strengthens and challenges our understanding of bonds and bonding. In turn, we will consider two complementary quantum approaches: VB theory and MO theory.

Valence Bond Theory

In valence bond theory the key assumption, as with Lewis structures above, is that electron pairs are localized on a single atom (one centre) or between two atoms (two centres). Electron pairs

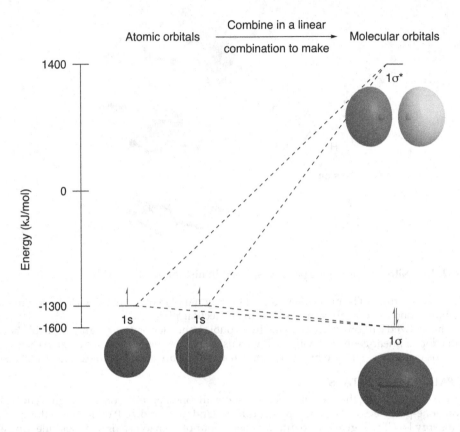

Figure 2.21 σ-orbital mixing in dihydrogen (H_2). Orbital images and energy values calculated at the CCSD(T)/cc-pVTZ level of theory and basis set.

localized to a single centre are called non-bonding electrons, or lone pairs in the language of Lewis structures. Electron pairs localized between two centres are called a shared pair or a bond pair. Recall that when a bond forms, the formation of a bond is the result of stabilization (lowering the energy) of the electrons involved (Figure 2.21). For our first example of VB theory, let us consider the simplest molecule dihydrogen (H_2). Here (Figure 2.21) we have two hydrogen atoms; each hydrogen has one electron located in the 1s-orbital. Bonding is the result of the overlap between two singly occupied atomic orbitals to form two new molecular orbitals.[23] The lower energy orbital is a bonding (stabilizing) orbital with orbital density between the two atoms. The higher energy orbital is an antibonding (destabilizing) orbital with a node between the two atoms. Since the electron density is along the internuclear axis, we call these orbitals σ orbitals. The bonding orbital is termed σ and the antibonding orbital is labelled as σ*.

Now, with our molecular orbital diagram, we add electrons following the rules we saw for orbital filling: the Aufbau principle (electrons are filled from the lowest energy orbital to the highest), Hund's rule (equal energy orbitals are filled singly before being doubly filled), and the Pauli exclusion principle (at most there can be two spin-paired electrons per orbital). If we put the two hydrogen electrons into the diagram in Figure 2.21, then, we can see that the electrons are now lower in energy (with the hydrogen atoms combined) than they were on two separate hydrogen atoms. This lowering of electron energy when shared by the two hydrogen atoms is what we call a bond.

At this point, we can introduce a formal definition for bond order, that is the number of bonds between two atoms. The bond order (p_{rs}) between two atoms (r and s) is half the difference between the number of electrons in bonding orbitals and the number of electrons in antibonding (*) orbitals. For dihydrogen, $p_{HH} = \frac{1}{2}(2 - 0) = 1.0$, that is a single bond exists between the two atoms of dihydrogen. In contrast, dihelium would have the same molecular orbital energy diagram but there are now four electrons that fill the σ and σ*-orbitals. For dihelium $p_{HeHe} = \frac{1}{2}(2 - 2) = 0$, no

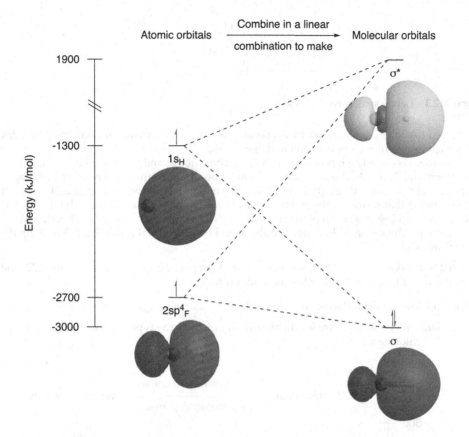

Figure 2.22 σ-orbital mixing in hydrogen fluoride (HF). Orbital images and energy values calculated at the CCSD(T)/cc-pVTZ level of theory and basis set.

bond exists between the two atoms of dihelium and no stable molecule forms when two helium atoms combine.

Problem 2.12 Which has a larger ionization energy (E_i) H or H_2. Explain your answer.

Problem 2.13 The dihelium cation (He_2^+) is a stable polyatomic ion with a dissociation energy (E_d) of 238 kJ/mol. Using Figure 2.21, explain why He_2^+ is stable in contrast to unstable He_2.

Single σ bonds are generally the lowest energy bond that can form between two atoms. Also, the σ-orbital density lies along the internuclear axis and is spherically symmetric, which means that σ bonds can freely rotate. Before considering other types of bonding, we will first look at σ bonding in a slightly more complicated example: hydrogen fluoride.

The bonding in HF comes from the overlap of the hydrogen 1s-orbital with a valence fluorine orbital. The valence orbitals of fluorine are 2s and $2p_x$, $2p_y$, and $2p_z$. When a main group element with p-orbitals bonds, the valence s orbital hybridizes (mixes) to some extent with the p-orbitals. The nature of the hybridization depends on the molecular shape and the energy of the electrons involved in the two atoms that bond. For HF, the 2s-orbital mixes with the $2p_z$-orbital to make two sp hybrid orbitals one that is 78.5% s and 21.5% p (s^4p) and the other that is 21.5% s and 78.5% p (sp^4) (Figure 2.22). The higher energy sp^4-orbital will mix with the hydrogen 1s orbital to produce the σ bond. The remaining orbitals on fluorine – the s^4p hybrid orbital and the unhybridized $2p_x$- and $2p_y$-orbitals – will contain the fluorine lone pairs. Since the lone pairs are in two different types of orbitals, the three lone pairs are non-equivalent energetically.

When looking at a Lewis structure, any single bond between two atoms is a σ bond. If two atoms are connected by a double (or triple bond), then there exists both a σ bond and a π bond (or π bonds) between the two atoms. Let's consider the bonding in ethene (Figure 2.23). Each carbon

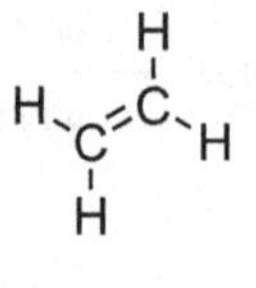

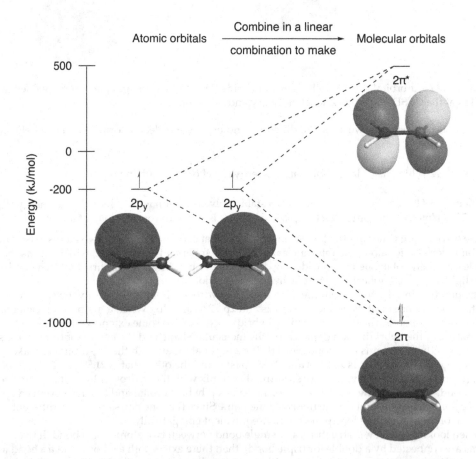

Figure 2.23   Lewis structure of ethylene.

is connected to each hydrogen atom by a σ bond. The two carbon atoms are connected by a double bond, which means there is one σ bond and one π bond.

   Because there is already a σ bond between the carbon atoms, and the σ bond orbital lies along the internuclear axis, the π bond must form through some other type of orbital overlap. Specifically, π bonds form through the facial overlap of parallel p-orbitals. Figure 2.24 shows the p-orbital mixing that results in the creation of a π bond. Note that a singular p-orbital has orbital density above and below a nodal plane at the atom. This means that a singular π bond, which is formed from p-orbitals, has orbital density above and below a nodal plane that exists along the internuclear axis.

Problem 2.14 Unlike σ bonds, π bonds cannot rotate. Compare Figure 2.24 with Figure 2.22, and provide an explanation of why π bonds cannot rotate.

Problem 2.15 Answer the following questions about dichlorine (Cl_2).

   a. Draw a Lewis structure for dichlorine and identify the type of bond that exists between the chlorine atoms.

Figure 2.24   π-orbital mixing in ethene. Orbital images and energy values calculated at the CCSD(T)/cc-pVTZ level of theory and basis set.

b. Dichlorine (Cl_2) has a dissociation energy (E_d) of 243 kJ/mol, but when one electron is added (Cl_2^-) the dissociation energy decreases to 122 kJ/mol.

  a. Where (into which molecular orbital) does this added electron go?

  b. Why does this decrease the dissociation energy?

c. The bond lengths for these two are 201 pm and 264 pm. Which bond length goes with which molecular entity? Explain.

d. What would adding one more electron do to the bond energy and the bond length? Explain.

Valence bond theory compliments our chemical sense that we can glean from Lewis structures while also grounding, validating, and deepening our understanding of bonding from a true

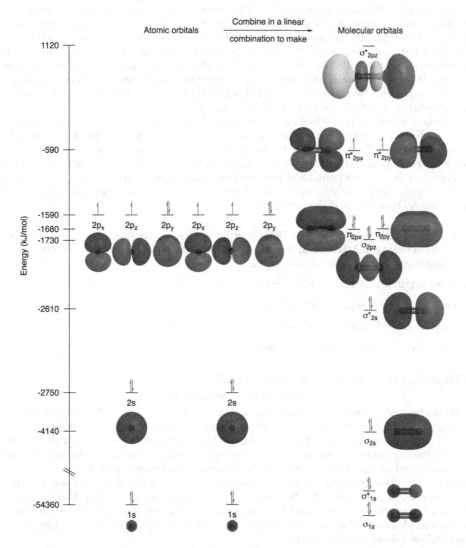

Figure 2.25 Full molecular orbital treatment of the bonding in dioxygen showing the combination of atomic orbitals (*left*) to produce molecular orbitals (*right*). Orbital images and energy values calculated at the CCSD(T)/cc-pVTZ level of theory and basis set. Note that the interaction of the electron spin and the orbital angular momentum leads to slightly (<100 kJ/mol) different energy values for filled or partially filled orbitals that are degenerate when unfilled: p, d, and f. The importance and effects of spin-orbit coupling are beyond the scope of this book and the values in the diagrams shown here do not include the effects of spin-orbit coupling.

quantum mechanical perspective. We will now consider molecular orbital theory, an alternative – but not contradictory – approach that provides us with a new and different way to consider bonding.

Molecular Orbital Theory

Molecular orbital theory is a quantum mechanical theory that does not approach chemical structure with Lewis structures in mind. Instead, molecular orbital theory was based on developing a quantum framework for molecular bonding based on spectroscopy – the interaction of light and molecules.[24] Given this different starting point, MO theory does not assume that electrons are localized, either to one atom or between two atoms, nor does it assume that only valence electrons are relevant to bonding. In fact, molecular orbital theory considers the interaction of all orbitals in bonding, and electrons can be delocalized over the entire molecule.

Figure 2.25 shows the molecular orbital diagram for dioxygen (O_2). This diagram is more complicated than the orbital diagrams we saw for VB theory because we are considering the entire set of orbitals and their combinations. There are some observations worth noting. First, as we saw in VB theory, bonding orbitals (those notes denoted with a *) are lower in energy than the atomic orbitals that combine to produce them. For example, the $2p_z$-orbitals (–1590 kJ/mol) combine to produce the σ_{2pz} bonding orbital (–1730 kJ/mol). Second, s-orbitals form from s-orbitals and from directly overlapping p-orbitals, while π bonds form from perpendicular, facially overlapping p-orbitals. Also, as we saw above, σ bonds (–1730 kJ/mol) are lower in energy than π bonds (–1680 kJ/mol). Finally, the bond order for dioxygen is 2.0 ($p_{OO} = \frac{1}{2}(10 - 8) = 2.0$).

Since molecular orbital theory looks at all the orbital interactions for a system, rather than the overlap of individual overlapping orbitals, one result is more immediately clear from the MO diagram (Figure 2.25): the ground state for dioxygen has two unpaired electrons. This means that dioxygen is paramagnetic, it will align with a magnetic field, in contrast to dihydrogen (Figure 2.21) where the electrons are paired and the molecule is diamagnetic, it will align against a magnetic field.[25] With deeper analysis, valence bond theory can also explain the paramagnetism of dioxygen, but MO theory can do so more readily.

Problem 2.16 Use MO theory to explain the following bond length data, in picometers (pm), and dissociation energy (E_d) data (in kJ/mol). Can you provide a trend for how bond order relates to bond length and bond dissociation energy?

O_2^+ bond length = 112 pm; E_d = 625 kJ/mol

O_2 bond length = 121 pm; E_d = 498 kJ/mol

O_2^- bond length = 128 pm; E_d = 439 kJ/mol

O_2^{2-} bond length = 149 pm; E_d = 213 kJ/mol

ELECTRON DELOCALIZATION (CONJUGATION)
Contributing Structures

We will now turn our attention to delocalization. Let's start by considering formate (HCO_2^-). Calculation of the number of bonds and unshared electrons shows that formate should have four bonds and ten unshared electrons. Carbon is the least electronegative element so it will go in the centre with the oxygen atoms and hydrogen atoms surrounding it. Drawing in the bonds, lone-pair electrons, and formal charges, we could draw formate as shown in Figure 2.26.

We have a model – Lewis structures are model structures of the real molecular species – in hand for formate. The model indicates that the two oxygen atoms are non-equivalent. One oxygen is negatively charged, the other is neutral. One oxygen is making a double bond to carbon, and one is making a single bond, which we would expect to show up as differences in bond lengths since double bonds are shorter than single bonds.

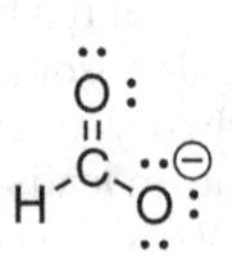

Figure 2.26 A Lewis structure for formate (HCO_2^-).

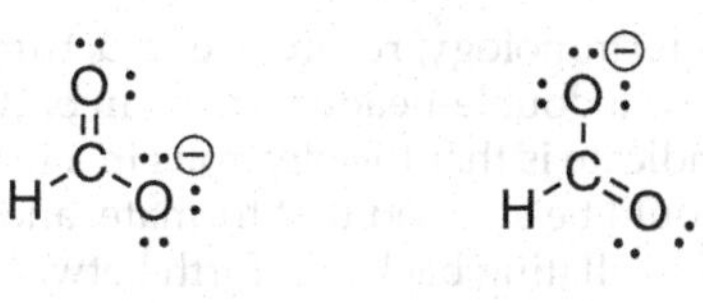

Figure 2.27 Valid Lewis structures of formate.

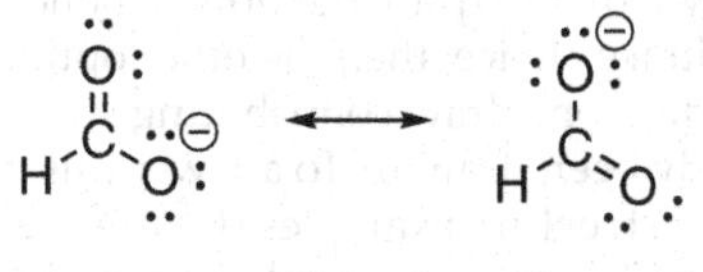

Figure 2.28 Two contributing structures of formate.

As with any good scientific model, it must be tested against empirical results. X-ray crystallographic analysis of formate shows that the two oxygen atoms are equivalent: the C–O bond lengths are both 127 pm.[26] What does this say about our Lewis structure model of formate? These results tell us that the original Lewis structure model has limitations. Specifically, the assumption that all electrons are localized, that is they are either on a single atom as a lone pair or they are between two atoms as a bond. If we go back and consider the Lewis structure that was drawn in Figure 2.26, it was arbitrarily decided, by the author, that the double bond would be drawn between the top oxygen atom and the carbon; however, it would be equally acceptable to draw either of the Lewis structures shown in Figure 2.27.

Both Lewis structures in Figure 2.27 are equally correct to draw (in that they all follow the rules of drawing good Lewis structures). Each Lewis structure differs from the other only in terms of where we drew the lone-pair electrons and the double bond. However, each of these Lewis structures is – by itself – a bad depiction of formate. In these situations, where multiple Lewis structures are acceptable to draw for a molecule, to get an accurate depiction of the molecule we must consider all of them. We call all the valid Lewis structures for a single molecule contributing

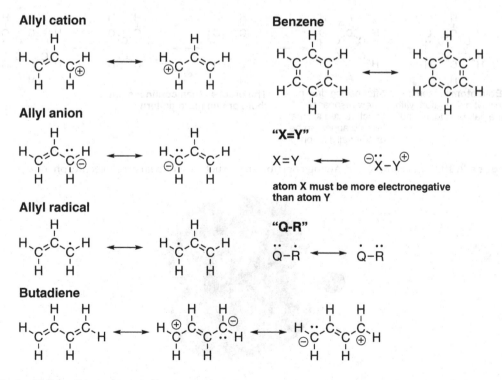

Figure 2.29 Common contributing structure patterns.

structures or, following Pauling's terminology, resonance structures.[27,28] To show their relationship as contributing structures, we draw a double-headed arrow in between each (Figure 2.28).[29]

What contributing structures indicate is that the electrons in a molecule are spread among multiple atoms, delocalized. It should be stressed that formate, and any molecule for which there are contributing structures, is not oscillating back and forth between the different forms, each Lewis structure is itself a poor approximation by chemists to describe a more complex reality. Contributing structures must be included whenever an arbitrary choice is made in a Lewis structure drawing, here the choice of whether to draw the double bond between the top oxygen or the right oxygen. If you make an arbitrary choice, then the other option must be drawn and considered.

Now the question arises, how does one draw contributing structures for a molecule (assuming a good Lewis structure has already been drawn)? To answer this question, let us examine several examples of common contributing structure examples (Figure 2.29).

The common examples in Figure 2.29 are then used as a template and looked for in larger, more complicated molecules. The contributing structure is drawn for that pattern and the process is then repeated as in Figure 2.30.

A note needs to be made on hybridization and conjugation. If an atom is involved in conjugation, then the sp^{SN-1} mnemonic can give incorrect results. To apply the sp^{SN-1} mnemonic for a particular atom, the contributing structure where the atom in question is making the most bonds must be considered. The hybridization is then assigned. That hybridization is then the correct hybridization for that atom, regardless of the contributing structure.

Problem 2.17 Draw all contributing structures for each of the following.

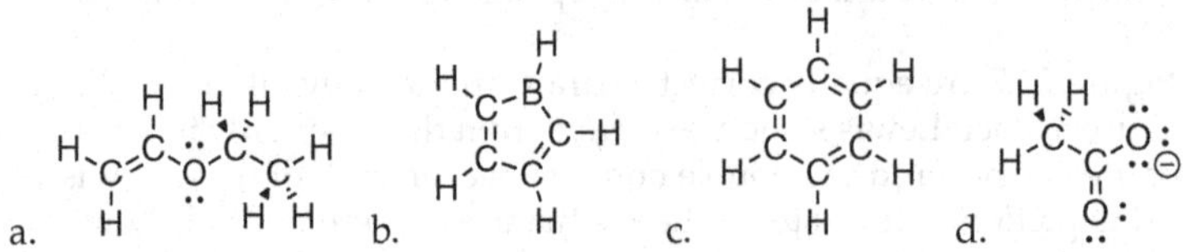

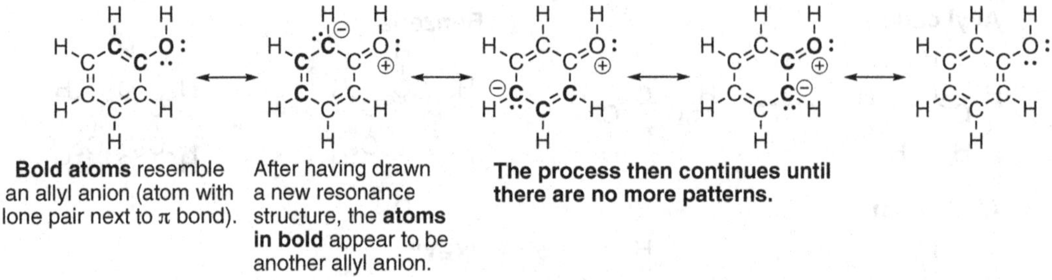

Bold atoms resemble an allyl anion (atom with lone pair next to π bond).

After having drawn a new resonance structure, the **atoms in bold** appear to be another allyl anion.

The process then continues until there are no more patterns.

Figure 2.30 An example of drawing contributing structures for a larger molecule, phenol.

Figure 2.31 π bonding orbital between carbon and the right oxygen atom.

Now, when drawing contributing structures, not every structure is valid nor does every contributing structure matter equally. Some basic guidelines for contributing structures:

■ The relative position of atoms cannot change between contributing structures.

■ Atom hybridization does not change between contributing structures.

■ Total formal charge cannot change between contributing structures.

■ Only lone-pair and bond electrons may differ between contributing structures.

In terms of ranking the importance of contributing structures, one should consider the following only: contributing structures where all atoms have a full octet are of higher priority than those contributing structures where all atoms do not have a full octet. [30]

Quantum Descriptions Of Delocalization

Delocalization is best described through quantum mechanics as the result of orbital overlap across multiple atoms. Valence bond theory describes delocalization in terms of conjugation. That is, like Lewis structures, VB theory assumes that electrons are localized on one atom or between two atoms. From this perspective, then, delocalization is viewed as the overlap of adjacent π electrons with an existing π bond or π system. We will reconsider the delocalized π bonding in formate (HCO_2^-), Figure 2.28. A π bond exists between the central carbon atom and the right oxygen atom (Figure 2.31), which is made up of the overlap of a carbon p-orbital and an oxygen p-orbital.

In formate, the left oxygen atom donates a p-orbital lone pair that is accepted by the π^*_{CO}-orbital (Figure 2.32). This donor–acceptor interaction creates a partial π bond between the left oxygen

Figure 2.32   A $n_O \to \pi^*_{CO}$ donor–acceptor interaction between carbon and the left oxygen atom.

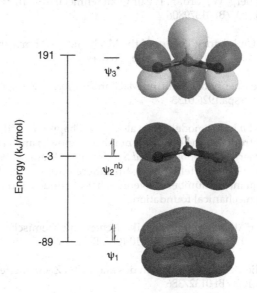

Figure 2.33 π molecular orbital diagrams for formate.

atom and the carbon atom. And because electron density is added to the π^*_{CO}-orbital, the amount of π bonding between the carbon atom and the right oxygen atom is reduced. This donor–acceptor interaction is called conjugation. We show this conjugation with contributing structures (Figure 2.28).

When considered through the lens of molecular orbital theory, the bonding in formate is the result of three p_z-orbitals overlapping (one each from carbon and the two oxygen atoms) to produce three molecular orbitals ($\psi_1 - \psi_3$), Figure 2.33. The lowest energy orbital (ψ_1) is a bonding orbital that shows a two-electron bond extending across all three atoms. The distribution of electrons across more than two atoms is called delocalization. The other occupied orbital (ψ_2) is non-bonding (nb), which shows electron density on the two end oxygen atoms and no bonding between any of the three atoms.

NOTES

1. The earliest drafts of this chapter were originally based on a primer produced by Dr Brian Esselman at UW-Madison.

2. The use of emission spectra to identify elements was first reported by Charles Wheatstone (Wheatstone, C. On the Prismatic Decomposition of Electrical Light. *Br. Assoc. Adv. Sci., Report*, **1835**, 11–12.) and Angstrom first demonstrated the spectra lines of gases (Angstrom, A.J. Otiska undersökningar. *Kungl. Svenska Vetenskapsakad. Handl* **1852**, 40, 339–366.)

3. Rydberg, J.R. Recherches sur la Constitution des Spectres D'Émission des Éléments Chimiques. *Kungl. Svenska Vetenskapsakad. Handl.* **1888–1889**, 23 (11), 1–155.

4. Thomson, J.J. XL. Cathode Rays. *Lond. Edinb. Dublin Philos. Mag. J. Sci.* **1897**, 44 (269), 293–316. DOI: 10.1080/14786449708621070

5. Bohr, N. Der Bau der Atome und die physikalischen und chemischen Eigenschaften der Elemente. *Z. Physik*, **1922**, 9 (1), 1–67. DOI: 10.1007/BF01326955

6. i) Heisenberg, W. Über quantentheoretische Umdeutung kinematischer und mechanischer Beziehungen. *Z. Phys.*, **1925**, 33 (1), 879–893. DOI: 10.1007/BF01328377

   ii) Born, M.; Jordan, P. Zur Quantenmechanik. *Z. Phys.*, **1925**, 34 (1), 858–888. DOI: 10.1007/BF01328531

   iii) Born, M.; Heisenberg, W.; Jordan, P. Zur Quantenmechanik. II. *Z. Phys.*, **1926**, 35 (8–9), 557–615. DOI: 10.1007/BF01379806

7. Schrödinger, E. An Undulatory Theory of the Mechanics of Atoms and Molecules. *Phys. Rev.*, **1926**, 28 (6), 1049–1070. DOI: 10.1103/PhysRev.28.1049

8. Dirac, P.A.M. On the Theory of Quantum Mechanics. *Proc. R. Soc. Lond. A*, **1926**, 112 (762), 661–677. DOI: 10.1098/rspa.1926.0133

9. Bohr had introduced n in his model of the atom (see Chapter 6). Arthur Sommerfeld refined Bohr's model and introduced the quantum numbers k (now ℓ) and m (now m_ℓ) to try to refine Bohr's model (Sommerfeld, A. *Atombau und Spektrallinien*. Braunschweig: Friedr. Vieweg & Sohn, **1921**.). The Bohr-Sommerfeld model still suffered from unresolved contradictions and paradoxes, but the quantum number convention has continued and been given a more rigorous, truly quantum mechanical foundation.

10. Bohr, N. Der Bau der Atome und die physikalischen und chemischen Eigenschaften der Elemente. *Z. Physik*, **1922**, 9 (1), 1–67. DOI: 10.1007/BF01326955

11. i) Pauli, W. Über die Gesetzmässigkeiten des anomalen Zeemaneffektes. *Z. Phys.*, **1923**, 16, 155–164. DOI: 10.1007/BF01327386

ii) Pauli, W. Über den Einfluss der Geschwindigkeitsabhängigkeit der Elektronenmasse auf den Zeemaneffekt. *Z Phys.*, **1925**, *31*, 373–385. DOI: 10.1007/BF02980592

iii) Pauli, W. Über den Zusammenhang des Abschlusses der Elektronengruppen im Atom mit der Komplexstruktur der Spektren. *Z. Phys.*, **1925**, *31*, 765–783. DOI: 10.1007/BF02980631

12. Hund, F. Zur Deutung verwickelter Spektren, I. *Z. Phys.*, **1925**, *33*, 345–371. DOI: 10.1007/BF01328319

13. Lewis, G.N. The Atom and the Molecule. *J. Am. Chem. Soc.*, **1916**, *38* (4), 762–785. DOI: 10.1021/ja02261a002

14. Langmuir, I. The Arrangement of Electrons in Atoms and Molecules. *J. Am. Chem. Soc.*, **1919**, *41* (6), 868–934. DOI: 10.1021/ja02227a002

15. i) The data most widely cited today are the revised Pauling values given by Allred: Allred, A.L. Electronegativity values from thermochemical data. *J. Inorg. Nucl.*, **1961**, *17* (3–4), 215–221. DOI: 10.1016/0022-1902(61)80142-5

ii) Tellurium value from: Huggins, M.L. Bond energies and polarities. *J. Am. Chem. Soc.*, **1953**, *75* (17), 4123–4126. DOI: 10.1021/ja01113a001.

iii) Noble gas values from: Allen, L.C.; Huheey, J.E. The definition of electronegativity and the chemistry of the noble gases. *J. Inorg. Nucl. Chem.*, **1980**, *42* (10), 1523–1524. DOI: 10.1016/0022-1902(80)80132-1

iv) All other values with only one decimal place are from: Gordy, W.; Thomas, W.J.O. Electronegativity of the Elements. *J. Chem. Phys.*, **1956**, *24* (2), 439–444. DOI: 10.1063/1.1742493

16. Coulomb, C.A. Premier mémoire sur l'éctricité et le magnétisme. *Histoire de l'Academie Royale des Sc.*, **1785**, 569–577.

17. i) Lewis, G.N. The Atom and the Molecule. *J. Am. Chem. Soc.*, **1916**, *38* (4), 762–785. DOI: 10.1021/ja02261a002

ii) Lewis, G.N. The Magnetochemical Theory. *Chem. Rev.*, **1924**, *1* (2), 231–248. DOI: 10.1021/cr60002a003

iii) Lewis, G.N. The Chemical Bond. *J. Chem. Phys.*, **1933**, *1* (1), 17–28. DOI: 10.1063/1.1749214

18. Formal charge comes from Langmuir's concept of residual charge: Langmuir, I. Types of Valence. *Science.* **1921**, *54*, 59–67. DOI: 10.1126/science.54.1386.59

19. Lavoisier, A.L. *Elements of Chemistry*; Creech: Edinburgh, 1790; pp. 159–172. Latimer, W.M. *The Oxidation States of the Elements and their Potentials in Aqueous Solution;* Prentice-Hall: New York, NY, 1938; p vii.

20. Reed, A. E.; Weinstock, R. B.; Weinhold, F. Natural population analysis. *J. Chem. Phys.* **1985**, *83* (2), 735–746. DOI: 10.1063/1.449486

21. Glendening, E. D.; Badenhoop, J. K.; Reed, A. E.; Carpenter, J. E.; Bohmann, J. A.; Morales, C. M.; Landis, C. R.; Weinhold, F. A. *NBO 6.0*, Theoretical Chemistry Institute: University of Wisconsin, Madison, 2013.

22. Gillespie, R.J.; Nyholm, R.S. Inorganic stereochemistry. *Q. Rev. Chem. Soc.*, **1957**, *11*, 339–380. DOI: 10.1039/QR9571100339

23. i) The original development of VB theory: Heitler, W.; London, F. Wechselwirkung neutraler Atome und homöopolare Bindung nach der Quantenmechanik. *Z. Phys.*, **1927**, *44* (6–7), 455–472. DOI: 10.1007/BF01397394

ii) A modern VB system for quantum mechanical calculations: E. D. Glendening, J, K. Badenhoop, A. E. Reed, J. E. Carpenter, J. A. Bohmann, C. M. Morales, P. Karafiloglou, C. R. Landis, and F. Weinhold, Theoretical Chemistry Institute, University of Wisconsin, Madison (2018).

24. i) Hund, F. Zur Deutung einiger Erscheinungen in den Molekelspektren, *Z. Phys.*, **1926**, *36* (9–10), 657–674. DOI: 10.1007/BF01400155

ii) Mulliken, R.S. The Assignment of Quantum Numbers for Electrons in Molecules. I. *Phys. Rev.*, **1928**, *32* (2), 186–222. DOI: 10.1103/physrev.32.186

iii) Lennard-Jones, J.E. The Electronic Structure of Some Diatomic Molecules, *Trans. Faraday Soc.*, **1929**, *25*, 668–686. DOI: 10.1039/TF9292500668

25. i) "Paramagnetic". *IUPAC. Compendium of Chemical Terminology*, 2nd ed. (the "Gold Book"). Compiled by A. D. McNaught and A. Wilkinson. Blackwell Scientific Publications, Oxford (1997). Online version (2019–) created by S. J. Chalk. DOI: 10.1351/goldbook.P04404

ii) "Diamagnetic". IUPAC. *Compendium of Chemical Terminology*, 2nd ed. (the "Gold Book"). Compiled by A. D. McNaught and A. Wilkinson. Blackwell Scientific Publications, Oxford (1997). Online version (2019–) created by S. J. Chalk. DOI: 10.1351/goldbook.d01668

26. Zachariasen, W.H. The crystal structure of sodium formate, $NaHCO_2$. *J. Am. Chem. Soc.*, **1940**, *62* (5), 1011–1013. DOI: 10.1021/ja01862a007

27. This concept is most often termed "resonance" following the name given by Pauling:

i) Pauling, L. The nature of the chemical bond. III. The transition from one extreme bond type to another. *J. Am. Chem. Soc.*, **1932**, *54* (3), 988–1003. DOI: 10.1021/ja01342a022

ii) Pauling, L.; Sherman, J. The nature of the chemical bond. VI. The Calculation from Thermochemical Data of the Energy of Resonance of Molecules Among Several Electronic Structures. *J. Chem. Phys.*, **1933**, *1* (8), 606–617. DOI: 10.1063/1.1749335

28. Ingold's development of similar ideas, along with his development of electron-pushing arrows, led him to term this phenomenon mesomerism: Ingold, C.K. Principles of an electronic theory of organic reactions. *Chem. Rev.*, **1934**, *15* (2), 225–274. DOI: 10.1021/cr60051a003

29. The double-headed arrow notation was introduced in Germany: Eistert, B. Zur Schreibweise chemischer Formeln und Reaktionsabläufe. *Ber. Dtsch. Chem. Ges.*, **1938**, *71* (2), 237–240. DOI: 10.1002/cber.19380710206

30. Suidan, L.; Badenhoop, J.K.; Glendening, E.D.; Weinhold, F. Common textbook and teaching misrepresentations of Lewis structures. *J. Chem. Educ.*, **1995**, *72* (7), 583–586. DOI: 10.1021/ed072p583

3 Alkanes, Cycloalkanes, and Nomenclature Fundamentals

Hydrocarbons – molecules made up of hydrogen and carbon atoms – are the traditional feedstock for organic chemistry. Hydrocarbons come from the extraction of coal, petroleum, and natural gas, nonrenewable feedstocks. Possible renewable feedstocks include biofuels from algae or biomass or using lignocellulose biomass directly (xylochemistry).[1] Hydrocarbons, whether from oil or biofuels, are a highly reduced feedstock, and so synthetic organic chemistry of hydrocarbons primarily involves selective oxidation (removing hydrogen atoms or adding bonds to N, O, F, P, S, Cl, Br, and I atoms). Lignocellulose is a highly oxygenated feedstock, and so organic chemistry starting from lignocellulose primarily involves selective reduction.

Alkanes (Table 3.1) are the first type of hydrocarbon that we will consider. Alkanes are saturated hydrocarbons, that is there is no possibility of adding any more hydrogens to the structure and have a general formula of C_nH_{2n+2}. Alkanes are also referred to as aliphatic hydrocarbons or as paraffins. Note, that the *-ane* suffix in the name denotes the alk*ane* class of compounds.

SKELETAL STRUCTURE

It is at this point that we will work on understanding the skeletal structures (Table 3.1), which are common throughout organic chemistry. As can be seen, even relatively simple hydrocarbons can involve a significant amount of drawing time to show a correct Lewis structure. The skeletal notation is used as a systematic shorthand notation. For skeletal structures, the following conventions are followed:

1. Carbon atoms are not shown unless necessary.

2. Hydrogen atoms on carbon atoms are not shown. Unless otherwise indicated (by formal charges or a single dot) there are assumed to be enough hydrogen atoms to fulfil the valence of carbon.

3. All other atoms are shown including all hydrogen atoms on non-carbon atoms (heteroatoms).

4. The lines indicating the bonds between atoms are shown. It is assumed that there is a carbon atom at the end of each line and at each intersection.

5. Multiple bonds are shown with multiple lines.

Problem 3.1 Convert each Lewis structure into a skeletal structure

a.

b.

c.

d.

e.

DOI: 10.1201/9781003479352-3

Table 3.1 The Name, Formula, and Structure of the First Ten Linear Alkanes

IUPAC Name	Molecular Formula	Condensed Structure	Lewis Structure	Skeletal Structure
methane	CH_4	CH_4	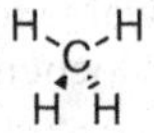 *(Lewis structure)*	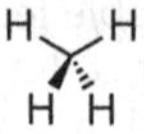 *(skeletal structure)*
ethane	C_2H_6	CH_3CH_3	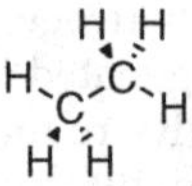 *(Lewis structure)*	*(skeletal structure)*
propane	C_3H_8	$CH_3CH_2CH_3$	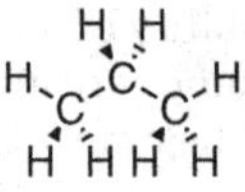 *(Lewis structure)*	 *(skeletal structure)*
butane	C_4H_{10}	$CH_3(CH_2)_2CH_3$	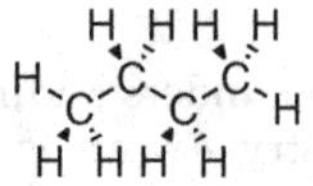 *(Lewis structure)*	 *(skeletal structure)*
pentane	C_5H_{12}	$CH_3(CH_2)_3CH_3$	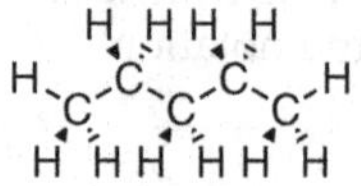 *(Lewis structure)*	 *(skeletal structure)*
hexane	C_6H_{14}	$CH_3(CH_2)_4CH_3$	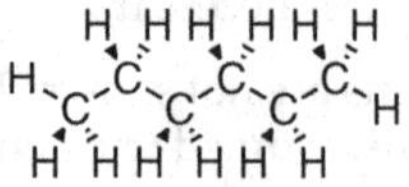 *(Lewis structure)*	 *(skeletal structure)*
heptane	C_7H_{16}	$CH_3(CH_2)_5CH_3$	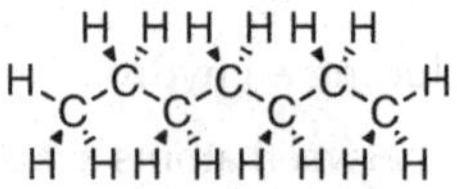 *(Lewis structure)*	 *(skeletal structure)*
octane	C_8H_{18}	$CH_3(CH_2)_6CH_3$	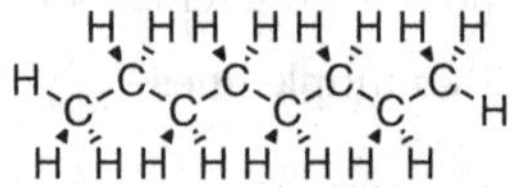 *(Lewis structure)*	 *(skeletal structure)*
nonane	C_9H_{20}	$CH_3(CH_2)_7CH_3$	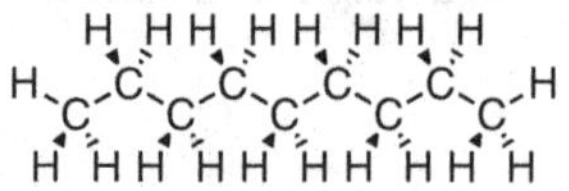 *(Lewis structure)*	 *(skeletal structure)*
decane	$C_{10}H_{22}$	$CH_3(CH_2)_8CH_3$	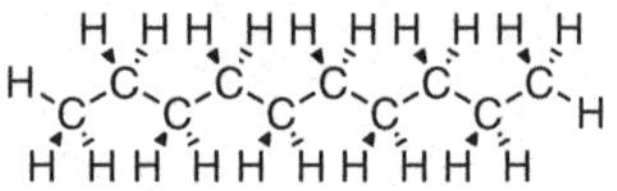 *(Lewis structure)*	 *(skeletal structure)*

Problem 3.2 Convert each skeletal structure into a Lewis structure

a.  *(skeletal structure)*

b. 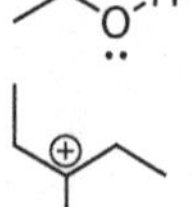 *(skeletal structure)*

c. 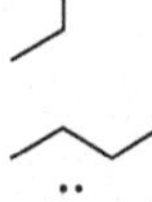 *(skeletal structure)*

d. 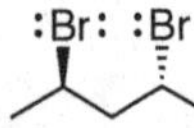 *(skeletal structure)*

e. *(skeletal structure)*

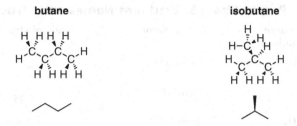

Figure 3.1 Linear and branched isomers of C_4H_{10}.

Not all alkanes (or all organic molecules) are straight chains. Alkanes can have branched structures too. This ability to constitute the same number of carbons and hydrogens into differently structured molecules gives rise to **constitutional isomerism**. There are two possible isomers with the formula C_4H_{10} (Figure 3.1).

Problem 3.3 Draw all the possible constitutional isomers for the formula C_5H_{12}.

IUPAC NOMENCLATURE

Given the vast number of isomers and structures that one might imagine for organic structures, keeping track of all these structures could quickly become confusing. As such, a systematic means of naming organic compounds is required. The naming convention for organic molecules is that developed by the International Union of Pure and Applied Chemistry (IUPAC).[2] For this chapter, the focus will be on the basics of nomenclature. In later chapters, additional layers of complexity will be added. The structure of any IUPAC name for organic molecules is as follows.

stereochemistry identifier – locant/substituents – parent – suffix

Stereochemistry identifier provides stereochemical designations (Chapter 6).
Locant/substituents indicates the location of and name or types of groups attached to the parent change.
The **parent** chain is the longest, continuous chain of carbon atoms (which contain the functional group if one is present), which gives the greatest number of substituents.
A **suffix** is an ending that indicates the highest priority functional group present in the molecule.

The first piece of information that one needs to convert a name to a structure or a structure to a name is the number of carbons in the longest chain (Table 3.2).

The -ane ending which indicates that the molecule is an alkane. Other suffixes will be introduced as other types of molecules are introduced. If there is a carbon chain that is a substituent on the parent chain, the suffix -yl is used (Table 3.3).

For a structure with halogen atoms fluorine, chlorine, bromine, or iodine on the parent chain they are listed as substituents and given the substituent names fluoro-, chloro-, bromo-, iodo-.

Table 3.2 Parent Chain Names and the Number of Carbons Represented

Parent	Number of carbon atoms in chain
meth-	1
eth-	2
prop-	3
but-	4
pent-	5
hex-	6
hept-	7
oct-	8
non-	9
dec-	10

Table 3.3   Parent Chain Substituent Names and Structures

IUPAC name (abbreviation)	Trivial Name	Skeletal structure
methyl (Me)	methyl (Me)	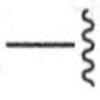
ethyl (Et)	ethyl (Et)	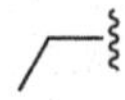
propyl (Pr)	propyl (Pr)	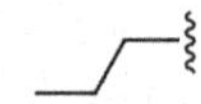
1-methylethyl	isopropyl (^{i}Pr)	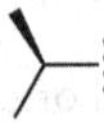
butyl (Bu)	n-butyl	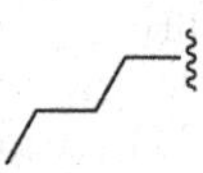
1-methylpropyl	sec-butyl (^{s}Bu)	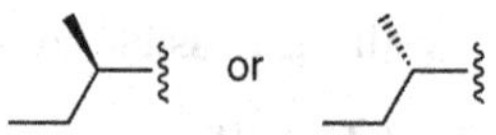 or
2-methylpropyl	isobutyl (^{i}Bu)	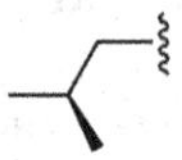
1,1-dimethylethyl	tert-butyl (^{t}Bu)	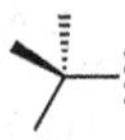

Note the squiggle at the end of the line indicates the attachment point onto the parent chain

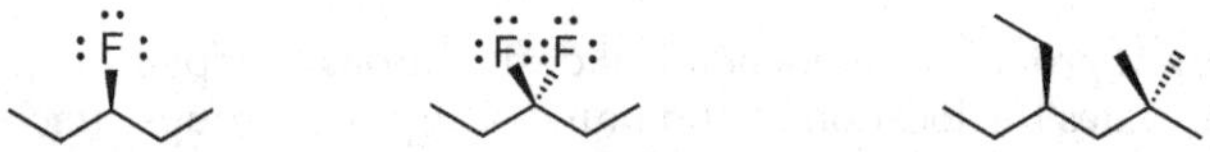

3-fluoropentane   **3,3-difluoropentane**   **4-ethyl-2,2-dimethylhexane**

Figure 3.2   Example skeletal structures and IUPAC names for 3-fluoropentane (*left*), 3,3-difluoropentane (*right*), and 4-ethyl-2,2-dimethylhexane (*right*).

If there are multiple of the same group then there is a prefix that indicates how many repeating groups there are: di (two), tri (three), tetra (four), penta (five), hexa (six), hepta (seven), octa (eight), nona (nine), and deca (ten).

The parent chain is numbered to give i) the greatest number of substituents, and ii) the smallest locant total. Substituents are listed alphabetically ignoring prefixes (*sec, tert*, di, tri, tetra).

Problem 3.4 Provide names for the following structures.

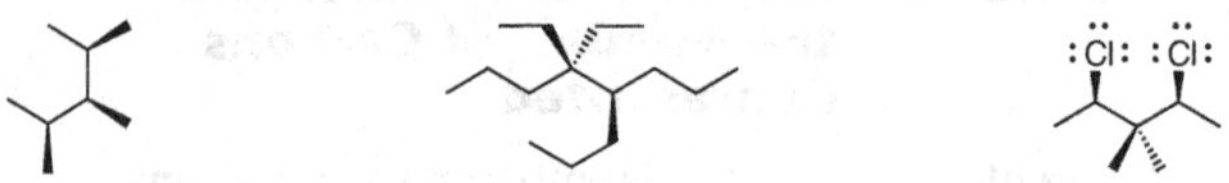

Problem 3.5 Convert each name to a correct structure.

5-isobutylnonane   **3,3-diiodo-2,4,4,6-tetramethylheptane**

CYCLOALKANES

Cycloalkanes (general formula C_nH_{2n}) are alkanes that have a ring structure (Figure 3.3). Since they have one less equivalent of hydrogen (notice the formula does not have the $2n+2$ but just $2n$ and two hydrogen atoms is defined as one equivalent), cycloalkanes are said to possess one unit of unsaturation. The cycloalkanes that will be considered in this chapter are shown in Figure 3.3.

As might be guessed by considering their small size, cyclopropane and cyclobutane are highly strained molecules. One way of demonstrating strain is with the enthalpy of formation ($\Delta_f H°$, Table 3.4). This is a measure of the amount of energy given off (–) or taken in (+) to form that molecule from its elements (here C(s, graphite) and H_2(g)). Where cyclopropane and cyclobutene have

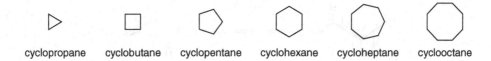

cyclopropane   cyclobutane   cyclopentane   cyclohexane   cycloheptane   cyclooctane

Figure 3.3   Name and skeletal structure of the first six cycloalkanes.

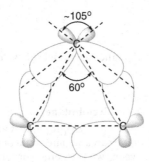

Figure 3.4   Bond-angle compression in cyclopropane that leads to "banana bonds" and high strain energy.

Table 3.4   Cycloalkane, Standard Enthalpy of Formation, and Standard Enthalpy of Formation Per Carbon Atom for the First Six Cycloalkanes

Molecule[3,4,5,6]	Number of carbon atoms	$\Delta_f H°$ (kJ/mol)	$\Delta_f H°$/C atom (kJ/mol)
cyclopropane	3	53.3	17.8
cyclobutane	4	26.9	6.7
cyclopentane	5	−105.5	−21.1
cyclohexane	6	−157.7	−26.3
cycloheptane	7	−157.9	−22.6
cyclooctane	8	−169.9	−21.2

positive $\Delta_f H°$ values, this means that they are less stable than graphite and hydrogen. As more carbon atoms are added to the rings, the rings become more stable until cyclohexane, which is the most stable small ring with the most negative $\Delta_f H°$ per carbon atom. Cyclopropane through cyclohexane will be considered each in turn.

Cyclopropane

Cyclopropane suffers from severe angle strain (that is compressing sp³ atoms with preferred 109.5° bond angles down to 60° bond angles). The bonds in cyclopropane are distorted and their bent shape is occasionally referred to as "banana bonds" (Figure 3.4).

In addition, cyclopropane (Figure 3.5) also suffers from what are called eclipsing interactions, that is C–H bonds on neighbouring carbons are parallel to one another, which causes electron-electron repulsion between the bonding orbitals.

Figure 3.5   Skeletal structure depiction and 3D model of cyclopropane.

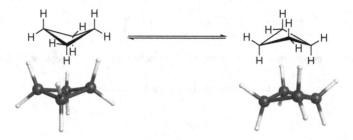

Figure 3.6 Skeletal structure depiction and 3D models of cyclobutane conformation interconversion.

Cyclobutane

Cyclobutane is only slightly less strained than cyclopropane. It also suffers from angle strain and eclipsing strain; however, cyclobutane can alleviate some of this by puckering (Figure 3.6). Cyclobutane rapidly interconverts between two puckered conformations.[7]

Cyclopentane

Cyclopentane has significantly less strain than cyclobutane as all the angle strain has been eliminated; however, planar cyclopentane would still suffer from destabilizing, eclipsing interactions. As such, cyclopentane rapidly interconverts (Figure 3.7), between different envelope (the carbon atoms roughly approximate the shape of an envelope with the flap open) and half-hair conformations. Because of the rapidity of this process and the relatively low barrier to interconversion, this process is known as pseudorotation.[6]

Cyclohexane

Cyclohexane has a very high-energy planar conformation and so cyclohexane adopts two possible conformations the chair and twist-boat (Figure 3.8). Given that there are roughly 1×10^5 chair conformations for every one twist-boat, it will be assumed that all cyclohexane molecules exist as chairs. Given their importance, more time will be dedicated to the cyclohexane chair conformation below.

Hyperconjugation

In the discussion of cycloalkanes above, we discussed their conformational interconversions. Alkanes and cycloalkanes by nature of the rotationally symmetric σ bonds, that is the bonds can be rotated 360° and the nature of the bond does not change, can adopt different conformations. Attention will now be given to those different conformations, the names thereof, and how to draw these conformations. The essential difference among different (cyclo)alkane isomers is whether bonds are eclipsed – which is destabilizing – or staggered – which is not destabilizing. These two possibilities can be visualized (Figure 3.9) most clearly for the simplest alkane, ethane.

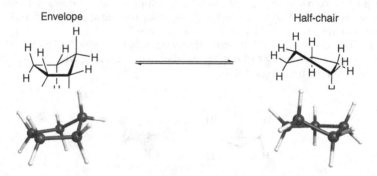

Figure 3.7 Skeletal structure depiction and 3D models of cyclopentane conformations and their interconversion.

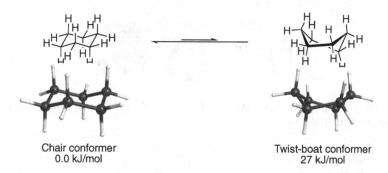

Chair conformer
0.0 kJ/mol

Twist-boat conformer
27 kJ/mol

Figure 3.8   Skeletal structure depiction, 3D models, and relative energies (in kJ/mol) of cyclohexane conformations and their interconversion.

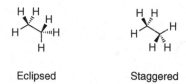

Eclipsed

Staggered

Figure 3.9   Structure and name of ethane rotomers.

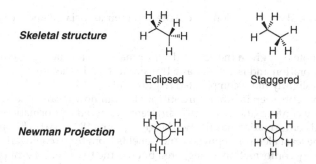

Skeletal structure

Eclipsed

Staggered

Newman Projection

Figure 3.10   Comparison of skeletal depictions and Newman projections of ethane rotational conformers (rotomers).

As depicted in skeletal structures (Figure 3.9), this difference loses some of its significance. To get a better grasp of the differences, these same structures will be redrawn as Newman projections (Figure 3.10).

To draw a Newman projection, one sights along a particular bond (in Figure 3.10, the C–C bond is being looked down), then the atoms are drawn coming off from the front atom (the vertex of the three bonds in the middle) and off from the back atom (the circle drawn behind). Even though in a true eclipsed conformation, the bonds are directly overlapping, in a Newman projection they are drawn slightly offset for clarity. At this point, a new term needs to be introduced, the dihedral angle. This is the angle between two atoms, where the vertex is made of two overlapping atoms, for ethane it is the angle between the two hydrogen atoms where the two carbon atoms form the vertex. In eclipsed ethane, the dihedral angle is 0°, while in staggered ethane it is 60°.

The eclipsed conformation is higher in energy because of the electron–electron repulsion between closely spaced bonds. Staggered ethane does not suffer from this repulsion as much. Also, the staggered conformation is stabilized by hyperconjugation.[8] That is, some of the electron density of the σ_{C-H} orbital donates into the accepting neighbour σ^*_{C-H} (Figure 3.11).

Together, the staggered conformation is stabilized by hyperconjugation and reduced electron-electron repulsion, while the eclipsed conformation is destabilized by electron-electron repulsion and a concomitant lack of hyperconjugation stabilization. This leads to an energy profile for C–C bond rotation in ethane that looks like Figure 3.12.

Figure 3.11 Hyperconjugation – partial donation of σ_{C-H} electron density into the empty σ^*_{C-H} – in the staggered ethane rotomer.

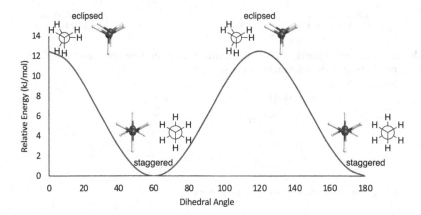

Figure 3.12 Rotational conformations of ethane and their associated energy values.

It is important to note that given the eclipsed conformation is an energy maximum, it exists for a single molecular vibration and is defined as a transition state. In moving from ethane to butane the picture becomes slightly more complicated (Figure 3.13)

The same basic concepts seen in ethane are still at play, but now it can be seen that there are two different eclipsed conformations and two different staggered conformations. The lowest energy staggered conformation has the two biggest groups 180° apart. This anti-staggered conformation has the least electron repulsion and benefits from hyperconjugation stabilization. The next lowest-energy conformation is staggered but has the two methyl groups 60° apart and is called *gauche*. The *gauche* conformation has hyperconjugation stabilization but suffers from more electron–electron repulsion because of the closer proximity of the large methyl groups. The

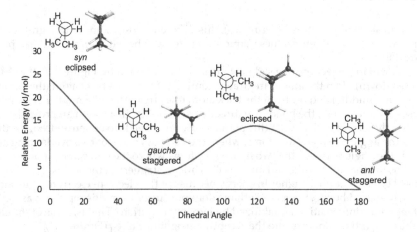

Figure 3.13 Rotational conformations of butane and their associated energy values.

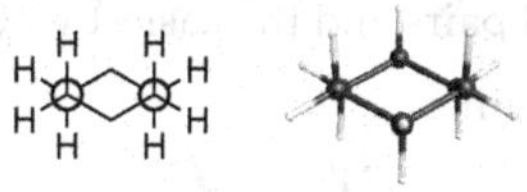

Figure 3.14 Newman projection and 3D model of chair cyclohexane.

high-energy transition states (eclipsed conformations) also come in two kinds: eclipsed and totally (or *syn*) eclipsed. The lower energy eclipsed has C–H bonds eclipsing C–Me bonds and suffers from electron-electron repulsion and has no hyperconjugation stabilization. The totally eclipsed (or *syn*-eclipsed) conformation has the maximum amount of electron-electron repulsion because of the significant electron-electron repulsion between eclipsing methyl groups.

Problem 3.6 Provide the four possible (two staggered and two eclipsed) conformations (as Newman projections) for 2,2,5,5-tetramethylhexane, when sighting down the carbon 3-carbon 4 bond.

Cyclohexane

The different conformations of cycloalkanes were discussed briefly above. Now, the conformations of cyclohexane will be considered in more detail. As mentioned, it will be assumed that all cyclohexane molecules exist as a chair conformer. If the Newman projection for chair cyclohexane is considered, one reason for the stability of this conformation is the fully staggered nature of all the bonds (Figure 3.14).

But cyclohexane is not locked into a single chair conformation and can, and in fact rapidly does, interconvert (Figure 3.15).

Given the importance of chair cyclohexane, it is important to be able to draw the conformation correctly, interconvert between chair conformations correctly, and interconvert between planar depictions and chair conformations.

When drawing chair cyclohexane, there are several rules for correctly drawing a chair:

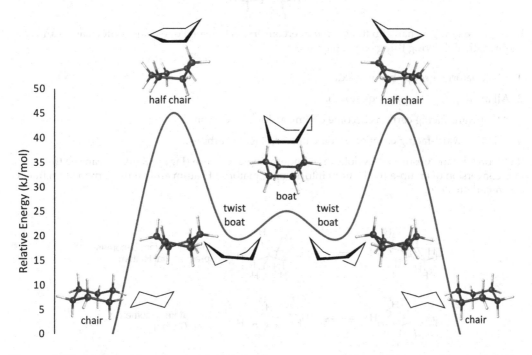

Figure 3.15 Rotational conformations (skeletal structures and 3D models) of cyclohexane and their associated, relative energy values (in kJ/mol).

1. The C–C bonds must be in parallel pairs and the easiest way to draw.

   a. Draw the end bonds as parallel lines.

   b. Then draw a longer set of C–C bonds starting from the inward-facing ends of the lines you drew previously.

   c. Finally, complete the chair with a final pair of parallel C–C bonds.

2. The equatorial C–H bonds must be parallel to the C–C bonds that make up the ring.

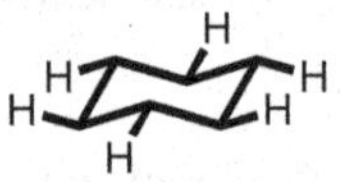

3. Each carbon atom must have one axial and one equatorial C–H bond. Axial bonds go straight up from up carbons and straight down from down carbons.

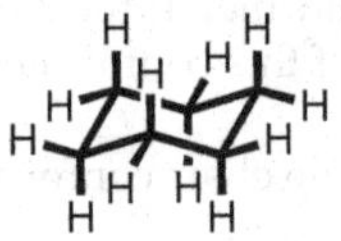

Thus, can one draw an appropriate chair. If the initial parallel lines had angled to the left rather than the right, then that would be an equally appropriate chair-flipped chair.

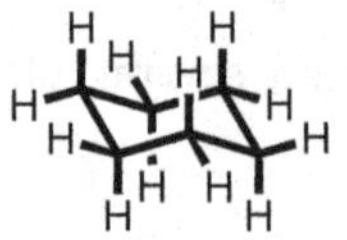

When undergoing a chair flip (that is the interconversion between one chair conformation and another), the following things must happen:

1. All equatorial groups become axial.

2. All axial groups become equatorial.

3. All upward-facing carbons become downward-facing carbons.

4. All downward-facing carbons become upward-facing carbons.

If we consider unsubstituted cyclohexane undergoing a ring flip (Figure 3.16), we can see this interconversion of an up-axial H atom into an up-equatorial H atom and an up-carbon atom into a down-carbon atom.

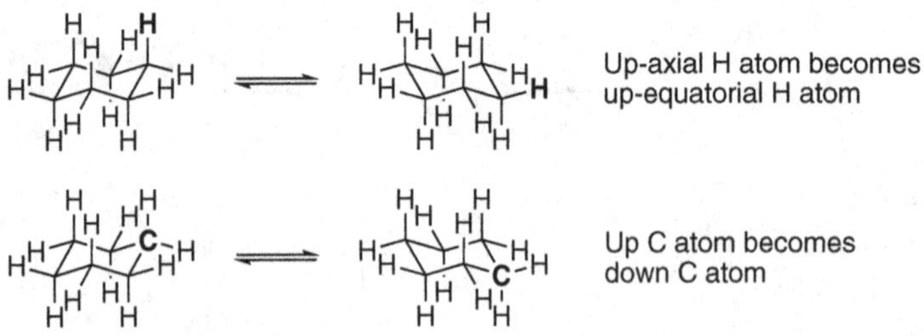

Up-axial H atom becomes up-equatorial H atom

Up C atom becomes down C atom

Figure 3.16 Ring flip showing the interconversion of axial and equatorial positions for a hydrogen atom in a skeletal structure (*top*) and the interconversion of up and down positions for a carbon atom in a skeletal structure (*bottom*).

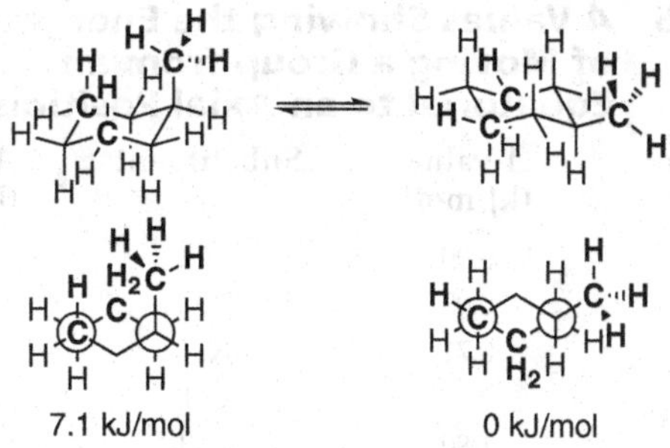

7.1 kJ/mol   0 kJ/mol

Figure 3.17 Ring flip of methylcyclohexane with the relative energies (in kJ/mol) associated with the axial and equatorial methyl group. The axial methyl group suffers from 1,3-diaxial inter-actions and *gauche* interactions, atoms bolded.

Now for unsubstituted cyclohexane, this is seemingly a pointless exercise; however, in look-ing at methylcyclohexane, significant differences between the two conformations can be seen in Figure 3.17.

In one chair conformation, the methyl group is axial and in one conformation the methyl group is equatorial. The axial methyl group is disfavoured because of steric repulsion between the methyl group and the other axial groups (here hydrogen atoms) in what is called 1,3-diaxial inter-actions (bold Figure 3.17) and the *gauche* interactions that occur (bold in Figure 3.17). In general, it is preferable for groups to be equatorial rather than axial and if there are two groups then the larger group is preferably in the equatorial position.

A measure of the preference for groups to be equatorial versus axial is the *A*-value table (a small excerpt is provided here).[9] This table lists the energetic cost (in kJ/mol) associated with moving a group from the equatorial position into an axial position (the positive *A* values mean energy must be added to accomplish this movement).

Finally, attention will be given to interconverting between planar cyclohexane depictions and chair conformations. To do this, take the planar cyclohexane:

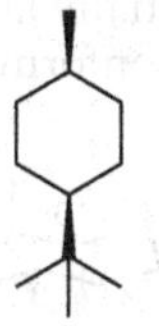

and number the atoms in the ring (where you start is arbitrary).

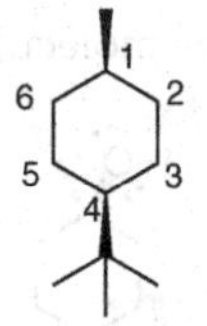

Then, draw a chair cyclohexane and number the atoms in the chair (again it is arbitrary where one starts).

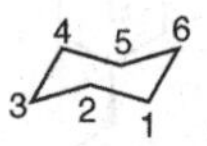

Then put the groups on the chair cyclohexane using Table 3.6.

Following this table, the methyl on carbon 1 is an up-group (wedge) on a down-carbon and so it should be equatorial. The *tert*-butyl group is an up-group (wedge) on an up-carbon and so it should be axial.

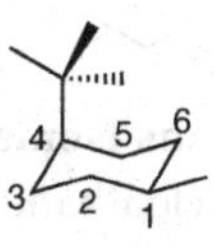

Table 3.5 A Values Showing the Energetic Cost of Moving a Group from an Equatorial to an Axial Position

Substituent	A-value (kJ/mol)	Substituent	A-value (kJ/mol)
–H	0.00	–SH	3.77
–D	0.03	–CO_2H	5.65
–CN	0.71	–NH_2	6.69
–F	0.63	–CH_3	7.11
–Cl	1.80	–C_2H_5	7.32
–Br	1.58	–$CH(CH_3)_2$	9.25
–I	1.80	–CF_3	8.78
–OCH_3	2.51	–Ph	12.55
–OH	3.64	–$C(CH_3)_3$	18.53

Table 3.6 Relationship between the Type of Substituent, Type of Carbon, and Axial or Equatorial

Type of substituent	Type of carbon	Axial or Equatorial
Wedge (up)	Up	Axial
Dash (down)	Up	Equatorial
Wedge (up)	Down	Equatorial
Dash (down)	Down	Axial

Then assess the chair that has been drawn. Is it the lowest energy? In reference to the A-value table, Table 3.5, the *tert*-butyl group is substantially larger (A-value 18.53 kJ/mol) than the methyl (7.11 kJ/mol) and so this is the higher energy conformation. Thus, a chair flip is necessary to produce the lowest-energy chair conformation:

Now the larger group is equatorial, this is the best chair depiction of this molecule.

Problem 3.7 Convert each of these cyclohexane molecules to the lowest-energy chair conformation.

Cis-Trans Isomerism

Due to the reduced degrees of freedom for cycloalkanes (compared to linear alkanes), cycloalkanes also have another type of isomerism: *cis–trans* isomerism (*cis* and *trans* are the first **stereochemistry identifiers**). Groups that are *trans* are on opposite faces of a ring (one wedge

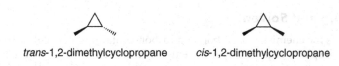

trans-1,2-dimethylcyclopropane   cis-1,2-dimethylcyclopropane

Figure 3.18 *Trans-* and *cis-* stereoisomers of 1,2-dimethylcyclopropane.

one dash) and groups that are *cis* are on the same face of a ring. Figure 3.18 shows *trans-* and *cis-*1,2-dimethylcyclopropane.

Problem 3.8 For each of the structures below, identify the substituents as having a *cis* or a *trans* relationship.

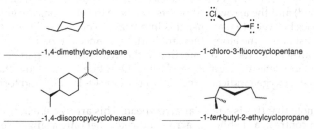

_____-1,4-dimethylcyclohexane   _____-1-chloro-3-fluorocyclopentane

_____-1,4-diisopropylcyclohexane   _____-1-*tert*-butyl-2-ethylcyclopropane

NOTES

1. Stubba, D.; Lahm, G.; Geffe, M.; Runyon, J.W.; Arduengo III, A.J.; Opatz, T. Xylochemistry – Making Natural Products Entirely from Wood. *Angew. Chem. Int. Ed.*, **2015**, *54* (47), 14187–14189. DOI: 10.1002/anie.201508500

2. Favre, H.A.; Powell, W.H. *Nomenclature of Organic Chemistry: IUPAC Recommendations and Preferred Names 2013*; The Royal Society of Chemistry, 2014. DOI: 10.1039/9781849733069

3. Knowlton, J.W.; Rossini, F.D. Heats of combustion and formation of cyclopropane. *J. Res. Natl. Bur. Stand.*, **1949**, *43* (2), 113–115. DOI: 10.6028/jres.043.013

4. Kaarsemaker, S. and Coops, J. Thermal Quantities of Some Cycloparaffins. Part III. Results of Measurements. *Recl. Trav. Chim. Pays-Bas*, **1952**, *71* (3), 261–276. DOI: 10.1002/recl.19520710307

5. Spitzer, R.; Huffman, H.M. The Heats of Combustion of Cyclopentane, Cyclohexane, Cycloheptane and Cyclooctane. *J. Am. Chem. Soc.*, **1947**, *69* (2), 211–213. DOI: 10.1021/ja01194a006

6. Legon, A.C. Equilibrium conformations of four- and five-membered cyclic molecules in the gas phase: determination and classification. *Chem. Rev.*, **1980**, *80* (3), 231–262. DOI: 10.1021/cr60325a002

7. Legon, A.C. Equilibrium conformations of four- and five-membered cyclic molecules in the gas phase: determination and classification. *Chem. Rev.*, **1980**, *80* (3), 231–262. DOI: 10.1021/cr60325a002

8. i) Weinhold, F. A New Twist on molecular shape. *Nature*, **2001**, *411*, 539–541. DOI: 10.1038/35079225
   ii) Pophristic, V.; Goodman, L. Hyperconjugation not steric repulsion leads to the staggered structure of ethane. *Nature*, **2001**, *411*, 565–568. DOI: 10.1038/35079036

9. i) Hirsch, J.A. Table of Conformation Energies – 1967. In *Topics in Stereochemistry*; John Wiley & Sons, Ltd, 1967, pp. 199–222. DOI: 10.1002/9780470147108.ch4
   ii) Jensen, F.R.; Bushweller, C.H. Conformational Preferences in Cyclohexanes and Cyclohexenes; HART, H.; KARABATSOS, G.J., Eds.; Advances in Alicyclic Chemistry; Elsevier, 1971, Vol. 3, pp. 139–194. DOI: 10.1016/B978-0-12-001303-6.50008-2

4 Arrow Pushing and Solvent

Organic chemistry is the chemistry of carbon and carbon-containing molecules. To understand the reactions that take place, organic chemistry employs a curved-arrow notation. For a tool that is so central to modern organic chemistry, the history of this notation is worth noting here. It was first developed in the 1920s by Robert Robinson[1] and early 1930s by Christopher Ingold,[2] but was not immediately popular. It took about a decade to work its way into textbooks (1938)[3] and teaching and research papers (1940).[4] Curved-arrow notation has been computationally demonstrated to have a link to the transformation of molecular orbitals as a reaction takes place.[5]

CURVED-ARROW NOTATION

The curved-arrow notation is based on a simple premise: the arrow shows the movement of electrons. To do this, the arrow starts where the electrons start (on a radical electron, on lone-pair electrons, or on a bond) and the arrowhead points to where the electrons are going (typically to an atom or to the space between two atoms). It is important to note that the electrons must always stay attached to at least one of the atoms they are initially connected to.

In Figure 4.1, we can see that the arrow starting at the hydroxide lone-pair electrons and moving to the hydrogen atom indicates the formation of a new O–H bond. The arrow starting at the O–H bond and moving to the oxygen indicates the cleavage of the O–H bond and formation of a new oxygen lone pair.

It is important to note that this is a regular arrowhead. This arrow type indicates that two electrons are moving. In contrast, a "fishhook" arrow indicates that only a single electron is moving. This is the type of arrow that is used to show radical reaction mechanisms (Figure 4.2).

Note in Figure 4.2 the different shape of the arrowheads and the lack of formal charges in this radical process. Here there are two arrows starting at the O–H bond (each arrow can only take one electron into account and there are two electrons per bond). One arrow shows the formation of a new oxygen radical, while the other arrow meets up with the radical arrow starting from the radical electron, which shows the formation of a new O–H bond.

REACTION MECHANISMS

Above, only relatively simple examples were considered. On the next several pages, a longer mechanism will be shown, and the arrows explained for each step. The final mechanism will then be provided, thereby showing how a complicated mechanism builds up from these individual steps.

First, we consider the reactants and products and take inventory of what has changed (Figure 4.3) in terms of the electrons: bonds and lone pairs.

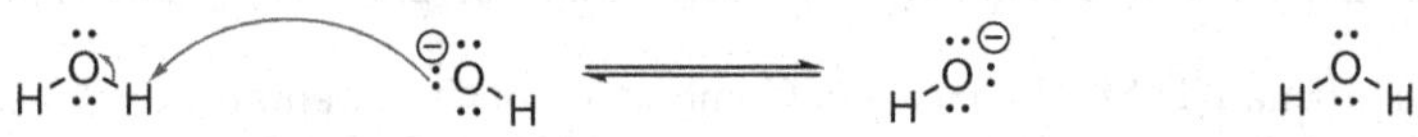

Figure 4.1 Curved arrows showing the movement of electrons in the reaction of water and hydroxide.

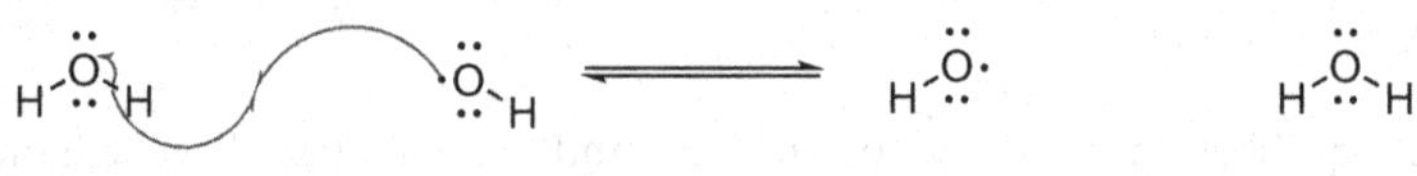

Figure 4.2 Curved fishhook arrows showing the movement of electrons in the reaction of water and a hydroxyl radical.

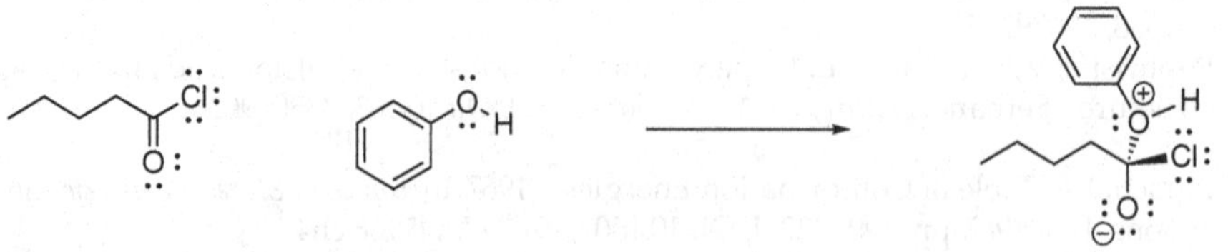

Figure 4.3 A mechanistic step with changes in bonds and lone pair.

DOI: 10.1201/9781003479352-4

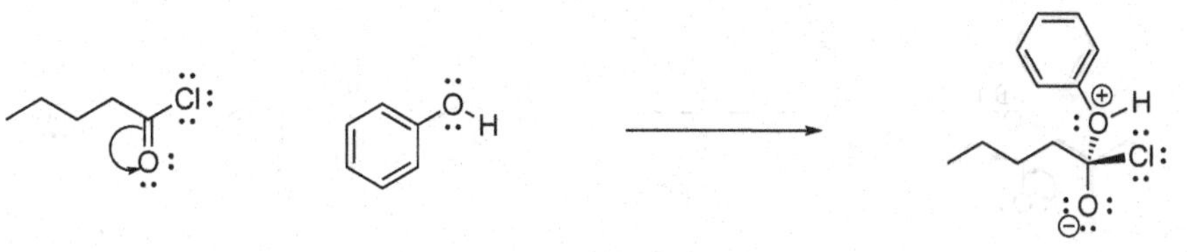

Figure 4.4 A mechanistic step with changes in bonds and lone pair and a curved arrow showing the changes to the CO double bond and oxygen lone pair.

In Figure 4.3, there are three changes to highlight:

1. The carbon atom and oxygen atom of the acid chloride have gone from being doubly bonded to singly bonded (loss of π bond).

2. The acid chloride oxygen atom has gained a lone pair and formal negative charge.

3. The phenol oxygen atom has lost an electron lone pair and is now making a bond to what had been the carbonyl carbon atom, in the process gaining a formal positive charge.

To account for these changes, we use curved arrows. The first point noted (1.) was the breaking of the C=O π bond and its localization (2.) as a lone pair on the oxygen atom. As such, we draw an arrow starting from the π bond and moving to the oxygen atom (Figure 4.4).

The third point (3.) noted was the loss of the phenol oxygen atom lone pair and the formation of a new C–O bond between the phenol oxygen atom and the acid chloride carbon atom. As such, we draw an arrow from the phenol oxygen atom lone pair to the carbonyl carbon atom showing the formation of that bond from the oxygen atom lone pair (Figure 4.5).

Together, then, Figure 4.5 shows the appropriate arrows to show the changes shown in Figure 4.3. The reaction, however, is not done and the second mechanistic step is shown in Figure 4.6.

In Figure 4.6, there are two changes to highlight:

1. The formally negative oxygen atom has lost a lone pair and is making a new carbon-oxygen π bond. The oxygen is now no longer formally negative.

2. The chlorine atom is no longer making a bond to carbon and has a new lone pair and negative formal charge.

The appropriate arrows that we would need to show for the formation of the C=O π bond starts at the oxygen atom lone pair and moves to the space between the carbon and oxygen atoms (Figure 4.7).

The single arrow in Figure 4.7, however, does not account for the C–Cl bond cleavage and so a second curved arrow is used to show the C–Cl bond localizing as a lone pair on the chlorine atom

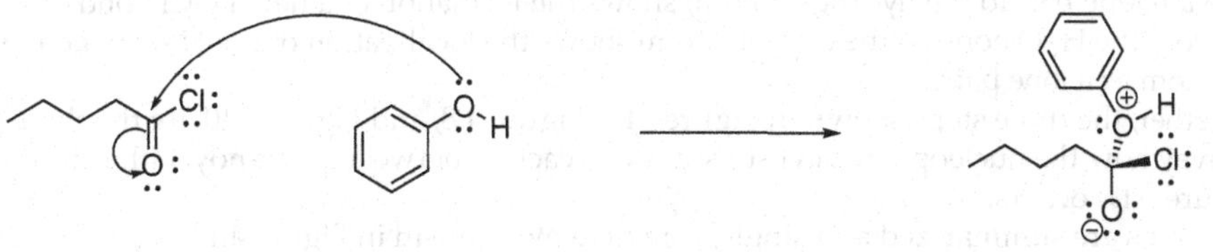

Figure 4.5 A mechanistic step with changes in bonds and lone pair with curved arrows that account for these changes.

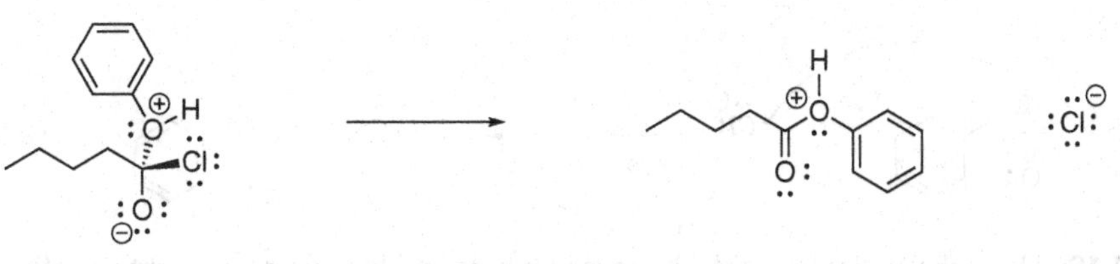

Figure 4.6 A mechanistic step with changes in bonds and lone pair.

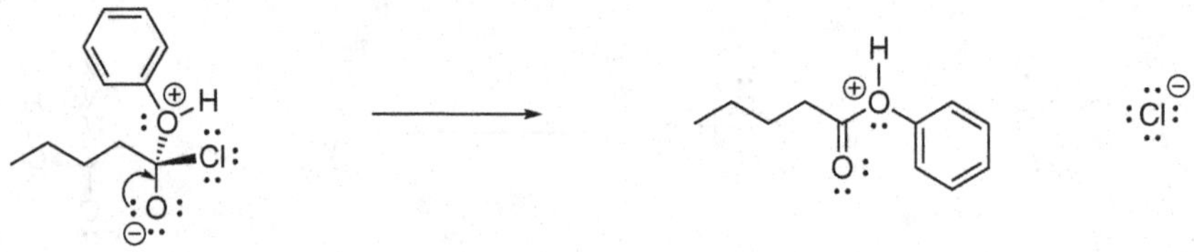

Figure 4.7 A mechanistic step with changes in bonds and lone pair and a curved arrow showing the changes to the CO bond and the oxygen atom lone pair.

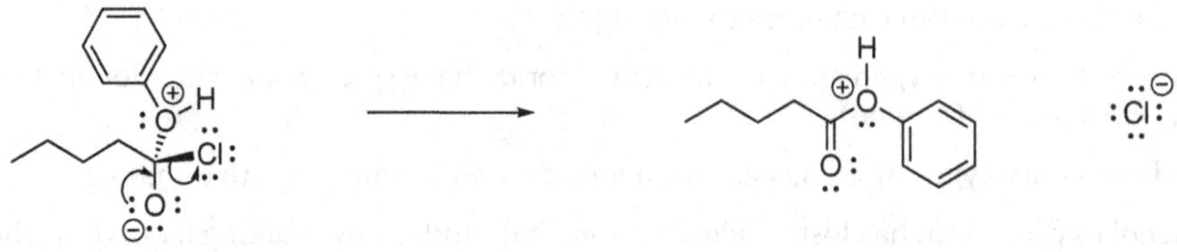

Figure 4.8 A mechanistic step with changes in bonds and lone pair with curved arrows that account for these changes.

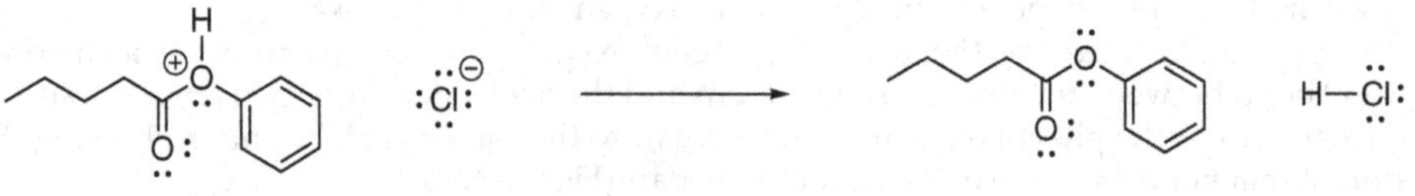

Figure 4.9 A mechanistic step with changes in bonds and lone pair.

and forming the chloride ion (Figure 4.8). Together these two arrows account for the changes in the mechanistic step shown in Figure 4.6.

Finally, the last mechanistic step involves the transfer of a hydrogen atom from oxygen to chlorine (Figure 4.9).

The first step is to, again, take stock of any changes:

1. The free chloride ion has lost a lone pair (and consequently a negative formal charge) and has made a new bond to hydrogen.

2. The formally positive oxygen atom gains a new lone pair (and thus loses its formal positive charge), which comes from the O–H bond breaking.

To account for these changes, two arrows are needed (Figure 4.10). The first arrow (from the chloride ion lone pair to the hydrogen atom) shows the formation of a new H–Cl bond. The second arrow (from the H–O bond to the oxygen atom) shows the localization of the H–O bond onto the oxygen atom as a lone pair.

Altogether, the three steps shown in Figures 4.5, Figure 4.8, and Figure 4.10 are the three steps that convey how the nucleophilic acyl substitution reaction between pentanoyl chloride and phenol (Figure 4.11) occurs.

These steps are summarized as a single, complete mechanism in Figure 4.12.

The specific mechanism shown in Figure 4.12 is something that will not be seen until Chapter 21. But given the starting materials and products for each step, you can draw the arrows necessary to show that change now.

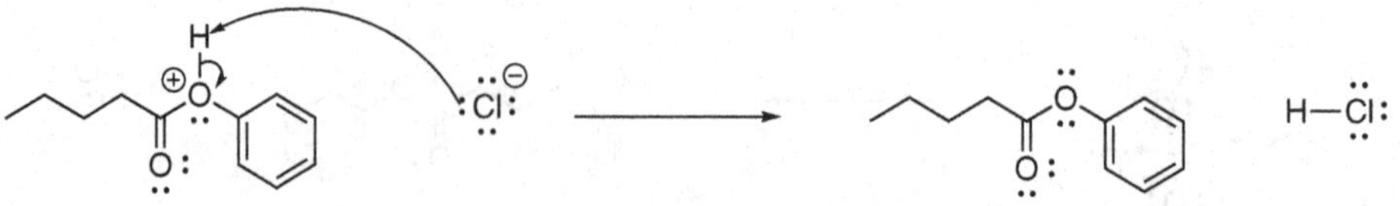

Figure 4.10 A mechanistic step with changes in bonds and lone pair highlighted with curved arrows that account for these changes.

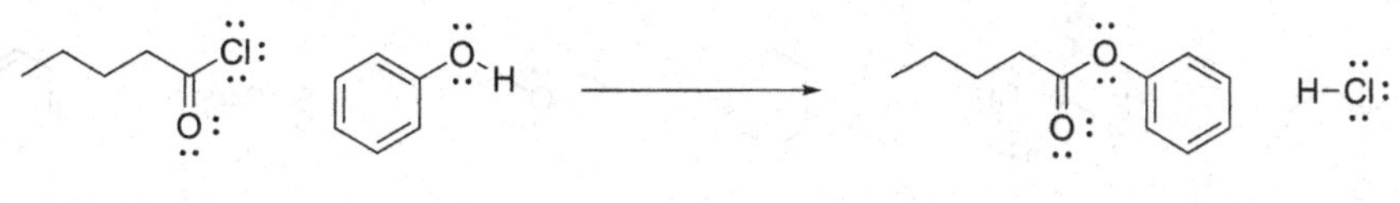

Figure 4.11 Nucleophilic acyl substitution reaction of pentanoyl chloride and phenol.

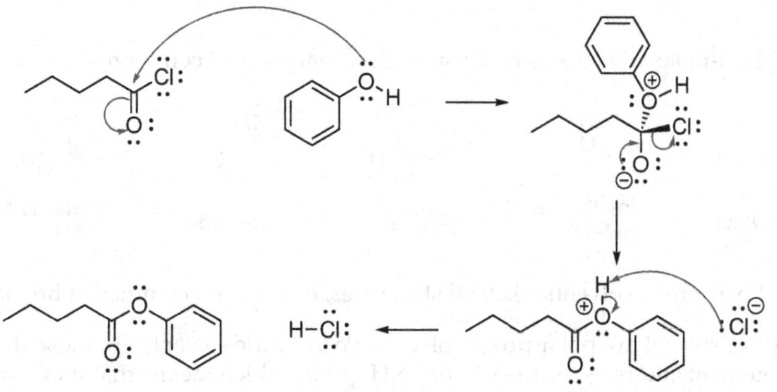

Figure 4.12 The three-step mechanism showing the movement of electrons as curved arrows.

Problem 4.1 For each of the following reactions, provide the appropriate arrows necessary to show the interconversion of starting material to product.

a.

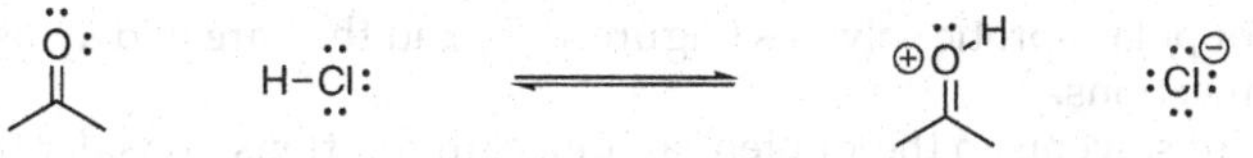

b. 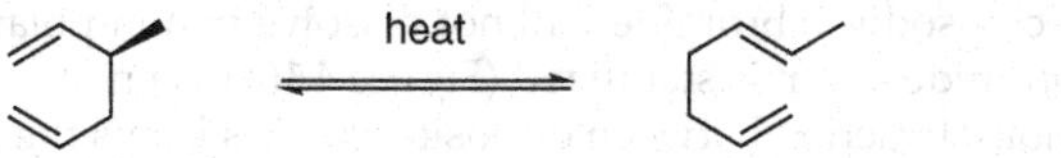   heat

c.

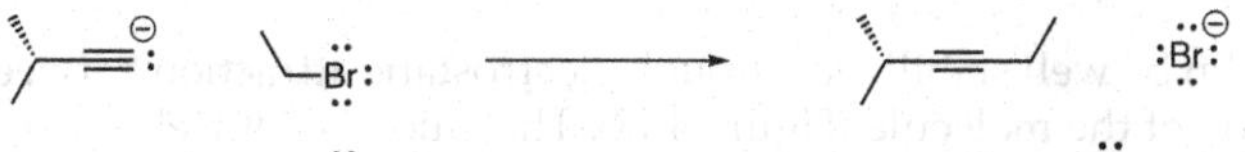

d.

e.

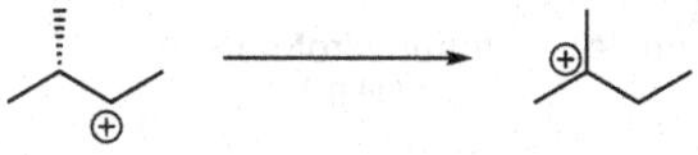

f.

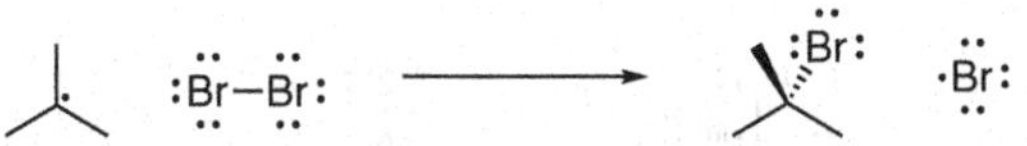

SOLVENTS

Solvents play an important role in chemistry. Solvents are the liquid medium in which reactions typically take place. In a chemical equation, the solvent is usually noted below the arrow (along with temperature and time of reaction). Because solvents can affect the reaction outcome it is important to note the solvent and understand its properties. There are three types of solvents: non-polar, polar protic, and polar aprotic. These categories are based on the intermolecular forces (IMFs) present in each.

The first type of solvent are non-polar solvents. These solvents only interact through dispersion interactions and have either a dipole of zero or only a relatively weak dipole moment. Examples of common non-polar solvents are hexane, carbon tetrachloride, benzene, diethyl ether, carbon disulfide, and toluene (Figure 4.13). Non-polar solvents are most used in radical reactions (which proceed through non-polar intermediates) and in the purification of organic chemicals.

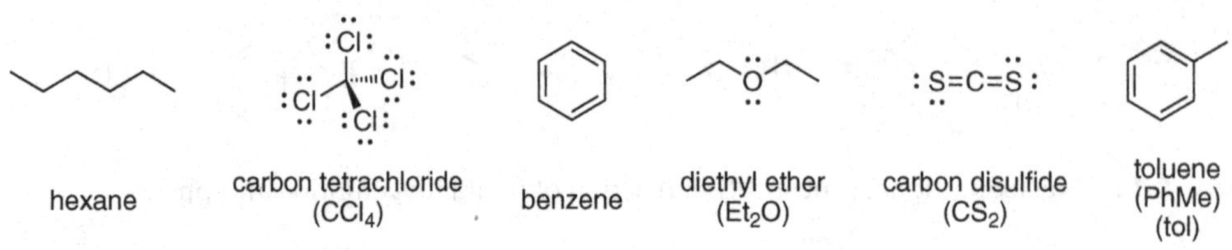

Figure 4.13 Non-polar solvents' skeletal structures, names, and common abbreviations.

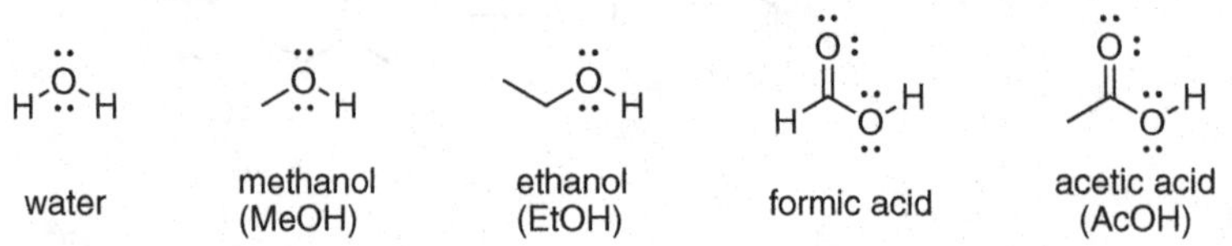

Figure 4.14 Polar protic solvents' skeletal structures, names, and common abbreviations.

The next type of solvent are polar protic solvents. Polar protic solvents are those that have a strong dipole moment and possess an –OH or –NH group, which means that they can hydrogen bond. Most commonly these are water, alcohol, or carboxylic acid solvents (Figure 4.14). Polar protic solvents are important in substitution and elimination reactions as they can solvate (surround and shield) both anions and cations.

The last type of solvent are polar aprotic solvents. These also possess a strong dipole moment, but do not possess either –OH or –NH groups and so cannot hydrogen bond. Since they do not hydrogen bond, polar aprotic solvents can solvate cations but cannot well solvate anions. There is a wide range of common examples of polar aprotic solvents (Figure 4.15), and they are widely used in many types of reactions and purifications.

To understand the effect of different solvents in their potential impact in reactions, consider the dissolution of sodium bromide. The ionic species sodium bromide will not dissolve in non-polar solvents. Neither the sodium cation nor the bromide anion is stabilized (Figure 4.16) by a non-polar solvent (like hexane) and so the coulombic attraction between opposite charges keeps the ions together as a solid.

In polar aprotic solvents, cations can be well stabilized through electrostatic attraction between the cation and electronegative portion of the molecule (Figure 4.17). The anion is not well

Figure 4.15 Polar aprotic solvents' skeletal structures, names, and common abbreviations.

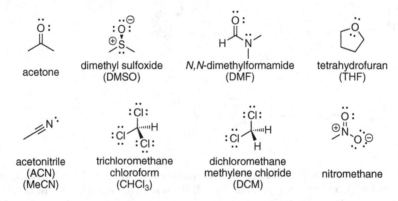

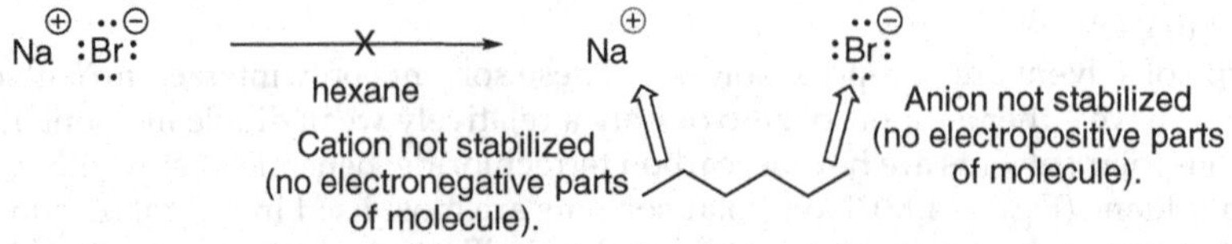

Figure 4.16 Non-polar solvents do not stabilize cations or anions.

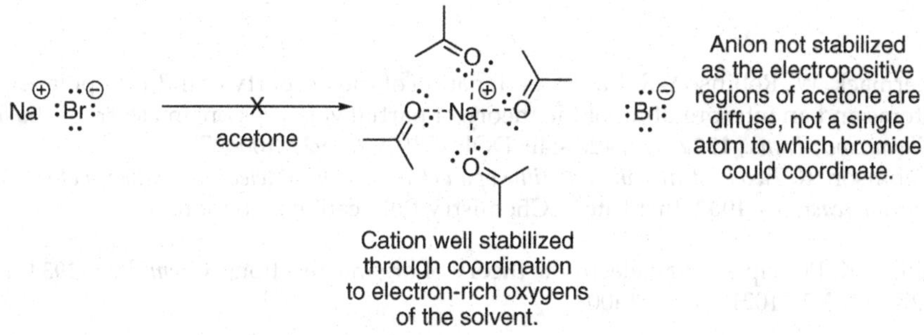

Figure 4.17 Polar aprotic solvents stabilize cations but do not stabilize anions.

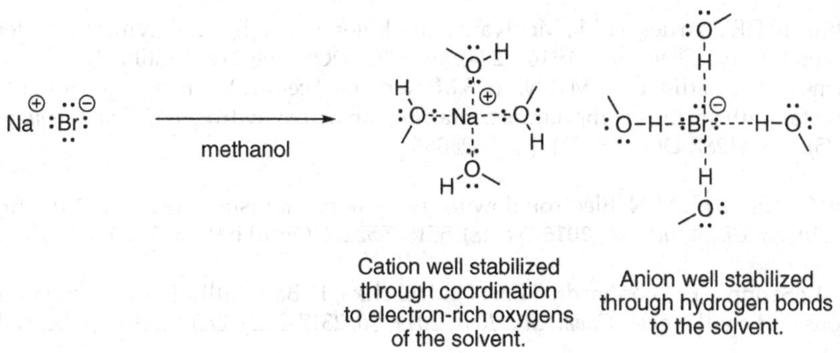

Figure 4.18 Polar protic solvents stabilize cations and anions.

stabilized for most polar aprotic solvents, like acetone, and so typically the unshielded anion strongly attracts the cation. Thus, sodium bromide does not dissolve in most polar aprotic solvents (DMF and DMSO can dissolve salts to a moderate extent).

Sodium bromide does dissolve in polar protic solvents: water (94 g/mL at 25 °C), methanol (17 g/mL at 25 °C), ethanol (2 g/mL at 25 °C), and formic acid (19 g/mL at 25 °C). This is because both the anion and cation are stabilized by coordination with the solvent (oxygen-rich atoms to the cation and hydrogen bonds to the anion, Figure 4.18). Ion coordination effectively shields the ions and coulombic attraction is overcome.

SOLVENTS AND GREEN CHEMISTRY

While solvents provide a liquid medium for conducting reactions, ensuring thorough and efficient mixing of the components, organic solvents are a special challenge in terms of both green chemistry and sustainability overall. The use of organic solvents increases the material cost of any reaction that uses them. In addition, organic solvents are frequently used in the purification of the final desired product, further increasing the material cost of a reaction. Finally, organic solvent waste is remediated through incinerating, which contributes to production of greenhouse gases.

There are several methods for avoiding the use of organic solvents. The first is to run a reaction neat, without solvent. This is not a universally applicable method but is feasible for a small handful of reactions. The second is to use high-speed ball milling to mix the chemicals inside of a rapidly rotating drum filled with metal balls that mix and break up the chemicals present in the drum.[6] The last method is to use water as a solvent.[7] Currently, none of these methods is a one-size-fits-all replacement for using organic solvents, but work is ongoing to further develop these greener methods of running reactions.

NOTES

1. i) Kermack, W.; Robinson, R. LI. An explanation of the property of induced polarity of atoms and an interpretation of the theory of partial valencies on an electronic basis. *J. Chem. Soc., Trans.*, **1922**, *121*, 427–440. DOI: 10.1039/CT9222100427
   ii) Robinson, R. *Two Lectures on an outline of an electrochemical (electronic) theory of the cause of organic reactions.* **1932**, Institute of Chemistry Publications: London.

2. Ingold, C. K. Principles of an electronic theory of organic reactions. *Chem. Rev.*, **1934**, *15* (2), 225–274. DOI: 10.1021/cr60051a003

3. Johnson, J.R. Modern electronic concepts of valence. In *Organic Chemistry: An Advanced Treatise*. Gilman, H., Ed.; John Wiley & Sons Inc.: New York, 1938; Volume II; pp. 1595–1711.

4. i) Johnson, P.R.; Barnes, H.M.; McElvain, S.M. Ketene Acetals. IV. Polymers of ketene diethylacetal[1] *J. Am. Chem. Soc.*, **1940**, *62* (4), 964–972. DOI: 10.1021/ja01861a071
   ii) Barnes, H.; Kundiger, D.; McElvain, S.M. Kentene Acetals. V. The reaction of ketene diethylacetal with various compounds containing an active hydrogen. *J. Am. Chem. Soc.*, **1940**, *62* (5), 1281–1287. DOI: 10.1021/ja01862a086

5. Knizia, G.; Klein, J.E.M.N. Electron flow in reaction mechanisms – revealed from first principles. *Angew. Chem. Int. Ed.*, **2015**, *54* (18), 5518–5522. DOI: 10.1002/anie.201410637

6. Stolle, A.; Szuppa, T.; Leonhardt, S.E.S.; Ondruschka, B. Ball milling in organic synthesis: solutions and challenges†. *Chem. Soc. Rev.*, **2011**, *40*, 2317–2329. DOI: 10.1039/C0Cs00195C

7. Boerner, L.K. Can organic chemists cut waste by switching to water? *Chemical and Engineering News*, 17 July, 2023. *101* (23).

5 Radical Reactions

This chapter will provide an introduction to organic radical chemistry. Radical reactions are unique as they involve species with single, unpaired electrons. Following his work on the synthesis of tetraphenylmethane,[1] organic radicals were first reported by Moses Gomberg in 1900.[2] Finding the work particularly interesting, Gomberg tried, in a follow-up publication, to lay claim to the entire field of organic radical chemistry by stating: "I beg to reserve this field for further work".[3]

RADICAL HALOGENATION

In this chapter, we will consider radical halogenation reactions, where a hydrogen atom is replaced with a halogen atom. To orientate and guide our work, consider the following series (Figure 5.1) of radical halogenations of isobutane.

In the following pages, several questions will attempt to be answered: 1) how does radical halogenation occur? 2) why is there a difference in outcome depending upon which halogen is used? 3) what can be gleaned from understanding radical halogenation in terms of understanding factors that dictate the outcome of reactions generally?

Initiation

Radical reactions, like radical halogenation, rely upon the formation of radicals (that is odd-electron species, also known as open-shell species). Typically, radicals can be generated through the application of heat (Δ) or light (hν) to molecules with weak bonds (Figure 5.2).

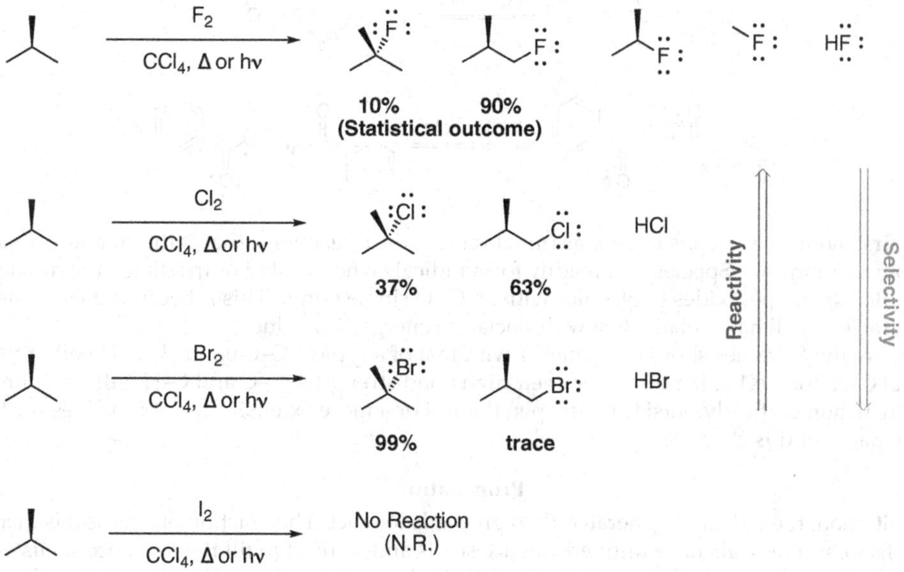

Figure 5.1 Reactions of isobutane with fluorine, chlorine, bromine, and iodine.

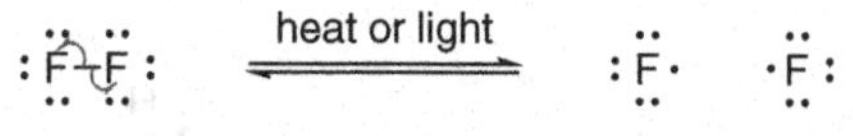

Figure 5.2 Homolysis of the F–F bond to yield two fluorine atoms (radicals).

DOI: 10.1201/9781003479352-5

Table 5.1 Dissociation Energy (E_d) Values for Common Radical Initiators

Chemical species	E_d (kJ/mol)
F–F	159
Cl–Cl	243
Br–Br	192
I–I	151
$H_3CO–OCH_3$	159

Problem 5.1 Draw the arrows showing homolysis for the following.

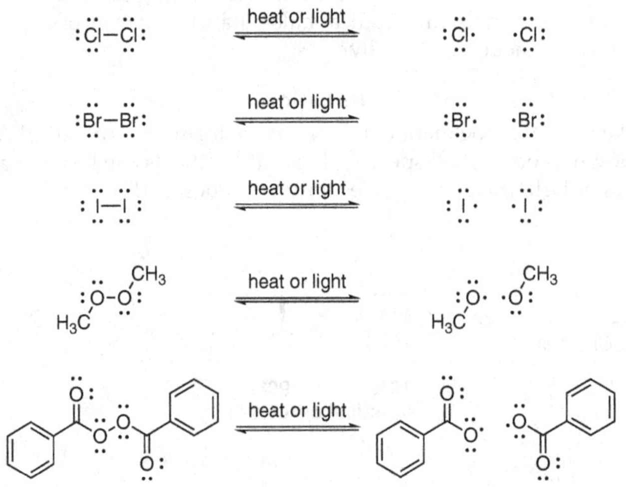

This first, homolytic step is known as initiation in radical reactions, that is the creation of radicals from non-radicals. Species that readily form radicals when heated or irradiated are dihalogens (F_2, Cl_2, Br_2, I_2) and peroxides (molecules with an O–O single bond). This is because these molecules (Table 5.1) all have relatively low dissociation energy (E_d) values.

Compare the E_d values shown in Table 5.1 with that of a typical C–C bond (363 kJ/mol) or that of a typical C–H bond (418 kJ/mol). Given their high bond strengths, C–C and C–H will not spontaneously break homolytically outside of an open flame. For a more extensive list of E_d values see the last two pages of this chapter.

Propagation

After initiation, the radical(s) generated then proceeds to react. The reaction of a radical is dictated by two factors: 1) radicals have unfilled valence shells and so react to fill their valence shells and 2) exothermic processes are favoured over endothermic steps. The reaction of a radical with a non-radical species is called a propagation step. One of the most common propagation steps is hydrogen atom abstraction. The radical species generated from the initiation step abstracts a hydrogen atom, meaning both the atom and one electron from the bond (Figure 5.3).

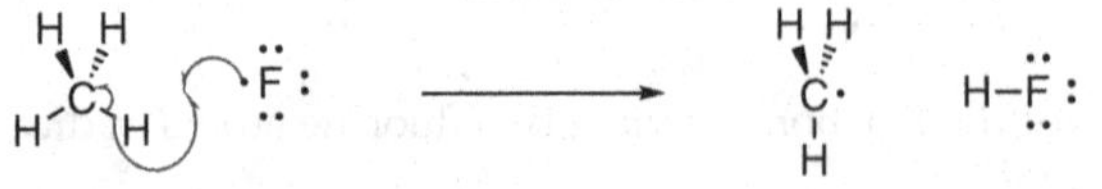

Figure 5.3 A propagation step where a fluorine atom abstracts a hydrogen atom.

Problem 5.2 Draw arrows to show the hydrogen atom abstraction in each case.

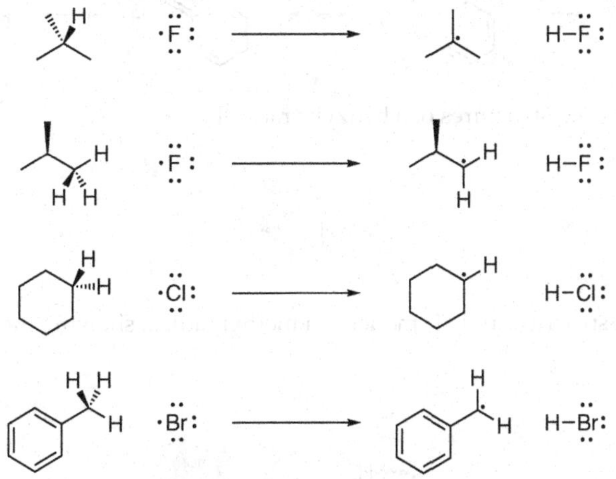

Propagation allows us to understand the reaction outcomes in Figure 5.1. As noted in the figure, fluorine is so reactive that it is entirely non-selective; chlorine is very reactive and gives only modest selectivity; bromine is very commonly used because it is very selective; and iodine is so selective that it is entirely unreactive. Why is there a difference in outcome depending upon which halogen is used? This can be rationalized if one considers the energetics of the different C–H bonds and halogens reacting. First, consider the strength of the different types of C–H bonds (Figure 5.4) that can be broken (a more extensive list of E_d values is available at the end of this chapter).

The trend in E_d values (Figure 5.4) is the result of two factors affecting the stabilities of the carbon radicals: delocalization and hyperconjugation. Delocalization (Chapter 2) is the spreading out of electron density over more than two atoms. This distribution of the unpaired electron over multiple atoms (Figure 5.5 and Figure 5.6) renders delocalization-stabilized radicals significantly more stable

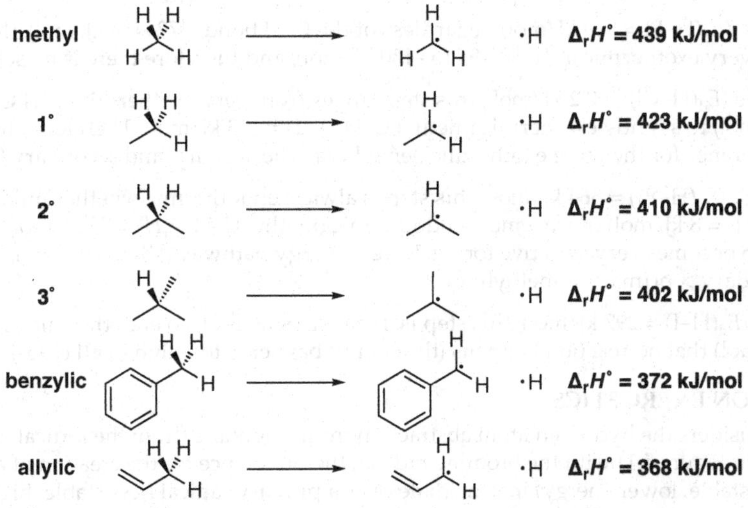

Figure 5.4 The dissociation energy values for different types of C–H bonds.

Figure 5.5 Contributing structures of an allylic radical.

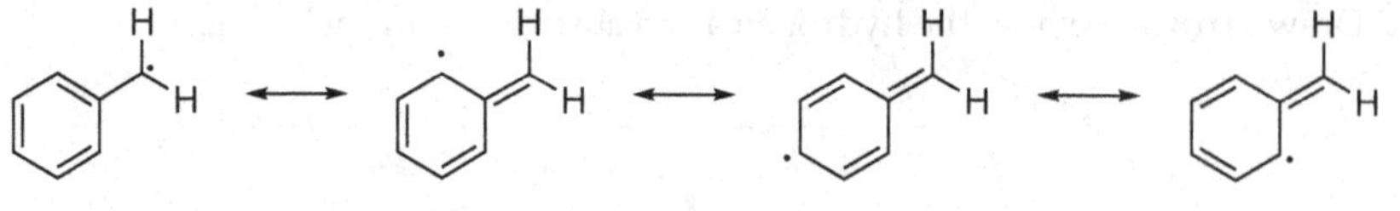

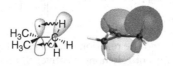

Figure 5.6 Contributing structures of a benzylic radical.

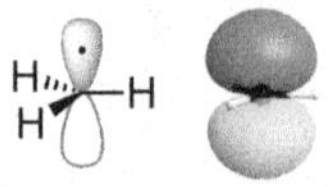

Figure 5.7 Skeletal structure and 3D model of a methyl radical showing the partially filled p-orbital.

Figure 5.8 Skeletal structure and 3D model of a *tert*-butyl radical showing the partially filled p-orbital and hyperconjugation stabilization from one of the neighbouring C–H σ bonds.

than similarly substituted, non-delocalization stabilized carbocations. In the Figure 5.4 trend, a primary allylic radical is 50 kJ/mol more stable than a primary, non-delocalization stabilized radical.

Hyperconjugation (Chapter 3) is the partial donation of electron density from filled σ bonding orbitals into partially filled orbitals. Increasing substitution of alkyl groups around the radical increases its stability, which is why radical stability trends tertiary > secondary > primary > methyl.

The $\Delta_r H°$ for the hydrogen abstraction step can be calculated by taking the E_d for the bond broken (C–H) and subtracting the E_d for the bond formed (H–X, Table 5.1).

For fluorine (E_d(H–F) = 569 kJ/mol), regardless of the C–H bond (372–439 kJ/mol), this step is always very exothermic ($\Delta_r H°$ = –197 to –130 kJ/mol), and therefore there is no selectivity.

For chlorine (E_d(H–Cl) = 452 kJ/mol), this step ranges from very exothermic (allyl C–H, $\Delta_r H°$ = –79 kJ/mol) to slightly exothermic (methyl C–H, $\Delta_r H°$ = –13 kJ/mol). Therefore, there is a modest preference for the more exothermic benzylic/allylic, tertiary, and secondary C–H bonds.

For bromine (E_d(H–Br) = 364 kJ/mol), this step is always endothermic whether mildly so (allyl C–H, $\Delta_r H°$ = 8 kJ/mol) or extremely endothermic (methyl C–H, $\Delta_r H°$ = 75 kJ/mol). As such, this step becomes very selective for the lowest energy pathway (allylic/benzylic >> tertiary > secondary > primary > methyl).

For iodine (E_d(H–I) = 297 kJ/mol), this step is, regardless of C–H, so endothermic ($\Delta_r H°$ = 75 to 142 kJ/mol) that no reaction happens (the energy barrier is too high in all cases).

PROPAGATION ENERGETICS

If one now considers the hydrogen atom abstraction from isobutane (from the radical halogenation example in Figure 5.1) with the bromine radical, the preference for the creation of a tertiary radical (more stable, lower energy) intermediate over a primary radical (less stable, higher energy) intermediate can now be understood in terms of radical stability. Consider a reaction coordinate diagram (Figure 5.9) that shows the energy pathways involved. The tertiary radical is more stable and lies below the primary radical in terms of energy. Since the tertiary radical is lower in energy, we infer that the pathways (curved lines) leading to the different intermediates are similarly lower for the tertiary radical compared to the primary radical.[4] Altogether, the lower energy intermediate is produced faster meaning that this will give rise to the major product.

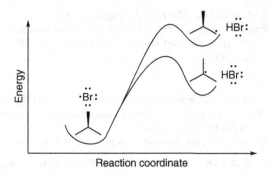

Figure 5.9 A reaction coordinate diagram showing the different energy pathways leading to the higher energy (less stable) primary radical intermediate and the lower (more stable) tertiary radical intermediate.

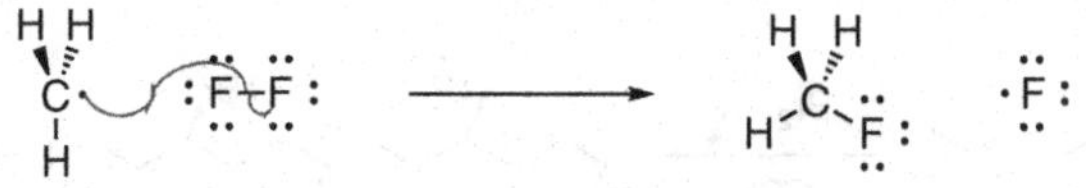

Figure 5.10 A propagation step where a carbon radical abstracts a fluorine atom from difluorine.

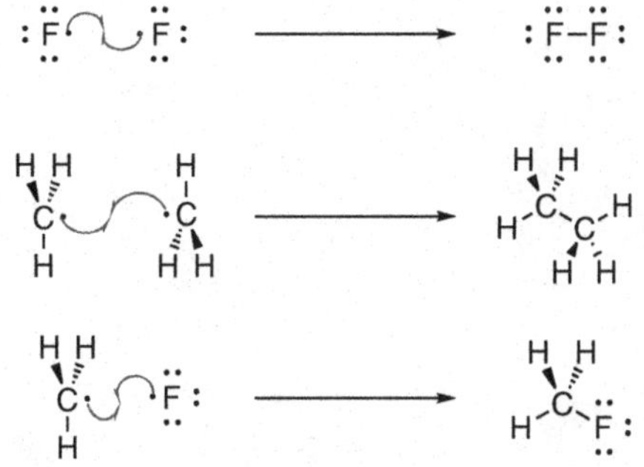

Figure 5.11 Termination steps involving the combination of different radicals present in solution.

Problem 5.3 Predict the most likely hydrogen atom abstraction intermediate(s) in each case.

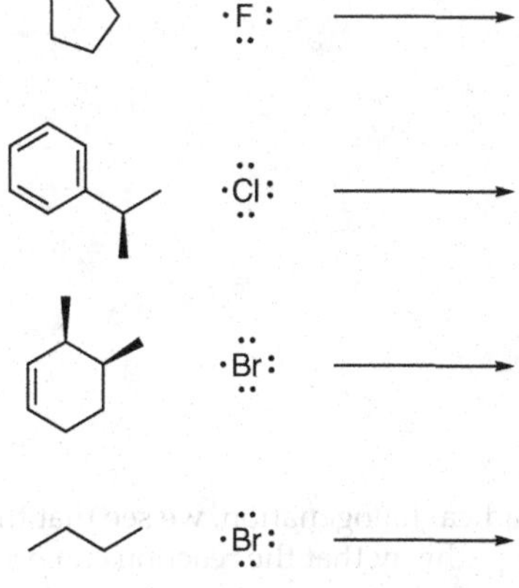

After formation of the carbon radical, the next propagation step is halogen-atom abstraction (Figure 5.10). This regenerates a halogen radical, and this cycle continues (hydrogen atom abstraction by a halogen radical, then halogen-atom abstraction by a carbon radical, then hydrogen atom abstraction by a halogen radical, etc.).

TERMINATION

Finally, a termination step, that is making non-radicals from radicals (Figure 5.11), only occurs once most of the reagents have been used up. Any radical intermediates in solution partner up and make a bond. This terminates the radical process.

Each step of a radical halogenation mechanism has now been individually investigated.

Problem 5.4 Draw a mechanism showing the initiation, propagation, and termination steps for the following radical halogenation reaction. Note that CCl_4 is the solvent (a common solvent for radical reactions) and that there are two products (one with a dashed bromine and one with a wedged bromine). Two stereoisomer products can be produced during radical reactions because the planar, carbon radical intermediate can add the halogen to either face with equal probability.

Problem 5.5 Predict the major product for each radical halogenation reaction. Rank each reaction in order of speed (that is rank the reactions 1–6 with 1 being the fastest and 6 being the slowest).

From the above investigation of radical halogenation, we see that the outcome is driven by energetics—what is the lowest energy pathway that the reaction can proceed along? This same maxim

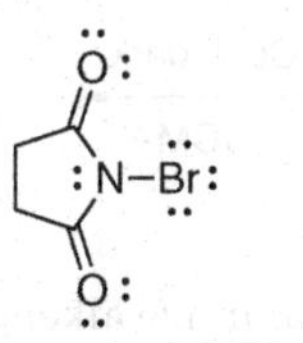

N-bromosuccinimide
(NBS)

Figure 5.12 Structure of *N*-bromosuccinimide (NBS).

will hold for many other organic reactions, and it will therefore be imperative to understand the mechanism of a reaction to understand which pathway is lower in energy and why.

RADICAL HALOGENATION AND GREEN CHEMISTRY

So far, we have considered radical bromination reactions only using dibromine. Dibromine is, however, is a noxious, volatile liquid that is both a hazardous and challenging reagent to work with. Instead of using dibromine, *N*-bromosuccinimide (NBS, Figure 5.12) is commonly used. From an atom economy standpoint, dibromine is the better reagent. NBS, however, is safer – while still an irritant it is not fatal if inhaled like bromine is – and easier to handle as a solid reagent. Several different mechanisms for radical bromination with NBS are presented in the literature. It is commonly accepted that NBS is a means of generating low concentrations of Br_2 in solution. The mechanism for the formation of Br_2 from NBS is not necessary to cover here. The radical substitution mechanism (proceeding from the low-level concentration of Br_2) then is the same as was seen above.

Other, safer methods of radical bromination also involve generating low levels of dibromine or bromine atoms *in situ*. The first involves using an aqueous solution of sodium bromide (NaBr) and sodium bromate ($NaBrO_3$) to generate dibromine with added hydrogen chloride (HCl).[5] The second – HBr, H_2O_2, H_2O – mimics the chemistry of naturally occurring bromoperoxidase enzymes and generates bromine atoms following initiation by hydrogen peroxide with water as a solvent.[6]

ALLYLIC HALOGENATION

Allylic substitution (Figure 5.13), substitution of C–H bonds next to alkenes, proceeds very rapidly (in comparison to non-allylic substitution) because the allylic radical that is generated is stabilized by delocalization.

However, as shall be seen in Chapter 12, alkenes can react with bromine to produce vicinal dibromides (Figure 5.14 the mechanism of and understanding of this reaction will be studied in Chapter 12, at this point, only be aware that this is a possible outcome).

The question, then, is how to achieve selectivity between allylic substitution and electrophilic addition. One way is to rigorously control the reaction conditions. Radical, allylic substitution is a radical process and so radical formation is favoured by running the reaction at higher temperatures, with light and using an inert, non-polar solvent (like carbon tetrachloride, Figure 5.15). The non-polar solvent disfavours the polar reaction pathway seen in electrophilic addition.

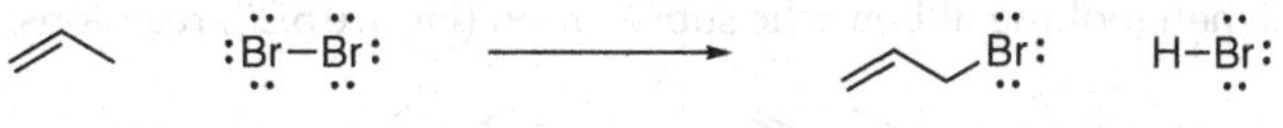

Figure 5.13 Radical bromination of propene.

Figure 5.14 Addition of bromine to propene.

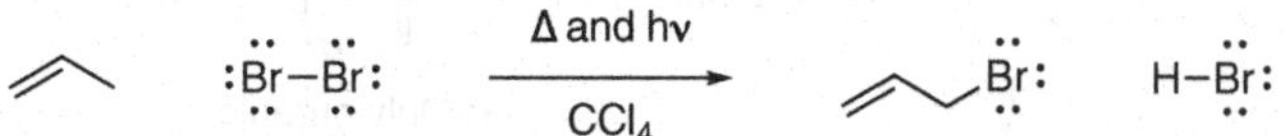

Figure 5.15 Conditions designed to maximize allylic radical substitution.

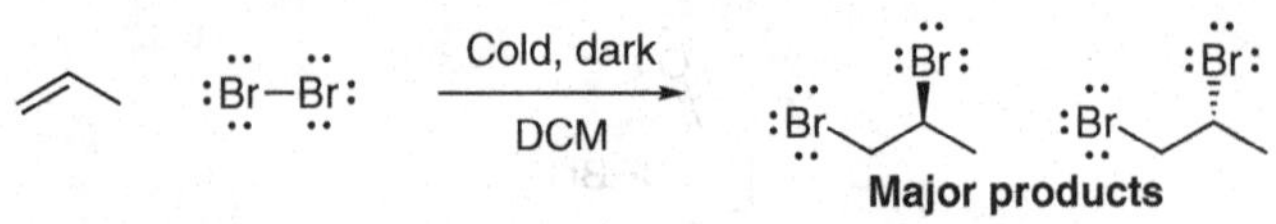

Figure 5.16 Conditions designed to maximize alkene addition.

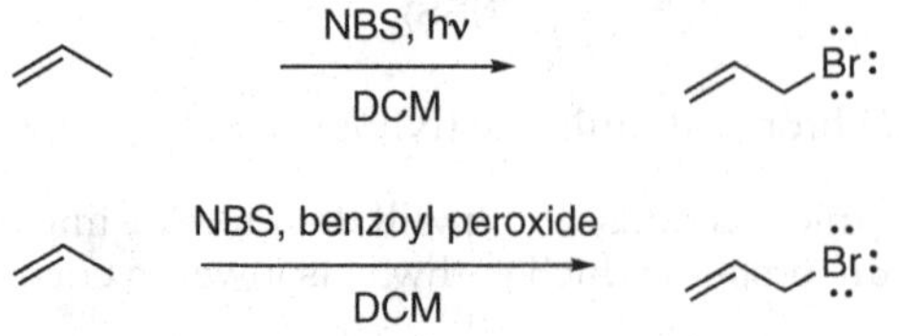

Figure 5.17 NBS substitution of propene.

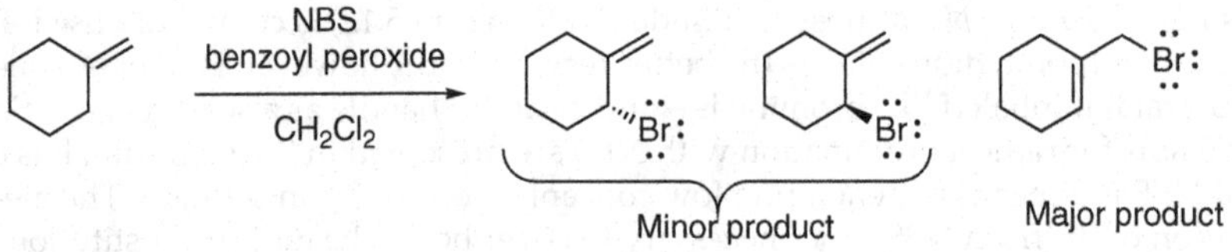

Figure 5.18 NBS bromination of an asymmetric alkene.

Electrophilic addition is not a radical process and so the formation of radicals is limited by conducting the reaction at lower temperatures and in the dark (Figure 5.16). Moreover, the reaction proceeds via an ionic mechanism, so a more polar solvent (dichloromethane, abbreviated as DCM) is used to favour this polar pathway.

Alternatively, allylic substitution can be favoured by using N-bromosuccinimide, what is referred to as the Wohl–Zielger bromination.[7] The concentration of dibromine generated will lead to radical substitution, but it is too low for Br_2 to add to the alkene and form a vicinal dibromide. NBS can be used alone with light to generate dibromine or with benzoyl peroxide as the initiator (Figure 5.17).

While the mechanism is the same as was seen above, allylic substitution does have one new wrinkle. That is, for asymmetric alkenes (Figure 5.18), multiple products can be produced.

This is because the radical intermediate that is produced has radical character at two different carbons (recall the delocalization of radicals in allylic systems).

And so, substitution can happen at both carbons. The major product will be that species which has the most substituted double bond (Figure 5.19).

BENZYLIC SUBSTITUTION

It should be noted that while allylic substitution reactions can involve rearrangements, that this is not a consideration when looking at benzylic substitution (Figure 5.20) reactions.

Figure 5.19 Contributing structures of the radical intermediate.

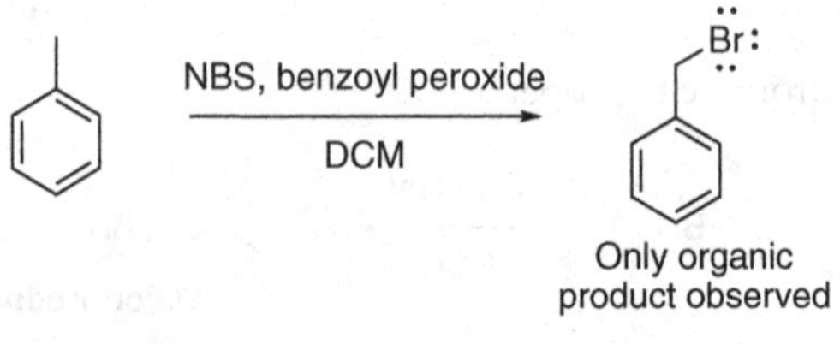

Figure 5.20 Benzylic substitution with NBS.

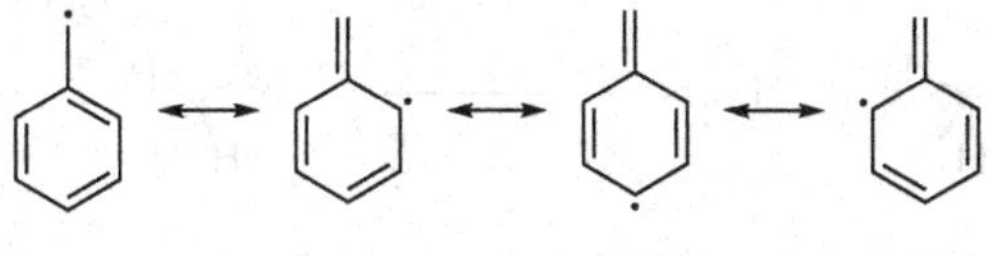

Figure 5.21 Contributing structures of the benzylic radical.

Here, although there are four carbons that display radical character (Figure 5.21), which stabilizes the radical intermediate, substitution does not occur on the ring. This would disrupt a unique feature of benzene (its aromaticity, a topic that will be investigated in Chapter 25) which is highly disfavoured.

Problem 5.6 Utilizing this information, predict the major products for the reactions below.

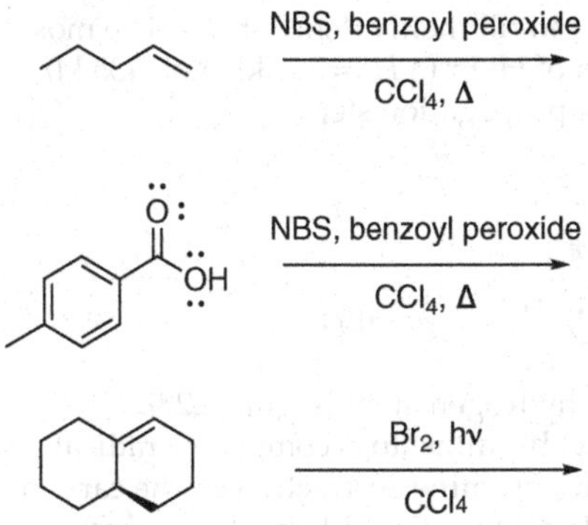

RADICAL ADDITION TO ALKENES

The final reaction (Figure 5.22) that shall be considered in this chapter is the addition of HBr to alkenes with peroxides.

This particular outcome is driven entirely by the presence of peroxides. As shall be seen in Chapter 12, a different product is obtained when there are no peroxides present. It should also be noted that HF, HCl, and HI do not engage in this type of chemistry.

Alone, neither HBr nor isobutene would, under most normal laboratory circumstances, produce radicals through heating or irradiation with light. As such, peroxides (in the example in Figure 5.22, diethyl peroxide) are used as a radical initiator (that is a molecule that will readily form radicals with heat or light and will initiate the radical chain process). The proclivity of peroxides to form radicals when exposed to light is why hydrogen peroxide solutions are typically sold in light-limiting opaque or brown bottles.

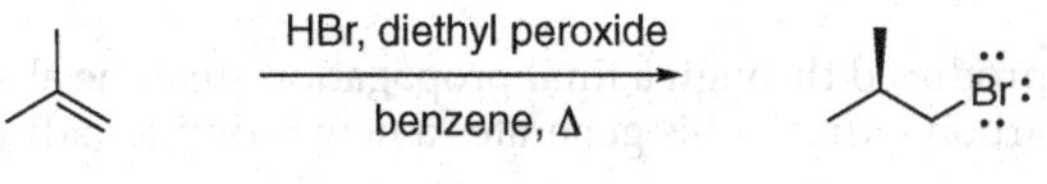

Figure 5.22 Radical addition of HBr to isobutene.

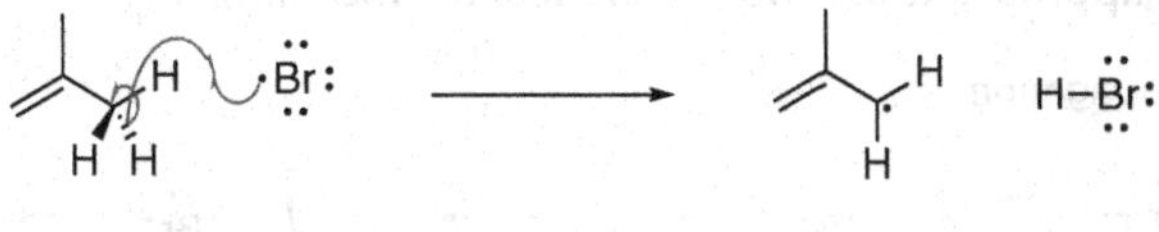

Figure 5.23 H–atom abstraction.

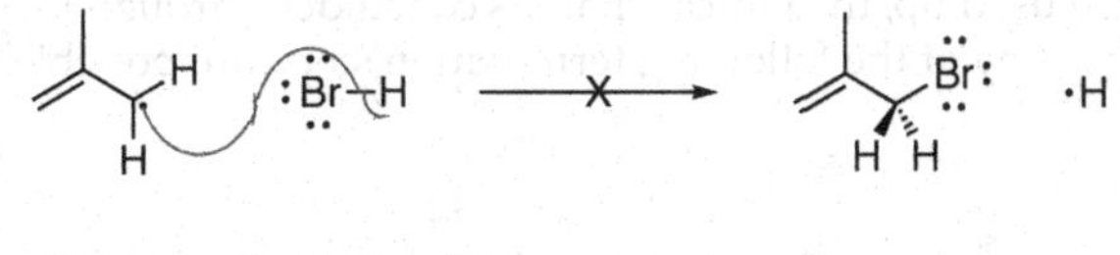

Figure 5.24 Propagation step that is too high energy to occur.

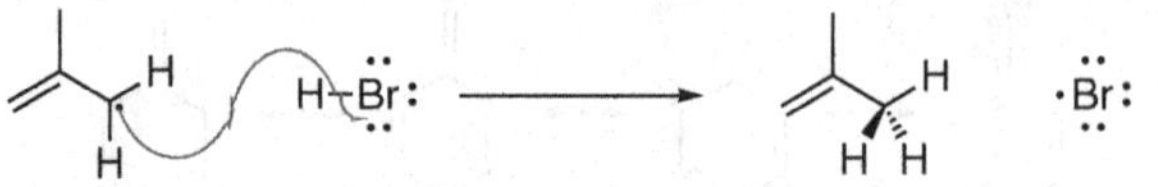

Figure 5.25 H–atom abstraction from HBr.

Problem 5.7 Draw the arrows to show the initiation step.

Initiation

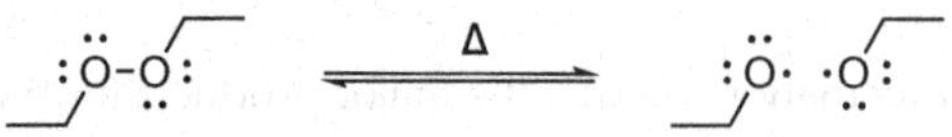

The alkoxy radicals generated then react (as was seen above) to fill their valence shells. The most energetically favourable atom to abstract is the hydrogen atom of H–Br ($\Delta_r H° = 364$ kJ/mol–435 kJ/mol = –71 kJ/mol). This generates a bromine radical in the first propagation step.

Problem 5.8 Draw the arrows to complete the mechanism.

Propagation

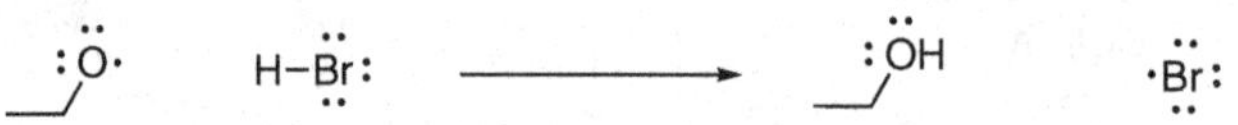

The bromine radical can then proceed to abstract the allylic hydrogen atom (Figure 5.23).

However, the carbon radical that is generated cannot abstract bromine (to accomplish a radical bromination, Figure 5.24). As the only source of bromine is HBr, bromine abstraction by the carbon radical would generate a free hydrogen radical, which energetically ($\Delta_r H° = 71$ kJ/mol) is never going to occur.

Therefore, when the allylic radical forms, the only pathway available to it is hydrogen atom abstraction (Figure 5.25) from H–Br.

While the step in Figure 5.26 can and does occur, this is obviously not a productive pathway as the result is where the reaction started (with a bromine radical and isobutane). Therefore, the productive pathway is addition of the bromine radical to the alkene. In the second propagation step, the bromine adds to the less substituted end of the alkene (due to steric repulsion) and generates the more stable carbon radical.

Problem 5.9 Draw the arrows to complete the mechanism.

Propagation

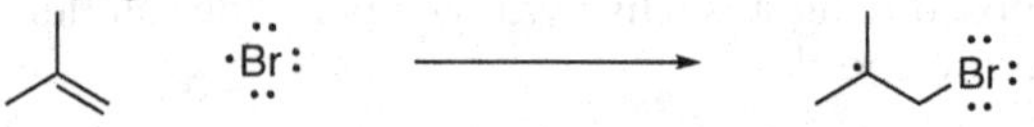

Finally, the product is produced through a final propagation step, the abstraction of a hydrogen atom from H–Br by the carbon radical. This generates a new bromine radical, which then starts a new radical cycle.

Problem 5.10 Draw the appropriate arrows to complete the mechanism.

Propagation

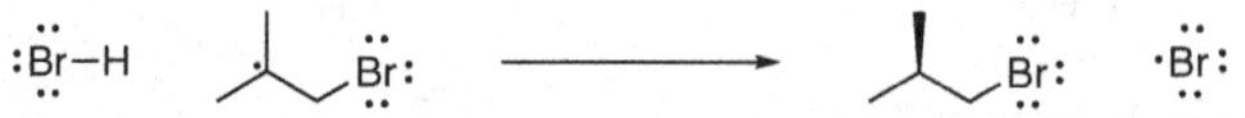

Once most reagents are used up, the radical chain is concluded through low-probability radical–radical termination steps. Any of the following termination steps are possible.

Problem 5.11 Draw the arrows to complete the mechanism.

Termination

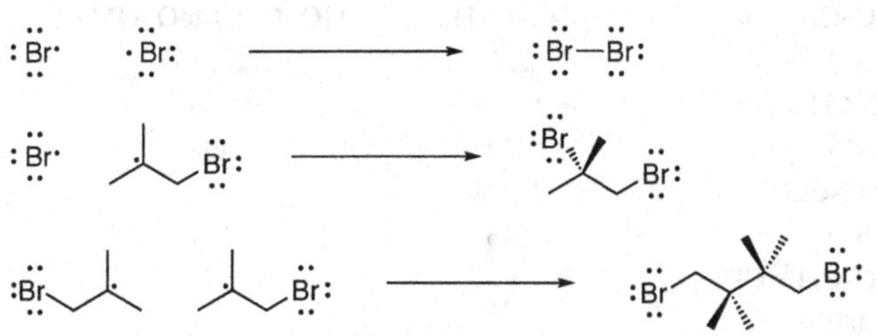

Problem 5.12 Predict the major product(s) for the following reaction.

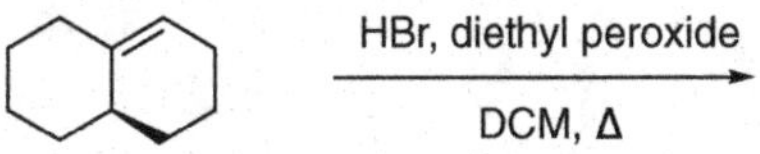

GENERAL PRACTICE PROBLEMS

Problem 5.13 Provide the reagent(s) necessary to affect the transformation.

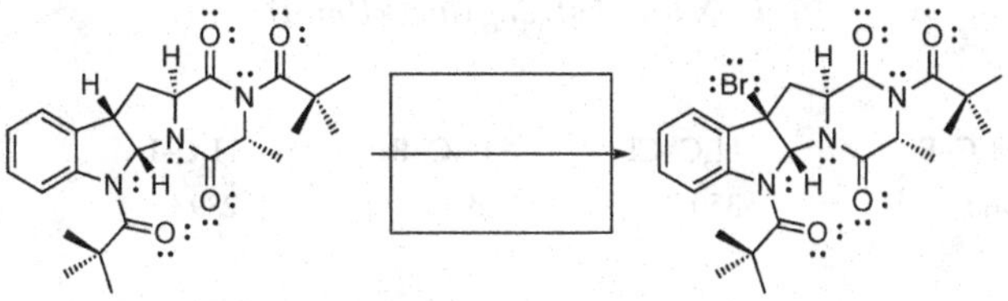

Problem 5.14 Predict the product(s).

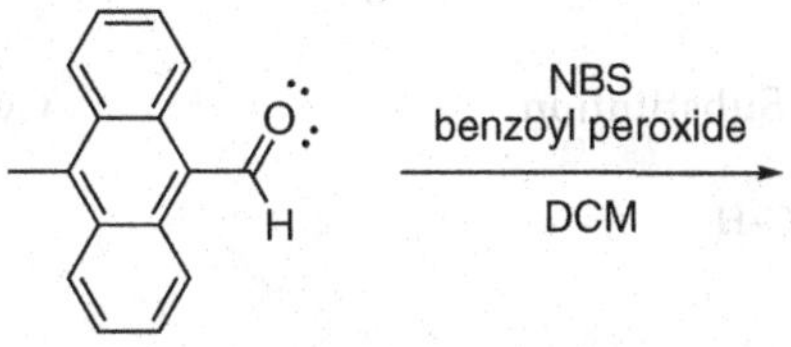

Dissociation energy values $(E_d)^8$
A–H bond strengths (kJ/mol)

H–H				
436	H$_3$C–H	H$_2$N–H	HO–H; MeO–H	F–H
	439	453	499; 440	569
			HS–H	Cl–H
			382	431
				Br–H
				364
				I–H
				297

A–A bond strengths (kJ/mol)

$H_3CH_2C–CH_2CH_3$	$H_2N–NH_2$	HO–OH; MeO–OMe	F–F
363	277	213; 159	159
$H_3CH_2C–CH=CH_2$	HN=NH	O=O	Cl–Cl
418	452	498	243
$H_3CH_2C–C≡CH$	N≡N		Br–Br
519	946		192
$H_2C=HC–CH=CH_2$			I–I
489			151
HC≡C–C≡CH			
628			
$H_2C=CH_2$			
728			
HC=CH			
960			

C–X bond strengths (kJ/mol)

$H_3C–F$	$H_3C–Cl$	$H_3C–Br$	$H_3C–I$
460	350	294	239

C–H bond strengths (kJ/mol)

Changes in Substitution				Cycloalkanes	Changes in Hybridization
	$H_3C–H$				$CH_3CH_2–H$
	439			445	421
$CH_3CH_2–H$	$(H_2C=CH)CH_2–H$	$PhCH_2–H$	$HC≡CCH_2–H$		$H_2C=CH–H$
421	369	370*	372	405	465
$(CH_3)_2CH–H$	$(H_2C=CH)_2CH–H$	$Ph_2CH–H$	$(HC≡C)_2CH–H$		Ph–H
411	321	354	324	400	472
$(CH_3)_3C–H$	$(H_2C=CH)_3C–H$	$Ph_3C–H$	$(HC≡C)_3C–H$		HC≡C–H
400	288	339	280		567

*Regardless of substitution on the phenyl ring, E_d changes by no more than 8 kJ/mol for benzylic C–H bonds.

NOTES

1. Gomberg, M. Tetraphenylmethan. *Ber. Dtsch. Chem. Ges.*, **1897**, *30* (2), 2043–2047. DOI: 10.1002/cber.189703002177

2. Gomberg, M. An instance of trivalent carbon: triphenylmethyl. *J. Am. Chem. Soc.*, **1900**, 22 (11), 757–771. DOI: 10.1021/ja02049a006

3. Gomberg, M. On trivalent carbon. *J. Am. Chem. Soc.*, **1901**, *23* (7), 496–502. DOI: 10.1021/ja02033a015

4. i) Leffler, J.E. Parameters for the description of transition states. *Science*, **1953**, *117*, 340–341. DOI: 10.1126/science.117.3039.340
   ii) Hammond, G.S. A correlation of reaction rates. *J. Am. Chem. Soc.*, **1955**, *77* (2), 334–338. DOI: 10.1021/ja01607a027

5. Adimurthy, S.; Ranu, B.C.; Ramachandraiah, G.; Ganguly, B.; Ghosh, P.K. Bromide-bromate Couple of Varying Ratios for Bromination, Vicinal Functionalisation and Oxidation in a Clean Manner. *Curr. Org. Synth.*, **2013**, *10* (6), 864–884.

6. Podgoršek, A.; Stavber, S.; Zupan, M.; Iskra, J. Free radical bromination by the H_2O_2–HBr system on water. *Tetrahedron Lett.*, **2006**, *47* (40), 7245–7247. DOI: 10.1016/j.tetlet.2006.07.109

7. i) Wohl, A. Bromierung ungesättigter Verbindungen mit N-Brom-acetamid, ein Beitrage zur Lehre vom Verlauf chemischer Vorgänge. *Ber. Dtsch. Chem. Ges. A/B*, **1919**, *52* (1), 51–63. DOI: 10.1002/cber.19190520109
   ii) Ziegler, K.; Schenck, G.; Krockow, E.W..; Siebert, A.; Wenz, A.; Weber, H. Die Synthese des Cantharidins. Liebigs Ann. **1942**, *551* (1), 1–79. DOI: 10.1002/jlac.19425510102

8. i) Batt, L.; Christic, K.; Milne, R.T.; Summers, A.J. Heats of formation of C_1–C_4 alkyl nitrities (RONO) and their RO–NO bond dissociation energies. *Int. J. Chem. Kinet.*, **1974**, *6*, 877–885. DOI: 10.1002/kin.550060610
   ii) Berkowitz, J.; Ellison, G.B.; Gutman, D. Three methods to measure RH bond energies. *J. Phys. Chem.*, **1994**, *98* (11), 2744–2765. DOI: 10.1021/j100062a009
   iii) Blanksby, S.J.; Ellison, G.B. Bond Dissociation Energies of Organic Molecules. *Acc. Chem. Res.*, **2003**, *36* (4), 255–263. DOI: 10.1021/ar020230d
   iv) Castelhano, A.L; Griller, D. Heats of formation of some simple alkyl radicals. *J. Am. Chem. Soc.*, **1982**, *104* (13), 3655–3659. DOI: 10.1021/ja00377a018
   v) Clark, K.B.; Culshaw, P.N.; Griller, D.; Lossing, F.P.; Simoes, M.J.A.; Walton, J.C. Studies of the formation and stability of pentadienyl and 3-substituted pentadienyl radicals. *J. Org. Chem.*, **1991**, *56* (19), 5535–5539. DOI: 10.1021/jo00019a012
   vi) Chase, M.W. Jr. *NIST-JANAF Thermochemical Table*, 4th Ed. *J. Phys. Chem. Data*, **1998**, Monograph 9.
   vii) *CRC Handbook of Chemistry and Physics*, 82d Ed. D.E. Lide, Ed.
   viii) Dobis, O.; Benson, S.W. Analysis of flow dynamics in a new, very low-pressure reactor. Application to the reaction: $Cl + CH_4 \rightarrow HCl + CH_3$. *Int. J. Chem. Kinet.*, **1987**, *19* (8) 691–708. DOI: 10.1002/kin.550190803
   ix) Ervin, K.M.; DeTuri, V.F. Anchoring the Gas-Phase Acidity Scale. *J. Phys. Chem. A*, **2002**, *106* (42), 9947–9956. DOI: 10.1021/jp020594n
   x) Menon, A.S.; Henry, D.J.; Bally, T.; Radom, L. Effect of substituents on the stabilities of multiply-substituted carbon-centered radicals[††]. *Org. Biomol. Chem.*, **2011**, *9*, 3636–3657. DOI: 10.1039/C1OB05196B
   xi) McMillen, D.F.; Golden, D.M.; Benson, S.W. Kinetics of the gas-phase reaction of cyclopropylcarbinyl iodide and hydrogen iodide. Heat of formation and stabilization energy of the cyclopropylcarbinyl radical[†]. *Int. J. Chem. Kinet.*, **1971**, *3*, 359–374. DOI: 10.1002/kin.550030406
   xii) McMillen, D.F.; Golden, D.M.; Benson, S.W. The rate of the gas phase iodination of cyclobutene. The heat of formation fo the cyclobutyl radical. *Int. J. Chem. Kinet.*, **1972**, *4*, 487–495. DOI: 10.1002/kin.550040503
   xiii) McMillen, D.F; Golden, D.M. *Ann.* Hydrocarbon Bond Dissociation Energies. *Rev. Chem.*, **1982**, *33*, 493–532. DOI: 10.1146/annurev.pc.33.100182.002425
   vii) *NIST Chemistry Webbook*, Mallard, W.G. Ed., http://webbook.nist.gov
   xiv) Parker, V.D.; Handoo, K.J.; Roness, F.; Tilset, M. Electrode potentials and the thermodynamics of isodesmic reactions. *J. Am. Chem. Soc.*, **1991**, *113* (20), 7493–7498. DOI: 10.1021/ja00020a007

xv) Pedley, J.B.; Naylor, R.D.; Kirby, S.P. *Thermochemical Data of Organic Compounds* Chapman and Hall: New York. 2nd Ed. 1986. DOI: 10.1007/978-94-009-4099-4

xvii) Russell, J.J.; Seetula, J.A.; Senkan, S.M.; Gutman, D. Kinetics and thermochemistry of the methyl radical: Study of the CH_3 + HCl reaction. *Int. J. Chem. Kinet.*, **1988**, *20*, 759–773. DOI: 10.1002/kin.550201002

xvi) Tsang, W. In *Energetics of Organic Free Radicals*; Simoes, M.J.A.; Greenberg, A.; Liebman, J.F., Eds.; Blackie Academic: New York, 1996, 22–58.

6 Stereochemistry

Organic chemistry has several different types of isomerism. In Chapter 3, constitutional isomers were discussed. Recall that constitutional isomers (Figure 6.1) are molecules that have the same molecular formula but a different connectivity (or constitution) of atoms.

CIS–TRANS ISOMERISM

In addition to constitutional isomers, another type of isomers are stereoisomers. Stereoisomers are molecules that have the same formula and the same connection of atoms; however, these isomers differ in terms of the arrangement of atoms in 3D space. Stereoisomers are also called configurational isomers. We have already encountered *cis-trans* isomerism, which is a type of stereoisomerism (Figure 6.2).

CHIRALITY

Another common form of stereoisomerism is chirality. To understand chirality we need to establish some definitions. The first is chirality itself. An object is chiral if it has a non-superimposable mirror image (Figure 6.3). This means that any chiral object does not possess a plane of symmetry nor a centre of symmetry. An object that posses a plane of symmetry or center of symmetry is termed achiral (Figure 6.4).

The two different-handed versions of a chiral object are called enantiomers. Enantiomers are stereoisomers that are non-superimposable mirror images of each other (Figure 6.5). Enantiomers have the same physical properties but differ in how they interact with other chiral species, like polarized light, enzymes, and so on.

Figure 6.1 The constitutional isomers of C_5H_{12}.

cis-1,4-dimethylcyclohexane   *trans*-1,4-dimethylcyclohexane

Figure 6.2 Stereoisomers of 1,4-dimethylcyclohexane.

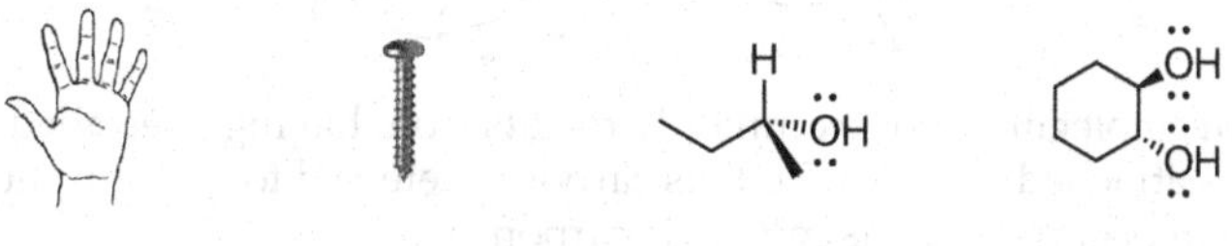

Figure 6.3 Examples of chiral objects (mirror images are non-superimposable).

plane of symmetry   planes of symmetry   plane of symmetry   plane of symmetry   planes of symmetry
                    center of symmetry   and
                                                                                 center of symmetry

Figure 6.4 Examples of achiral objects (mirror images are superimposable).

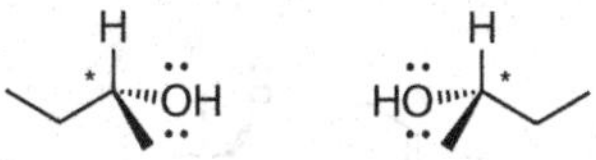

Figure 6.5 Examples of enantiomers (non-superimposable mirror image isomers) 2-butanol. Note on symbology and terminology, the starred carbons are termed chirality centres, stereocentres, or asymmetric carbons.

DOI: 10.1201/9781003479352-6

67

cis-1,4-dimethylcyclohexane   *trans*-1,4-dimethylcyclohexane

Figure 6.6   Examples of *cis*- and *trans*-1,4-dimethylcyclohexane diastereomers.

In contrast to enantiomers, diastereomers are stereoisomers that are not mirror images of each other (not necessarily chiral). Diastereomers (Figure 6.6) have different physical properties from each other.

Problem 6.1 Identify each pair of molecules as the same, enantiomers or diastereomers.

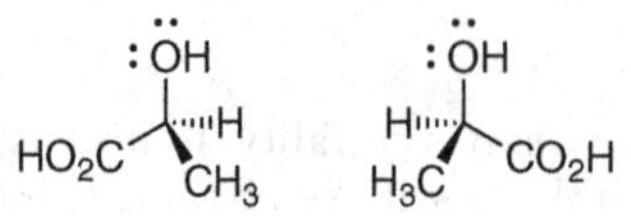

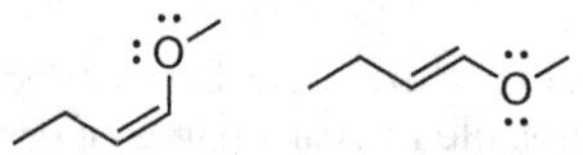

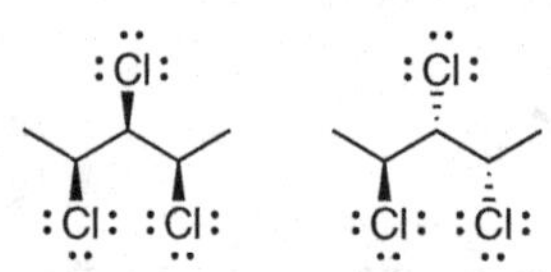

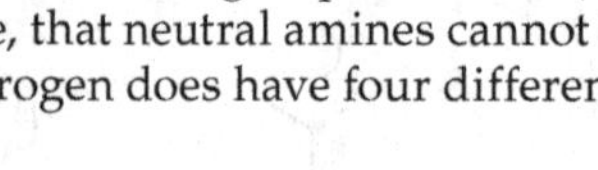

Chirality in organic molecules most commonly results from having a tetrahedral carbon with four different groups attached (Figure 6.7). This carbon is referred to as a chirality centre and can be referred to as a stereocentre or an asymmetric carbon.

Chirality centres, however, need not be carbons. Chirality centres, in organic chemistry, can be N, P, S, Si, Ge, or Sn atoms with four different groups attached (Figure 6.8).

It is worth noting at this juncture, that neutral amines cannot be chirality centres. While neutral amines are tetrahedral, and the nitrogen does have four different groups around it (in the example

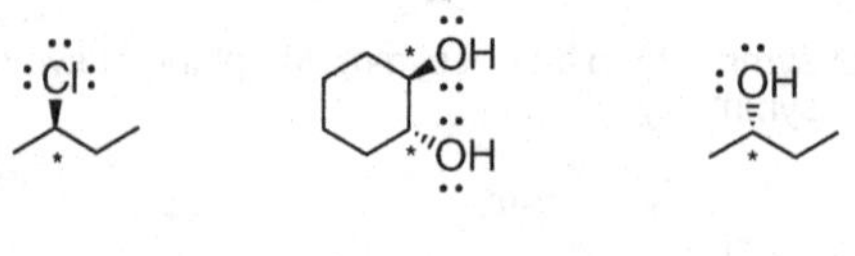

Figure 6.7   Examples of carbon-atom chirality centres.

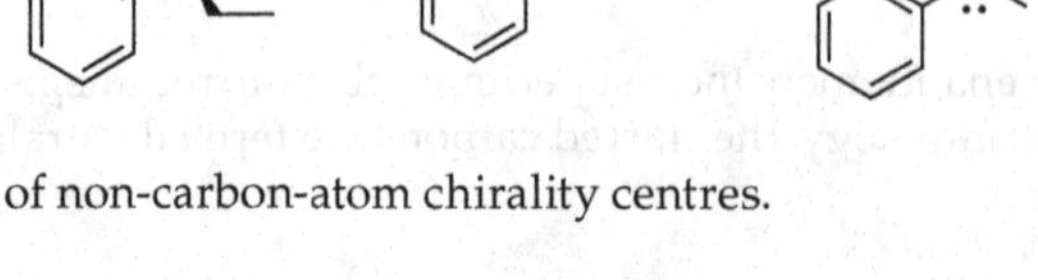

Figure 6.8   Examples of non-carbon-atom chirality centres.

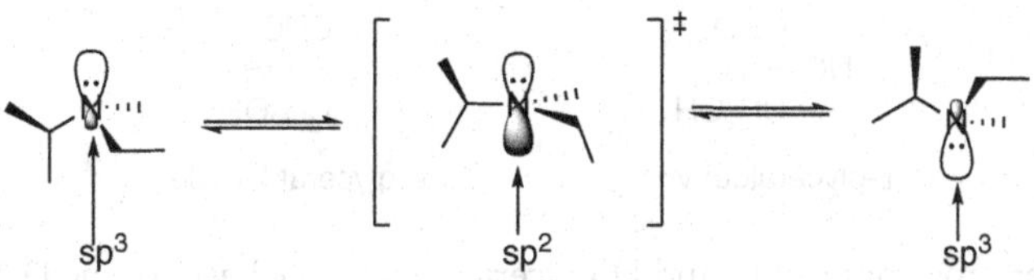

Figure 6.9 Example of nitrogen atom inversion, which prevents neutral amines from being chirality centres.

in Figure 6.9, one methyl, one ethyl, one isopropyl and one lone pair), amines rapidly undergo pyramidal inversion at room temperature.[1] This inversion rapidly interconverts and equilibrates the two different configurations at nitrogen (Figure 6.9).

Problem 6.2 Identify all the chirality centres (place a star next to each) in the following molecules.

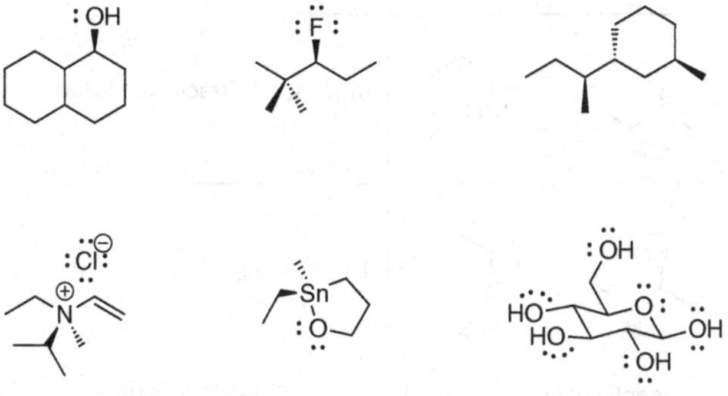

CONFORMATIONAL ISOMERISM

The last type of stereoisomerism that will be mentioned here is conformational isomerism. Conformational stereoisomers are different conformations of a molecule that, by definition, have different 3D configurations of atoms, butane for example (Figure 6.10).

Anti-butane and *gauche* butane are conformational stereoisomers of each other. Because these species are due to the rapid rotation of single bonds, however, their stereoisomerism has no consequence at room temperature and will not be considered further. Molecules which possess conformational stereoisomers that cannot rapidly interconvert at room temperature are called atropisomers. Atropisomerism is an advanced topic, which is beyond the scope of this textbook.

STEREOCHEMISTRY SYSTEMS

There are several systems of nomenclature for describing stereoisomers. Each will be introduced, but organic chemistry typically focuses on the *R/S* system. Please note that none of the following systems is at all related to any of the others.

Relative Stereochemistry: The D,L–System

This system is frequently used in biochemistry and biology. To apply this system, each molecule of interest is compared to glyceraldehyde (Figure 6.11). This system was originally based upon a conjecture made in 1891 by Hermann Emil Fischer.[2] His guess was proven correct by X-ray analysis in 1951.[3]

Very few molecules in organic chemistry lend themselves to the D,L–system. Students are likely to encounter this in biology or biochemistry, for example, D–glucose, D–mannose, and L–proline.

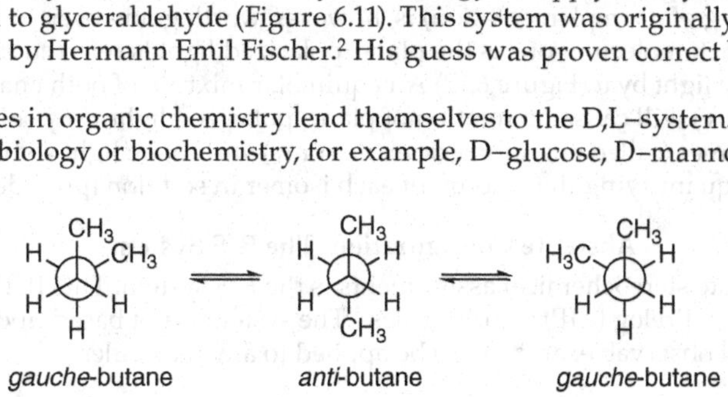

Figure 6.10 Conformational isomers of butane.

$$CHO$$
$$HO\!\!-\!\!H$$
$$CH_2OH$$

L-glyceraldehyde

$$CHO$$
$$H\!\!-\!\!OH$$
$$CH_2OH$$

D-glyceraldehyde

Figure 6.11 Fisher projections of L- and D-glyceraldehyde, the bases of the D, L–system.

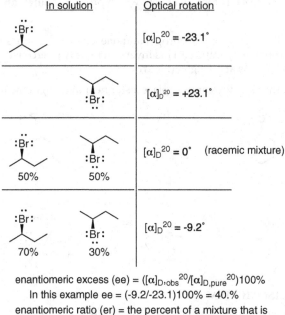

In solution | Optical rotation

$[\alpha]_D{}^{20}$ = -23.1°

$[\alpha]_D{}^{20}$ = +23.1°

$[\alpha]_D{}^{20}$ = 0° (racemic mixture)

50% 50%

$[\alpha]_D{}^{20}$ = -9.2°

70% 30%

enantiomeric excess (ee) = ($[\alpha]_{D,obs}{}^{20}/[\alpha]_{D,pure}{}^{20}$)100%
  In this example ee = (-9.2/-23.1)100% = 40.%
enantiomeric ratio (er) = the percent of a mixture that is
the dominant enantiomer.
  In this example, er = 70% *R*

Figure 6.12 Example of polarimetry of 2-bromobutane, the value for a for pure isomers and mixtures, and the enantiomeric excess and enantiomeric ratios of a 70/30 mixture.

Optical Rotation: The d(+), l(–) System

Molecules are analyzed by polarimetry, a technique that was developed over the course of the 19th century and integral to our development of concepts relating to chemical structure and reaction mechanisms.[4] Enantiomers rotate plane, polarized light by the same magnitude, but in equal and opposite directions. Therefore, each enantiomer will be assigned as either (+) that is dextrorotatory (*d*) or (–) that is levorotatory (*l*).

Given that each sample must be analyzed by a polarimeter (a pricey piece of equipment), this system is not particularly useful for analyzing a 2D Lewis structure. The mechanics of this process (and an understanding of its implementation) is, however, worth going over.

The amount that each molecule rotates the plane, polarized light is referred to as α. A pure sample will rotate the light by α (Figure 6.12) An equimolar mixture of both enantiomers (referred to as a racemic mixture) will cause no rotation of plane, polarized light. Any solution that has an excess of one enantiomer (called enantiomeric excess or *ee*) will rotate light some fraction of α. This provides a means of quantifying the amount of each isomer in solution (provided α is known).

Absolute Configuration: The R/S System

The system for absolute stereochemical assignments is the *R/S* system. This IUPAC system is based upon the Cahn–Ingold–Prelog (CIP) priority rules.[5] The system is not based upon any chemical property nor physical observable, and so can be applied to any molecule.

The CIP Priority Rules are:

1. a) Highest atomic number has highest priority, e.g., F > C

   b) If atomic number is equal, the highest pass has highest priority, e.g.: D (^{2}H) > H (^{1}H).
2. If the first atoms attached are equal, then consider the next atoms (in priority order) until the first point of difference is found.

Once all the groups attached to the chirality centre have been assigned a priority (1–4), one uses one's hands (Figure 6.13). The thumb is orientated towards the lowest priority group (4) and the fingers towards the highest (1). If the fingers curl in the correct priority order (1–3) with the right hand, it is an R (*rectus*) stereocentre. If the fingers curl in the correct priority order with the left hand, it is an S (*sinister*) stereocentre.

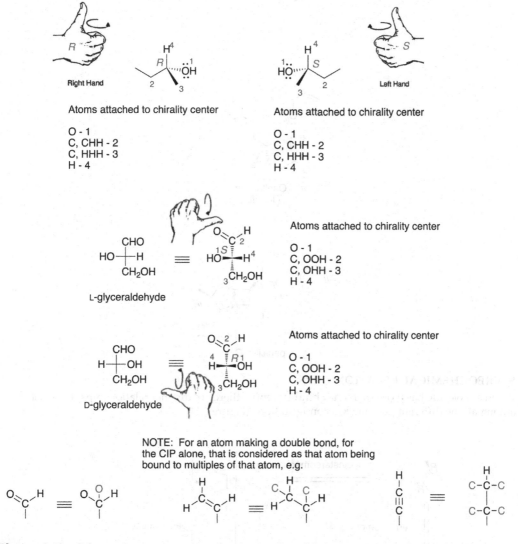

Atoms attached to chirality center

O - 1
C, CHH - 2
C, HHH - 3
H - 4

Atoms attached to chirality center

O - 1
C, CHH - 2
C, HHH - 3
H - 4

L-glyceraldehyde

Atoms attached to chirality center

O - 1
C, OOH - 2
C, OHH - 3
H - 4

D-glyceraldehyde

Atoms attached to chirality center

O - 1
C, OOH - 2
C, OHH - 3
H - 4

NOTE: For an atom making a double bond, for the CIP alone, that is considered as that atom being bound to multiples of that atom, e.g.

Figure 6.13 Schematic representation of assigning R and S designations with one's hands.

Problem 6.3 Identify all chirality centres and assign each with the correct *R/S* designation.

cis-2-methylcyclopentanol

naproxen

Oseltamivir
(Tamiflu)

penicillin G

STEREOCHEMICAL RELATIONSHIPS

When a molecule has more than one chirality centre, there are distinct relationships that exist among all the different possible *R/S* configurations (Figure 6.14).

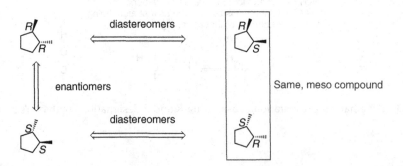

Figure 6.14 The three possible stereoisomers of 1,2-dimethylcyclopentane.

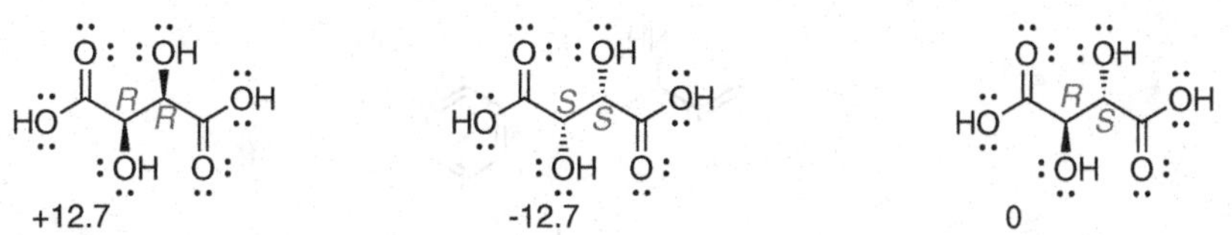

$[\alpha]_D^{20}$	+12.7	-12.7	0
MP (°C)	171-174	171-174	146-148
ρ (g/mL)	1.76	1.76	1.6
solubility	139 g/100 mL H_2O	139 g/100 mL H_2O	125 g/100 mL H_2O
pK_{a1}	2.98	2.98	3.23
pK_{a2}	4.34	4.34	4.82

Figure 6.15 Stereoisomers of tartaric acid and their physical properties of each.

Enantiomers have all opposite-assignment chirality centres and are non-superimposable mirror images.

Diastereomers have one or more identical chirality centres and one or more different chirality centres and are not mirror images.

If the molecule is identical to its mirror image and has chirality centres, it is a *meso* compound. A *meso* compound contains an internal symmetry element and is achiral. A simple test for *meso* compounds is to name the molecule and its mirror image. If the names are the same, then it is a *meso* compound.

Enantiomers differ from each other only in how they interact with other chiral molecules and chiral objects, like plane-polarized light. Diastereomers, on the other hand, have different physical properties from each other (Figure 6.15).

Problem 6.4 Here is a reaction that will be seen in Chapter 12, the acid-catalyzed hydration (addition of water) of an alkene. Correctly assign all chirality centres with an *R/S* designation. In the spaces provided, state the relationship that exists between each pair of stereoisomer products produced.

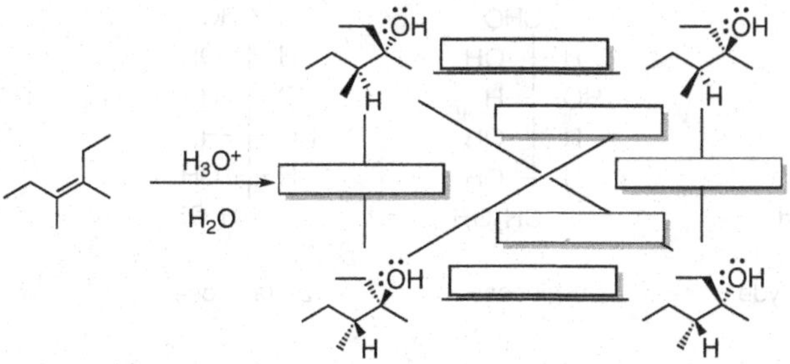

ENANTIOMERIC RESOLUTION

Typically, reactions of achiral starting materials produce a racemic mixture (that is an equal amount of both enantiomers) of products. Separating enantiomers is challenging because they do not have different physical properties (Figure 6.14). If the product of a reaction is acidic or basic, the enantiomers can be separated (or resolved) by adding a chiral acid or base and forming a diastereomeric salt (Figure 6.16).

Separating enantiomers is both materially and energetically costly. The difficulty of separating enantiomers has led to an ongoing focus on developing synthetic methods that are asymmetric, that is they preferentially produce only one stereochemical product. Selective, asymmetric processes are more atom economical and obviate the need for difficult chemical separations. Advanced organic chemistry courses focus heavily on stereoselective synthetic methods.

CHIRALITY IN BIOCHEMISTRY

Stereochemistry also plays an important role in biology. Living systems exhibit homochirality (that is the same chirality type) for proteins and sugars. For example, all naturally occurring amino acids are L-amino acids (Figure 6.17). All naturally occurring amino acids are D-sugars (Figure 6.18), which consider the chirality of the last (bottom-most) carbon atom in the chain.

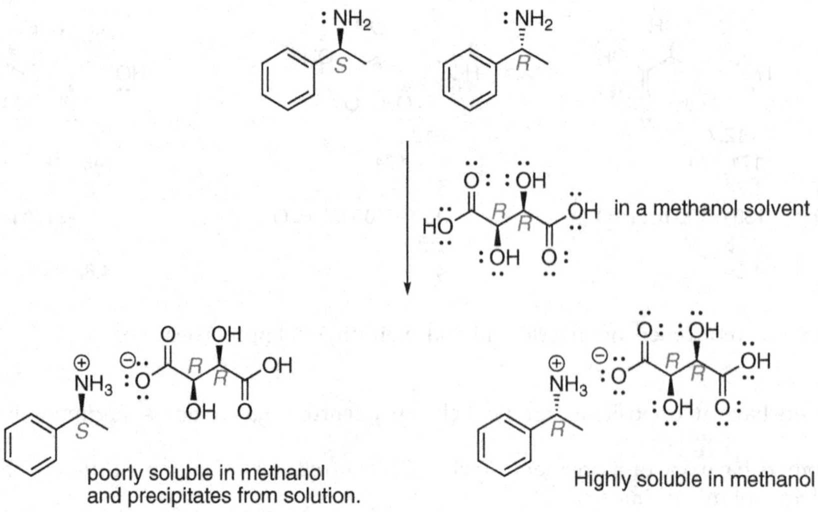

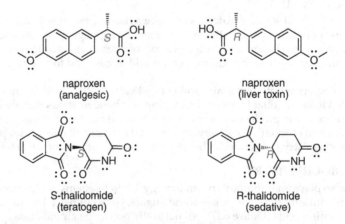

Figure 6.16 Chiral resolution of an amine using (2*R*,3*R*)-tartaric acid.

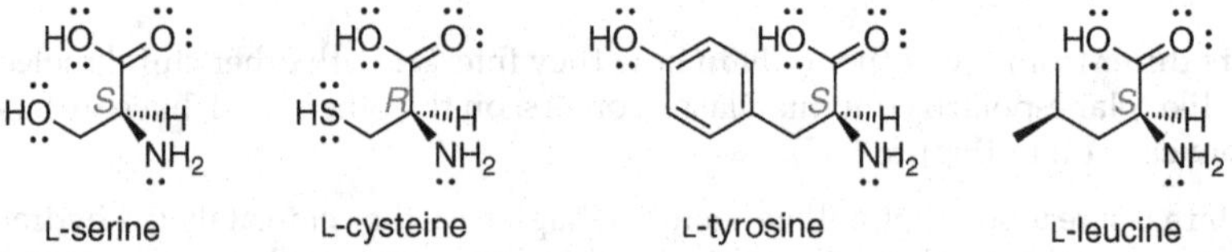

Figure 6.17 Homochirality among four example amino acids.

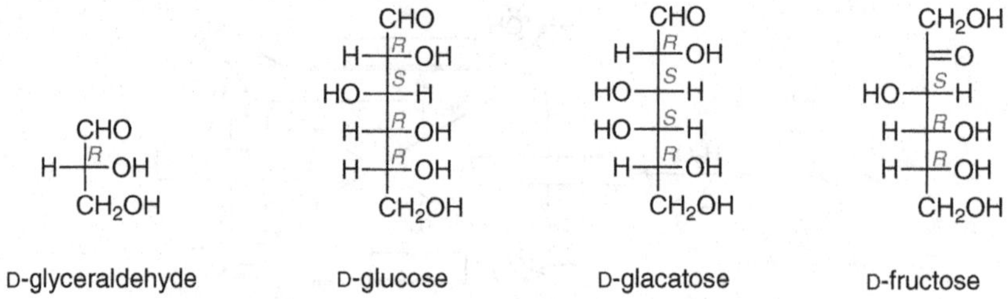

Figure 6.18 Homochirality among four example sugars.

Figure 6.19 Chiral chemicals and their physiological responses.

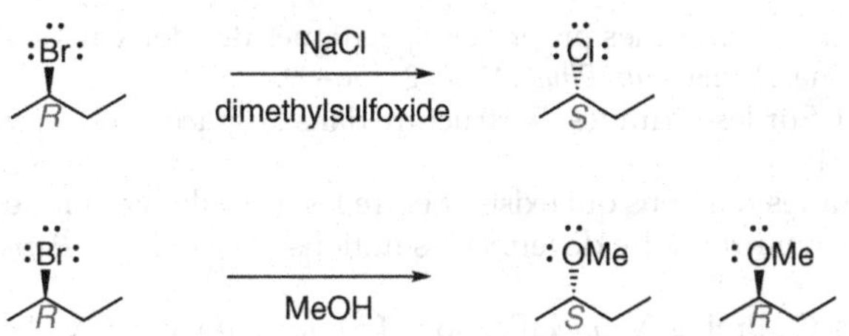

Figure 6.20 Different stereochemical outcomes observed in two different substitution reactions.

The origins of homochirality in living systems remain a mystery. Several possible hypotheses have been proposed. The author's favourite hypothesis is that the chiral dominance of L-amino acids and D-sugars on earth is due to meteorites seeding earth with molecules that were enantiomerically enriched by circularly polarized light in space.

Given the chiral nature of living systems, enantiomers can produce different physiological responses (Figure 6.19).

Paclitaxel (sold under the brand name Taxol) is a commonly used chemotherapy agent. The 3D shape of this molecule is directly responsible for its behaviour. Stereoisomers of paclitaxel do not exhibit the same properties.

Problem 6.5 Identify all chirality centres and assign each chirality centre its correct *R/S* designation.

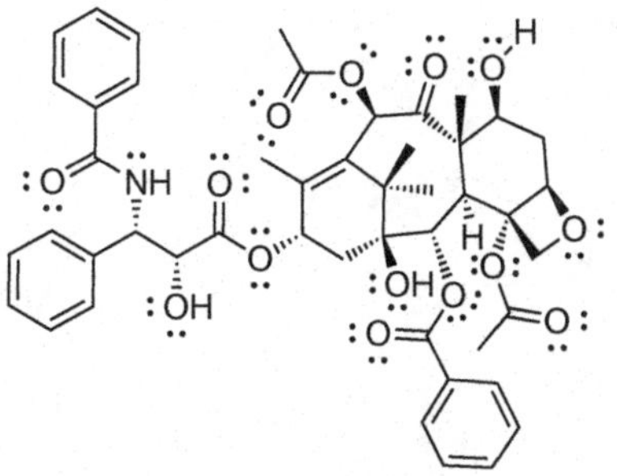

STEREOCHEMISTRY AND REACTION MECHANISMS

Looking ahead, consider the following two reactions (Figure 6.20). There are different stereochemical outcomes. This difference is a direct result of the mechanistic differences between the two reactions. Moving forward, the stereochemical outcome of a reaction will become a means of inferring different, likely pathways that take place in a reaction mechanism.

NOTES

1. Dennison, D.M. The Infra-Red Spectra of Polyatomic Molecules. Part II. *Rev. Mod. Phys.*, **1940**, *12* (3), 175–214. DOI: 10.1103/RevModPhys.12.175

2. i)   Fischer, E. Über die Configuration des Traubenzuckers und seiner Isomeren. I. *Ber. Dtsch. Chem. Ges.*, **1891**, 24 (1), 1836–1845. DOI: 10.1002/cber.189102401311
   ii) Fischer, E. Über die Configuration des Traubenzuckers und seiner Isomeren. II. *Ber. Dtsch. Chem. Ges.*, **1891**, 24 (2), 2683–2687. DOI: 10.1002/cber.18910240278

3. Bijvoet, J.M.; Peerdeman, A.F.; van Bommel, A.J. Determination of the absolute configuration of optically active compounds by means of X-rays. *Nature*, **1951**, *168* (4268), 271–272. DOI: 10.1038/168271a0

4. i)   Biot, J.B. Phenomene de polarization successive, observes dans des fluids homogenes. *Bull. Soc. Philom.*, **1815**, 190–192.
   ii) Biot, J.B. Extrait d'un mémoire sur les rotations que certaines substances impriment aux axes de polarization des rayons lumineux. *Ann. Chim. Phys.*, **1818**, *9*, 372–389.

iii) Pasteur, L. Recherches sur les propriétés spécifiques des deux acides qui composent l'acide racémique. *Ann. Chim. Phys.*, **1850**, *28*, 56–99.

iv) van't Hoff, J.H. Sur les formules de structure dans l'espace *Arch. Neerl. Sci. Exactes Nat.*, **1874**, *9*, 445–454.

v) Le Bel, J.-A. Sur les relations qui existent entre les formules atomiques des corps organiques et le pouvoir rotatoire de leurs dissolutions. *B. Soc. Chim. Paris.*, **1874**, *22*, 337–347.

5. Cahn, R.S.; Ingold, C.; Prelog, V. Specification of Molecular Chirality. *Angew. Chem. Int. Ed.*, **1966**, *5* (4), 385–415. DOI: 10.1002/anie.196603851

7 Infrared Spectroscopy

Spectroscopy is the study of how electromagnetic radiation interacts with atoms and molecules. Electromagnetic radiation consists of oscillating, perpendicular electronic and magnetic waves. These waves (Figure 7.1) have a certain wavelength (λ) and energy associated with them.

ELECTROMAGNETIC SPECTRUM

Recall from general chemistry and physics that electromagnetic radiation exists as a continuum of wavelengths and energies, which are divided into different regions (Figure 7.2).

SPECTROSCOPIC METHODS

The different types of electromagnetic radiation are useful for different types of spectroscopic studies.

X-rays (also called Röntgen rays in some European countries) are useful for analyzing the core electrons of atoms.

Ultraviolet and visible (UV/Vis) wavelengths are useful for analyzing the valence-electronic structure of atoms and molecules.

Infrared (IR) wavelengths is useful for looking at bond vibrations in molecules (and will be the focus of this chapter).

Microwaves (μW) are useful for studying the rotational states of molecules (useful for determining what molecules exist in space).

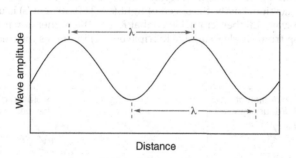

Figure 7.1 A simplified model of an electromagnetic wave (only showing one component, the electric wave). Where $E = h\nu = \dfrac{hc}{\lambda}$; $h = 6.625 \times 10^{-34}\,Js; c = 299792458\,m/s$.

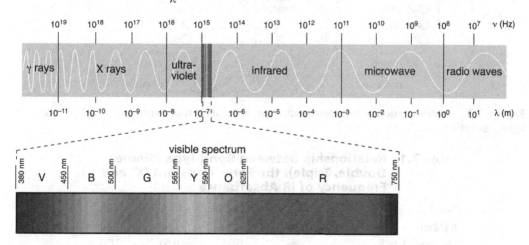

Figure 7.2 The electromagnetic spectrum.

DOI: 10.1201/9781003479352-7

Radio frequencies (RF) are useful for interrogating nuclear and electronic spin states, which will be analyzed in greater detail in Chapter 9 Nuclear Magnetic Resonance (NMR) spectroscopy.

All the spectroscopic types mentioned above are absorption spectroscopy techniques. In absorption spectroscopy, the appropriate type of radiation is generated, the radiation is passed through a sample and the absorbance of radiation from the spectrum is measured. Molecules do not absorb radiation continuously but absorb only discreet wavelengths (or energies) of radiation called photons. This is due to the quantum nature of the electronic, vibrational, rotational, and spin states being interrogated.

FUNDAMENTALS OF INFRARED SPECTROSCOPY

IR spectroscopy interrogates the nature of the vibrational states of bonds. Although bonds are drawn as discreet lines in Lewis structures, they can be better envisioned as springs linking two atoms. The atoms are in constant motion (stretching and bending the spring-like bond between them) and when an appropriate photon of light is added, this causes the spring to stretch farther (Figure 7.3).

Mathematically, the vibrational frequencies of bonds can be modelled as a simple, harmonic oscillator using Hooke's Law, Equation 7.1.[1]

$$v = \frac{1}{2\pi c}\sqrt{\frac{K}{\mu}}, \mu = \frac{m_1 m_2}{m_1 + m_2}$$ (Equation 7.1)

This equation calculates the frequency v, also called wavenumbers (cm^{-1}). K is the force constant, which can be thought of as the stiffness of the spring connecting the atom, for the bond in question (in N/m). Each bond has its own force constant. The term • μ is the reduced mass (in kg) of the two atoms (m_1 is mass of atom one in kilogram and m_2 is mass of atom two in kilogram) that make up the bond in question. Lastly, the speed of light ($c = 3.00 \times 10^{10}$ cm/s) is used to convert the frequency (in Hz) into wavenumber (cm^{-1}). Given that K is in the numerator, increasing the force constant (i.e. increasing from single to double to triple bonds) increases the frequency of the IR absorbance (Table 7.1).

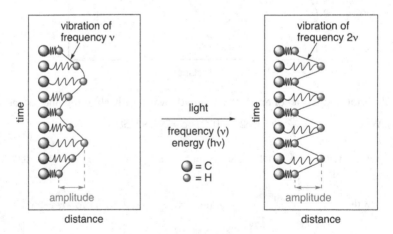

Figure 7.3 Representation of two atoms vibrating, bond shown as a spring, before and after light absorption.

Table 7.1 Relationship between Bond Type (Single, Double, Triple), the Force Constant (*K*) and Frequency of IR Absorbance

Bond	C–C	C=C	C≡C
K (N/m)[2]	450	970	1560
Frequency (cm^{-1})	1200	1650	2150

Table 7.2 **Relationship between Atom Hybridization (sp³, sp², and sp), the Force Constant (K) and Frequency of IR Absorbance**

Bond	C_{sp3}–H	C_{sp2}–H	C_{sp}–H
Fraction s character at carbon	25%	33%	50%
K (N/m)³	488	505	588
Frequency (cm⁻¹)	2996	3047	3288

Table 7.3 **Relationship between Reduced Mass (μ) and Frequency of IR Absorbance**

Bond	C–H	C–C	C–O	C–Cl	C–Br
μ (x10⁻²⁷ kg)	1.54	9.97	11.4	14.9	17.3
Frequency (cm⁻¹)	3000	1200	1100	750	650

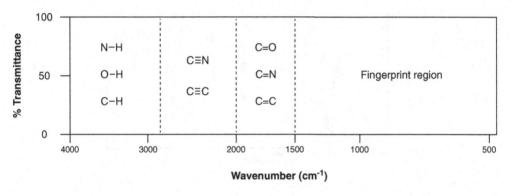

Figure 7.4 Typical organic chemistry IR spectral window and breakdown of the different regions.

There is also an increase in the force constant, K, when the %s-orbital character increases in a bond. For example, a C–H bond increases in strength as the carbon goes from sp³-hybridized to sp²-hybridized to sp-hybridized (Table 7.2).

Similarly, given that μ is in the denominator, increasing the reduced mass reduces the frequency (Table 7.3) of the IR absorbance (note K is not the same for all single bonds, but its variance is less significant than the changing masses in these examples).

Given the dependence upon the type of bond (single, double, or triple) and the masses of the atoms involved IR spectroscopy is useful for determining the different types of functional groups that are in different molecules. To help in determining the types of bonds in unknown organic molecules, tables listing the types of bonds and their typical IR frequencies have been compiled (a reference has been provided in Table 7.4). Typically, only the 1500–4000 cm⁻¹ region (diagnostic region) is considered and the region below 1500 cm⁻¹ (fingerprint region) is not considered. A general breakdown of the different regions of an IR spectrum is found in Figure 7.4.

Problem 7.1 A common technique for studying mechanisms is to substitute protium (¹H or H) with deuterium (²H or D) and study changes in reaction rates. If the C–H and C–D bonds both have similar force constants (assume 500. N/m) calculate the expected IR frequency of a C–D bond. The mass of a carbon atom is 1.9945x10⁻²⁶ kg and the mass of a deuterium atom is 3.3444x10⁻²⁷ kg.

IR SPECTROSCOPY PRACTICE

Problem 7.2 For each of the following spectra (structure of molecule on the spectrum), assign all significant peaks >1500 cm^{-1} using the simplified IR correlation table (Table 7.4).

a. Hexane

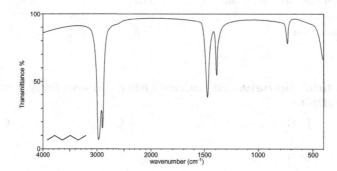

b. Hex-1-ene

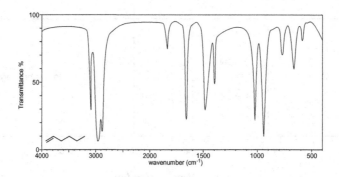

c. Hex-1-yne

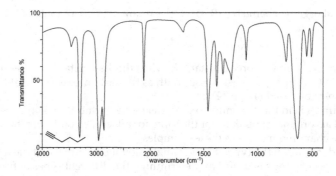

d. Hexan-1-ol

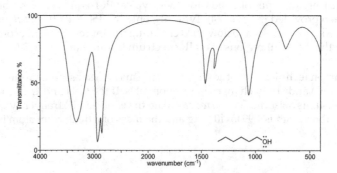

e. Acetophenone

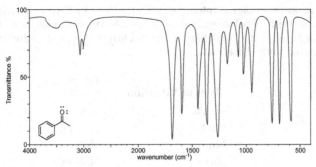

USING IR SPECTROSCOPY TO DIFFERENTIATE RELATED COMPOUNDS

Problem 7.3 Match each C_4H_8O isomer with its appropriate IR spectrum.

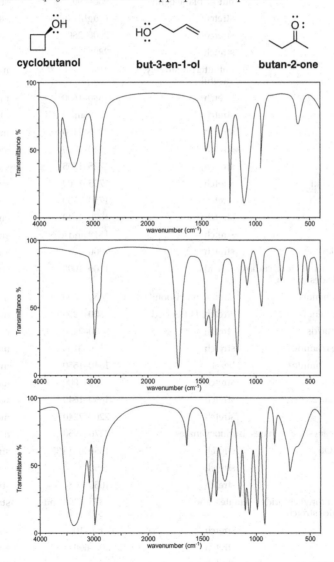

cyclobutanol **but-3-en-1-ol** **butan-2-one**

Table 7.4

Simplified IR Correlation Table. Signal Intensity Indicates Relative Percent Transmittance: Strong Corresponds to a Signal with 50% Transmittance or Less, Medium Corresponds to a Signal with 50% Transmittance or More, and a Weak Signal Shows Only a Marginal Reduction in Percent Transmittance Compared to the Baseline

Bond Type		Type of Vibration	Frequency (cm^{-1})	Intensity
C–H	Alkanes	stretch	3000–2850	strong
	–CH$_3$	bend	1450–1375	medium
	–CH$_2$–	bend	1465	medium
	Alkenes	stretch	3100–3000	medium
		out-of-plane bend	1000–650	strong
	Arenes	stretch	3150–3050	strong
		out-of-plane bend	900–690	strong
	Alkyne	stretch	roughly 3300	strong
	Aldehyde	stretch	2900–2800	weak
		stretch	2800–2700	weak
C–C	Alkane	not interpretatively useful	n/a	n/a
C=C	Alkene	stretch	1680–1600	medium, weak
	Aromatic	stretch	1600 and 1475	medium, weak
C≡C	Alkyne	stretch	2250–2100	medium, weak
C=O	Aldehyde	stretch	1740–1720	strong
	Ketone	stretch	1725–1705	strong
	Carboxylic acid	stretch	1725–1700	strong
	Ester	stretch	1750–1730	strong
	Amide	stretch	1670–1640	strong
	Anhydride	stretch	1810 and 1760	strong
	Acid chloride	stretch	1800	strong
C–O	Alcohols, ethers, esters, carboxylic acids, and anhydrides stretch		1300–1000	strong
O–H	Alcohols/phenols	stretch, no H-bond	3650–3600	medium
	Alcohols/phenols	stretch, H-bonded	3500–3200	medium, broad
	Carboxylic acids	stretch	3400–2400	medium, broad
N–H	1°,2° amines/amides	stretch	3500–3100	medium
	1°,2° amines/amides	bend	1640–1550	medium, strong
C–N	Amines	stretch	1350–1000	medium, strong
C=N	Imines/oximes	stretch	1690–1640	weak – strong
C≡N	Nitriles	stretch	2260–2240	medium
X=C=Y	Allenes, ketenes, isocyanates, isothiocyanates		2270–1950	medium, strong
N=O	Nitro (R-NO$_2$)	stretch	1550 and 1350	strong
S–H	Thiols	stretch	2550	weak
S=O	Sulfoxides	stretch	1050	strong
	Sulfones, sulfonyl chlorides, sulfates, and Sulfonamides stretch		1375–1300 and 1200–1140	strong
C–X	Fluorine	stretch	1400–1000	strong
	Chlorine	stretch	800–600	strong
	Bromine, iodine	stretch	<667	strong

NOTES

1. Hooke, Robert. Lectures de potentia restitutiva, or, Of spring: explaining the power of springing bodies: to which are added some collections. Printed for J. Martyn, London, 1678.

2. Robinson, E.A.; Lister, M.W. A linear relationship between bond orders and stretching force constants. *Can. J. Chem.*, **1963**, *41*, 2988–2995. DOI: 10.1139/v63-439

3. Jenkins, H.O. Bond energies, internuclear distances, and force constants in series of related molecules. *Trans. Faraday Soc.*, **1955**, *51*, 1042–1044. DOI: 10.1039/TF9555101042

8 Mass Spectrometry

Mass spectrometry is a method for analyzing the masses of ions and analyzing their fragmentation (if the ions are molecular). Note that for IR, the term spectroscopy (meaning the study of the interaction of electromagnetic radiation and matter) is used while for mass spectrometry the term spectrometry (meaning the quantitative measurement of the intensity of electromagnetic radiation) is used. Mass spectrometry, often called mass spec or simply MS, developed from canal ray tubes,[1] was first proposed in 1898,[2] and the first mass spectrum (of neon) was taken in 1913.[3] Mass spectrometry has been instrumental in demonstrating the existence of isotopes, providing a means for measuring isotopic decay (radiocarbon dating), and determining the origins – natural versus synthetic – of compounds. In our study of mass spectrometry, the data provided (the mass spectrum) will be utilized to provide the likely fragments within our molecules and it can provide a check on whether a structure that is proposed is likely or unlikely.

Typically, mass spectrometers are coupled with gas chromatography (GC, separates molecules based on boiling points) or liquid chromatography (LC, separates molecules based on polarity) units, which will separate molecules in each sample and ensure that the mass spectrum analyzed is for a single molecule. Electron-ionization mass spectrometers (EI-MS) are a common type of mass spectrometer and will be the only type considered here. An EI-MS does several things: 1) it converts neutral molecules (or atoms) into ions; 2) it separates the ions based on their mass-to-charge quotient (m/z); and 3) it measures the relative abundance of each type of ion.

Once a sample is analyzed, a mass spectrum is produced (Figure 8.1), the data for which is summarized in Table 8.1.

BASE PEAKS AND ISOTOPES

Several points should be made for the mass spectrum. First, the peak that corresponds to the molecular ion (Figure 8.2, $CH_4^{+\bullet}$) is called the parent peak or molecular ion peak. Second, the spectrometer then normalizes the spectrum to the tallest m/z peak, which is then called the base peak (here the molecular ion peak and base peak are the same, that is not always the case). Third, it should be noted that there is a peak at $m/z = 17$ [M+1], which comes from the 1% of molecules that have ^{13}C rather than ^{12}C. Table 8.1 contains a list of the common isotopes that will be seen in organic mass spectra.

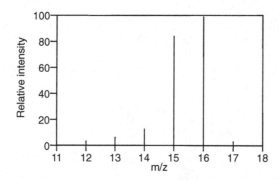

Figure 8.1 Mass spectrum of methane (CH_4). Melton, C.E.; Rosenstock, H.M. Collision-Induced Dissociations in the Mass Spectrum of Methane. *J. Chem. Phys.*, **1957**, *26* (3), 568–571. DOI: 10.1063/1.1743346

Table 8.1 Methane Mass Spectrum Peak Data

m/z	Relative intensity
12	1.3
13	3.8
14	9.4
15	83.3
16	100.0
17	1.1

DOI: 10.1201/9781003479352-8

Table 8.2   Elements Commonly Found in Organic Molecules, Their Isotopes, Percent Abundance, and Masses

Element	Relative atomic mass (u)	Isotope	Isotopic mass (u)	Relative abundance
Hydrogen	1.0080	^1H	1.007 83	100.00
		^2H	2.014 10	0.02
Carbon	12.011	^{12}C	12.0000	100.00
		^{13}C	13.0034	1.11
Nitrogen	14.007	^{14}N	14.0031	100.00
		^{15}N	15.0001	0.38
Oxygen	15.999	^{16}O	15.9949	100.00
		^{17}O	16.9991	0.04
		^{18}O	17.9992	0.20
Sulfur	32.06	^{32}S	31.9721	100.00
		^{33}S	32.9715	0.75
		^{34}S	33.9679	4.25
		^{36}S	35.9671	0.01
Chlorine	35.45	^{35}Cl	34.9689	100.00
		^{37}Cl	36.9659	31.58
Bromine	79.904	^{79}Br	78.9183	100.00
		^{81}Br	80.9163	97.28

σ-BOND FRAGMENTATION

Now, the origin of the different peaks in the methane mass spectrum will be analyzed. Upon ionization in an EI-MS, the molecule loses an electron. Electrons are lost preferentially from lone-pair > π-bond > σ-bond electrons. For methane, there are only σ-bonds and so upon ionization, methane ionizes to become a radical cation which fragments by separation of the radical cation C–H bond into radical and cation fragments as follows, note strange species are produced in the high-energy, vacuum environment of the spectrometer (Figure 8.2).

Alkanes follow a similar pattern, whereby a σ-bond loses an electron to form a radical cation alkane. One of the σ-bonds is then fragmented to give a radical and a cation. The cation can further fragment to give a cation and an ethylene. For linear alkanes, this process should be repeated iteratively to cleave off a methyl radical and an ethyl radical (Figure 8.3). Note that for branched alkanes, where it is possible to form secondary or tertiary radicals/cations, there is a preference for fragmentation at those C–C bonds. Linear alkanes lack this preference because any C–C bond that can be fragmented will only generate primary radicals/cations.

Note that if the mass spectrum for pentane (Table 8.3) is now analyzed, the peaks that were predicted from the Figure 8.3 analysis are present (those from the ethyl radical fragmentation are more intense signals, which is because of the higher stability of an ethyl radical compared to a

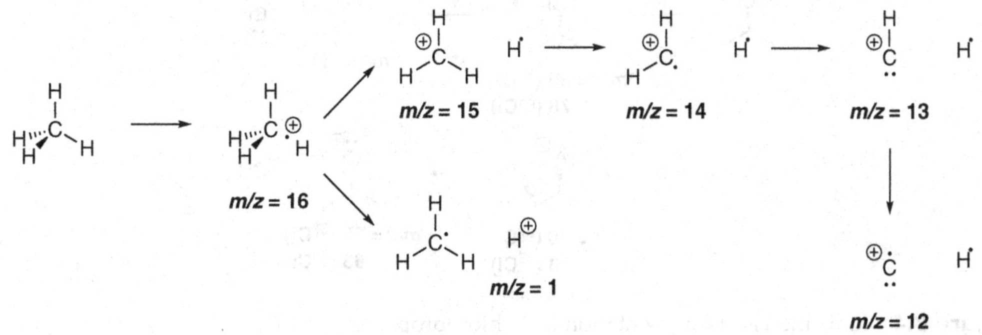

Figure 8.2   Breakdown of the methyl radical cation ($CH_4^{+\bullet}$) in the mass spectrometer to give the different observable *m/z* signals.

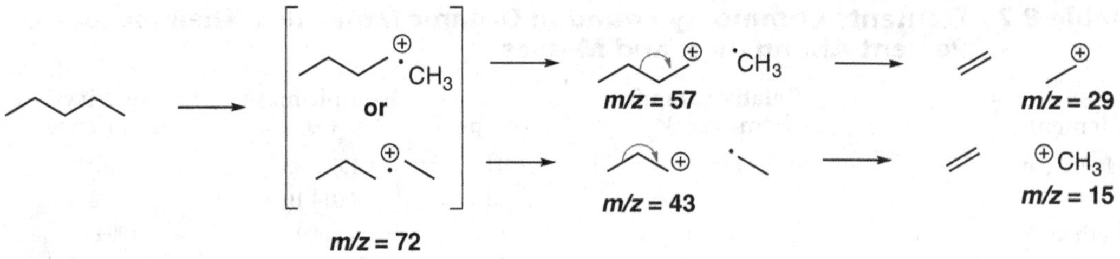

Figure 8.3 Ionization and fragmentation of pentane.

methyl radical. In addition, there are numerous other peaks, which are much harder to rationalize without covering many more – more complicated – fragmentation pathways.

ABC FRAGMENTATION – HALOCARBON FRAGMENTATION

Now 2-chloropropane will be analyzed (Figure 8.4). It should be noted that there is a peak at m/z 78 and 80 with a roughly 3:1 ratio (Table 8.4), which corresponds to the two isotopes of chlorine (^{35}Cl and ^{37}Cl, see Table 8.2) that exist in a roughly 3:1 ratio. In addition, 2-chloropropane forms a molecular ion by losing an electron from the chlorine lone pair. Chlorine, bromine, iodine, and alcohols (OH) then follow a similar pattern whereby they leave as a neutral group and leave behind a carbocation. The cation created by the loss of a chloride is the base peak. The 2-chloropropane can also cleave by formation of a C–Cl double bond and loss of an alkyl radical.

The second fragmentation pathway is a type of generalized fragmentation that occurs in EI-MS, where the radical (A) forms another bond to the atom it is attached to (B) and causes the B–C bond to break, generating a cation (+A=B) and radical (C·) (Figure 8.5).

Table 8.3 Pentane Mass Spectrum Peak Data for Significant Signals

m/z	Relative intensity
27	20
29	20
39	20
41	58
42	60
43	100
57	20
72	15

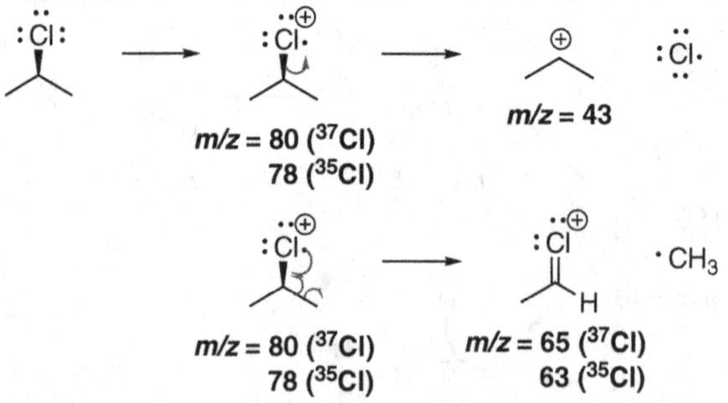

Figure 8.4 Ionization and fragmentation of 2-chloropropane.

Table 8.4 Pentane Mass Spectrum Peak Data for Significant Signals

m/z	Relative intensity
27	29
39	9
41	21
43	100
63	17
65	6
78	9
80	3

Figure 8.5 Fragmentation pattern commonly observed when an atom has a radical electron (A), after losing an electron upon ionization.

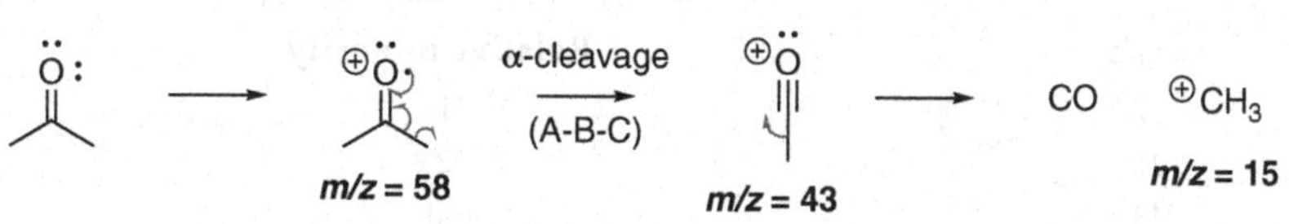

Figure 8.6 Ionization and fragmentation of acetone showing α cleavage, an example of A–B–C cleavage seen in Figure 8.5.

Table 8.5 Acetone Mass Spectrum Peak Data for Significant Signals

m/z	Relative intensity
15	23
42	9
43	100
58	63.8
59	3

CARBONYL FRAGMENTATION – ABC FRAGMENTATION AND MCLAFFERTY REARRANGEMENT

Now acetone will be investigated (Figure 8.6). Acetone loses an electron from one of the oxygen lone pairs to generate a radical cation ($m/z = 58$). The molecular ion then undergoes α-cleavage to generate an acylium ion ($m/z = 43$), the acylium ion is the base peak – a common occurrence for carbonyl-containing compounds. Finally, loss of CO creates a carbocation ($m/z = 15$). The peak data can be seen in Table 8.5.

Note this α-cleavage process is another example of the ⁺˙A–B–C → +A=B and C˙ (Figure 8.5). This fragmentation pathway (what is here referred to as the A–B–C pathway) is general for all radical cations.

Finally, pentan-2-one will be examined (Figure 8.7). The expected fragmentation pathways (as already seen above) occur to give an acylium ion (the base peak, Table 8.6) by α-cleavage, and

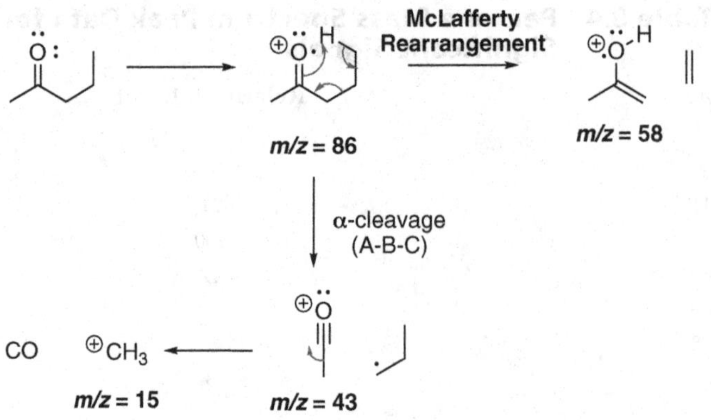

Figure 8.7 Ionization and fragmentation of pentane-2-one showing a cleavage and a McLafferty rearrangement.

Table 8.6 Pentan-2-One Mass Spectrum Peak Data for Significant Signals

m/z	Relative intensity
27	11
41	12
43	100
58	10
71	11
86	20

methyl cation (by loss of CO). 2-pentanone also has a chain of atoms long enough to undergo a McLafferty rearrangement to give a peak at $m/z = 58$ (Table 8.6).

Above, each of the major fragmentation pathways – with which you should be familiar – are exemplified: σ-bond cleavage, loss of a halogen atom, A–B–C cleavage, and McLafferty rearrangement.

MASS SPECTRUM PRACTICE

Problem 8.1 For each of the following compounds and m/z data, explain the major peaks based on likely fragmentation pathways. This is a useful skill as mass spectrometry serves frequently to confirm a structure rather than to determine one.

a. Propan-2-ol, Table 8.7

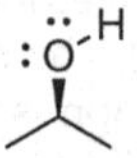

Table 8.7 Propan-2-ol Mass Spectrum Peak Data for Significant Signals

m/z	Relative intensity
19	5
27	12
29	6
41	7
43	10
45	100
59	4

b. Acetophenone, Table 8.8

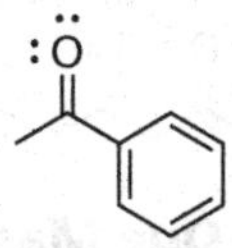

Table 8.8 Acetophenone Mass Spectrum Peak Data for Significant Signals

m/z	Relative intensity
15	3
43	13
51	23
77	73
105	100
120	26

c. Benzyl fluoride, Table 8.9

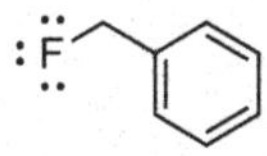

Table 8.9 Benzyl Fluoride Mass Spectrum Peak Data for Significant Signals

m/z	Relative intensity
39	8
51	8
83	13
91	83
109	100
110	55

d. Benzyl chloride, Table 8.10

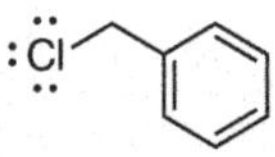

Table 8.10 Benzyl Chloride Mass Spectrum Peak Data for Significant Signals

m/z	Relative intensity
39	7
63	7
65	11
91	100
126	25
128	8

e. Hexan-2-one, Table 8.11

Table 8.11   Hexan-2-One Mass Spectrum Peak Data for Significant Signals

m/z	Relative intensity
15	4
27	8
29	15
41	14
43	100
57	16
58	50
85	6
100	8

NOTES

1. Goldstein, E. Ueber eine noch nicht untersuchte Strahlungsform an der Kathode inducierter Entladung. *Ann. Phys. (Berl.),* **1898,** *300* (1), 38–48. DOI: 10.1002/andp.18983000105

2. Wien, W. Untersuchungen über die electrische Entladung in verdünnten Gasen. *Ann. Phys. (Berl.),* **1898,** *301* (6), 440–452. DOI: 10.1002/andp.18983010618

3. Thompson, J.J. Bakerian Lecture: Rays of positive electricity. *Proc. R. Soc. Lond. A.,* **1913,** *89* (607), 1–20. DOI: 10.1098/rspa.1913.0057

9 Nuclear Magnetic Resonance[1,2]

Electrons have an inherent quantum mechanical property called spin. Recall that for an electron its spin can be $\pm\frac{1}{2}$, and these are often denoted by drawing ↑ and ↓. Atomic nuclei also have spin values that can vary over a much wider range. When placed into a magnetic field, the spin states of electrons and nuclei have different energies, which can be probed with electromagnetic radiation in the microwave and radio frequency regions of the spectrum. These techniques are called collectively called Magnetic Resonance Spectroscopy (MRS). The science of Electron Spin Resonance (ESR)/Electron Paramagnetic Resonance (EPR) will not be discussed here. The focus of this chapter will be Nuclear Magnetic Resonance (NMR) spectroscopy, which looks at the nuclei of atoms.

PHYSICAL BASIS OF THE NMR EXPERIMENT

Several atomic nuclei are NMR active. Some of the most useful nuclei are those that have nuclear spins of $\frac{1}{2}$, they are most useful because their spectra and analysis are relatively easy: ^1H, ^{13}C, ^{15}N, ^{19}F, and ^{31}P. Other nuclei have other spin values, which are harder to analyze, ^2H (spin = 1), ^{10}B (spin = 3), ^{11}B (spin = 3/2), and ^{14}N (spin = 1). Some nuclei common to organic molecules are entirely NMR inactive: ^{12}C, ^{16}O, ^{32}S. This chapter will look at the common spin-$\frac{1}{2}$ nuclei ^1H and ^{13}C.

In the absence of a magnetic field, nuclear spin states are randomly oriented. When an external magnetic field is applied (Figure 9.1), the spin states will align with or against the field, which creates different energy states. The stronger the magnetic field the larger the energy gap $\Delta\varepsilon_p$ (Figure 9.2) between spin states. This makes instruments with more powerful magnets – higher field strengths – more sensitive, which gives better results. It should be noted that even at very high

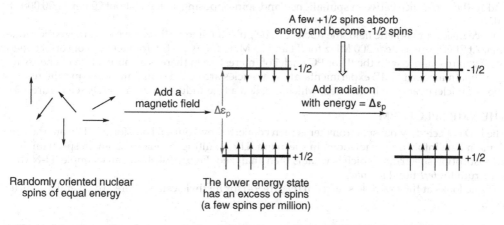

Figure 9.1 A general overview of the Magnetic Resonance Spectroscopy (MRS) experiment.

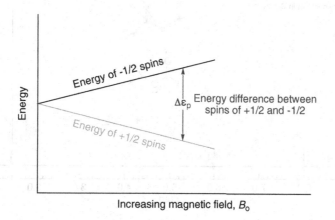

Figure 9.2 Effect of magnetic field strength (B_o) on the energy difference ($\Delta\varepsilon_p$) between spin states.

DOI: 10.1201/9781003479352-9

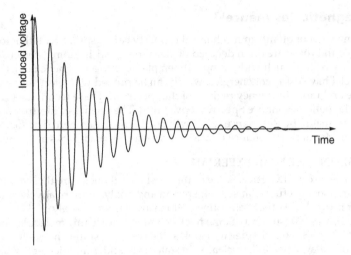

Figure 9.3 Example of the free-induction decay (FID) data captured by the NMR spectrometer.

field strengths, the difference in energy between the two spin states is much less than 5 kJ/mol and so the difference between spin-aligned and spin-opposing states is about 20 per 1,000,000 at 7.05 T.

Once inside a magnetic field, the nuclei can be pulsed using radio wave electromagnetic radiation. At 7.05 T, this is near 300 MHz for ^1H and 75 MHz for ^{13}C. If the frequency is of correct energy, we say in resonance with the ^1H or ^{13}C nuclei of interest, then there is an absorption of energy and spin flip. In modern NMR experiments, all frequencies are tested simultaneously and the relaxation of nuclei over time is detected, which is called a free-induction decay (FID): see Figure 9.3.

THE NMR SPECTRUM

The FID collected by the spectrometer is then converted by Fourier Transform (FT) from the time-domain data into a frequency-domain spectrum. The resulting frequency-domain spectrum is an absorption spectrum, which is what is then analyzed. Figure 9.4 shows an example ^1H-NMR spectrum for *tert*-butyl formate.

If one looks at the example spectrum (Figure 9.4) for methyl acetate, there are several points to note.

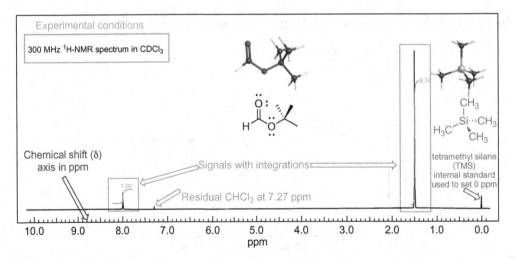

Figure 9.4 Annotated ^1H-NMR spectrum of *tert*-butyl formate.

1. The x-axis, or chemical shift (δ) axis, is in units of ppm (parts per million). This allows one to compare data from different frequency spectrometers by taking the frequency at which a signal occurs in Hz (above or below the spectrometer listed frequency) and dividing that frequency value by the frequency of the spectrometer in MHz.

2. There is a peak at 0 ppm, which is an internal standard. The agreed-upon standard for ^{1}H-NMR and ^{13}C-NMR is tetramethylsilane (TMS).

3. Each signal – two in this spectrum – has an integration (highlighted in blue), in NMR the integration values correspond to the number of hydrogens.

4. For this spectrum, the experimental conditions are listed above the spectra: a 300 MHz spectrometer, ^{1}H-NMR experiment, in $CDCl_3$ solvent.

5. The solvent used is deuterated chloroform ($CDCl_3$). A deuterated solvent is used because the solvent is there in a significantly higher amount than the sample to be analyzed. If $CHCl_3$ were used, the chloroform hydrogen nucleus signal would completely swamp the analyte signals. By using $CDCl_3$ (D (^{2}H) does not show up in ^{1}H-NMR) the solvent signal is suppressed. However, no solvent is completely deuterated (most are 99.9%+ deuterated) and so there is always a residual solvent peak (the undeuterated $CHCl_3$) in the spectrum.

SPECTROMETER PROPERTIES – FREQUENCY

What the various information in an NMR spectrum means will now be looked at in turn. The chemical shifts, positions on the x-axis in NMR spectra, arise from the motion of electrons around the atomic nuclei. This motion of electrons creates a magnetic field that opposes the applied field from the spectrometer. Different types of the same nuclei in a molecule are surrounded by different amounts of electron densities and so their chemical shifts vary.

The frequency at which ^{1}H nuclei, also referred to as protons, can be pulsed depends upon the NMR spectrometer's magnetic field strength (B). As such, at $B_o = 1.41$ T, hydrogen nuclei precess at a frequency of 60 MHz, at 2.35 T at 100 MHz, at 7.05 T at 300 MHz, and so on. As such, frequency (hertz) is the unit used to represent energy in NMR spectroscopy; however, the use of frequency for the x-axis is problematic, as the position of absorption signals is dependent upon the strength of the spectrometer.

For this reason, the distance between the reference, tetramethylsilane (TMS), and the position of a specific peak in the spectrum, the chemical shift, is reported in the dimensionless units of δ, which is the same on all spectrometers, regardless of field strength.

$$\delta = \frac{\text{frequency shift from TMS in Hz}}{\text{spectrometer frequency in MHz}} \qquad \text{(Equation 9.1)}$$

Given that the numerator of δ is in hertz and the denominator is in megahertz, the chemical shift is also referred to as ppm (part per million).

MOLECULAR PROPERTIES – CHEMICAL SHIFT

Because of the differences in electron density for different types of functional groups, hydrogen nuclei have well-defined regions in which their signals appear. These data have been assembled into chemical shift tables. Figure 9.5 provides a general list of chemical shift regions. Note, to the left in the spectrum is defined as downfield or deshielded which is hydrogen nuclei with less electron density, while to the right is defined as upfield or shielded which is hydrogen nuclei with more electron density.

Note that the chemical shift of an –OH hydrogen nucleus and R_2NH/RNH_2 signals are quite variable and depends upon several factors, including concentration of the sample, NMR solvent, and temperature. As seen in Figure 9.6, the –OH signal depends heavily upon the concentration of alcohol in the solvent (going from >5 ppm in neat (pure) ethanol to 1 ppm in very dilute solutions in CCl_4). It should also be noted that in dilute solutions the –OH hydrogen nucleus signal is typically broad and featureless. A common test for determining whether a given signal is, in fact, an –OH or R_2NH/RNH_2 hydrogen nucleus signal is to shake with a small amount of D_2O. In D_2O, there is rapid H/D exchange and the –OH or R_2NH/RNH_2 signal will disappear. Figure 9.6 shows the ranges for acidic, amine, phenol, alcohol, and thiol hydrogen nuclei.

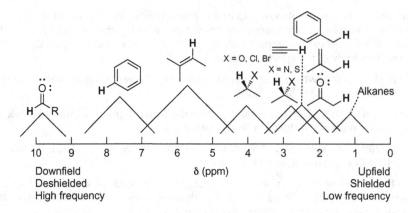

Figure 9.5 Different chemical shift values for different types of ^1H nuclei.

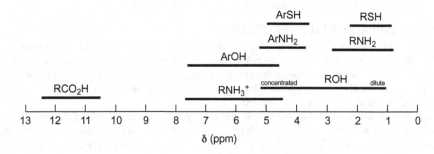

Figure 9.6 Chemical shift ranges for ^1H nuclei in –OH, –NH, and –SH functional groups.

MOLECULAR PROPERTIES – (IN)EQUIVALENT HYDROGENS AND NUMBER OF NMR SIGNALS

The number of signals in an NMR spectrum corresponds to the number of inequivalent nuclei, that is the number not related by symmetry. Consider methyl acetate (Figure 9.7), its spectrum contains two signals, one for each methyl group. The hydrogen nuclei on each methyl group in methyl acetate are unique from one another because the two methyl groups are in different chemical environments (one is attached to an oxygen atom, and one is attached to a carbonyl carbon).

Exploring this further, if one were to consider acetone and dimethyl carbonate (Figure 9.8), each of these molecules only has one signal. The two methyl groups in acetone are chemically equivalent. The two methyl groups in dimethyl carbonate are chemically equivalent.

methyl acetate

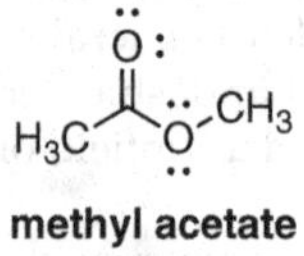

Figure 9.7 Structure of methyl acetate, an asymmetrical molecule.

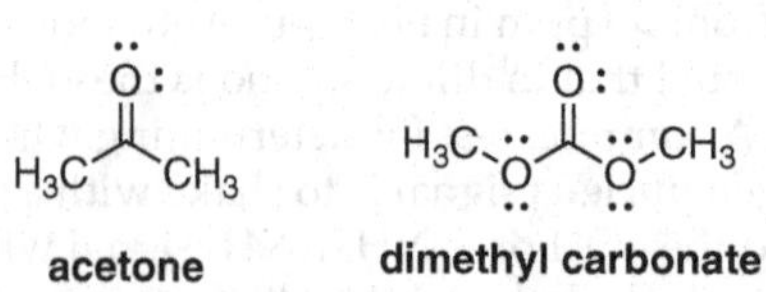

acetone   **dimethyl carbonate**

Figure 9.8 Structures of acetone and dimethyl carbonate, symmetrical molecules.

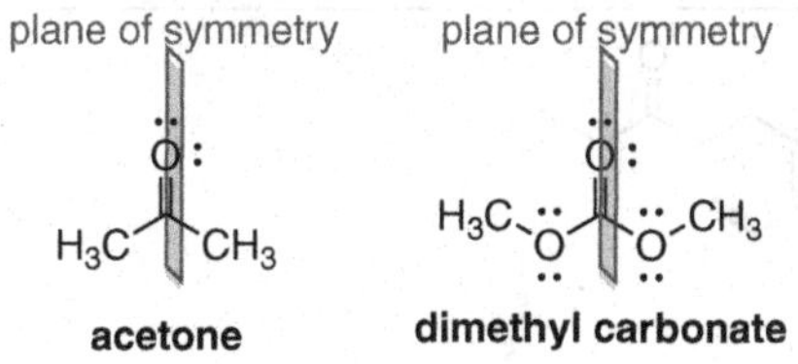

Figure 9.9 Planes of symmetry in acetone and dimethyl carbonate.

Put another way, the two methyl groups in acetone are related to each (Figure 9.9) other by a plane of symmetry (as are the two methyl groups of dimethyl carbonate).

It will be important to consider symmetry in molecules. In molecules without symmetry, each methyl (CH_3), methylene (CH_2) and methine (CH) will be chemically different. In molecules with symmetry, those methyls, methylenes, or methines related to one another by symmetry will be chemically equivalent and will show a single signal at the same chemical shift. Note: typically, the two hydrogen nuclei in methylene are equivalent to one another, unless there is a chiral centre in the molecule. When a molecule possesses even a single chiral centre, then all CH_2 hydrogen nuclei are non-equivalent (diastereotopic) and show different signals. These spectra quickly become very hard to interpret and are referred to as second order.

Problem 9.1 For each of the following molecules, predict the number of signals that would be expected in each ^{1}H-NMR spectrum. Examples:

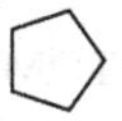

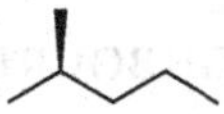

Practice Problems

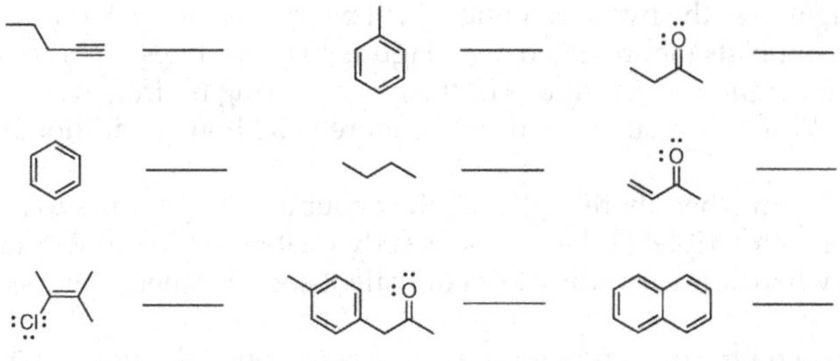

MOLECULAR PROPERTIES – NUMBER OF HYDROGEN ATOMS AND SIGNAL INTEGRATION

NMR possess a unique quality among all spectroscopic techniques, where the signal intensity (specifically the integral area for each signal) is proportional to the number of nuclei giving rise to that signal. Therefore, NMR spectra (unlike IR spectroscopy) are integrated (Figure 9.10).

In the Figure 9.10 spectrum, the integration values are reported as vertical displacements (in mm) for the integration curves. Given that the molecule formula is $C_9H_{10}O$, there are 10H in the molecule.

Total displacement: 26.5 + 11.8 + 16.2 = 54.5 mm

Thus, there is 10H/54.5 mm or 0.183H/mm

To determine the number of hydrogen nuclei that give rise to each signal, the integration value for each signal is multiplied the 0.183 H/mm:

(26.5 mm)(0.183H/mm) = 4.85, which rounds to 5H

(11.8 mm)(0.183H/mm) = 2.16, which rounds to 2H

(16.2 mm)(0.183H/mm) = 2.96, which rounds to 3H

The integration value – here a vertical displacement – gives the relative number of hydrogen nuclei. It is not possible to determine the absolute numbers without additional information, specifically the molecular formula. Sometimes a numerical value (whether the absolute area or a

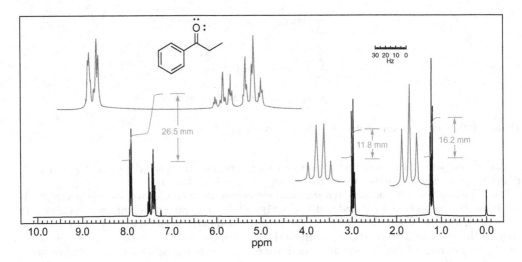

Figure 9.10 Integrated ^{1}H-NMR spectrum of 1-phenylpropan-1-one.

relative area) will be given for the integration, or sometimes, as in the example in Figure 9.10, the integration values' vertical displacement must be measured with a ruler. The number of hydrogen nuclei corresponding to each signal can be estimated by rounding to the nearest integer.

MOLECULAR PROPERTIES – NEIGHBOURING HYDROGEN ATOMS AND COUPLING

If two hydrogen nuclei have different chemical shifts and are within two bonds (*geminal* or *gem*, from Latin meaning twin), three bonds (*vicinal*, from Latin meaning neighbour), or occasionally four bonds (long-range) then the hydrogen nuclei will be coupled to each other. That is the signals will be split into doublets (abbreviated as d, Figure 9.11, two lines separated by the coupling constant *J*) due to two magnetic orientations of the neighbouring hydrogen nucleus (50% spin up and 50% spin down). When there are two, three, or more neighbours, additional splitting can be observed.

Regular multiplets form when there are equivalent coupling neighbours to a given hydrogen. The number of lines in a multiplet (Table 9.1) is directly related to the number of neighbours (and follows the *n*+1 rule, where *n* is the number of equivalent coupling neighbours), and the intensities follow Pascal's triangle.

The shorthand for each type of common multiplet is s = singlet, d = doublet, t = triplet, q = quartet, p = pentet. The appearance of each multiplet triplet through heptet appears in Figure 9.12 (doublet, *n* = 1 is in Figure 9.11). Note that octets and nonets can become hard to distinguish because the smallest, outside lines start to be hard to distinguish from the baseline.

The distance between each peak of a multiplet (in Hz) is referred to as the coupling constant (*J*). Determination of this distance provides information as to the type of neighbour (Table 9.2) that causes the splitting. In general, geminal (${}^2J$), vicinal (${}^3J$) and long-range (${}^4J$) couplings fit in the following ranges (the superscripted number is the number of bonds between coupled nuclei).

SIGNAL ANALYSIS

Given all the information covered so far, attention will be given to how one analyzes NMR spectra. Each NMR spectrum is a collection of individual multiplets and so the crux of NMR analysis is signal analysis. For example, let us analyze the multiplet in Figure 9.13.

Figure 9.11 Appearance of an ^{1}H-NMR doublet (d).

Table 9.1 **Relationship between the Number of Equivalent Coupling Neighbours, the Type of Multiplet, the Intensity of the Lines, and Common Structures That Give These Multiplet Signals**

Number equivalent coupling neighbours (n)	Intensities	Signal appearance	Example part structures
0	1	singlet	$X–CH_3$, $X–CH_2–CH_2–X$, C_6H_6
1	1 1	doublet	$X_2CH–CH_3$, $X_2CH–CHY_2$
2	1 2 1	triplet	$XCH_2–CH_3$, $X_2CH–CH_2–CHX_2$
3	1 3 3 1	quartet	$XCH_2–CH_3$, $XCH_2–CH_2–CHX_2$
4	1 4 6 4 1	pentet	$XCH_2–CH_2–CH_2X$, $CH_3–CH_2–CHX_2$
5	1 5 10 10 5 1	sextet	$CH_3–CH_2–CH_2X$, $CH_3–CHX–CH_2R$
6	1 6 15 20 15 6 1	heptet	$X–CH(CH_3)_2$, $(X–CH_2)_3CH$
7	1 7 21 35 35 21 7 1	octet	$CH–CH(CH_3)_2$
8	1 8 29 56 70 56 29 8 1	nonet	$XCH_2–CH(CH_3)_2$

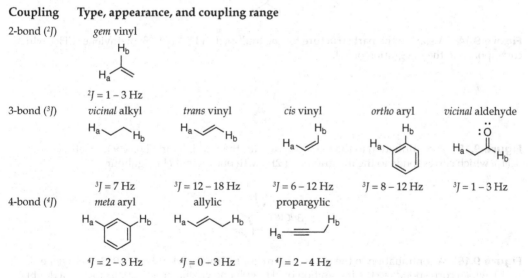

| triplet $n = 2$ | quartet $n = 3$ | pentet $n = 4$ | sextet $n = 5$ | heptet $n = 6$ |

Figure 9.12 Appearance of ^{1}H-NMR regular multiplets from $n = 2$ to $n = 6$.

Table 9.2 **Relationship between the Type of Neighbour and the Coupling Constant (J)**

Coupling	Type, appearance, and coupling range
2-bond (^2J)	*gem* vinyl

$^2J = 1 – 3$ Hz

| 3-bond (^3J) | *vicinal* alkyl | *trans* vinyl | *cis* vinyl | *ortho* aryl | *vicinal* aldehyde |

$^3J = 7$ Hz $^3J = 12 – 18$ Hz $^3J = 6 – 12$ Hz $^3J = 8 – 12$ Hz $^3J = 1 – 3$ Hz

| 4-bond (^4J) | *meta* aryl | allylic | propargylic |

$^4J = 2 – 3$ Hz $^4J = 0 – 3$ Hz $^4J = 2 – 4$ Hz

This has an integration of 2H, which means this is a methylene (CH_2). Start by drawing a part structure, a structure that fits the data for the multiplet and will represent a likely part of the whole final structure, Figure 9.14.

Then the peak multiplet is analyzed. This is a doublet, which means that there is one neighbour. Measuring the coupling constant (using the Hz scale provided) gives a coupling constant value of roughly 7 Hz, which means there is a single neighbouring, vicinal alkyl hydrogen nucleus. Add this to the part structure (Figure 9.15).

Finally, the chemical shift is considered. A chemical shift of 4.11 ppm is in the range (Figure 9.7) for hydrogen nuclei on a carbon next to an oxygen, bromine, or chlorine. So, this is added to the part structure (Figure 9.16).

Without a chemical formula to rule in or out Br, Cl, or O, this is the final part structure for this multiplet. Further analysis of the NMR spectrum would proceed by analyzing each multiplet in turn and independently coming up with a part structure for the signal. Then the final step is conversion of all the part structures one has come up with into one, complete, and final structure.

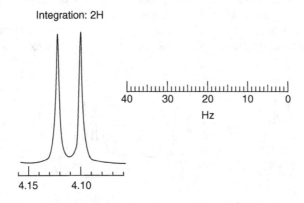

Integration: 2H

40   30   20   10   0
Hz

4.15   4.10

Figure 9.13 ^{1}H-NMR doublet with an integration of 2H at 4.11 ppm.

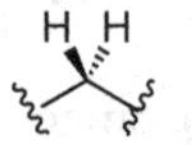

Figure 9.14 A start to the part structure for the multiplet in Figure 9.13: methylene (CH_2), which corresponds to the integration of 2H.

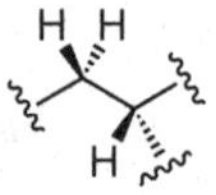

Figure 9.15 A continuation to the part structure for the multiplet in Figure 9.13: methylene (CH_2), which corresponds to the integration of 2H, with one vicinal H neighbour.

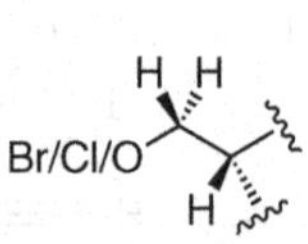

Figure 9.16 A continuation to the part structure for the multiplet in Figure 9.13: methylene (CH_2), which corresponds to the integration of 2H, with one vicinal H neighbour to give a doublet, and adjacent to a Br, Cl, or O atom, to account for the chemical shift.

Problem 9.2 Using the example from above, analyze each multiplet and draw a reasonable part structure.

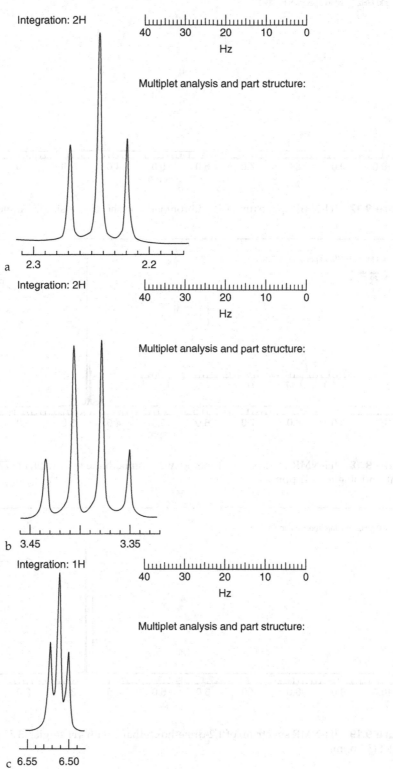

Integration: 2H

40   30   20   10   0
Hz

Multiplet analysis and part structure:

a   2.3   2.2

Integration: 2H

40   30   20   10   0
Hz

Multiplet analysis and part structure:

b 3.45   3.35

Integration: 1H

40   30   20   10   0
Hz

Multiplet analysis and part structure:

c 6.55   6.50

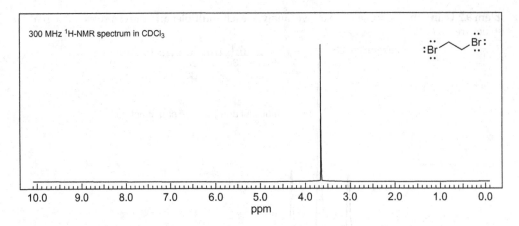

Figure 9.17 ¹H-NMR spectrum of 1,2-dibromoethane showing a single, singlet signal.

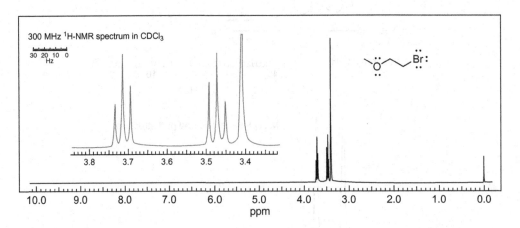

Figure 9.18 ¹H-NMR spectrum of 1-methoxy-2-bromoethane with triplet (3.72 ppm), triplet (3.45 ppm), and singlet (3.41 ppm).

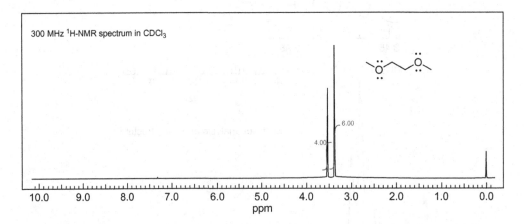

Figure 9.19 ¹H-NMR spectrum of 1,2-dimethoxyethane with 4H singlet (3.55 ppm) and 6H singlet (3.4 ppm).

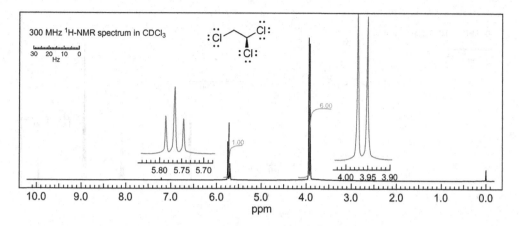

Figure 9.20 ¹H-NMR spectrum of 1,1,2-trichloroethane with 1H triplet (5.76 ppm) and 2H doublet (3.96 ppm).

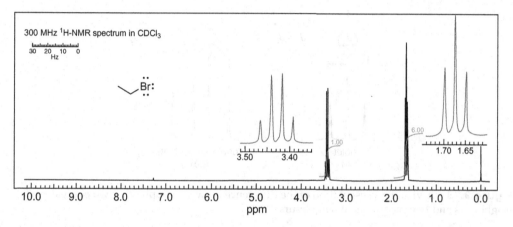

Figure 9.21 ¹H-NMR spectrum of 1-bromoethane with 2H quartet (3.44 ppm) and 3H triplet (1.67 ppm).

To help give the reader a sense of the relationship between structure and NMR spectrum, the next several pages, some ¹H-NMR spectra are shown that demonstrate common coupling patterns.

In the examples in Figures 9.17–9.22, only spectra where all the couplings to a hydrogen nucleus are the same (that is all neighbours that cause splitting were equivalent), so that simple multiplets like triplets, quartets, etc., are formed. There are, however, many cases where a hydrogen nucleus may be coupled to two non-equivalent neighbours by different coupling constants, which leads to a doublet of doublets rather than a triplet (Figure 9.23). This is most often observed in arenes (where a hydrogen nucleus has an ortho ($^3J = 8$ Hz) and a meta ($^4J = 2$ Hz) neighbour) and in alkenes (which have different coupling constants for *gem* ($^2J = 2$ Hz), *cis* ($^3J = 10$ Hz) and *trans* ($^3J = 15$ Hz) neighbours). Arenes and alkenes are also important exceptions to the generality that neighbouring hydrogen nuclei must be within three bonds to show coupling.

Consider Figure 9.24, which shows a splitting to non-equivalent neighbours in the spectrum for methyl *trans*-but-2-enoate. Readers can see a doublet of doublets and two doublets of quartets. Given all the information that has been covered so far, you can now analyze the spectrum in Figure 9.24 and assign each signal to the hydrogen nucleus (or nuclei) in the molecule that gives rise to that signal.

A brief mention needs to be given to NMR shorthand notation. It is very common not to include full NMR spectra in literature reports (given the size of each spectrum), but rather to give a shorthand listing of each signal, multiplet type, coupling constant(s), and its integration. Reconsider the spectrum of *trans*-but-2-enoate (Figure 9.24). The spectrum gives a lot of information but takes up a lot of room. The shorthand listing for this spectrum would be: ¹H NMR (300 MHz, CDCl₃) δ:

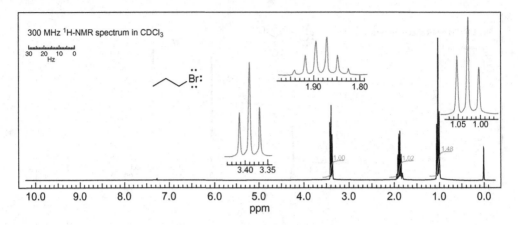

Figure 9.22 ¹H-NMR spectrum of 1-bromopropane with 2H triplet (3.39 ppm), 2H sextet (1.89 ppm), and 3H triplet (1.04 ppm).

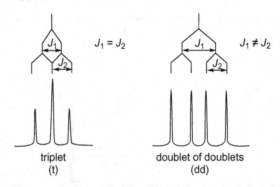

Figure 9.23 A visual representation of the difference between coupling to two equivalent neighbours and two inequivalent neighbours.

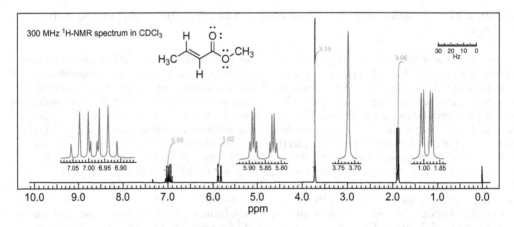

Figure 9.24 ¹H-NMR spectrum of methyl *trans*-but-2-enoate with 1H doublet of quartets (6.98 ppm, 16 Hz and 8 Hz coupling), 1H doublet of quartets (5.85 ppm, 8 Hz and 2 Hz coupling), 3H singlet (3.73 ppm), and 3H doublet of doublets (1.89 ppm, 16 Hz and 2 Hz coupling).

6.98 (dq, J = 16, 8 Hz, 1H), 5.85 (dq, J = 8, 2 Hz, 1H), 3.73 (s, 3H), 1.89 (dd, J = 16, 2 Hz, 3H). Both full spectra and shorthand notations for NMR data are commonly seen.

If one considers the spectrum for methyl *trans*-but-2-enoate (Figure 9.24), a question that may be asked is, "Why do the alkene hydrogen nuclei have such dramatically different chemical shifts?"

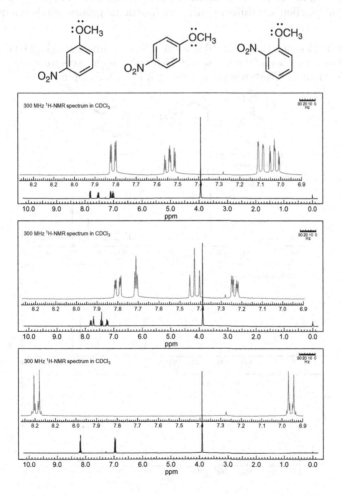

Figure 9.25 Contributing structures of methyl *trans*-but-2-enoate.

The answer is delocalization. Delocalization of electrons can lead to increased chemical shifts. First, if a positive formal charge is depicted at a certain location, that position will have lower electron density, that is it will be deshielded and will appear more downfield (to the left) in the NMR spectrum. Second, if a negative formal charge is depicted at a certain location, that position possesses higher electron density, that is it will be shielded and will appear more upfield (to the right) in the NMR spectrum. Consider the contributing structures for *trans*-but-2-enoate (Figure 9.25).

The alkene carbon farthest from the carbonyl has decreased electron density due to conjugated withdrawal by the carbonyl. This is the origin of that alkene hydrogen nuclei very downfield chemical shift (6.98 ppm, when alkenes typically appear 4.5–6.5 ppm).

Delocalization impacts on chemical shifts can also be starkly seen with substituted benzene rings. Unsubstituted benzene has a chemical shift of 7.3 ppm. Nitrobenzene, which is electron withdrawing, has chemical shifts of 8.20 ppm, 7.70 ppm, and 7.55 ppm, all significantly downfield of benzene. In contrast, aniline, with an electron-donating amino group, has chemical shifts all upfield of benzene: 7.21 ppm, 6.73 ppm, and 6.51 ppm.

Problem 9.3 Determine which ^{1}H-NMR spectrum corresponds to which isomer.

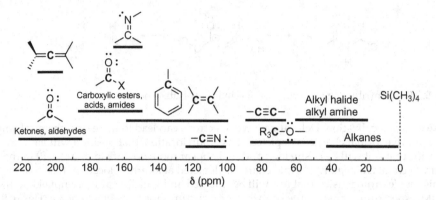

Figure 9.26 Different chemical shift values for different types of ^{13}C nuclei.

CARBON NMR

Another very useful NMR experiment in organic chemistry is ^{13}C-NMR. 99% of naturally occurring carbon is ^{12}C, which is NMR inactive. As such, there is only a 1% chance that any carbon is ^{13}C. The odds of two ^{13}C nuclei being next to each other is 0.01%. Given the low probability of ^{13}C nuclei neighbouring each other, the ^{13}C-^{13}C coupling is not typically observed and ^{13}C NMR shows singlets for each non-equivalent carbon (^{1}H-^{13}C coupling would be observed, but ^{13}C NMRs are run to decouple ^{1}H–^{13}C). Integration values for ^{13}C NMR are not determined as they are unreliable for several reasons. The typical ^{13}C chemical shifts for different functional groups are shown in Figure 9.26.

Problem 9.4 Shown below are the ^{13}C-NMR spectra of three isomers of $C_6H_{10}O$. Determine the expected numbers of carbons for each isomer. Determine which spectrum corresponds to which compound. Note that CDCl$_3$ shows up as a 1:1:1 triplet at 77.27 ppm.

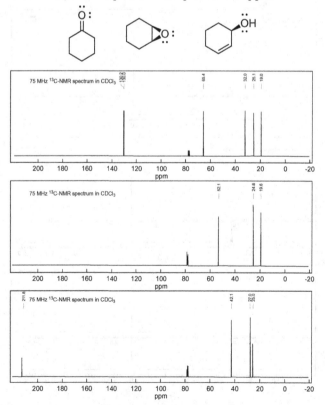

STRUCTURE DETERMINATION PRACTICE

For the following problems, a molecular formula, IR, MS, and NMR data will be provided. The task will be to determine the structure of each molecule by using all the spectroscopic information available. Also note, that it is helpful to determine the index of hydrogen deficiency (IHD), which tells the number of π bonds and/or rings in the molecule. To calculate the IHD, Equation 9.2 is used.

$$IHD = \frac{2C + 2 - H - X + N}{2}$$

(Equation 9.2)

C is the number of carbon atoms
H is the number of hydrogen atoms
X is the number of halogen atoms
N is the number of nitrogen atoms

Problem 9.5 Determine the structure of the molecule with the formula C_5H_8.

a. Determine the IHD:

b. Analyze its IR spectrum and label all significant absorptions above 1500 cm⁻¹.

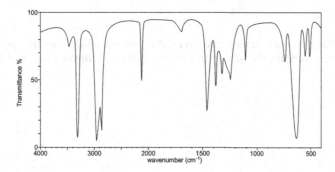

c. Analyze the ¹H-NMR spectrum and draw part structures for each signal. What are the peaks at 7.27 ppm and 0.0 ppm?

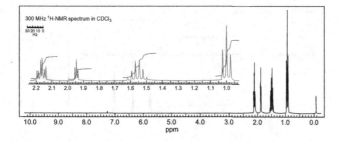

d. Analyze the ¹³C-NMR data, determine the number of non-equivalent carbons in the spectrum and propose any functional groups that may be present (based on chemical shift).
¹³C NMR (75 MHz, CDCl₃) δ: 84.5, 68.2, 22.0, 20.4, 13.4.

e. Propose a structure for this molecule. Given the mass spectrum data for the proposed structure, does the structure make sense with the fragmentation observed?

Structure:
C_5H_8 mass spectrum peak data.

m/z	Relative intensity
27	33
29	24
38	10
39	55
40	61
41	22
42	22
50	7
51	9
43	44
65	7
67	100
68	15

Problem 9.6 Modelling your analysis approach on Problem 9.5, analyze all the spectroscopic data and propose a structure for the compound whose molecular formula is $C_5H_9BrO_2$.

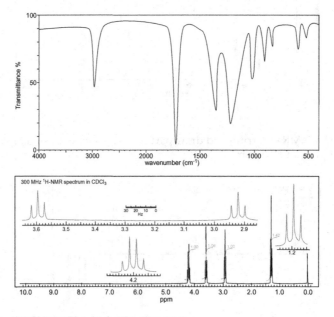

^{13}C NMR (75 MHz, CDCl$_3$) δ: 170.4, 61.0, 37.8, 25.9, 14.2.
$C_5H_9BrO_2$ mass spectrum peak data.

m/z	Relative intensity
27	57
28	18
29	100
45	20
55	12
73	28

101	19
107	37
109	36
135	29
137	28
152	50
184	50
180	3
182	3

Proposed structure:

Problem 9.7 Modelling your analysis approach on Problem 9.5, analyze all the spectroscopic data and propose a structure for the compound whose molecular formula is C_9H_9BrO.

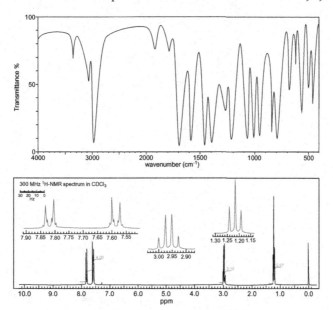

^{13}C NMR (75 MHz, $CDCl_3$) δ: 199.4, 135.7, 131.9, 129.6, 128.0, 31.7, 7.9.

C_9H_9BrO mass spectrum peak data.

m/z	Relative intensity
29	5
50	12
57	3
75	17
76	17
155	28
157	28
184	100
182	100
212	12
214	12

Proposed structure:

NOTES

1. This chapter is based on materials from the author's time as a student in Hans Reich's CHEM605 course at UW-Madison.

2. Now run by the American Chemical Society Division of Organic Chemistry but originally assembled by Hans Reich (Reich, H.J. Introducing JCE ChemInfo: Organic. *J. Chem. Educ.*, **2005**, *82* (3), 495–496. DOI: 10.1021/ed082p495), the Hans Reich collection is a great resource for spectroscopy data and information. NMR Spectroscopy: https://organicchemistrydata .org/hansreich/resources/nmr/?page=nmr-content%2F Accessed: 31 October 2023.

10 Acids and Bases

Arrhenius proposed one of the earliest acid–base definitions (in 1884)[1] in which acids are molecules that produce $H^+(aq)$ ions upon addition to water and bases are molecules that produce HO^- upon addition to water. Arrhenius's original definition needs to be amended to highlight what is now known about H^+ in aqueous solutions: $H^+(aq)$ does not exist as a freely occurring species, rather it exists (Figure 10.1) as an equilibrium mixture of hydronium (H_3O^+),[2] the Zundel cation ($H_5O_2^+$),[3] and the Eigen cation ($H_9O_4^+$).[4] Typically only the hydronium ion will be depicted.

While the Arrhenius definition of acids and bases is quite useful in water, it is a limiting definition that does not cover the full scope of acid–base chemistry. As such the primary focus will be on the Brønsted–Lowry and Lewis acid–base theories, both play a far more prominent role in organic chemistry.

BRØNSTED–LOWRY ACID–BASE THEORY

In Brønsted–Lowry acid–base theory (1923),[5] an acid is a molecule that is a H^+ (proton) donor and a base is a molecule that is a H^+ (proton) acceptor. In Brønsted–Lowry theory, acids and bases are defined, specifically, in relation to one another. Acids donate a proton to a base, and this produces a conjugate base while the base accepts the proton and becomes a conjugate acid (Figure 10.2).

It is common to use the shorthand notations **HA** for an acid and **:B** for a base. In a specific example of HCl reacting with water (Figure 10.3), the base (water) takes the proton (as indicated by the blue arrow from the oxygen lone pair to the H) from HCl (the bond from the Cl–H bond to the chlorine) to give chloride (Cl^-, conjugate base) and hydronium (H_3O^+, conjugate acid).

Problem 10.1 In each of the following problems, identify the acid, base, conjugate acid, and conjugate base. Also, draw the appropriate arrows showing the flow of electrons.

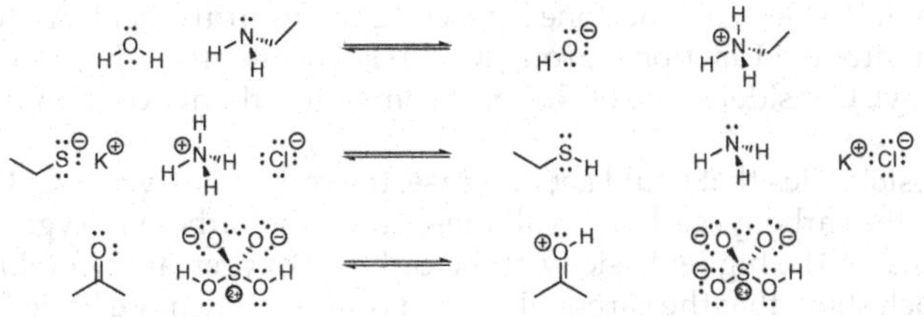

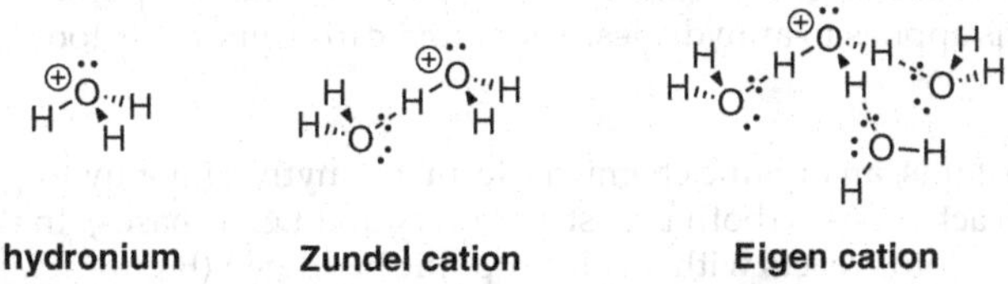

Figure 10.1 Structure of hydronium, Zundel, and Eigen cations.

acid + base ⇌ conjugate base + conjugate acid

HA **:B** **A⁻** **H-B⁺**

Figure 10.2 Terminology and symbology of a generic Brønsted–Lowry acid–base reaction.

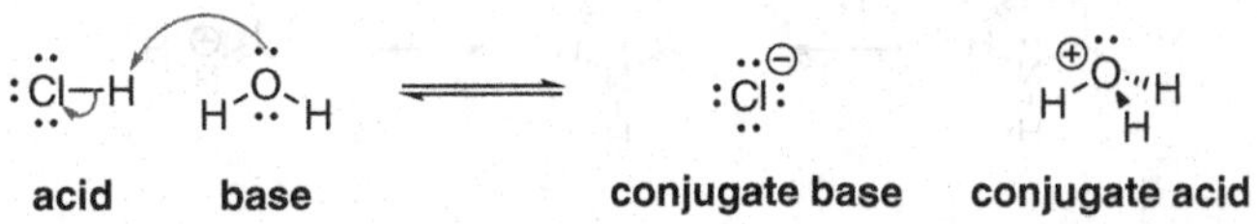

Figure 10.3 Electron-pushing arrows for a Brønsted–Lowry acid–base reaction with each species correctly identified.

DOI: 10.1201/9781003479352-10

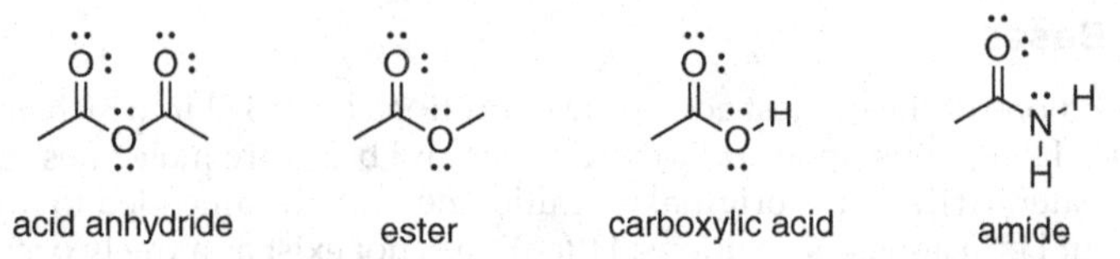

Figure 10.4 Common carbonyl functional groups.

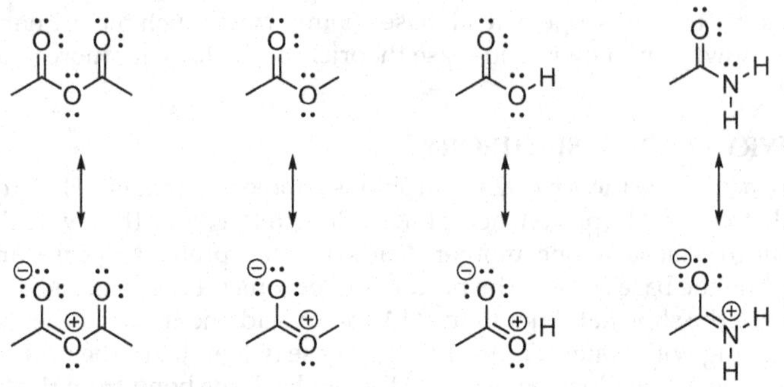

Figure 10.5 Contributing structures of common carbonyl functional groups.

CARBONYL BASES

In the final question in Problem 10.1, acetone is reacting with sulfuric acid. Acetone is an example of a molecule with a carbonyl functional group (C=O). There are many different functional groups that contain a carbonyl. Consider some of the most common carbonyl-containing functional groups (Figure 10.4).

There are two possible sites that could act as a base, the carbonyl oxygen (C=O) or the oxygen or nitrogen attached to the carbonyl carbon. In all these cases, the carbonyl oxygen is the much more basic site. The reasons for the higher basicity of the carbonyl oxygen are twofold. First, contributing structures for each show that the carbonyl oxygen is inherently more basic (Figure 10.5).

Moreover, once protonated at the carbonyl oxygen, the species is stabilized by delocalization of charge (Figure 10.6) while protonation at the non-carbonyl oxygen or nitrogen produces a species that does not benefit from charge delocalization/stabilization (the two possibilities here shown just for the amide, but this applies to anhydrides, esters, and carboxylic acids too).

ALKENE BASES

In addition to heteroatoms, an organic chemistry term for anything not hydrogen or carbon, acting as bases, π bonds can act as bases (both Brønsted–Lowry and Lewis bases). In these situations, the bond between carbon atoms reacts with an electrophilic hydrogen (Figure 10.7) or Lewis acid. The result is breaking of the C–C double bond (leaving just a single bond) and formation of a carbocation. As shall be seen, this process occurs to produce the carbocation with the most C–C bonds

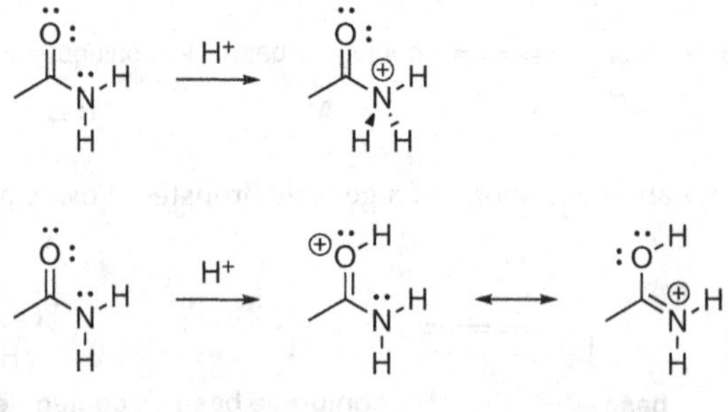

Figure 10.6 Different regiochemical protonation of an amide nitrogen atom (*top*), which generates a cation next to an electron-withdrawing carbonyl (destabilizing), and an amide carbonyl oxygen atom (*bottom*), which generates a delocalization-stabilized cation.

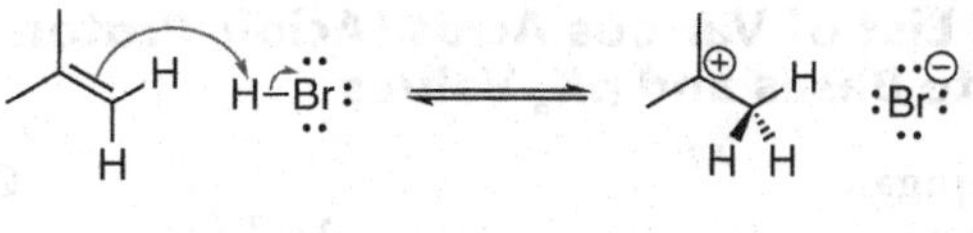

Figure 10.7 Protonation of an alkene.

preferentially. This reactivity trend for alkenes (molecules with C–C π bonds) will be a key focus in Chapter 12.

ACIDITY CONSTANTS

The quantitative measure of an acid's strength is determined by measuring the equilibrium concentration of acid, conjugate base, and protonated solvent (called a lyonium ion). In general chemistry, this process is only considered in water, where the lyonium ion is hydronium (Equation 10.1)

$$K_a = \frac{\left[H_3O^+\right]\left[A^-\right]}{[HA]}$$ (Equation 10.1)

Typically, the acidity constant (K_a) values for organic molecules are very small values, less than 1, which indicates predominantly unionized, and so pK_a values (pK_a = –logK_a) are considered. The smaller (or more negative) the pK_a the stronger the acid. The bigger the pK_a value the weaker the acid. Values for many molecules have been tabulated. A representative pK_a table (Table 10.1) is shown below.

Recall that the solvent plays an important role. While pK_a values are typically determined in water, pK_a values have also been determined in other solvents such as DMSO. DMSO cannot stabilize anions as effectively as water, and so pK_a values are typically higher. For example, the p$K_{a(water)}$ for acetic acid is 4.8, while the p$K_{a(DMSO)}$ for acetic acid is 12.6.

Problem 10.2 Rank the acids from most acidic to least acidic (1 = most acidic ⟶ 6 = least acidic).

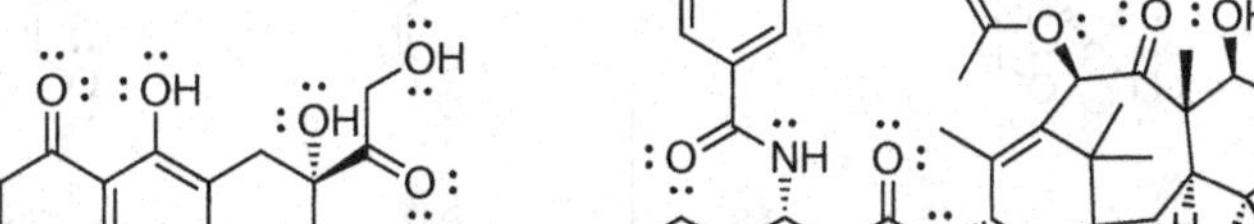

Problem 10.3 For each of the following molecules, identify the most acidic hydrogen and estimate its pK_a.

doxorubicin – chemotherapy

paclitaxel – chemotherapy

penicillin G – antibiotic

nepetalactone – ingredient in catnip

Table 10.1 Selected List of Various Acids (Acidic Proton in Red) with Their Conjugate Bases and pK_a Values

Acid	Conjugate base	pK_a	Acid	Conjugate base	pK_a
H—I	I⁻	−10	protonated amide	amide	−0.5
H—N≡ (protonated nitrile)	N≡	−10	H₃O⁺	H₂O	0[6]
H—Br	Br⁻	−9	trifluoroacetic acid	trifluoroacetate	0.2
protonated acetone (H—O⁺)	acetone (O)	−7.5	p-toluenesulfonic acid	p-toluenesulfonate	0.7
H—Cl	Cl⁻	−7	phosphoric acid H₃PO₄	H₂PO₄⁻	2.1
protonated thioether (H—S⁺)	thioether S	−7	H—F	F⁻	3.2
protonated acetic acid	acetic acid	−6.2	anilinium	aniline	4.6
protonated sulfuric acid	sulfuric acid	−5.2	acetic acid	acetate	4.8
protonated ether (H—O⁺)	ether	−3.8	pyridinium	pyridine	5.2
protonated alkene/isopropyl cation	alkene	−3	carbonic acid	bicarbonate	6.4
protonated alcohol (H—O⁺)	alcohol	−2.5	H—S—H (H₂S)	HS⁻	7.0
protonated nitro/nitric acid	nitrate	−1.4	2,4-pentanedione	enolate	9.0

(Continued)

Table 10.1 (Continued) Selected List of Various Acids (Acidic Proton in Red) with Their Conjugate Bases and pK_a Values

Acid	Conjugate base	pK_a	Acid	Conjugate base	pK_a
H—≡N:	⊖:≡N:	9.1			19.2
		9.2	H—≡—H	⊖:≡—H	25
		10			25
		10.2			25
		10.6			28
H—S:	:S:⊖	10.7	H—H	:H⊖	35
		14.0[54]			38
		15.7			40
		16			43
		16.7			44
		17			50

113

ACID–BASE EQUILIBRIA

In addition to knowing the relative strength of an acid, the pK_a values also allow one to determine the position of an equilibrium in an acid–base reaction. For example, take the acid–base reaction in Figure 10.8.

The pK_a values are provided for the acid (HCl, $pK_a = -7$) and conjugate acid (hydronium, $pK_a = -1.7$). To determine the equilibrium, the following formula is utilized (Equation 10.2)

$$K_c = 10^{(pK_{a,\text{conjugate acid}} - pK_{a,\text{acid}})}$$

(Equation 10.2)

Utilizing this formula for the reaction in Figure 10.8, $K_c = 10^{(-1.7--7)} = 10^{5.3}$. Recall that $K_c < 1$, reactants are favoured; $K_c = 1$, is a perfect equilibrium; $K_c > 1$, products favoured. Given that $K_c = 10^{5.3}$ for this reaction, this equilibrium heavily favours the products. In fact, for any reaction where $K_c \geq 10^5$, it will be considered that the reaction is essentially unidirectional (Figure 10.9).

Similarly, any reaction where $K_c \leq 10^{-5}$ will be a reaction that exclusively favours reactants (Figure 10.10).

Otherwise, if $10^{-5} \leq K_c \leq 10^5$ it is most appropriate to draw equilibrium arrows to show which side is favoured (Figure 10.11).

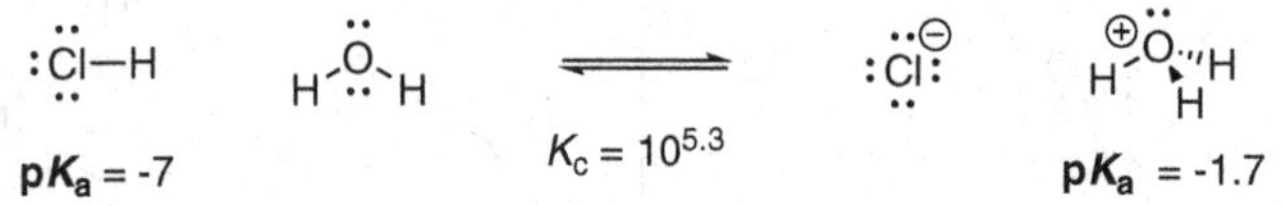

Figure 10.8 Example of using pK_a values to calculate an equilibrium constant.

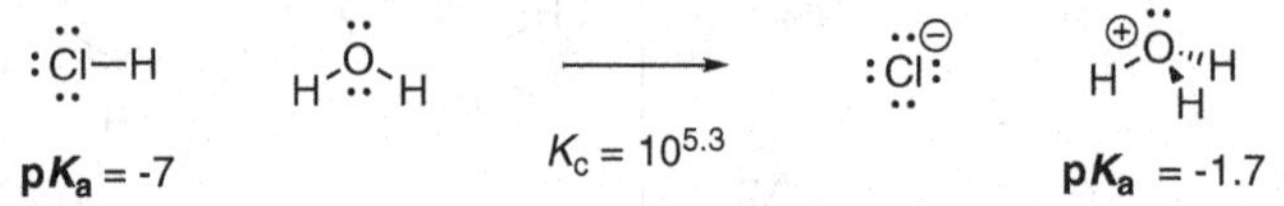

Figure 10.9 Example of a very product-favoured process

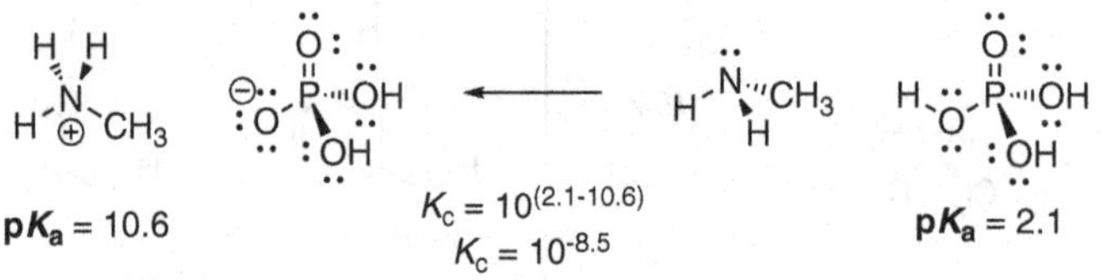

Figure 10.10 Example of a very reactant-favoured process

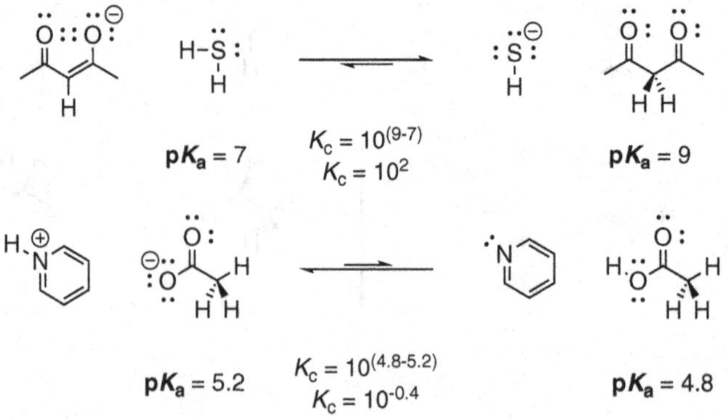

Figure 10.11 Examples of slightly product-favoured (*top*) and slightly reactant-favoured (*bottom*) processes.

Problem 10.4 For each of the following acid–base reactions, determine K_c and draw the correct arrow in the box provided.

Finally, trends in Brønsted–Lowry acidity–basicity will be investigated, that is how differences in atomic size, conjugation, inductive (polar) effects, and hybridization affect acidity and basicity.

1. Atomic size – as the atom bearing the acidic proton gets larger in size, the acidity increases: HF < HCl < HBr < HI. This is because the larger atoms (I is much larger than F) are more polarizable (it can smear the electrons over a bigger atomic volume) and better able to stabilize extra electrons. Typically, acidity increases as one goes down a group.

2. Conjugation – delocalizes electron density, making it more stable. Delocalization stabilization reduces basicity – by making the base more stable – and thus typically increases acidity of the conjugate acid.

3. Electronegativity – increased electronegativity of the atom with the acidic hydrogen leads to higher acidity. For example, the acidity of the second row in the periodic table follows the trend: $H_4C < H_3N < H_2O < HF$. As such, moving across the periodic table, acidity increases as the resulting anions are better able to bear the extra electrons.

4. Inductive (polar) effect – Placing highly electronegative atoms on a molecule tends to increase acidity (through inductively pulling electron density away from the base and making a more stable base and thus a stronger acid). Conversely, placing groups that have relatively low electronegativities on a molecule (carbon or silicon substituents) tends to increase the basicity (and reduce acidity) by increasing the energy of nearby lone-pair electrons.

5. Hybridization – acidity increases with great %s-character. Acetylene (C_2H_2, sp-hybridized carbon atoms) is more acidic than ethylene (C_2H_4, sp²-hybridized carbon atoms) is more acidic than ethane (C_2H_6, sp³-hybridized carbon atoms) with pK_a values of 25, 44 and 50, respectively. The greater the s-character, the lower in energy the orbital in which the lone-pair electrons reside. The lower the energy, the more stable the lone pair (less basic) and thus the more acidic the acid.

6. Charge – having positive charges makes molecules more acidic. Having negative charges makes molecules more basic.

Problem 10.5 Identify the strongest acid in each set of molecules. Explain what factor(s) make your selection the strongest acid.

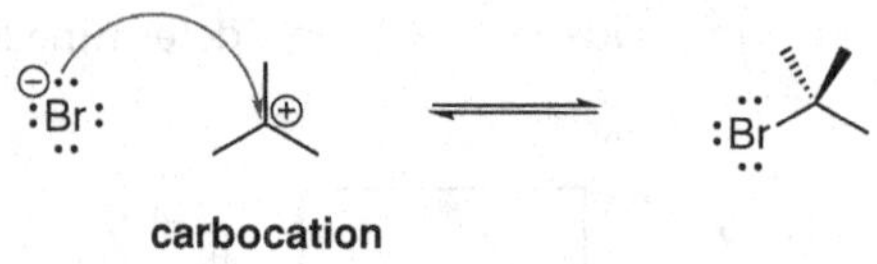

carbocation

Figure 10.12 Example of a Lewis acid–base reaction.

Problem 10.6 Identify the strongest base in each set of molecules. Rationalize your selection.

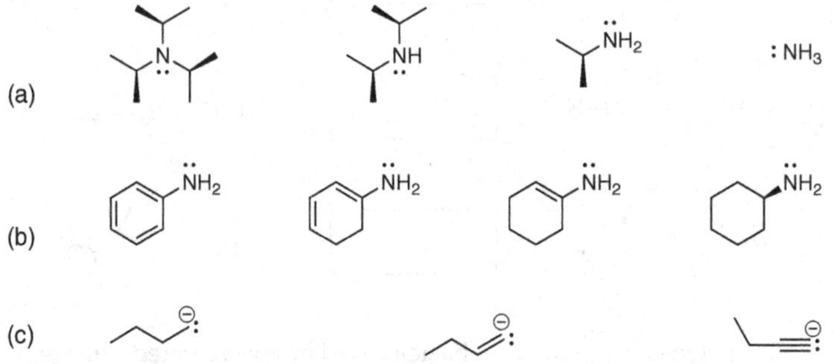

(a)

(b)

(c)

LEWIS ACID–BASE THEORY

Lewis acid–base theory[7] is the last theory that will be considered in this chapter, but it has perhaps the greatest significance for future studies. Specifically, recall that a Lewis acid is an electron-pair acceptor, and a Lewis base is an electron-pair donor. In organic chemistry, Lewis acid–base reactions are most of what makes up organic reactivity. It should be noted, however, that in organic chemist terms a Lewis acid is an electrophile and a Lewis base a nucleophile. Lewis bases (nucleophiles) match up reasonably well with Brønsted–Lowry bases (atoms or molecules with lone-pair electrons or extra electron density). Lewis acids (electrophiles) are molecules that are less familiar. Perhaps one of the most important electrophiles in organic chemistry – certainly one of the most well-studied – is the carbocation (Figure 10.12).

ACID–BASE CATALYSIS AND GREEN CHEMISTRY

An important use of acids and bases is as catalysts in (bio)synthetic chemistry. The added acid or base can add or remove a hydron, respectively, which alters the energy of the reaction and allows for transformations that would not be feasible without the added acid or base. Common examples include hydrolysis (breaking of bonds with water) and dehydration (removing water and forming new bonds). In the laboratory this is frequently accomplished with mineral acids (HCl, H_2SO_4, H_3PO_4) or Lewis acids (commonly transition metal cations or compounds containing a group 13 atom). Strong acids, like sulfuric acid, present a risk as hazardous chemicals. Weaker acids like phosphoric acid can be used to reduce risk, or solid acids have been developed and deployed that are recyclable, significantly reducing waste, improving atom economy, and reducing the chemical hazard of using a corrosive liquid.[8]

In cells, enzyme active sites contain charged amino acid side chains (arginine, histidine, lysine, aspartic acid, and glutamic acid), or Lewis basic side chains (serine, threonine, and cysteine), with water to effect these transformations. A wide number of enzymes use acid–base catalysis, but a common example is hydrolase enzymes, which can break (lyse) bonds using water. Enzymes themselves, including hydrolases, are also used in laboratory synthesis because of their high selectivity, the mild conditions in which they operate, and the rapidity with which they work.[9]

NOTES

1. Arrhenius, S. Thesis: Recherches sur la conductibilité galvanique des électrolytes. **1884**.

2. IUPAC has officially deprecated the use of hydronium, but it is so widely used that it will be used here. Officially H_3O^+ should be termed oxidanium or oxonium.

3. Zundel, G.; Metzger, H. Energiebänder der tunnelnden Überschuß-Protonen in flüssigen Säuren. Eine IR-spectroskopische Untersuchung der Natur der Gruppierungen $H_5O_2^+$. Z. *Phys. Chem. (N F)*., **1968**, *58* (5–6), 225–245. DOI: 10.1524/zpch.1968.58.5_6.225

4. Wicke, E.; Eigen, M.; Ackerman, T. Über den Zustand des Protons (Hydroniumions) in wäßriger Lösung. Z. *Phys. Chem. (N F)*. **1954**, *1* (5–6), 340–364. DOI: 10.1524/zpch.1954.1.5_6.340

5. i) Brönsted, J. N. Einige Bemerkungen über den Begriff der Säuren und Basen. *Recl. Trav. Chim. Pays-Bas.*, **1923**, *42* (8), 718–728. DOI: 10.1002/recl.19230420815
   ii) Lowry, T.M. The uniqueness of hydrogen. *J. Soc. Chem. Ind. (London).*, **1923**, *42* (3), 43–47. DOI: 10.1002/jctb.5000420302

6. Meister, E.C.; Willeke, M.; Angst, W.; Togni, A.; Walde, P. Confusing Quantitative Descriptions of Brønsted–Lowry Acid–Base Equilibria in Chemistry Textbooks – A Critical Review and Clarification for Chemical Educators. *Helv. Chem. Acta*, **2014**, *97* (1), 1–31. DOI: 10.1002/hlca.201300321

7. Lewis, G. N. *Valence and the structure of atoms and molecules.* The Chemical Catalog Co., Inc.: New York, New York, 1923.

8. Gong, S.; Liu, L.; Zhang, J.; Cui, Q. Stable and eco-friendly solid acids as alternative to sulfuric acid in the liquid phase nitration of toluene. *Process Saf. Environ. Prot.*, **2014**, *92* (6), 577–582. DOI: 10.1016/j.psep.2013.03.005

9. *Organic Synthesis Using Biocatalysis.* Animesh Goswami and Jon D. Stewart Eds. Elsevier: Oxford, **2015**. DOI: 10.1016/C2012-0-07124-4

11 Alkene Structure and Nomenclature

In this chapter, alkene structure and nomenclature will be introduced. Alkenes are the second type of hydrocarbon that will be considered in this book. Alkenes are unsaturated hydrocarbons; that is, there is the possibility of adding more hydrogens to the structure and have a general formula of C_nH_{2n} (for monoalkenes). Alkenes are also referred to as olefins. Alkynes are also unsaturated hydrocarbons, and they will be considered further in Chapter 13. Arenes (of which benzene is the best example) are another type of unsaturated hydrocarbon, which be considered in depth in Chapter 25 and Chapter 26; however, benzene rings will often be seen as substituents on alkanes, alkenes, and alkynes. When benzene is a substituent, it is called a phenyl group (Ph).

π BONDS

Alkenes contain a C=C functional group. The carbons making a double bond are sp²-hybridized atoms and are trigonal planar in geometry. The π bond (Figure 11.2) is made up of overlapping p-orbitals on adjacent carbon atoms.

CIS–TRANS ISOMERISM

The necessity of having in-phase, facial p-orbital alignment means that unlike σ bonds the π bond cannot rotate without breaking the π bond. Because π bonds cannot rotate alkenes, like cycloalkanes, exhibit *cis*, *trans* isomerism. *Cis* isomers (Figure 11.3) have two of the same groups on the same side and *trans* isomers have two of the same groups on the opposite side.

Cis isomers are higher in energy than *trans* isomers because of the increased steric strain of having two, large groups on the same side. *Cis* and *trans* isomers exhibit different physical properties, namely different melting points and boiling points (Figure 11.3).

IUPAC NOMENCLATURE

The IUPAC nomenclature system for alkenes builds upon the basics that were covered for alkanes/cycloalkanes. Alkenes are indicated in the IUPAC nomenclature system through the use of the suffix -ene. The simplest alkene is ethene ($CH_2=CH_2$) and the next simplest is propene ($CH_3CH=CH_2$). For longer-chain alkenes, the following IUPAC roles are followed:

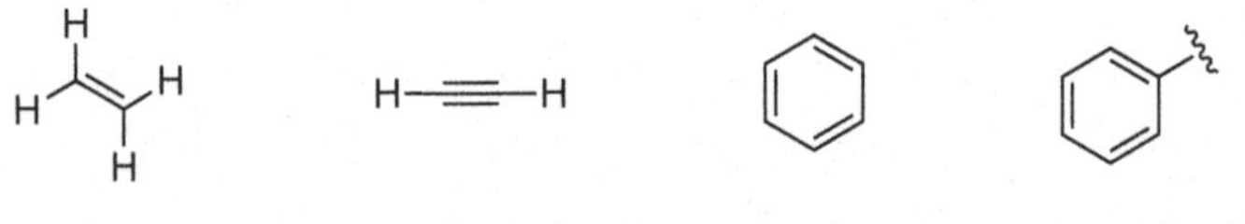

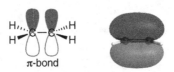

| ethene (ethylene) | ethyne (acetylene) | benzene | phenyl (Ph) |

Figure 11.1 Examples of unsaturated hydrocarbons and IUPAC names (common names or abbreviations in parentheses).

π-bond

Figure 11.2 Structure and 3D model of the π bond in ethene.

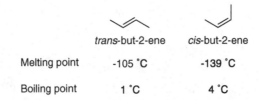

	trans-but-2-ene	*cis*-but-2-ene
Melting point	-105 °C	-139 °C
Boiling point	1 °C	4 °C

Figure 11.3 Structure of *trans*-but-2-ene and *cis*-but-2-ene and their physical properties.

DOI: 10.1201/9781003479352-11

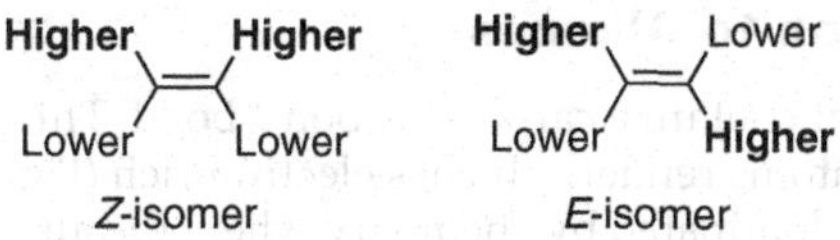

Z-isomer   E-isomer

Figure 11.4 Relationship between higher priority groups in a Z-isomer (*left*) and E-isomer (*right*).

1. Number the longest carbon chain that contains the double bond in the direction that gives the carbon atoms of the double bond the lowest possible numbers.

2. Indicate the location of the double bond by the number of its first carbon.

3. Name branched or substituted alkenes in a manner like alkanes.

4. Number the carbon atoms, locate and name substituent groups, locate the double bond, and name the main chain.

Examples

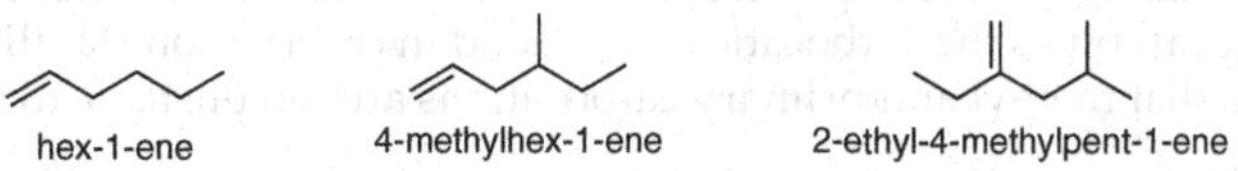

hex-1-ene   4-methylhex-1-ene   2-ethyl-4-methylpent-1-ene

Note: in the newest IUPAC rules (followed above), the locant for the –ene ending goes between the parent chain prefix and the –ene suffix. The older system (still commonly employed) puts the –ene locant before the parent name (in parentheses), e.g. 1-hexene instead of hex-1-ene.

Note: many alkenes have trivial names that are used almost exclusively instead of the recommended IUPAC names, e.g.: ethylene (ethene) and propylene (propene).

E/Z STEREOCHEMISTRY IDENTIFIERS

As mentioned above, different alkene configurations can be indicated by *cis* and *trans* stereochemical terms; however, the *cis* and *trans* terms only apply when the two groups are the same. The official IUPAC terminology employs the stereochemical designators Z- and E- (which come from the German terms *zusammen* (meaning together) and *entgegen* (meaning across or opposite)). To assign Z- and E-, the Cahn-Ingold-Prelog prioritization rules (recall from Chapter 5). If the highest priority groups on either end of the double bond are on the same side, it is the Z-isomer. If the highest priority groups are on opposite sides of the double bond, it is the E-isomer (Figure 11.4).

Cycloalkenes are named by IUPAC conventions following the above rules; however, the C=C is always numbered as positions 1 and 2 in the ring. For dienes (two alkene groups in a single molecule) and polyenes (multiple alkene groups in a single molecule), each alkene group must be given a locant and each alkene must be given an appropriate stereochemical designation.

Problem 11.1 Convert each IUPAC name to a skeletal structure and each structure to a name.

a. 1-fluorocyclopentene   b. (2Z,6E)-3-bromo-4,4-dimethylnona-2,6-diene

c.   d.

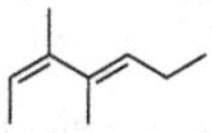

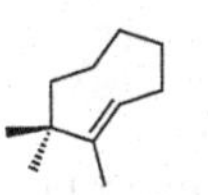

12 Electrophilic Addition to Alkenes

Alkenes are hydrocarbons that contain a carbon–carbon π bond. This region of extra electron density between the two carbon atoms renders alkenes electron-rich (Figure 12.1).

The reactivity of alkenes is dominated by chemistry where the alkene functional group reacts as a base or nucleophile. While possible, reactions where an alkene reacts as an electrophile are much less common.

ALKENES AS BRØNSTED–LOWRY BASES

One of the first reactions involving alkenes is the addition of H–X. As mentioned above, alkenes react as bases/nucleophiles. As a first step, the addition of HX proceeds via a Brønsted–Lowry, acid–base reaction between the mineral acid and the alkene base.

Problem 12.1 Draw the correct, electron-pushing arrows for the following Brønsted–Lowry acid–base reaction.

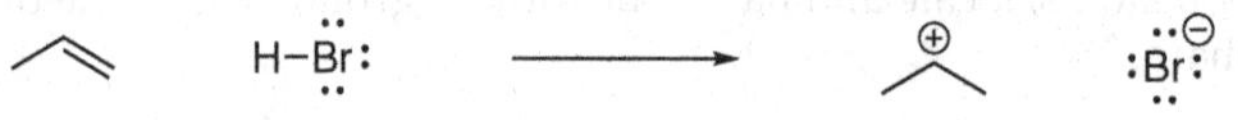

Note that the above reaction proceeds preferentially to add H+ to the terminal carbon, thus leaving the more substituted carbon as the carbocation. The trend in carbocation stability follows the trend in Figure 12.2.[1] Note, that methyl and primary carbocations are too unstable to form under normal reaction conditions.

The above trend (Figure 12.2) in carbocation stabilities is the result of two factors: conjugation and hyperconjugation. Delocalization (see Chapter 2) is the spreading out of charge and electron density over more than two atoms. This distribution of positive charge over multiple atoms renders delocalization-stabilized carbocations significantly more stable than similarly substituted, non-delocalization-stabilized carbocations. In the above trend, primary allylic carbocation is 120 kJ/mol more stable than a primary, non-delocalization-stabilized carbocation. Hyperconjugation (see Chapter 3) is the partial donation of electron density from filled σ bonding orbitals into unfilled orbitals. Increasing the substitution of alkyl groups around the carbocation increases its stability, which is why carbocation stability trends tertiary > secondary >> primary >>> methyl.

ENERGETICS OF ASYMMETRIC ALKENE PROTONATION – MARKOVNIKOV ADDITION

Reconsidering the protonation of propene with hydrogen bromide, the preference for the creation of a secondary carbocation (more stable, lower energy) intermediate over a primary carbocation (less stable, higher energy) intermediate can now be understood in terms of carbocation stability. Now if a reaction coordinate diagram is considered, the secondary carbocation lies below the primary carbocation in terms of energy. The Hammond-Leffler postulate is then used to infer that the transition state leading to the secondary carbocation is lower in energy than the transition state leading to the primary carbocation. [2]

Figure 12.1 Orbital depiction of an alkene π-bond (*left*) in ethylene and electrostatic potential map of ethylene (red corresponds to high electron density).

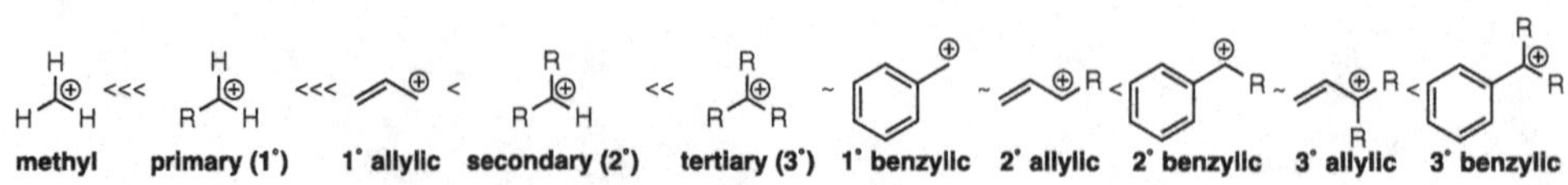

methyl   primary (1°)   1° allylic   secondary (2°)   tertiary (3°)   1° benzylic   2° allylic   2° benzylic   3° allylic   3° benzylic

Figure 12.2 Relative stabilities of different carbocation types. Each < represents a 40 kJ/mol energy difference. R stands for alkyl substituents.

DOI: 10.1201/9781003479352-12

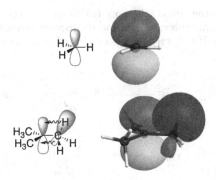

Figure 12.3 Orbital depiction of the empty p-orbital in the methyl carbocation (*top*) not stabilized by hyperconjugation and of the *tert*-butyl carbocation (*bottom*), which is stabilized by hyperconjugation (orbital overlap and electron donation from a neighboring filled σ_{CH} orbital).

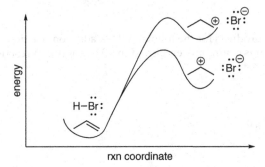

Figure 12.4 A reaction coordinate diagram showing the different energy pathways leading to the higher energy (less stable) primary carbocation intermediate and the lower (more stable) secondary carbocation intermediate.

Altogether, the lower energy intermediate is produced faster meaning that this will give rise to the major product. The final product is the result of the Lewis base (nucleophile) bromide adding to the Lewis acid (electrophile) carbocation.

Problem 12.2 Draw the arrow needed to complete this mechanism.

The result is the addition of hydrogen and bromide to the two carbons of the alkene. The reaction proceeds – because of carbocation stability – to place the hydrogen at the less substituted carbon (carbon with more hydrogens) and the nucleophile at the more substituted carbon (carbon with fewer hydrogens). This tendency was first reported by Vladimir Markovnikov (also spelt Markownikoff per German transliteration rules) in 1870.[3] To honour his work this regiochemical outcome is called Markovnikov addition.

Problem 12.3 Draw arrows for the Brønsted–Lowry acid–base reaction of HX with various alkenes.

Problem 12.4 Draw arrows for the Lewis acid–base reaction of each carbocation with a halide. Note that the carbocation is a planar intermediate and the incoming nucleophile will attack either face with equal probability. If a chiral centre is created, then a racemic mixture is produced.

Problem 12.5 Predict the products for the following H–X addition reactions. Dichloromethane (DCM) is a solvent that is commonly employed in alkene addition reactions.

CARBOCATION REARRANGEMENTS

Carbocations are unique intermediates because they can undergo rearrangements.[4] That is if a neighbouring group (typically hydrogen or alkyl substituents) can move with its electrons and produce a more stable carbocation, then that rearrangement will occur. Carbocation rearrangement is always something that needs to be considered.

Problem 12.6 Draw the arrows necessary to complete the following carbocation rearrangements. (If stuck on a problem, drawing the implicit hydrogen atoms is a good step.)

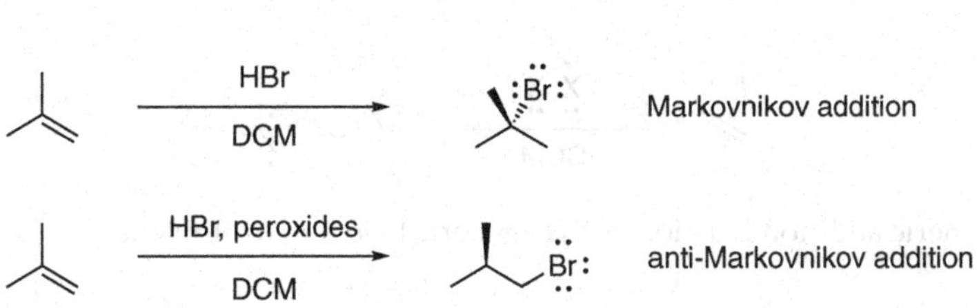

Figure 12.5 Predict the products for the following H–X addition reactions. Dichloromethane (DCM) is a solvent that is commonly employed in alkene addition reactions.

Problem 12.7 Predict the product for each of the following reactions.

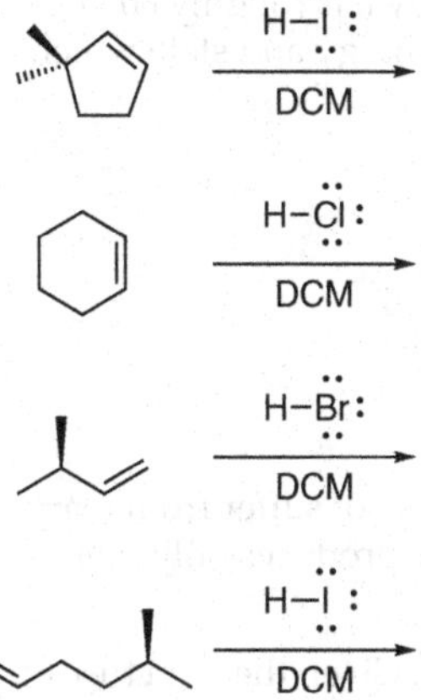

ANTI-MARKOVNIKOV ADDITION OF HYDROGEN BROMIDE WITH PEROXIDES

Recall from Chapter 5 that HBr in the presence of peroxides will add to alkenes to place the H atom on the carbon with fewer H atoms (more substituted carbon) and the Br on the carbon with more H atoms (less substituted carbon). This anti-Markovnikov addition occurs because of a difference in mechanism (radical versus ionic) and is unique to HBr, other hydrogen halides do not show a reversal in regiochemistry with the addition of peroxides. This peroxide effect for HBr was conclusively demonstrated by Morris S. Kharasch.[5] Kharasch's work disproved other hypotheses linking the difference in outcome to temperature, light, and variations in solvent.

HYDRATION OF AN ALKENE

While the products are different, the reaction of water/acid with an alkene and alcohols/acid with an alkene are mechanistically very similar to the reaction of HX with an alkene.

Problem 12.8 Propose a mechanism for the following reactions.

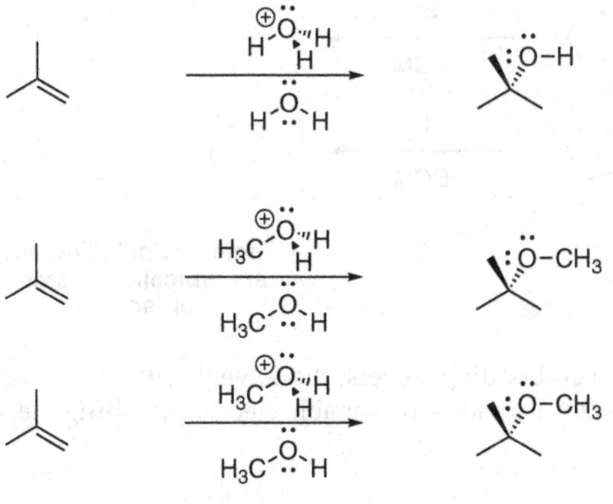

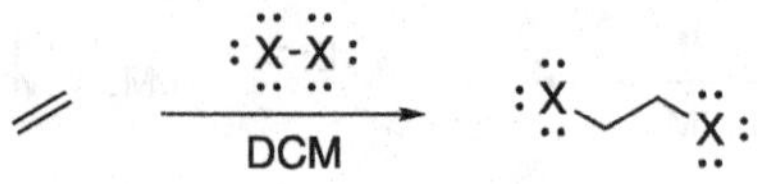

Figure 12.6 Generic addition reaction of X_2 (a generic halogen) to ethylene.

DIHALOGEN ADDITION TO AN ALKENE

Alkenes can also react with dihalogens (X_2). While radical allylic substitution (Chapter 5) is always possible, judicious choice of reagents helps to ensure that the reactions proceed through addition to alkene (Figure 12.6), which proceeds in a manner that is similar and dissimilar to HX addition.

Problem 12.9 Provide arrows for the following reactions. Note that pathway i. is roughly 60 kJ/mol higher in energy.[6] Provide an explanation why pathway ii. is lower in energy and state which pathway is more likely to occur.

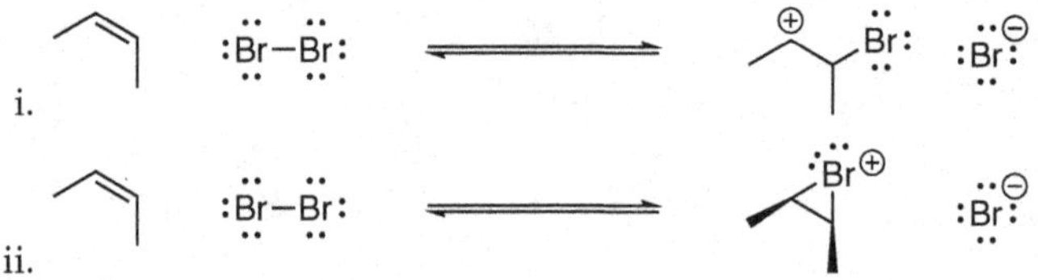

While it is a cation, no atom lacks a full octet and so the halonium ion does not suffer from rearrangements like carbocations. Opening a halonium ion with a nucleophile produces only *trans* products.

Problem 12.10 Provide electron-pushing arrows for the following reaction. Given the structure of the halonium ion, explain why only trans products are produced.

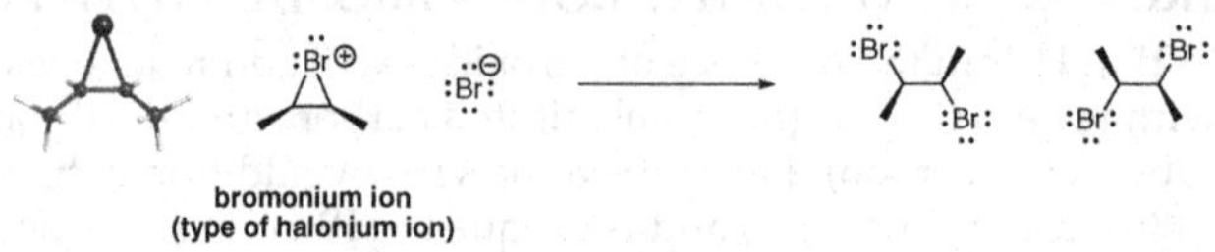

bromonium ion
(type of halonium ion)

Problem 12.11 Predict the product(s) for each reaction.

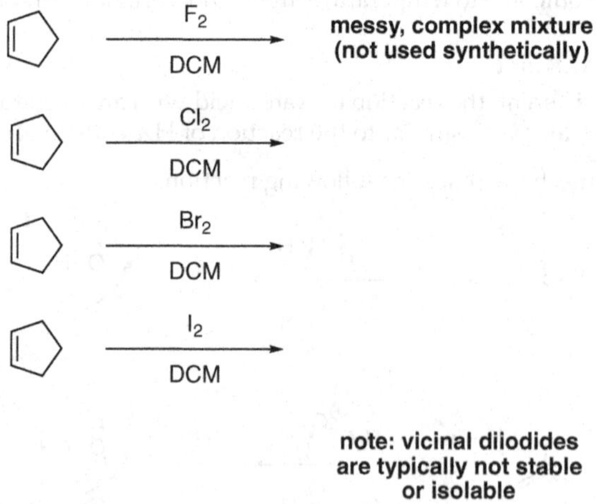

note: vicinal diiodides
are typically not stable
or isolable

In the presence of a hydrogen-bonding solvent, the solvent will attack the halonium ion intermediate rather than the halide (X^-). Provide a reasonable electron-pushing mechanism to account for the products formed.

Problem 12.12 Give two reasons why the solvent acts as the nucleophile and attacks the halonium rather than the halide.

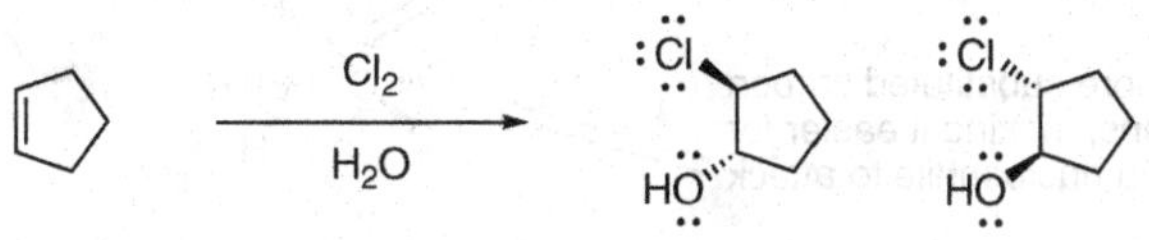

Problem 12.13 Predict the product for each of the following reactions.

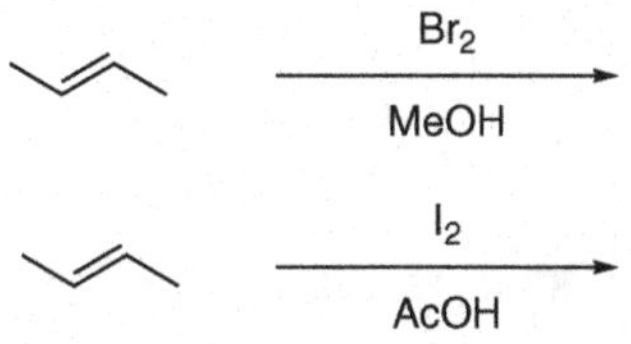

In addition to utilizing chlorine, bromine, or iodine in aprotic and protic solvents, various interhalogens (ClF, BrCl, ICl, and IBr) and the pseudohalides IN_3 can also be utilized in a similar fashion (Figure 12.7). Here the less electronegative atom forms the halonium ion intermediate and the other atom (or N_3^- group for IN_3) is added as a nucleophile.

In symmetrically substituted halonium ions, nucleophilic attack at either side occurs with equal probability. For asymmetric halonium ions, attack occurs preferentially at the more substituted side, the side with more non-hydrogen groups (Figure 12.8).

Problem 12.14 Predict the product for each reaction.

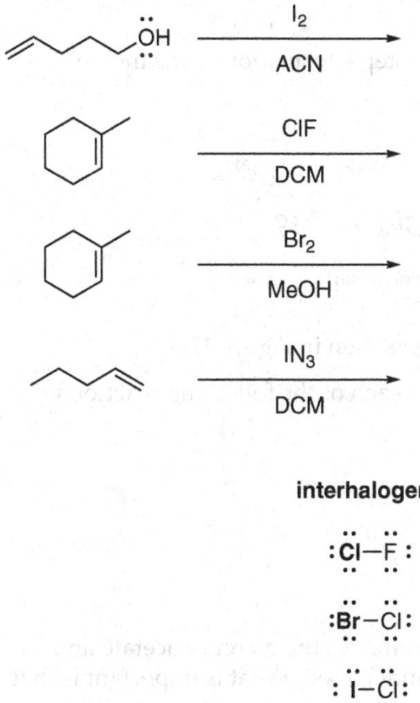

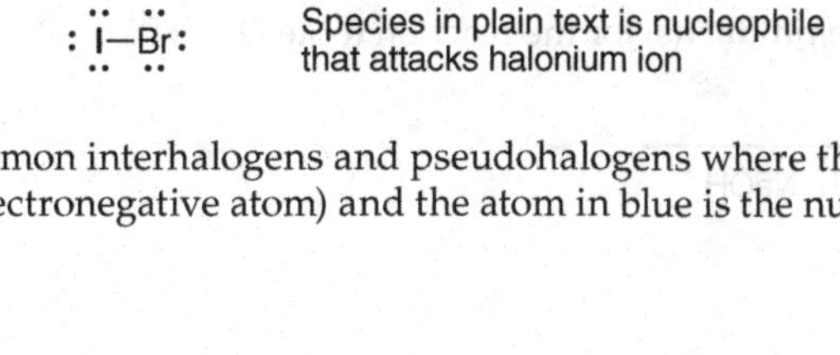

interhalogens

pseudohalogen

Atom in bold forms halonium ion

Species in plain text is nucleophile that attacks halonium ion

Figure 12.7 List of common interhalogens and pseudohalogens where the atom in red forms the halonium ion (the less electronegative atom) and the atom in blue is the nucleophile that opens the halonium ion.

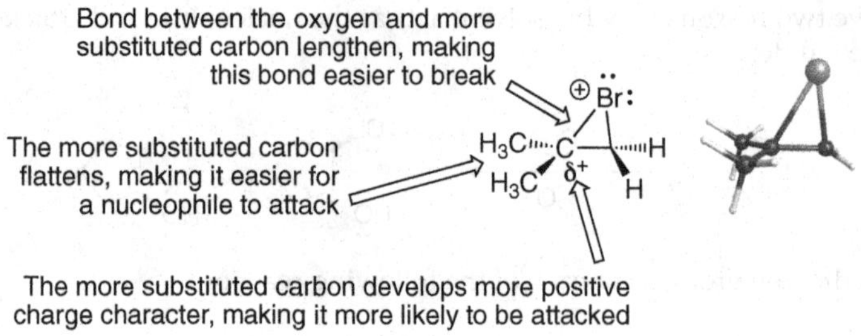
Bond between the oxygen and more substituted carbon lengthen, making this bond easier to break
The more substituted carbon flattens, making it easier for a nucleophile to attack
The more substituted carbon develops more positive charge character, making it more likely to be attacked
Br
H₃C
H₃C
C
H
H
δ+

Figure 12.8 Structure of an asymmetric bromonium ion.

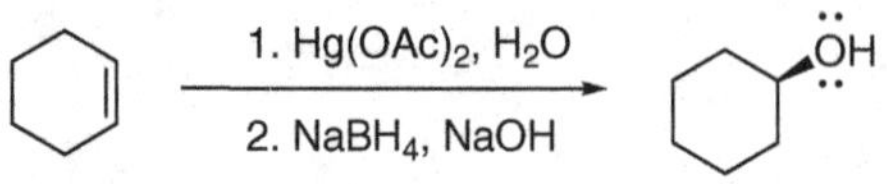
1. Hg(OAc)₂, H₂O
2. NaBH₄, NaOH
OH

Figure 12.9 Example oxymercuration-reduction of cyclohexene to give cyclohexanol.

MARKOVNIKOV HYDRATION WITHOUT REARRANGEMENT – OXYMERCURATION-REDUCTION

Oxymercuration-reduction reactions are synthetically useful because they produce alcohols and ethers without going through a carbocation intermediate and therefore avoid potential rearrangements.[7] The reaction is very similar to the addition of halogen (X_2) in a polar, protic solvent (Figure 12.9).

Problem 12.15 Provide the electron-pushing arrows for the first step – formation of the mercurinium ion.

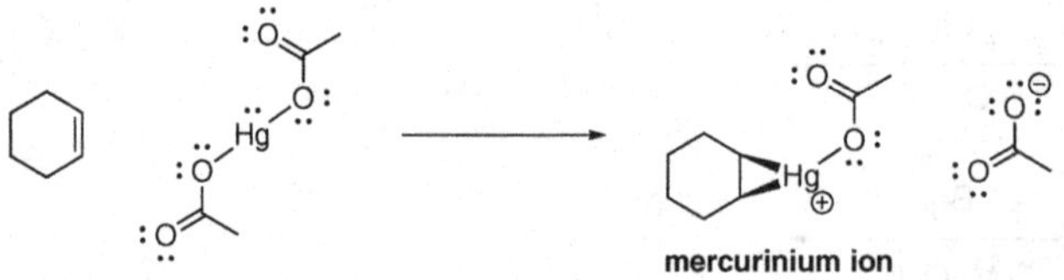
Hg
Hg
mercurinium ion

The mercurinium ion behaves just like the halonium ion that was seen in Figure 12.8.

Problem 12.16 Given this information, predict the product(s) for each of the following reactions.

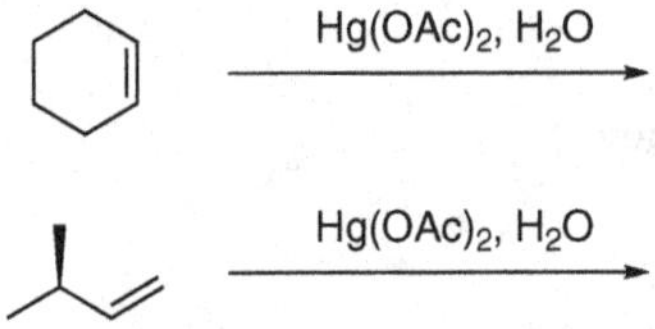
Hg(OAc)₂, H₂O
Hg(OAc)₂, H₂O

The final step in oxymercuration-reduction (2. $NaBH_4$, NaOH) removes the mercury acetate and replaces it with a hydrogen atom. The mechanism is not fully understood. What is important is that the stereochemistry is scrambled.

Problem 12.17 Provide the product(s) for the reaction (note D = ^2H).

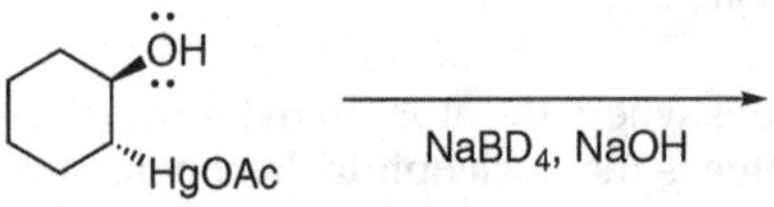
OH
HgOAc
NaBD₄, NaOH

Problem 12.18 Provide the product(s) for each reaction.

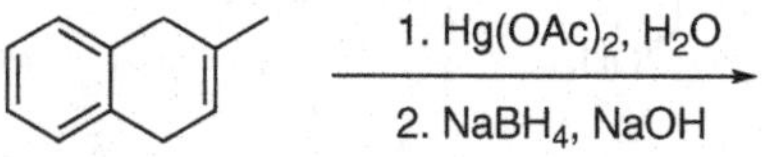

$$\text{1. Hg(OAc)}_2, \text{H}_2\text{O}$$
$$\text{2. NaBH}_4, \text{NaOH}$$

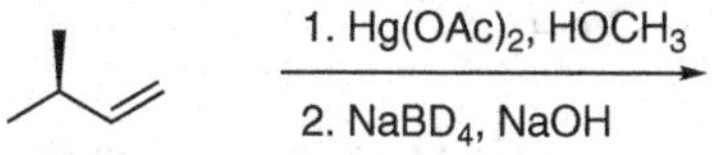

$$\text{1. Hg(OAc)}_2, \text{HOCH}_3$$
$$\text{2. NaBD}_4, \text{NaOH}$$

Oxymercuration-reduction is an early example of a synthetic process that offers improvements – no rearrangements – over prior synthetic methodologies, i.e.: acid-catalyzed hydration. The oxymercuration-reduction reaction also follows a mechanistic pattern that is common among introductory organic reactions (the mercurinium ion reacts in the same manner as halonium ions and as shall be seen in protonated epoxides). The use of a mercury reagent, however, presents a chemical hazard, which has led to chemists developing newer and safer hydration methods including the Mukaiyama hydration, which uses a cobalt catalyst, oxygen gas, and phenylsilane (PhSiH$_3$).[8] It is worth highlighting that there will be other examples throughout this book where an older synthetic method is shown. This is for two reasons. The first reason is that the reactions shown provide insight into core organic mechanisms and help to cement important reactivity patterns. The second reason is that newer reactions often employ transition metal catalysts, the understanding of which requires further study of inorganic and organometallic chemistry.

ANTI-MARKOVNIKOV HYDRATION

Hydroboration oxidation is a third method for producing alcohols.[9] In contrast to the previous two (H$_3$O$^+$/H$_2$O and oxymercuration-reduction), hydroboration oxidation produces alcohols with opposite (anti-Markovnikov) regiochemistry (Figure 12.10).

BH$_3$ is added in a single concerted step and the outcome is determined by steric repulsion (the large BH$_2$ group orients towards the less substituted side of the alkene) during the addition. This step determines the regiochemistry of the final product.

Problem 12.19 Draw the arrows for this first step (the transition state is shown for this reaction).

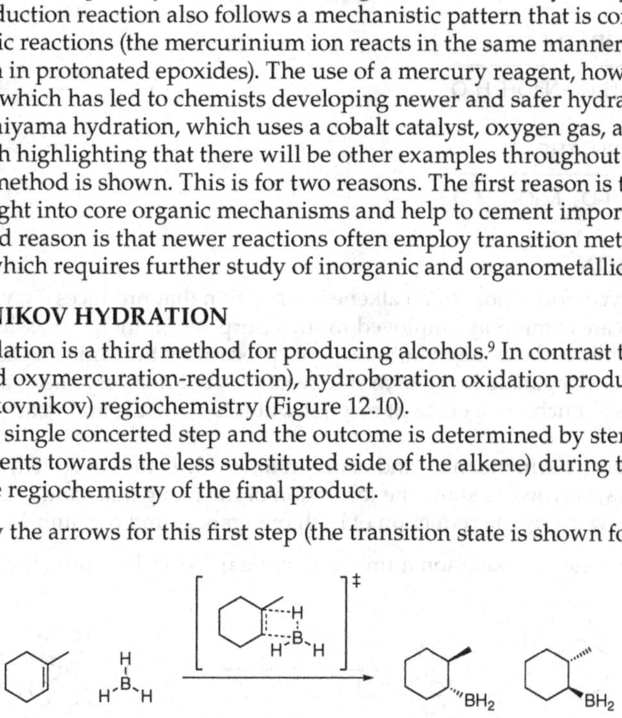

The second step of the reaction proceeds by a mechanism that is beyond the scope of this book. The important point is that the C–B bond stereochemistry is retained when it is converted into a C–O bond.

Problem 12.20 Predict the product for each of the following oxidation steps.

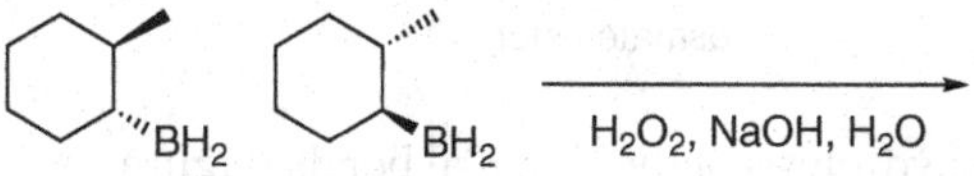

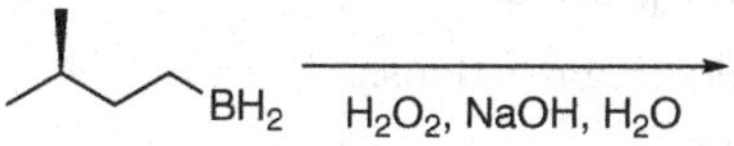

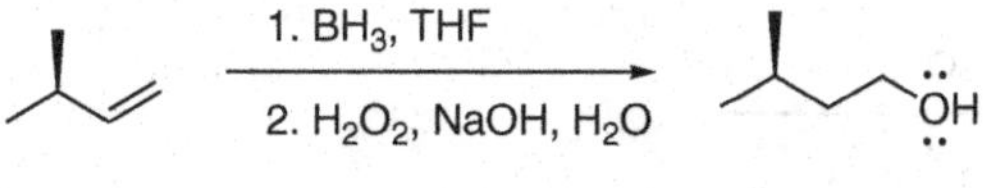

$$\text{H}_2\text{O}_2, \text{NaOH}, \text{H}_2\text{O}$$

$$\text{H}_2\text{O}_2, \text{NaOH}, \text{H}_2\text{O}$$

$$\text{1. BH}_3, \text{THF}$$
$$\text{2. H}_2\text{O}_2, \text{NaOH}, \text{H}_2\text{O}$$

Figure 12.10 Example hydroboration oxidation of 3-methylbutene to give 3-methylbutanol.

127

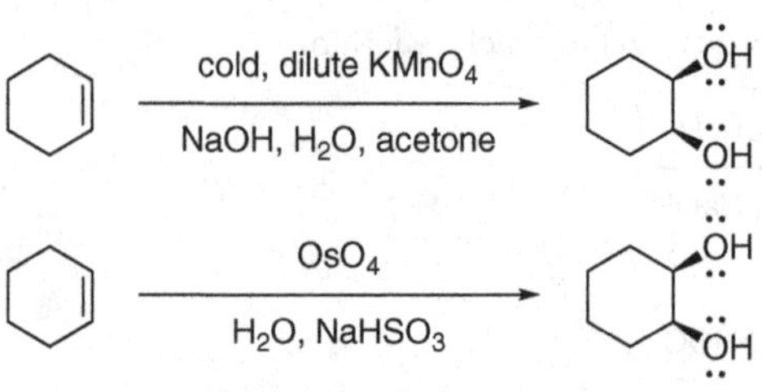
cold, dilute KMnO4
NaOH, H2O, acetone
OH
OH
OsO4
H2O, NaHSO3
OH
OH

Figure 12.11 Example of dihydroxylation reactions of cyclohexene with $KMnO_4$ (*top*) and OsO_4 (*bottom*).

Problem 12.21 Predict the final product(s) for each reaction.

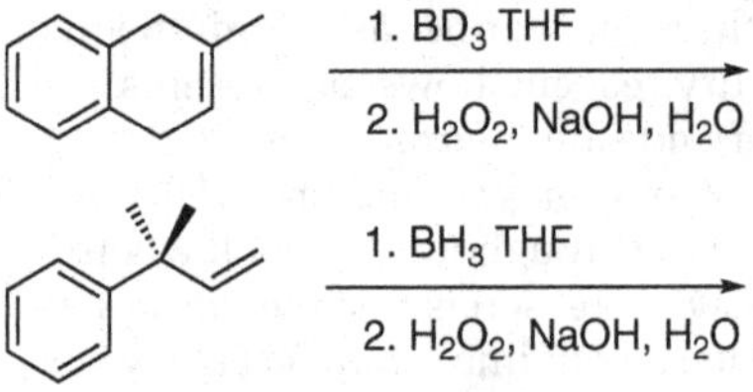
1. BD3 THF
2. H2O2, NaOH, H2O
1. BH3 THF
2. H2O2, NaOH, H2O

DIHYDROXYLATION

Dihydroxylation (glycol formation) of an alkene is a reaction that produces *cis*, vicinal diols (Figure 12.11). Two reagents are commonly employed for this purpose: $KMnO_4$ (potassium permanganate)[10] and OsO_4 (osmium tetroxide).[11] Both reagents proceed by the same mechanism and produce the same product. Permanganate is prone to side reactions, but it is very cheap and easy to handle. Osmium tetroxide is a much more efficient reagent, but it is a volatile and toxic solid that is also very expensive.

The first step for both permanganate and osmium tetroxide is the concerted, *syn* addition across the double bond. Draw arrows to show the formation of the manganate and osmate esters. The reaction is driven forward by the reduction of both manganese and osmium.

Problem 12.22 Determine the oxidation number of manganese and osmium both before and after the reaction.

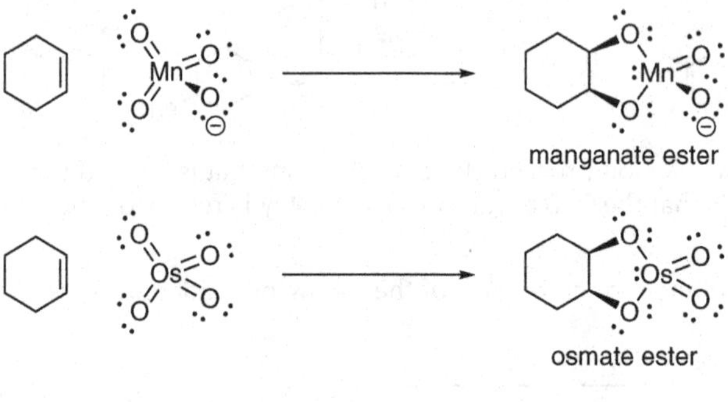
Mn
Mn
manganate ester
Os
Os
osmate ester

The final step of each dihydroxylation reaction is the hydrolysis of the O–metal bonds (Figure 12.12). The mechanism for this bond cleavage is beyond the scope of this textbook. Note, however, that the C–O bond stereochemistry is unaffected by the process.

Problem 12.23 Predict the products for each of the following reactions.

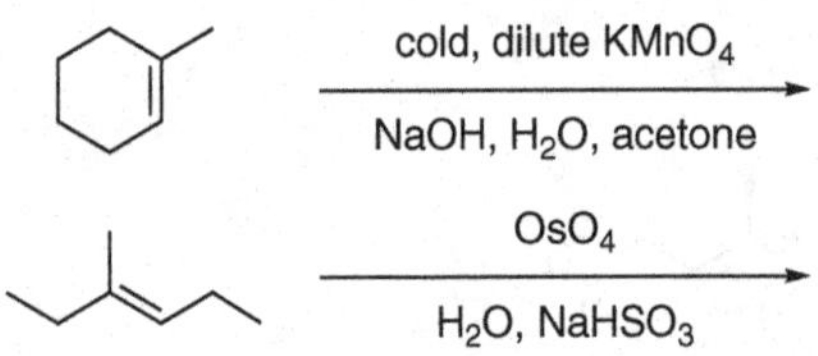
cold, dilute KMnO4
NaOH, H2O, acetone
OsO4
H2O, NaHSO3

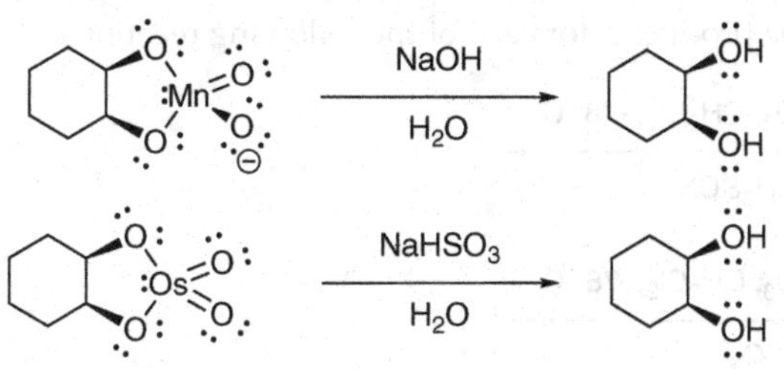

Figure 12.12 Cleavage of the manganate and osmate ester intermediates.

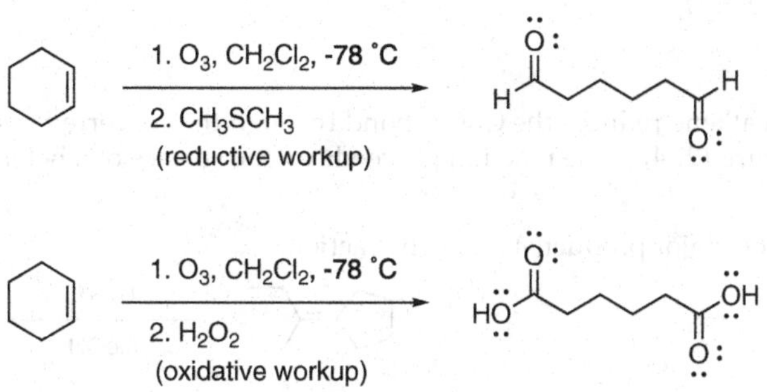

Figure 12.13 Example of ozonolysis reactions of cyclohexene under reductive (*top*) and oxidative (*bottom*) conditions.

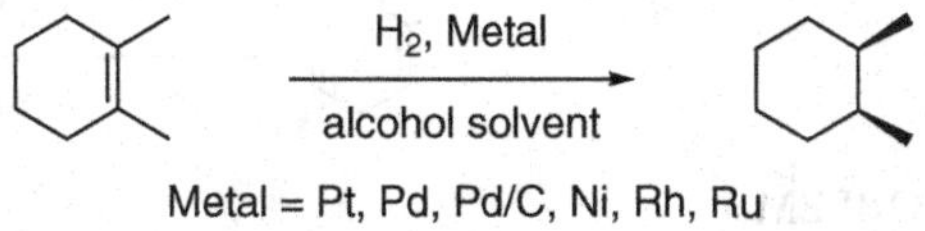

Metal = Pt, Pd, Pd/C, Ni, Rh, Ru

Figure 12.14 Example reduction reaction of 1,2-dimethylcyclohexene.

OZONOLYSIS

Ozonolysis, Figure 12.13, is a unique reaction because it breaks both the π-bond and the σ-bond of an alkene and produces two carbonyl functional groups.[12] Ozonolysis has two possible workup conditions that can lead to different outcomes depending upon the presence (or absence) of hydrogens on the alkene carbons.

Problem 12.24 Draw all electron-pushing arrows for the first step of the ozonolysis reaction. Note the workup steps proceed by mechanisms that are beyond the scope of this book.

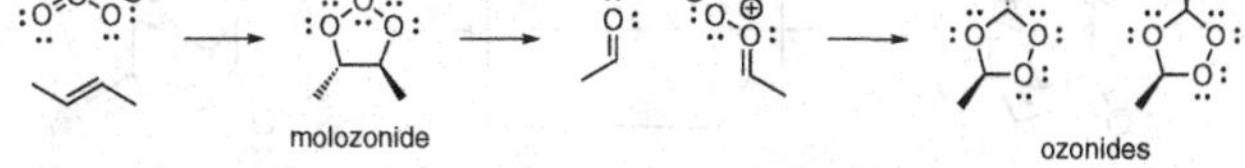

molozonide   ozonides

A reductive workup (Me$_2$S, Zn and HCl, or PPh$_3$) will leave any protons attached to the alkene carbons as hydrogens, aldehyde products. An oxidative workup (H$_2$O$_2$) will turn any protons attached to the alkene carbons to –OH's, carboxylic acid products.

Problem 12.25 Predict the product(s) for each of the following reactions.

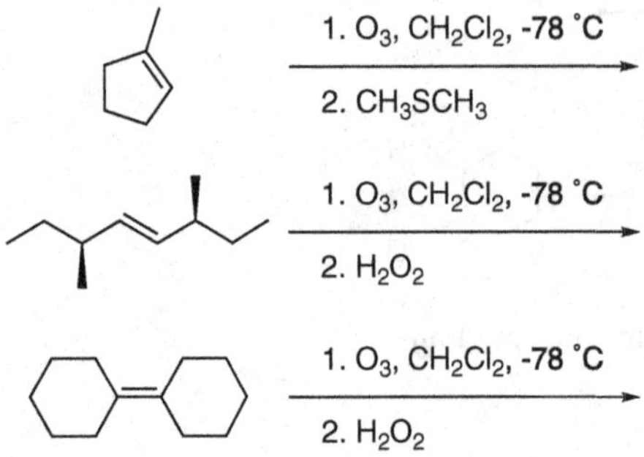
1. O3, CH2Cl2, -78 °C
2. CH3SCH3
1. O3, CH2Cl2, -78 °C
2. H2O2
1. O3, CH2Cl2, -78 °C
2. H2O2

HYDROGENATION

Hydrogenation of an alkene reduces the C=C π-bond through the concerted, *syn* addition of two hydrogen atoms (Figure 12.14).[13] The reaction proceeds on the surface of a heterogeneous, metal catalyst.

Problem 12.26 Give the major product(s) for each reaction.

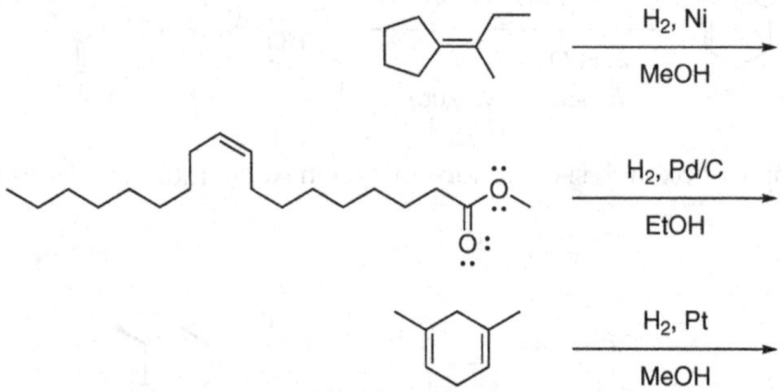
H2, Ni
MeOH
H2, Pd/C
EtOH
H2, Pt
MeOH

GENERAL PRACTICE PROBLEMS

Problem 12.27 Provide the reagent(s) necessary to complete the reaction and propose a reasonable explanation for why the alkene near the carbonyl does not react.

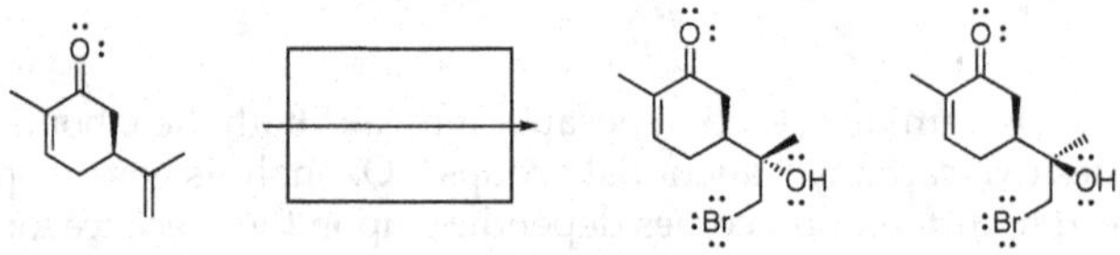
OH
Br
OH
Br

Problem 12.28 Provide the reagent(s) necessary to complete the reaction.

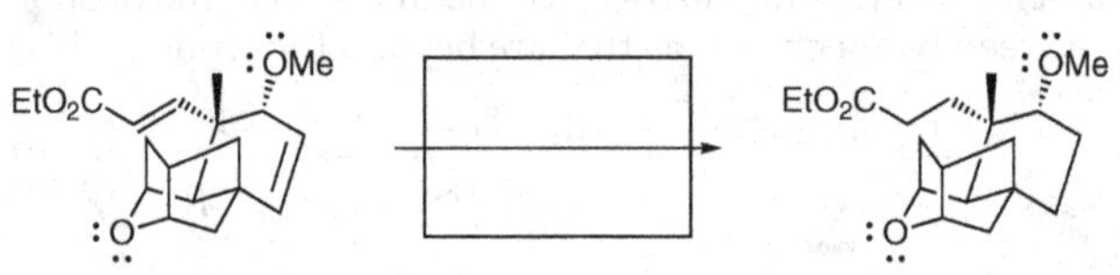
EtO2C
OMe
EtO2C
OMe

Problem 12.29 Predict the product(s).

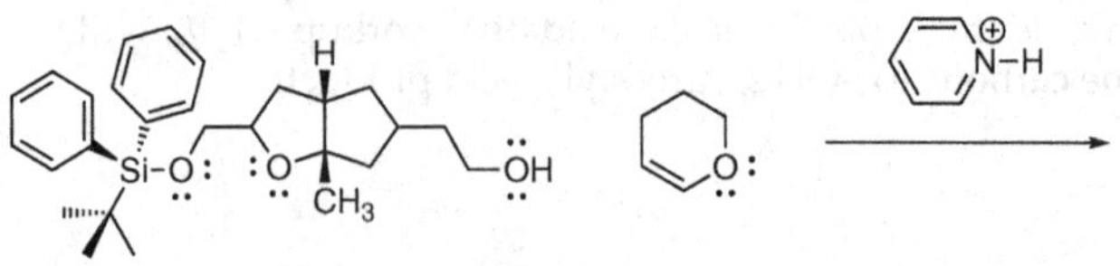
H
Si-O
CH3
OH
N-H

NOTES

1. Calculated relative stability trend, which is in excellent agreement with experimental gas-phase hydride affinities: Anslyn, E.V.; Dougherty, D.A. Strain and Stability in *Modern Physical Organic Chemistry*. University Science Books: Sausalito, CA, 2006. pp. 65–145.

2. i) Leffler, J.E. Parameters for the description of transition states. *Science*, **1953**, *117*, 340–341. DOI: 10.1126/science.117.3039.340
   ii) Hammond, G.S. A correlation of reaction rates. *J. Am. Chem. Soc.*, **1955**, *77* (2), 334–338. DOI: 10.1021/ja01607a027

3. Markownikoff, W. I. Ueber die Abhängigkeit der verschiedenen Vertretbarkeit des Radicalwasserstoffs in den isomeren Buttersäuren. *Liebigs Ann.*, **1870**, *153* (2), 228–259. DOI: 10.1002/jlac.18701530204

4. Meerwein, H.; van Emster, K. Über die Gleichgewichts-Isomerie zwischen Bornylchlorid, Isobornylchlorid und Camphen-chlorhydrat. *Ber. Dtsch. Chem. Ges.*, **1922**, *55* (8), 2500–2528. DOI: 10.1002/cber.19220550829

5. Kharasch, M.S.; Mayo, F.R. The peroxide effect in the addition of reagents to unsaturated compounds. I. The addition of hydrogen bromide to allyl bromide. *J. Am. Chem. Soc.*, **1933**, *55* (6), 2468–2496. DOI: 10.1021/ja01333a041

6. Berman, D.; Anicich, V.; Beauchamp, J. Stabilities of isomeric halonium ions $C_2H_4X^+$ (X = chlorine, bromine) by photoionization mass spectrometry and ion cyclotron resonance spectroscopy. General considerations of relative stabilities cyclic and acyclic isomeric onium ions. *J. Am. Chem. Soc.*, **1979**, *101* (5), 1239–1248. DOI: 10.1021/ja00499a032

7. Hofmann, K.; Sand., J. Ueber das Verhalten von Mercurisalzen gegen Olefine. *Ber. Dtsch. Chem. Ges.* **1900**, *33* (1), 1340–1353. DOI: 10.1002/cber.190003301231

8. Shigeru, I.; Teruaki, M. A New Method for Preparation of Alcohols from Olefins with Molecular Oxygen and Phenylsilane by the Use of Bis(acetylacetonato) cobalt(II). *Chem. Lett.*, **1989**, *18* (6), 1071–1074. DOI: 10.1246/cl.1989.1071

9. Brown, H. C.; Zweifel, G. A stereospecific cis hydration of the double bond in cyclic derivatives. *J. Am. Chem. Soc.*, **1959**, *81* (1), 247–247. DOI: 10.1021/ja01510a059

10. Kekulé, A.; Anschütz, R. Ueber Tanatar's Bioxyfumarsäure. *Ber. Dtsch. Chem. Ges.*, **1880**, *13* (2), 2150–2152. DOI: 10.1002/cber.188001302213

11. Milas, N.; Terry, E. Oxidation of fumaric and of maleic acid to tartaric acid. *J. Am. Chem. Soc.*, **1925**, *47* (5), 1412–1418. DOI: 10.1021/ja01682a028

12. i) Schönbein, C.F. Ueber die langsame Verbrennung des Aethers. *Ber. Verh. Nat. Ges. Basel*, **1847**, *7*, 4–6.
    ii) Schönbein, C.F. Ueber das Verhalten des Ozons zum oelbildenden Gas. *Ber. Verh. Nat. Ges. Basel*, **1847**, *7*, 7–9.

13. Sabatier, P.; Senderens, J.-B. Action du nickel sur l'éthylène. Synthèse de l'éthane. *C. R. Acad. Sci.*, **1897**, *124*, 1358–1360.

13 Reactions of Alkynes

Alkynes are molecules that contain a carbon–carbon triple bond (two, perpendicular carbon–carbon π bonds). This region of extra electron density between the two carbon atoms renders alkynes, like alkenes, electron-rich (Figure 13.1).

The reactivity of alkynes is much richer than that of alkenes, as alkynes can act both as an electrophile (terminal alkyne C–H bonds can be deprotonated) and as a nucleophile (the π bonds donate electron density to an electrophile). Both types of reactivity will be seen throughout this chapter.

ACETYLIDE IONS

Alkynes, unlike alkenes and alkanes, are weak acids (pK_a of ethyne = 25). Recall from Chapter 10 that this stems from the higher per cent s-character of the carbon, which renders the resulting conjugate base lower in energy, more stable, and therefore less basic.

This allows for the deprotonation of terminal alkynes (Figure 13.3) with strong bases (typically NaH or $NaNH_2$) to produce sodium acetylides.

The acetylide anion can be employed in nucleophilic substitution reactions (S_N2 reactions, the details of which will be a topic in Chapter 14) with primary alkyl halides. In this manner, terminal alkynes can be extended into larger, internal alkynes.[1]

Figure 13.1 Orbital depiction of the two, perpendicular alkyne π bonds (*left*) in ethyne and electrostatic potential map of ethyne (red corresponds to high electron density and blue corresponds to low electron density).

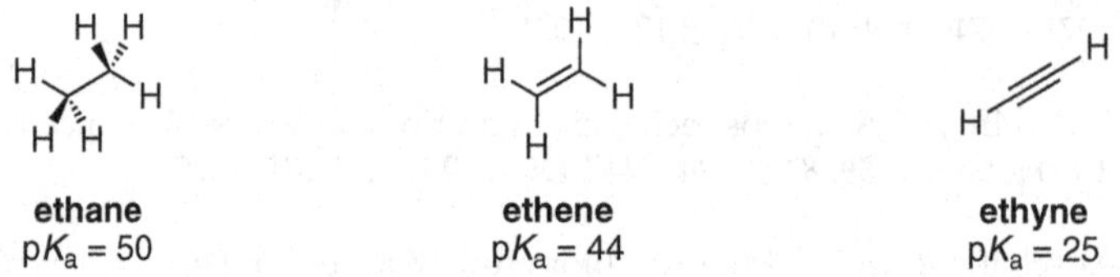

ethane
$pK_a = 50$

ethene
$pK_a = 44$

ethyne
$pK_a = 25$

Figure 13.2 Acidity of different hydrocarbons.

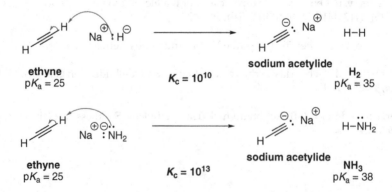

ethyne
$pK_a = 25$

$K_c = 10^{10}$

sodium acetylide

H_2
$pK_a = 35$

ethyne
$pK_a = 25$

$K_c = 10^{13}$

sodium acetylide

NH_3
$pK_a = 38$

Figure 13.3 Deprotonation of alkynes with hydride (*top*) and amide (*bottom*) bases.

DOI: 10.1201/9781003479352-13

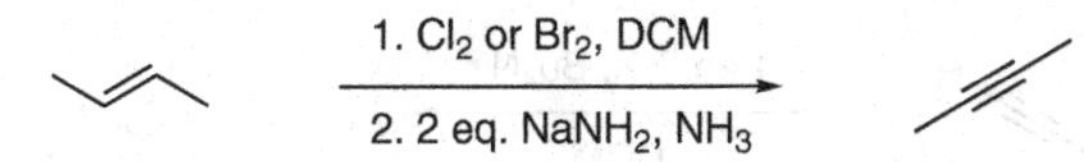

1. Cl₂ or Br₂, DCM

2. 2 eq. NaNH₂, NH₃

Figure 13.4 Formation of an alkyne from an alkene through halide addition and double elimination.

Problem 13.1 Draw the arrows to complete the mechanism.

Problem 13.2 Predict the product(s) for each reaction below.

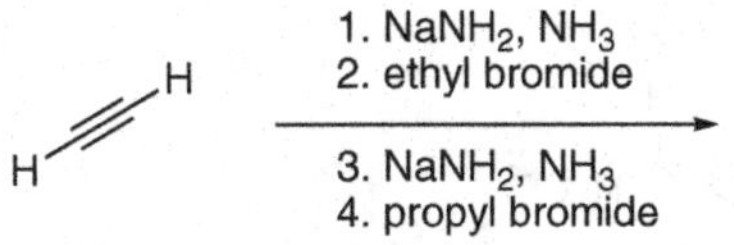

1. NaNH₂, NH₃
2. ethyl bromide

3. NaNH₂, NH₃
4. propyl bromide

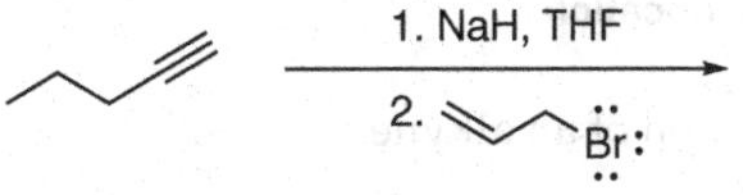

1. NaH, THF

2.

SYNTHESIS OF ALKYNES

As discussed above, larger alkynes can be synthesized from terminal alkynes. Alkynes can also be prepared from alkenes through what is called an elimination reaction, H–X is eliminated from the reaction to form a π bond (Figure 13.4). Here two successive elimination reactions occur to form the two, perpendicular π bonds of the alkyne). First, a vicinal dihalide is created by adding Cl₂ or Br₂ to an alkene (for a review of this mechanism, see Chapter 11). Excess strong base (sodium amide, NaNH₂, is most employed) is added to cause the elimination reactions (Chapter 14).

In the elimination mechanism, the strong base NaNH₂ acts as a base to remove a proton on a carbon next to the carbon with the halogen. As the proton is abstracted, the C–H bonding pair of electrons becomes the new carbon-carbon π bond and the halogen leaves. This same process then proceeds a second time to create the second carbon-carbon π bond.

Problem 13.3 Draw the appropriate arrows to complete the double-elimination mechanism.

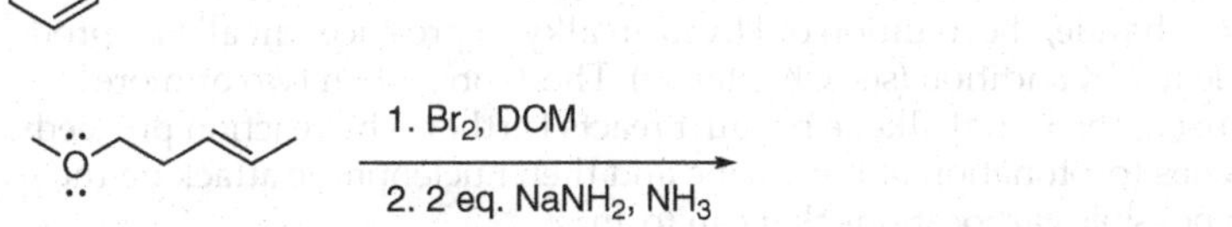

Problem 13.4 Predict the final product for each reaction.

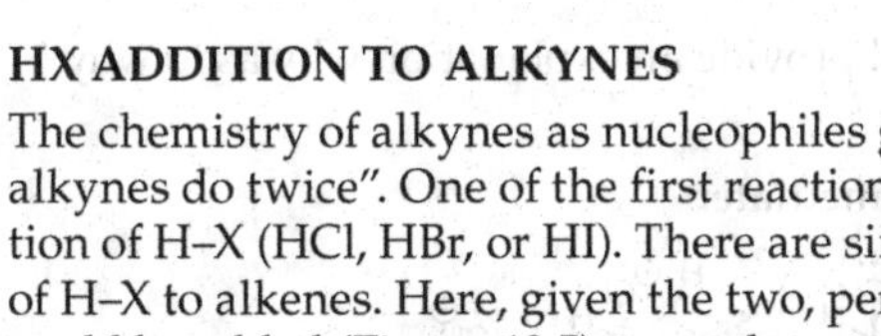

1. Cl₂, DCM

2. 2 eq. NaNH₂, NH₃

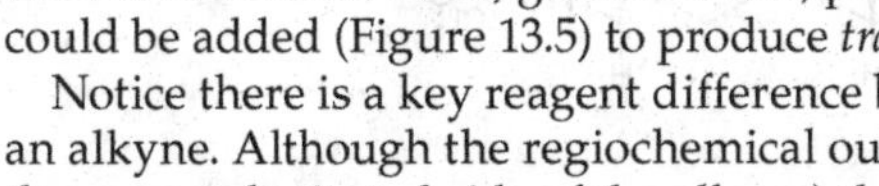

1. Br₂, DCM

2. 2 eq. NaNH₂, NH₃

HX ADDITION TO ALKYNES

The chemistry of alkynes as nucleophiles generally follows the precept, "Whatever alkenes do, alkynes do twice". One of the first reactions involving alkynes reacting as nucleophiles is the addition of H–X (HCl, HBr, or HI). There are similarities and dissimilarities to the comparable addition of H–X to alkenes. Here, given the two, perpendicular π bonds, one or two equivalents of H–X could be added (Figure 13.5) to produce *trans*-vinyl halides or *geminal* dihaloalkanes, respectively.

Notice there is a key reagent difference between the addition of HX to an alkene and HX to an alkyne. Although the regiochemical outcome is still Markovnikov (the halogen ends up at the more substituted side of the alkyne) the reaction requires the addition of extra halide (X⁻). Typically, this is accomplished by adding the tetrabutylammonium (Bu₄N⁺) halide. This is because

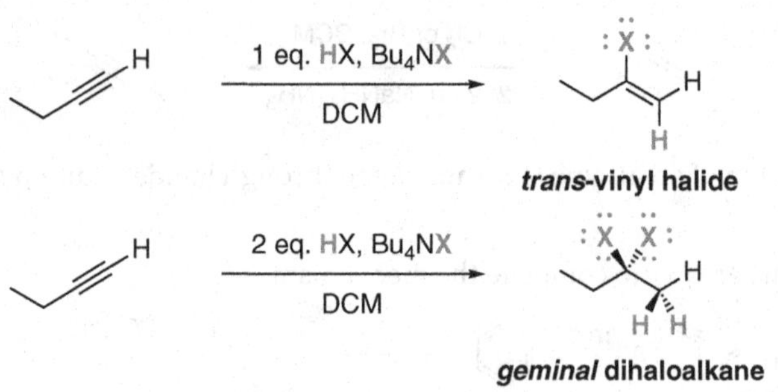

Figure 13.5 Addition of one equivalent (*top*) and two equivalents (*bottom*) of HX to but-1-yne.

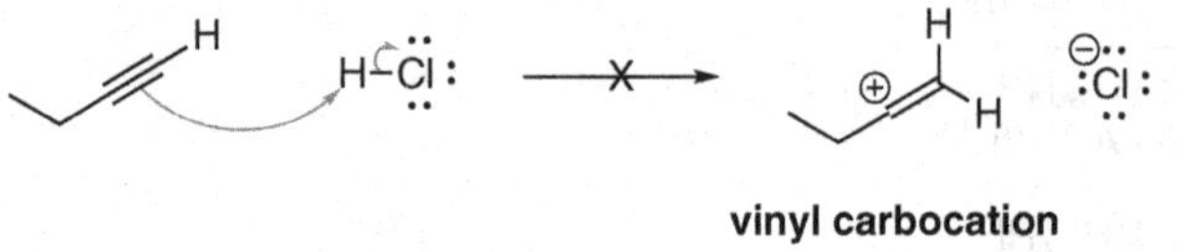

vinyl carbocation

Figure 13.6 High-energy (disfavoured) pathway in the protonation of an alkyne.

regular sodium or potassium salts are not soluble in DCM, while the Bu_4NX salts are organic-solvent soluble sources of halide (X^-). The expected protonation of the alkynes (following the example of alkene protonation, Figure 13.6) does not occur because the resulting vinyl carbocation is too unstable (vinyl carbocations have stabilities like that of primary carbocations[2]).

As such, the reaction then proceeds via a termolecular (three species coming together in a single step) reaction with the addition of chloride (from Bu_4NCl) and hydrogen (from HCl) proceeding to give *trans*-specific addition.[3]

Problem 13.5 Draw the arrows to complete the mechanism.

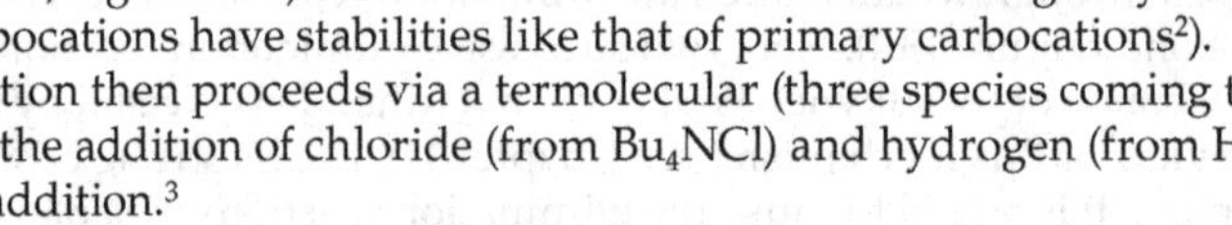

Problem 13.6 Predict the major product(s) for each of the following.

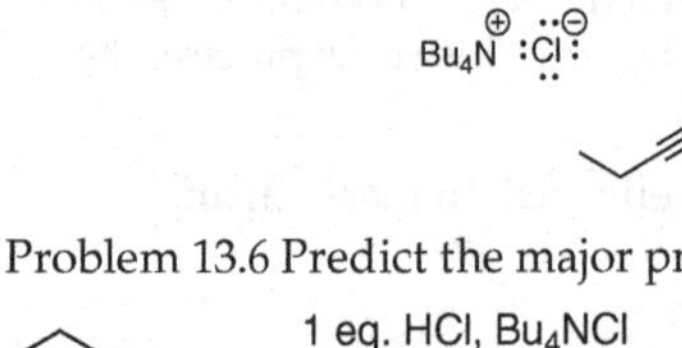

In addition to needing additional halide, the addition of HX to an alkyne produces an alkene product. Alkenes are still susceptible to HX addition (see Chapter 12). Therefore, when two or more equivalents of HX/Bu_4NX are used, the initial alkene product reacts further. The reaction proceeds as was seen previously for alkenes (protonation of the alkene and then nucleophilic attack on the carbocation), but there are two possible carbocations that can form.

Problem 13.7 Provide the arrows for each protonation and provide an explanation why as to why one outcome is disfavoured, and one is favoured.

Protonation Pathway Leading to a Higher-Energy Intermediate:[4]

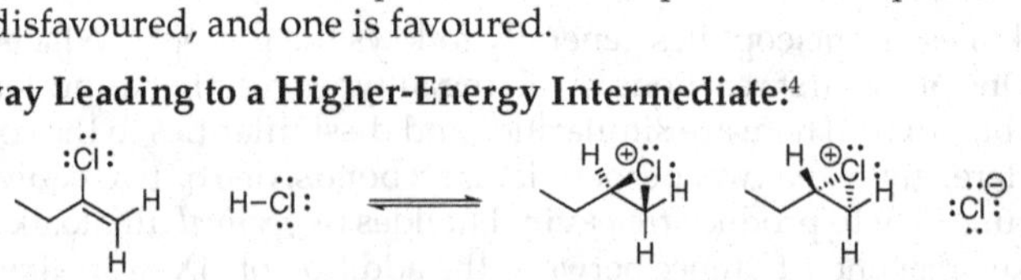

Protonation Pathway Leading to a Lower-Energy Intermediate:

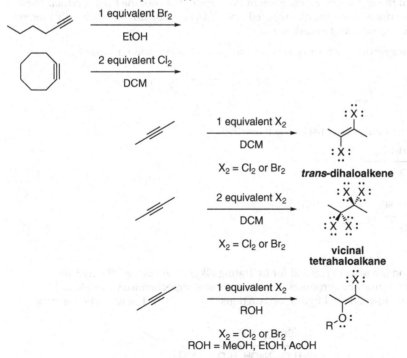

Explanation

The final step of the reaction is attack of the halide on the alkyl carbocation to produce a *geminal* dihalide.

Problem 13.8 Draw the arrows for this last step in the mechanism.

Problem 13.9 Predict the product(s) for each reaction.

2 eq. HCl, Bu₄NCl
———————————
DCM

2 eq. HBr, Bu₄NBr
———————————
DCM

DIHALOGEN ADDITION TO ALKYNES

The addition of X_2 to an alkyne (Figure 13.7) proceeds via a nearly identical mechanistic pathway as was seen for alkene + X_2 reactions. Again, two outcomes are possible depending on the equivalents of X_2 added. For the addition of one equivalent, the reaction produces a *trans*-dihaloalkene. For the addition of two equivalents, the reaction produces vicinal tetrahaloalkanes.

Problem 13.10 Predict the product(s) for each reaction.

1 equivalent Br₂
———————————
EtOH

2 equivalent Cl₂
———————————
DCM

1 equivalent X_2
———————————
DCM

X_2 = Cl₂ or Br₂   ***trans*-dihaloalkene**

2 equivalent X_2
———————————
DCM

X_2 = Cl₂ or Br₂   **vicinal tetrahaloalkane**

1 equivalent X_2
———————————
ROH

X_2 = Cl₂ or Br₂
ROH = MeOH, EtOH, AcOH

Figure 13.7 Addition of one equivalent (*top*) and two equivalents (*middle*) of X_2 and one equivalent of X_2 in a protic solvent (*bottom*) to but-2-yne.

135

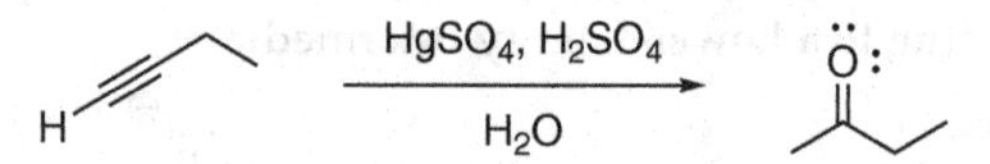

Figure 13.8 Hydration of a but-1-yne with sulfuric acid and mercury(II) sulfate.

HYDRATION OF ALKYNES

Hydration of alkynes continues the trend of mirroring previous alkene chemistry, while introducing interesting differences. For example, while alkenes can be readily hydrated with strong, mineral acids, like H_2SO_4, in water, alkynes cannot be hydrated under the same conditions due to the relatively high instability of vinyl cations. As such, the hydration of alkynes necessitates the addition of a water-soluble Lewis acid $(HgSO_4)$[5] along with sulfuric acid and water (Figure 13.8). The reaction with mercury(II) sulfate is presented here for the same reason as oxymercuration-reduction in Chapter 12, it cements some fundamental reactivity patterns, but this reaction suffers from the same drawback (it utilizes a toxic mercury compound) and the added disadvantage that it uses strongly acidic conditions. Other, safer Lewis acids are more commonly used today, including gold(III) compounds.[6] It should be noted that instead of a diol product being produced, a ketone is the final product.

Problem 13.11 Provide the electron-pushing arrows for the mechanism. Here the entire mechanism is presented. Note any similarities to previous mechanisms.

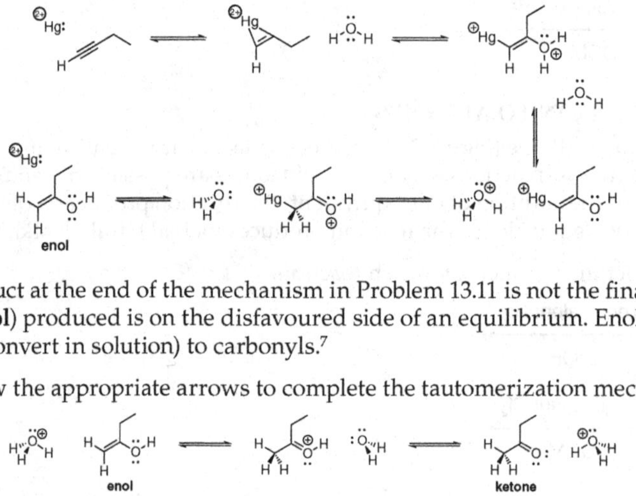

Note that the product at the end of the mechanism in Problem 13.11 is not the final product. The enol (alkene-alcohol) produced is on the disfavoured side of an equilibrium. Enols rapidly tautomerize (that is interconvert in solution) to carbonyls.[7]

Problem 13.12 Draw the appropriate arrows to complete the tautomerization mechanism.

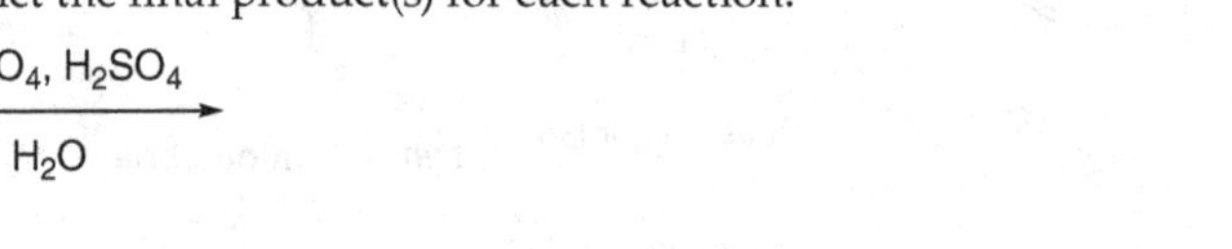

Problem 13.13 Predict the final product(s) for each reaction.

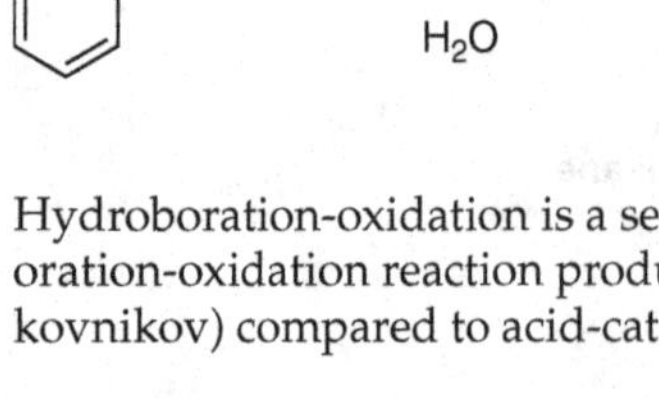

Hydroboration-oxidation is a second method for hydrating alkynes. As before, the hydroboration-oxidation reaction produces products with the opposite regiochemistry (anti-Markovnikov) compared to acid-catalyzed hydration reactions (Figure 13.9). The regiochemistry of

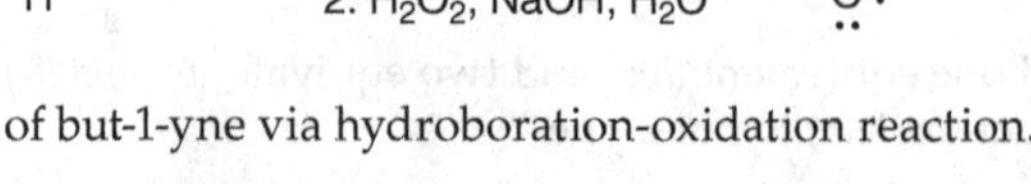

Figure 13.9 Hydration of but-1-yne via hydroboration-oxidation reaction.

H
B
9-borabicyclo[3.3.1]nonane
9-BBN
H–B
O
catecholborane
(HBcat)
H–B
disiamylborane
(HB(sia)2)

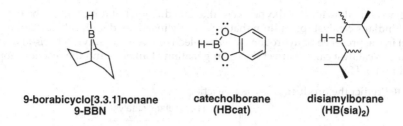

Figure 13.10 Bulky boranes used to produce anti-Markovnikov hydrations of an alkyne.

hydroboration-oxidation reactions is driven by steric repulsion (see Chapter 11) and alkynes, due to their linear shape, are generally less sterically encumbered. As such, to achieve good selectivity, bulkier boranes (Figure 13.10) are used with alkynes.[8] The key points of the mechanism were seen in the reaction of borane with alkenes: hydrogen and borane are added simultaneously to the same face with the larger borane on the less substituted carbon, and subsequent oxidation with peroxide converts the C–BR_2 group into a C–OH. Finally, just like the reaction of an alkyne with mercury(II) sulfate and sulfuric acid, the initial enol product rapidly tautomerizes to a carbonyl.

Problem 13.14 Predict the final product(s) for each reaction.

1. HB(sia)2, THF
2. H2O2, NaOH, H2O
1. HB(sia)2, THF
2. H2O2, NaOH, H2O

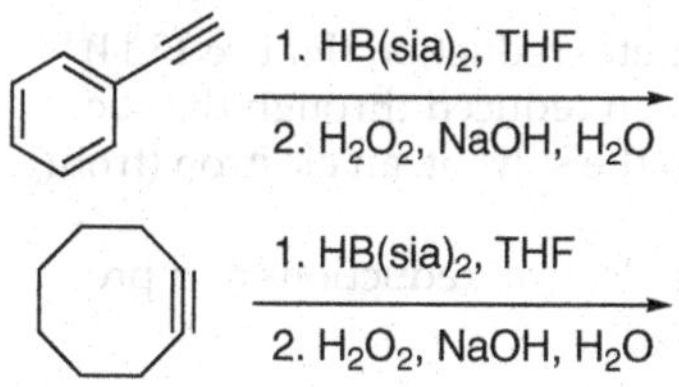

REDUCTION OF ALKYNES

Just as with alkenes, alkynes can be reduced with the addition of H_2 (Figure 13.11). Reduction of an alkyne can selectively produce *cis*-alkenes, *trans*-alkenes, or alkanes depending upon conditions. Hydrogenation[9] of an alkyne reduces the two C≡C π bonds and converts the alkyne to an alkane. The reaction proceeds on the surface of a heterogeneous, metal catalyst and the hydrogen atoms are added in two, successive *syn* additions.

Problem 13.15 Predict the product(s) for each reaction.

H2, Ni
ethanol
H2, Pd/C
methanol

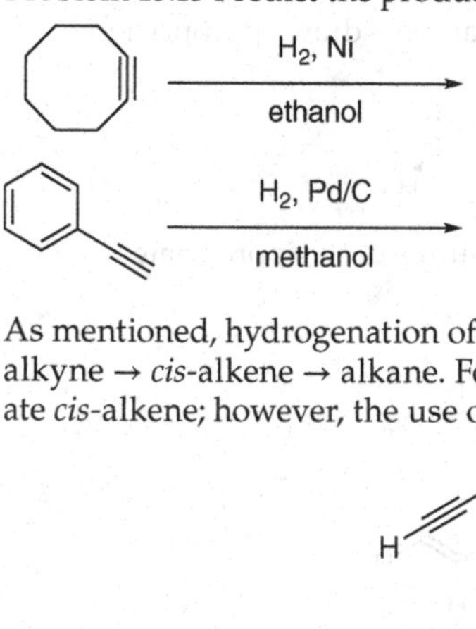

As mentioned, hydrogenation of an alkyne produces alkanes through two, successive reductions: alkyne → *cis*-alkene → alkane. For most catalysts, the reaction is too rapid to stop at the intermediate *cis*-alkene; however, the use of a poisoned catalyst (a catalyst spiked with additives to make it

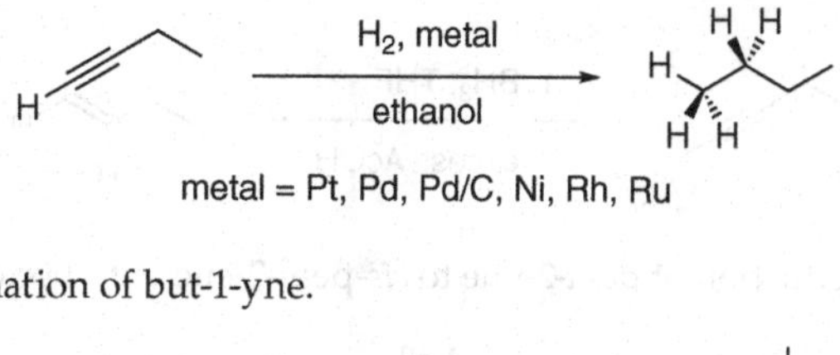

H
H2, metal
ethanol
H H
H
H H
metal = Pt, Pd, Pd/C, Ni, Rh, Ru

Figure 13.11 Hydrogenation of but-1-yne.

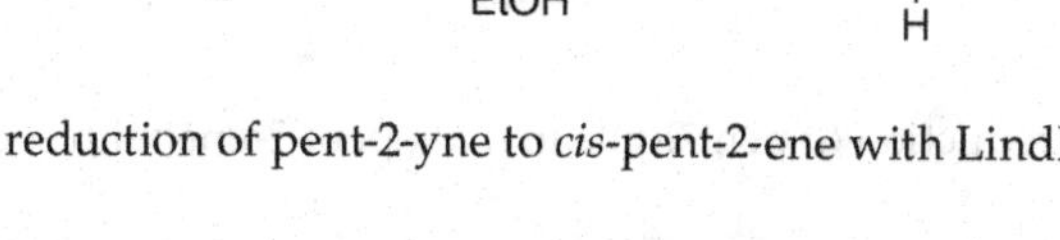

H2, Lindlar's catalyst
EtOH
H
H

Figure 13.12 Selective reduction of pent-2-yne to *cis*-pent-2-ene with Lindlar's catalyst.

less reactive) would allow for the alkyne → *cis*-alkene reduction but not catalyze the reduction to the alkane. Lindlar's catalyst (palladium that is partially inactivated with calcium carbonate and lead acetate) in the presence of hydrogen gas will selectively convert alkynes to *cis*-alkenes (Figure 13.12).[10] Lindlar's catalyst cannot affect the hydrogenation of alkenes and the reaction stops after the first addition of H_2.

Problem 13.16 Predict the product(s) for each reaction.

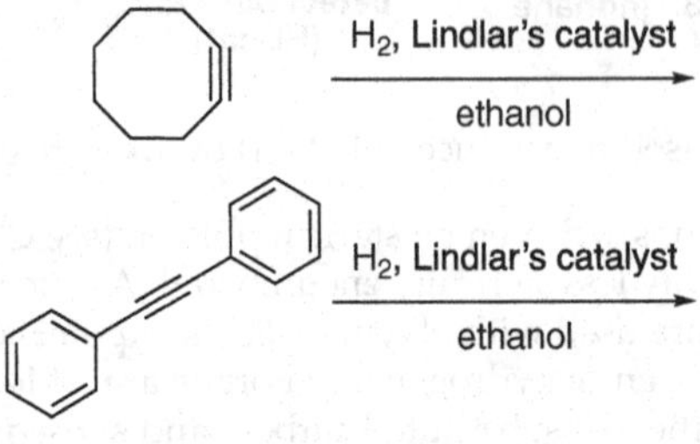

Cis-alkenes can also be produced selectively (Figure 13.13) by using borane (BH_3).[11] Borane adds across one of the π-bonds as expected (see Chapter 12). The addition of excess acetic acid (AcOH) then replaces the boron with a hydrogen atom. The mechanism for the C–B → C–H transformation is beyond the scope of this chapter.

Trans-alkenes are produced in a reaction known as a dissolving metal reduction (Figure 13.14).[12] Here sodium metal is dissolved in liquid ammonia. The alkyne is then reduced through the successive transfer of an electron (from the sodium), a proton (H^+) from the solvent, an electron (from another sodium) and a proton (from another solvent molecule).

The mechanism (in Problem 13.17) is a series of successive, single-electron reductions and protonations of the resulting anions.

Problem 13.17 Answer the following questions about the mechanism as directed.[13]

Step one: Single-electron reduction of the alkyne by sodium to generate a radical anion with *E* geometry. While the radical anion can interconvert between *E* and *Z* isomers, the *E* isomer is more stable and therefore the favoured isomer during the reaction. Draw appropriate electron-pushing arrows to complete the mechanism.

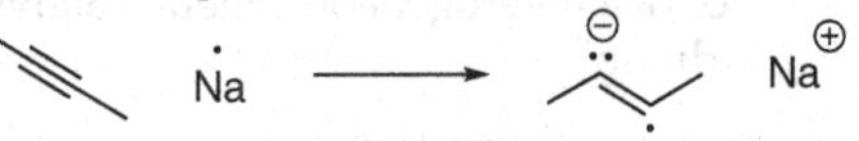

Step two: Deprotonation of the ammonia solvent by the radical anion – draw appropriate electron-pushing arrows to complete the mechanism.

Step three: Single-electron reduction of the vinyl radical by sodium – draw appropriate electron-pushing arrows to complete the mechanism.

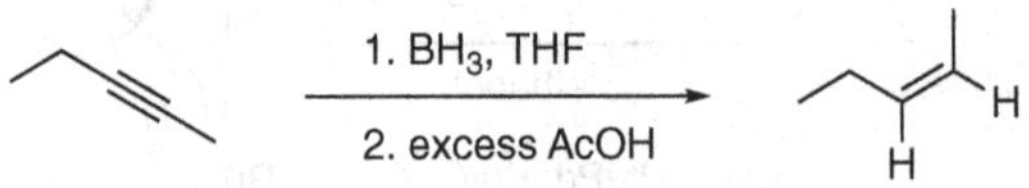

Figure 13.13 Selective reduction of pent-2-yne to *cis*-pent-2-ene with borane.

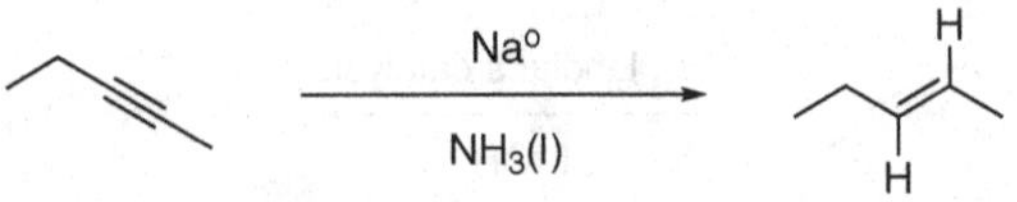

Figure 13.14 Selective reduction of pent-2-yne to *trans*-pent-2-ene with sodium and ammonia.

138

Step four: Deprotonation of the ammonia solvent by the vinyl anion – draw appropriate electron-pushing arrows to complete the mechanism.

Problem 13.18 Predict the product(s) for each reaction below.

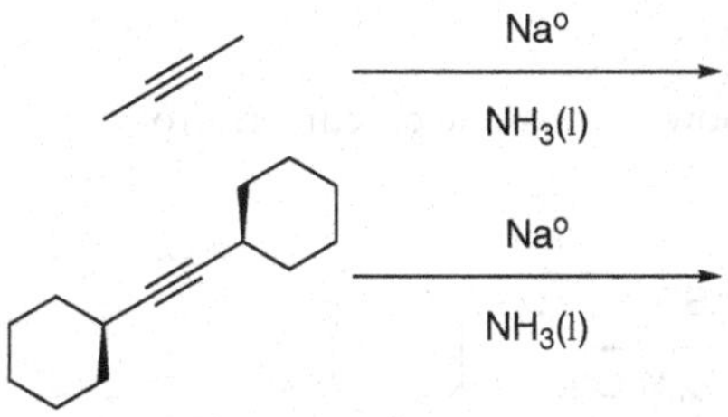

$$\text{Na}^\circ$$
$$\xrightarrow{\hspace{3cm}}$$
$$\text{NH}_3(l)$$

$$\text{Na}^\circ$$
$$\xrightarrow{\hspace{3cm}}$$
$$\text{NH}_3(l)$$

RETROSYNTHESIS

Retrosynthesis is a way of thinking about designing a synthesis.[14] If one is provided with a target molecule, then the question is, "Knowing what I know about organic reaction chemistry, what possible precursor could this molecule have been created from?" Then once all possible precursors have been evaluated, the question is then, "From this molecule what possible precursors could have been started with?" Retrosynthesis uses its own arrow to indicate that the relationship between molecules is the molecule after the arrow is a synthetic precursor of the molecule before the arrow (Figure 13.16). The question is then repeated until one has worked backwards to a reasonable starting material. Consider the following example in Figure 13.15.

As the problem solver, the task then is to try and establish a way of synthesizing cyclooctanone from *cis*-cyclooctene. Retrosynthetic analysis will be employed. At this point, the only way that has been seen to make carbonyls is from alkynes. Cyclooctyne would be proposed as a synthetic precursor (Figure 13.16).

Cyclooctyne, however, is not *cis*-cyclooctene, so the analysis continues (Figure 13.17). Two processes have been seen for making alkyne functional groups: from acetylides and from alkenes, both are considered.

At this point, the retrosynthetic analysis has provided with a route that gets from the available precursor to the target molecule. Using this route, a forward synthesis is proposed in Figure 13.18.

Figure 13.18 is not the only possible forward synthesis; the synthesis shown in Figure 13.19 is also a reasonable proposal based on Figure 13.17.

Target **Available starting material**

Figure 13.15 Desired synthetic target and available starting material from which to synthesize it.

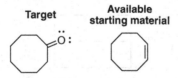

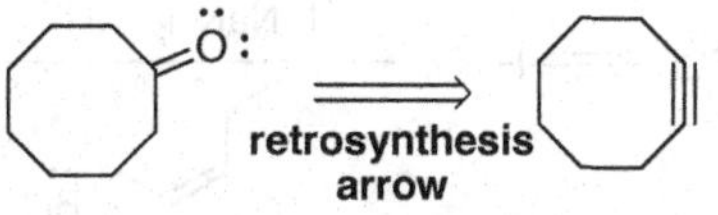

Figure 13.16 A retrosynthetic step showing a possible precursor (as indicated by the ⇒ arrow), from which we could make the molecule on the left.

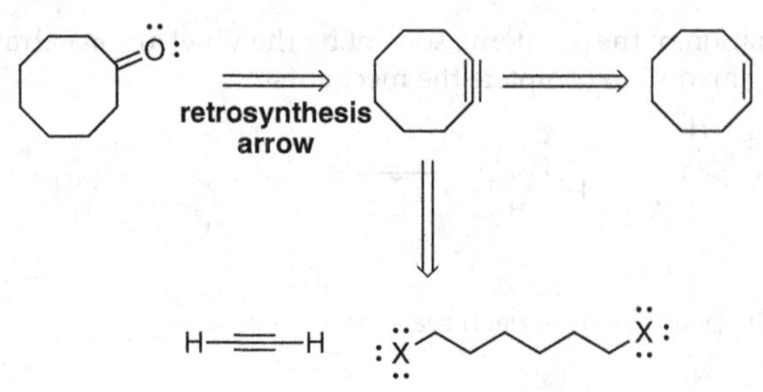

Figure 13.17 An expansion of the retrosynthetic analysis showing possible precursors, from which cyclooctyne could be synthesized.

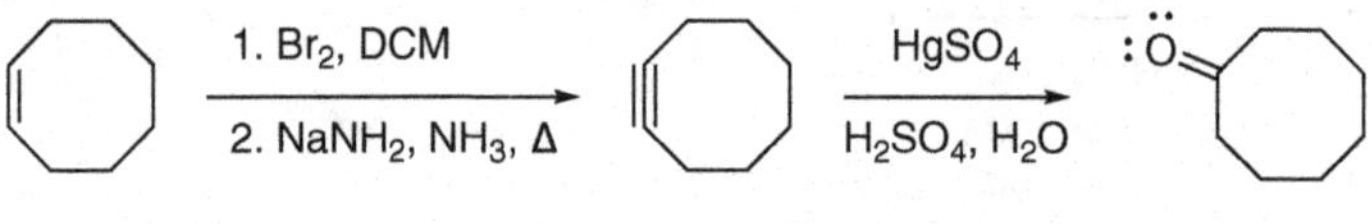

Figure 13.18 A forward synthesis based on the retrosynthetic analysis in Figure 13.17.

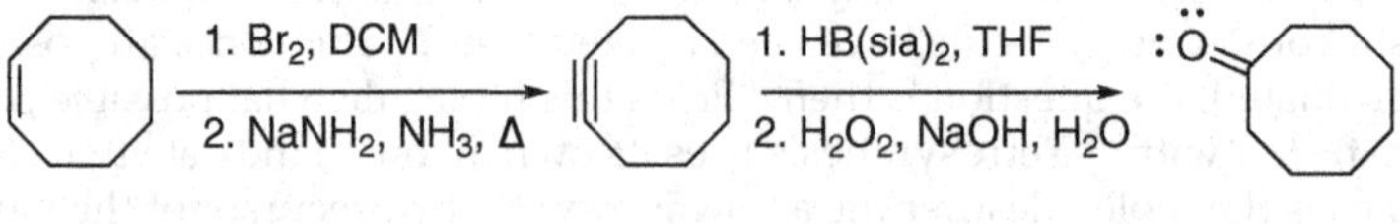

Figure 13.19 A forward synthesis based on the retrosynthetic analysis in Figure 13.17.

Problem 13.19 Propose a full retrosynthetic analysis (on the next page) for making the 1-heptanol from acetylene and 1-bromopentane.

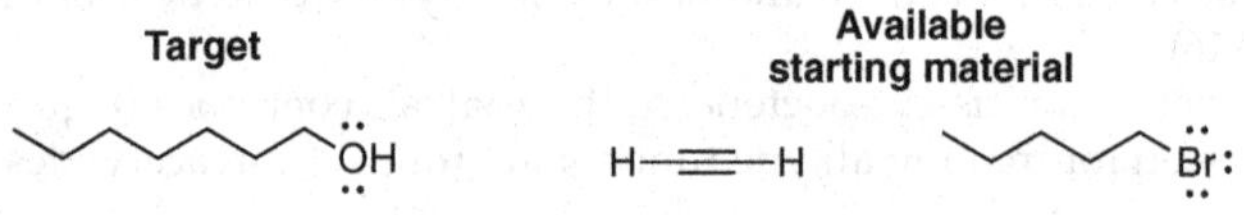

Target	Available starting material

Retrosynthetic analysis:

Now propose a forward synthesis based upon the retrosynthetic analysis:

GENERAL PRACTICE PROBLEMS

Problem 13.20 Provide the reagent(s) necessary to complete the reaction.

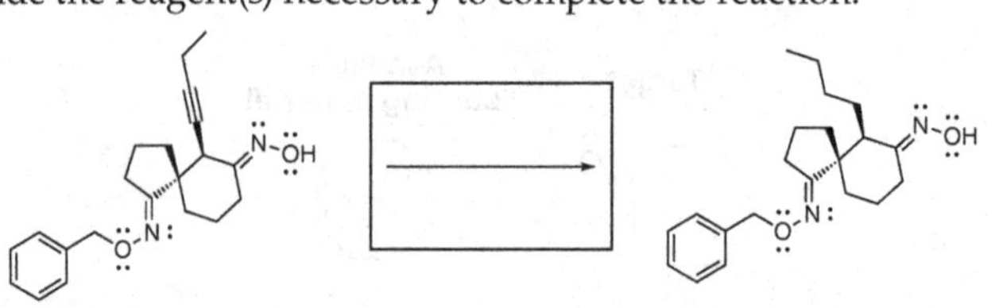

Problem 13.21 Predict the product(s).

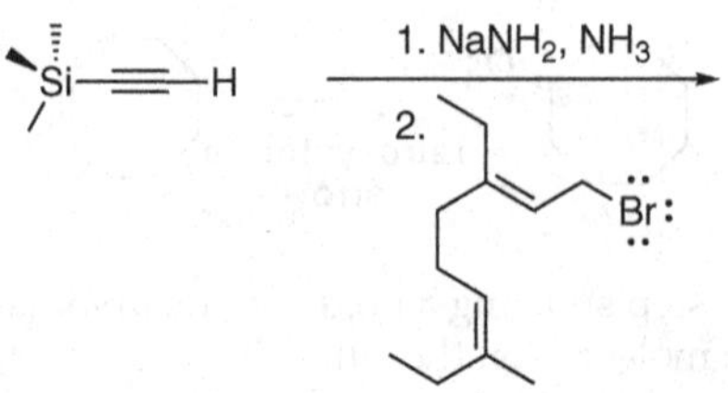

NOTES

1. Jones, E.R.H. Acetylene and acetylenic compounds in organic synthesis. *J. Chem. Soc.*, **1950**, 754–761. DOI: 10.1039/JR9500000754

2. Calculated relative stability, which is in excellent agreement with experimental gas-phase hydride affinities: Anslyn, E.V. Dougherty, D.A. Strain and Stability in *Modern Physical Organic Chemistry*. University Science Books: Sausalito, CA, 2006. pp. 65–145.

3. i) Fahey, R.C.; Lee, D.J. Polar additions to olefins and acetylenes. IV. Evidence for synchronous C-H and C-Cl bond formation in the *trans* addition of hydrogen chloride to 3-hexyne. *J. Am. Chem. Soc.*, **1967**, *89* (11), 2780–2781. DOI: 10.1021/ja00987a076
   ii) Weiss, H.M.; Touchette, K.M. Bromide assisted addition of hydrogen bromide to alkynes and allenes. *J. Chem. Soc., Perkin Trans. 2*, **1998**, 1523–1528. DOI: 10.1039/A703569A

4. Berman, D.; Anicich, V.; Beauchamp, J. Stabilities of isomeric halonium ions $C_2H_4X^+$ (X = Cl, Br) by photoionization mass spectrometry and ion cyclotron resonance spectroscopy. General considerations of the relative stabilities of cyclic and acylic isomeric onium ions. *J. Am. Chem. Soc.*, **1979**, *101* (5), 1239–1248. DOI: 10.1021/ja00499a032

5. Vogt, R.R.; Nieuwland, J.A. The role of mercury salts in the catalytic transformation of acetylene into acetaldehyde, and a new commercial process for the manufacture of paraldehyde. *J. Am. Chem. Soc.*, **1921**, *43* (9), 2071–2081. DOI: 10.1021/ja01442a010

6. Fukuda, Y.; Utimoto, K. Effective transformation of unactivated alkynes into ketones or acetals with gold(III) catalyst. *J. Org. Chem.*, **1991**, *56* (11), 3729–3731. DOI: 10.1021/jo00011a058

7. In general, alcohols on double bonds rapidly tautomerize into the corresponding carbonyl. This was originally known as Erlenmeyer's rule following his discovery of this trend: 1) Erlenmeyer, E. Ueber Phenylbrommilchsäure. *Ber. Dtsch. Chem. Ges.*, **1880**, *13* (1), 305–310. DOI: 10.1002/cber.18800130187. 2) Erlenmeyer, E. Verhalten der Glycerinsäure und der Wainsäure gegen wasserentziehende Substanzen. *Ber. Dtsch. Chem. Ges.*, **1881**, *14* (1), 320–323. DOI: 10.1002/cber.18810140173

8. Brown, H.C.; Gupta, S.K. Catecholborane (1,3,2-benzodioxaborole) as a new, general monohydroboration reagent for alkynes. A convenient synthesis of alkeneboronic esters and acids from alkynes via hydroboration. *J. Am. Chem. Soc.*, **1972**, *94* (12), 4370–4371. DOI: 10.1021/ja00767a072

9. Sabatier, P.; Senderens, J.-B. Action du nickel sur l'éthylène. Synthèse de l'éthane. *Comptes Rendus Acad. Sci. Paris*, **1897**, *124*, 1358–1360.

10. Lindlar, H. Ein neuer Katalysator für selective Hydrierungen. *Helv. Chim. Acta*, **1952**, *35* (2), 446–450. DOI: 10.1002/hlca.19520350205

11. Corey, E.J.; Kang, J. Short, stereocontrolled syntheses of irreversible eicosanoid biosynthesis inhibitors. 5,6-, 8,9-, and 11,12-dehydroarachidonic acid. *Tetrahedron Lett.*, **1982**, *23* (16), 1651–1654. DOI: 10.1016/s0040-4039(00)87181-6

12. i) Kraus, C.; White, G. Reactions of strongly electropositive metals with organic substances in liquid ammonia solution. I. Preliminary investigations. *J. Am. Chem. Soc.*, **1923**, *45* (3), 768–777. DOI: 10.1021/ja01656a032

    ii) Campbell, K.N.; Eby, L.T. The preparation of higher *cis* and *trans* olefins. *J. Am. Chem. Soc.*, **1941**, *63* (1), 216–219. DOI: 10.1021/ja01846a050

13. Damrauer, R. Dissolving metal reduction of acetylenes: A computational study. *J. Org. Chem.*, **2006**, *71* (24), 9165–9171. DOI: 10.1021/jo061583j

14. Corey, E.J. Robert Robinson Lecture. Retrosynthetic thinking – essentials and examples. *Chem. Soc. Rev.*, **1988**, *17*, 111–133. DOI: 10.1039/CS9881700111

14 Nucleophilic Substitution and β-Elimination Reactions

Initially, let us consider the reaction of fluoride with hydrogen bromide (Figure 14.1). In the Brønsted-Lowry terminology, fluoride is a base that is protonated by hydrogen bromide, the acid. This produces hydrogen fluoride, the conjugate acid, and bromide, the conjugate base. Utilizing pK_a values, the favorability (K_c) of the reaction can be calculated. Here $K_c = 10^{12.2}$. The reaction is favoured in the forward reaction because bromide is a weaker base (is a more stable anion due to its larger atomic size) than fluoride.

The same reaction can also be described using Lewis acid-base terminology (Figure 14.2): fluoride is a Lewis base or nucleophile (Nuc) that attacks hydrogen of hydrogen bromide, the electrophile or Lewis acid. This produces hydrogen fluoride, the substitution product, and bromide, the leaving group (LG).

Regardless of the terminology employed to describe the reaction, the reaction occurs in a bimolecular step: The fluoride lone pair donates its electrons to the H–Br σ* orbital, which breaks the H–Br bond (Figure 14.3). This orbital alignment requires that the fluoride approach the backside of the H–Br bond and, by filling the σ* orbital.

This concerted, bimolecular reaction proceeds through a single transition state where the fluoride-hydrogen bond forms as the hydrogen-bromide bond breaks (Figure 14.4).

This type of reaction is called a S_N2 reaction. The nomenclature breaks down as follows: S stands for substitution, $_N$ for nucleophilic, and 2 for bimolecular. This is a common type of reaction that can occur with carbon, phosphorous and sulfur electrophiles too. This chapter will focus on carbon electrophiles such as methyl bromide (Figure 14.5).

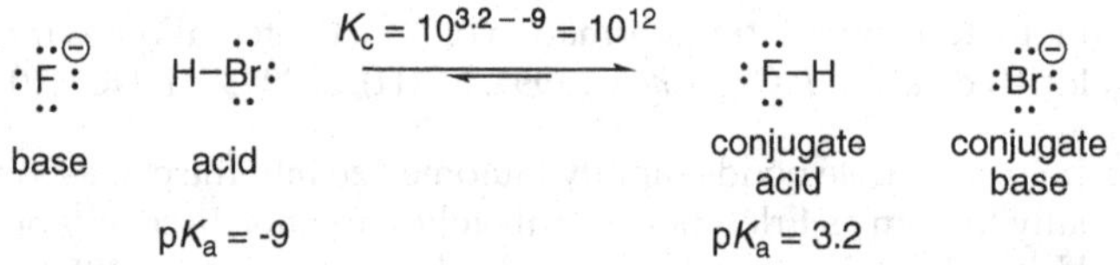

Figure 14.1 Reaction of fluoride with hydrogen bromide with Brønsted-Lowry acid-base terminology.

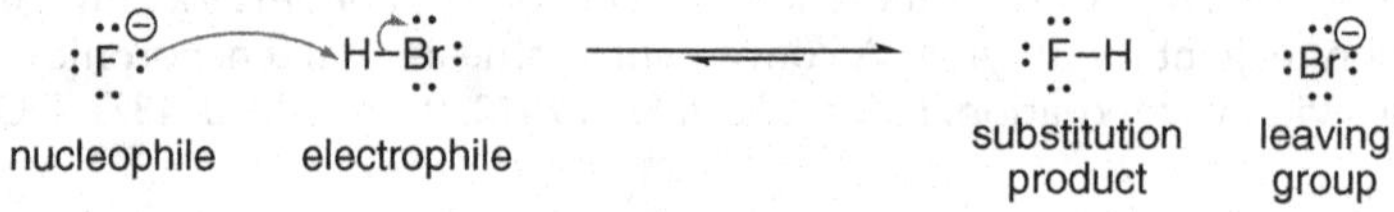

Figure 14.2 Reaction of fluoride with hydrogen bromide with Lewis acid-base terminology.

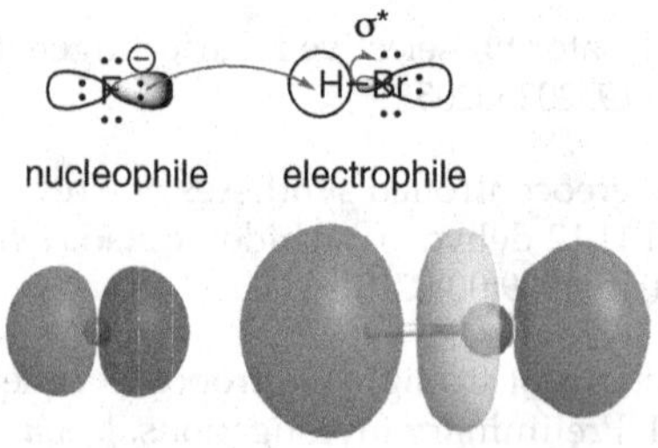

Figure 14.3 Structure and 3D model of fluoride lone-pair orbital donating its electrons to σ^*_{HBr}.

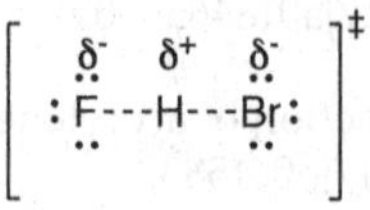

Figure 14.4 Transition state showing the F–H bond making and H–Br bond breaking.

DOI: 10.1201/9781003479352-14

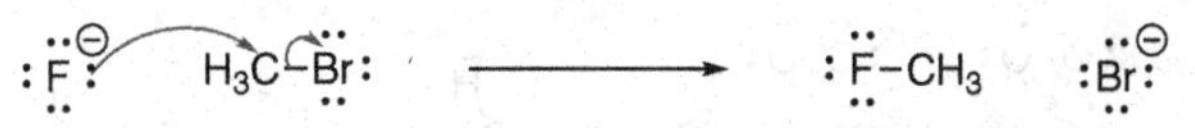

$$:\ddot{F}: \quad H_3C-\ddot{Br}: \quad \longrightarrow \quad :\ddot{F}-CH_3 \quad :\ddot{Br}:$$

Figure 14.5 S_N2 reaction between fluoride and methyl bromide.

Problem 14.1 Draw the arrows for all the following S_N2 reactions showing some of the most common nucleophiles.

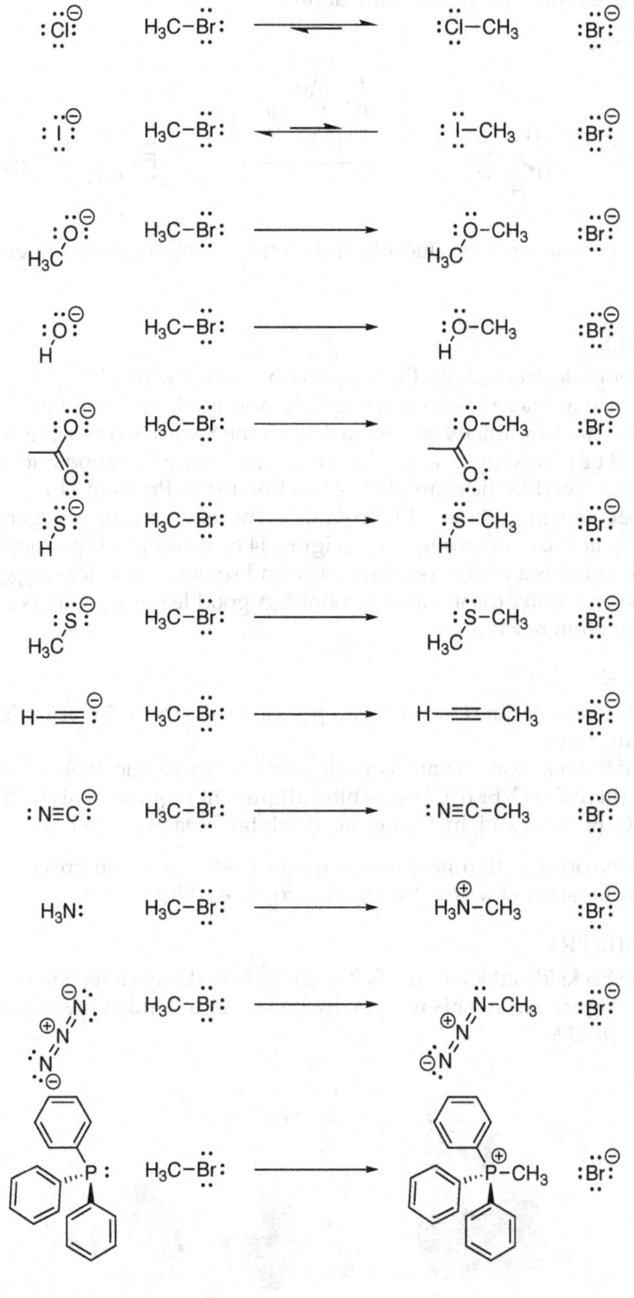

$$:\ddot{Cl}: \quad H_3C-\ddot{Br}: \quad \rightleftharpoons \quad :\ddot{Cl}-CH_3 \quad :\ddot{Br}:$$

$$:\ddot{I}: \quad H_3C-\ddot{Br}: \quad \rightleftharpoons \quad :\ddot{I}-CH_3 \quad :\ddot{Br}:$$

$$\underset{H_3C}{:\ddot{O}:} \quad H_3C-\ddot{Br}: \quad \longrightarrow \quad \underset{H_3C}{:\ddot{O}-CH_3} \quad :\ddot{Br}:$$

$$\underset{H}{:\ddot{O}:} \quad H_3C-\ddot{Br}: \quad \longrightarrow \quad \underset{H}{:\ddot{O}-CH_3} \quad :\ddot{Br}:$$

$$:\ddot{O}: \quad H_3C-\ddot{Br}: \quad \longrightarrow \quad :\ddot{O}-CH_3 \quad :\ddot{Br}:$$

$$\underset{H}{:\ddot{S}:} \quad H_3C-\ddot{Br}: \quad \longrightarrow \quad \underset{H}{:\ddot{S}-CH_3} \quad :\ddot{Br}:$$

$$\underset{H_3C}{:\ddot{S}:} \quad H_3C-\ddot{Br}: \quad \longrightarrow \quad \underset{H_3C}{:\ddot{S}-CH_3} \quad :\ddot{Br}:$$

$$H-\!\!\!\equiv\!\!\!: \quad H_3C-\ddot{Br}: \quad \longrightarrow \quad H-\!\!\!\equiv\!\!\!-CH_3 \quad :\ddot{Br}:$$

$$:N\equiv C: \quad H_3C-\ddot{Br}: \quad \longrightarrow \quad :N\equiv C-CH_3 \quad :\ddot{Br}:$$

$$H_3N: \quad H_3C-\ddot{Br}: \quad \longrightarrow \quad \overset{\oplus}{H_3N}-CH_3 \quad :\ddot{Br}:$$

$$H_3C-\ddot{Br}: \quad \longrightarrow \quad :\ddot{Br}:$$

$$P: \quad H_3C-\ddot{Br}: \quad \longrightarrow \quad \overset{\oplus}{P}-CH_3 \quad :\ddot{Br}:$$

Attention will now be paid to key aspects of the S_N2 reaction, specifically the nature of the nucleophile, the leaving group, the nature of the carbon electrophile, and the reaction conditions.

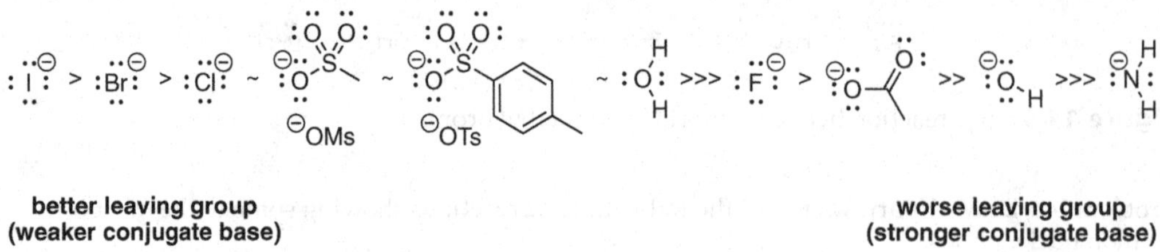

Figure 14.6 Relative order of leaving group ability.

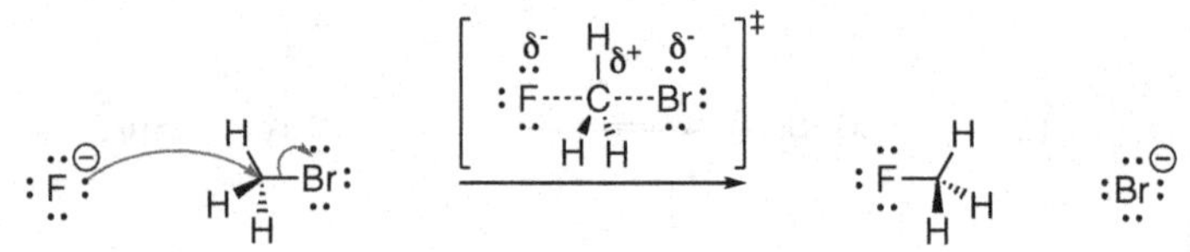

Figure 14.7 SN2 reaction between fluoride and methyl bromide showing pentavalent transition state.

S$_N$2 NUCLEOPHILES

In terms of the nucleophile, the S$_N$2 reaction requires a strong nucleophile. Nucleophilicity is a kinetic term, that is strong nucleophiles react quickly, and weak nucleophiles react slowly. Several different, quantitative nucleophilicity scales exist, with the Swain-Scott[1] being the most common. Herein emphasis will be placed upon a qualitative understanding of good nucleophiles, of which an extensive list is presented in the example S$_N$2 reaction list in Problem 14.1.

In the example reactions in Problem 14.1, bromide is the leaving group. In general, good leaving groups are the conjugate bases of strong acids (Figure 14.6). Good leaving groups are halides (I⁻, Br⁻, and Cl⁻; F⁻, although it is a halide, is rather basic and so not a good leaving group) and pseudo-halides (⁻OMs, ⁻OTs) and water (protonated alcohols). A good leaving group is necessary for any substitution or elimination reaction.

BACKSIDE ATTACK

Just as was seen in the case of fluoride and hydrogen bromide, the S$_N$2 mechanism with carbon occurs in a bimolecular step.

Over the course of the reaction, the nucleophile lone pair (here fluorine) donates its electrons to the C–LG σ* orbital (here the C–Br σ*). This orbital alignment requires that the fluoride approach the backside of the C–Br bond and, by filling the σ* orbital, breaks the C–Br bond (Figure 14.8).

Problem 14.2 Given the orbital alignment necessary for the S$_N$2 reaction provide an explanation for the relative rates of reaction,[2] with different electrophiles (Figure 14.9).

S$_N$2 STEREOCHEMISTRY

A direct effect of the backside attack in the S$_N$2 reaction is that any stereochemistry is inverted during the reaction (Figure 14.10). This is formally known as a Walden inversion after Paul Walden who discovered this in 1896.[3]

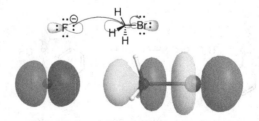

Figure 14.8 Structure and 3D model of fluoride lone-pair orbital donating its electrons to the σ*$_{CBr}$ orbital.

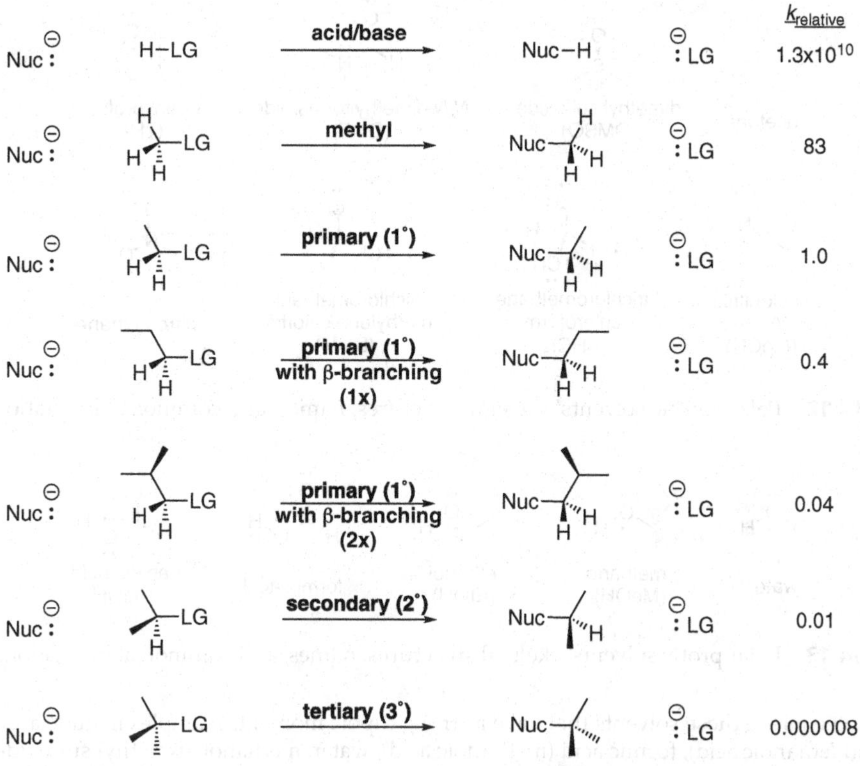

Figure 14.9 Relative rates of S_N2 substitution with different electrophiles.

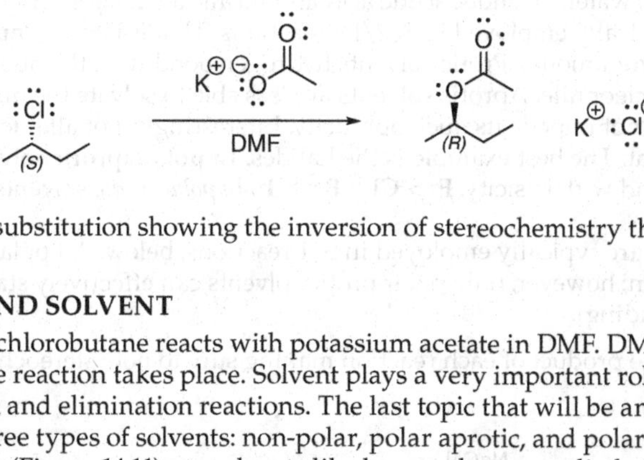

Figure 14.10 S_N2 substitution showing the inversion of stereochemistry that occurs.

S_N2 REACTIONS AND SOLVENT

In Figure 14.10, (S)-2-chlorobutane reacts with potassium acetate in DMF. DMF is a solvent, the medium in which the reaction takes place. Solvent plays a very important role in the competition between substitution and elimination reactions. The last topic that will be analyzed is the role of solvent. There are three types of solvents: non-polar, polar aprotic, and polar protic.

Non-polar solvents (Figure 14.11) are solvents like hexane, benzene, diethyl ether, carbon tetrachloride, and carbon disulfide. Non-polar solvents are useful in radical reactions (which proceed through non-polar intermediates) but not very useful in substitution or elimination reactions.

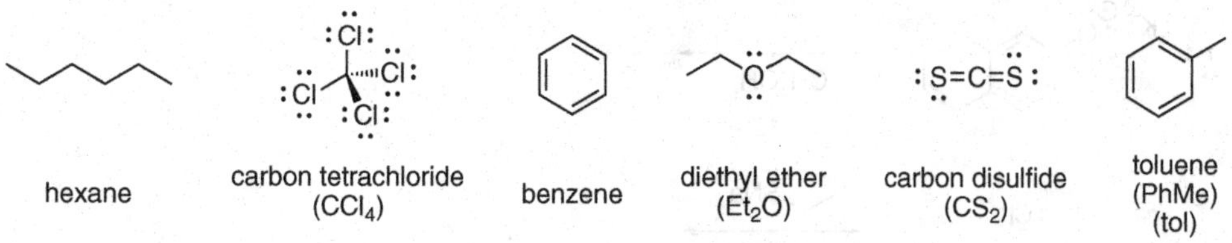

Figure 14.11 Non-polar solvents' skeletal structures, names, and common abbreviations.

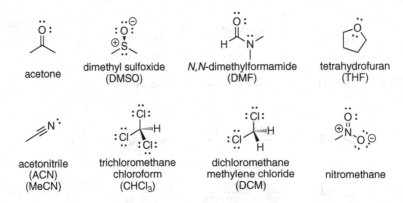

Figure 14.12 Polar aprotic solvents' skeletal structures, names, and common abbreviations.

water   methanol   ethanol   formic acid   acetic acid
         (MeOH)   (EtOH)   (AcOH)

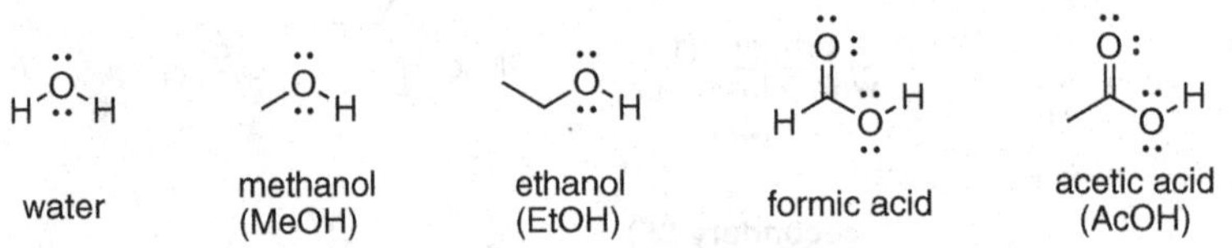

Figure 14.13 Polar protic solvents' skeletal structures, names, and common abbreviations.

Polar solvents are those solvents that have a strong dipole moment, examples include acetone, acetic acid (ethanoic acid), formic acid (methanoic acid), water, methanol, dimethyl sulfoxide (DMSO), N,N-dimethylformamide (DMF), tetrahydrofuran (THF), acetonitrile (ACN), dichloromethane (DCM). Polar solvents can be further classified as aprotic (those that are not hydrogen-bond donors: acetone, DMSO, DMF, THF, ACN, DCM, Figure 14.12) and protic (those that are hydrogen-bond donors: water, alcohols, acetic acid and formic acid, Figure 14.13).

Polar aprotics are typically employed in S_N2/E2 reactions. That is because nucleophiles are typically electron-rich and/or anionic. Protic solvents hydrogen bond with the nucleophile and hinder its ability to act as a nucleophile. Aprotic solvents are less able to solvate the nucleophile, rending it "naked" and thus do not impede its nucleophilicity. Interestingly, not all nucleophiles are slowed down to the same extent. The best example is the halides. **In polar aprotic** solvents, the nucleophilicity of the halides trend with basicity, **F^- > Cl^- > Br^- > I^-**. *In polar protic* solvents, the trend reverses, *I^- > Br^- > Cl^- > F^-*.

Polar protic solvents are typically employed in S_N1 reactions, below. All polar solvents can help stabilize the carbocation; however, only polar protic solvents can effectively stabilize the anion (through hydrogen bonding).

Problem 14.3 Predict the product of each reaction making sure to note stereochemistry where appropriate.

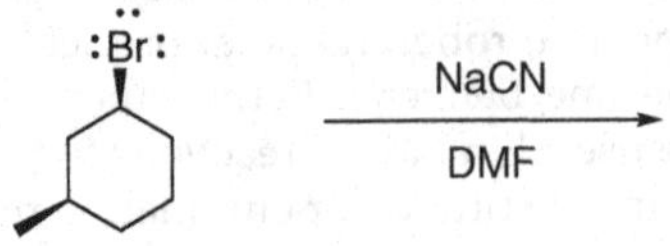

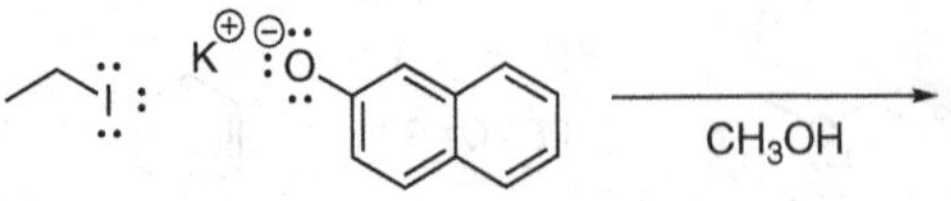

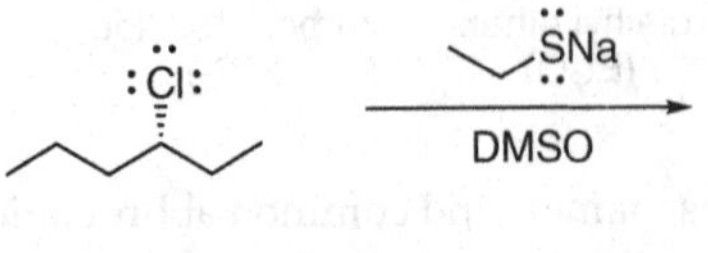

BIMOLECULAR SUBSTITUTION AND ELIMINATION COMPETITION

Consider the following series of reactions (Figure 14.14) with fluoride as a nucleophile, a strong and weakly basic (pK_a(HF) = 3.2) nucleophile, with different carbon electrophiles. Fluoride readily substitutes bromide on all carbon electrophiles except for the tertiary alkyl bromide. Under normal circumstances, tertiary carbon electrophiles do not undergo S_N2 reactions.

Now contrast the Figure 14.14 examples of reactions, where fluoride (a strong and weakly basic nucleophile) is used, with the reactions in the list below (Figure 14.15). Here methoxide is used as the nucleophile.[4] Both fluoride and methoxide are strong nucleophiles (they react quickly), but unlike fluoride methoxide is a strongly basic nucleophile (pK_a $HOCH_3$ is 16).

For methyl, primary, and primary with a single β-branch carbon electrophiles, the strongly basic nucleophile methoxide reacts by the expected S_N2 pathway. For primary with two β-branches, secondary, and tertiary carbon electrophiles, the rate of the S_N2 reaction slows down (see page 173) and so strongly basic nucleophiles react by what is called an E2 mechanism, a type of β-elimination. Strongly basic nucleophiles are generally defined as those nucleophiles whose conjugate acids have pK_a values greater than 11.

β–Elimination is the competing counterpart to nucleophilic substitution. All nucleophiles have a lone pair of electrons and can thus act as Brønsted-Lowry bases. In a β–elimination reaction, the nucleophile removes a proton from a β-carbon forming a double bond after loss of the leaving group (Figure 14.16).[5]

When possible, elimination proceeds to give the most stable alkene product (the most substituted alkene product). This outcome was determined empirically by Alexander Mikhaylovich Zaitsev in 1875,[6] and it is called Zaitsev's (in older texts spelt Saytzeff following German transliteration rules) rule in his honour.

The E2 reaction is closely related in mechanism to the S_N2 mechanism, and as pointed out above, in competition with the S_N2 reaction. A strongly basic nucleophile and the carbon electrophile react in a single, concerted step to produce an alkene, leaving group anion (LG⁻) and conjugate acid of our basic nucleophile (Figure 14.17). The E2 nomenclature stands for elimination, bimolecular (bimolecular because there are two molecules involved in the slow step, here the only step).

For methyl and primary carbon electrophiles, the rate of substitution is faster than elimination. For secondary and tertiary electrophiles (or primary with lots of β-branching) the rate of substitution is diminished and E2 reactions typically occur when strongly basic nucleophiles such as alkoxides (RO⁻) and amides (R_2N^-) are used. E2 elimination can be favoured for primary, unbranched carbon electrophiles if a bulky base like *tert*-butoxide is used. The E2 elimination is also selective, when possible, for the more substituted alkene product (Zaitsev's rule).

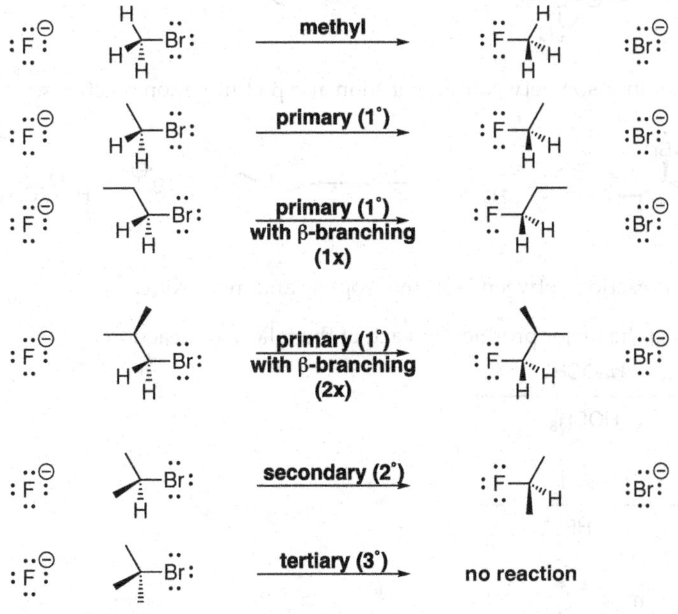

Figure 14.14   Reaction of fluoride with different carbon electrophiles.

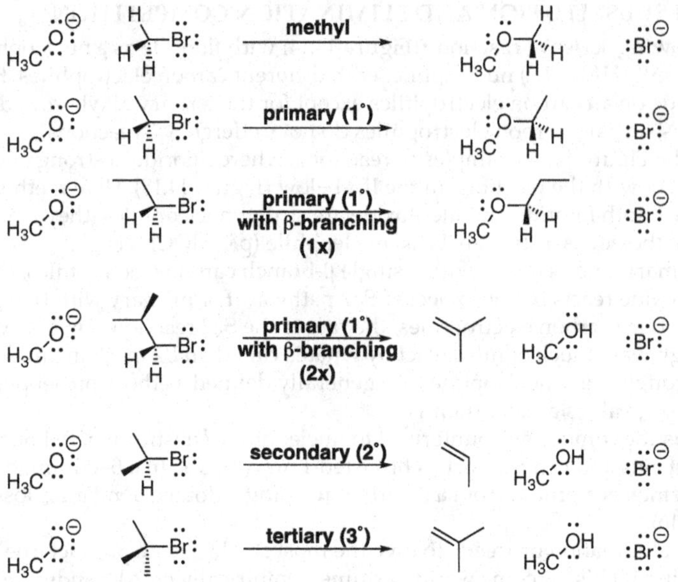

Figure 14.15 Reaction of methoxide with different carbon electrophiles.

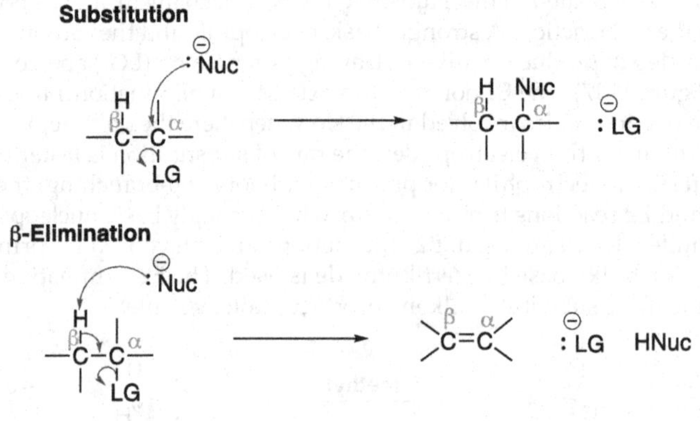

Substitution

β-Elimination

Figure 14.16 Comparison between substitution and β-elimination reactions.

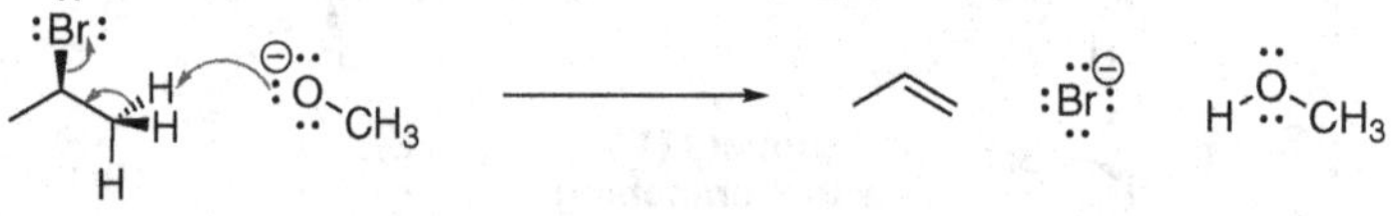

Figure 14.17 E2 reaction between 2-bromopropane and methoxide.

Problem 14.4 Predict the major product for each of the following reactions.

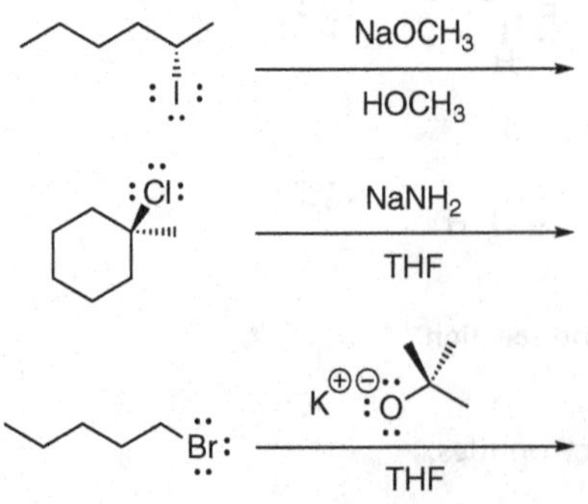

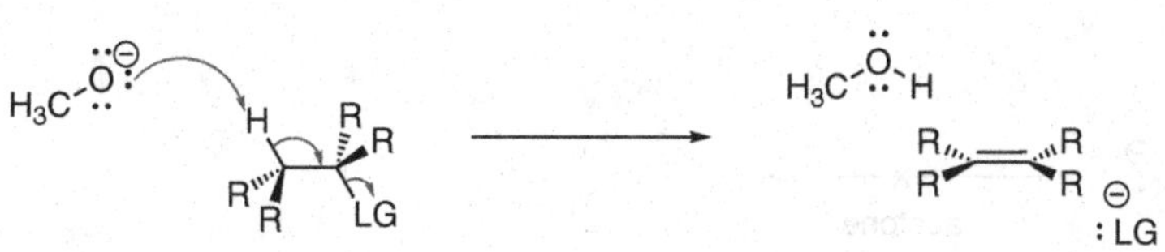

Figure 14.18 The anti-coplanar arrangement of H–C–C–LG required in an E2 reaction.

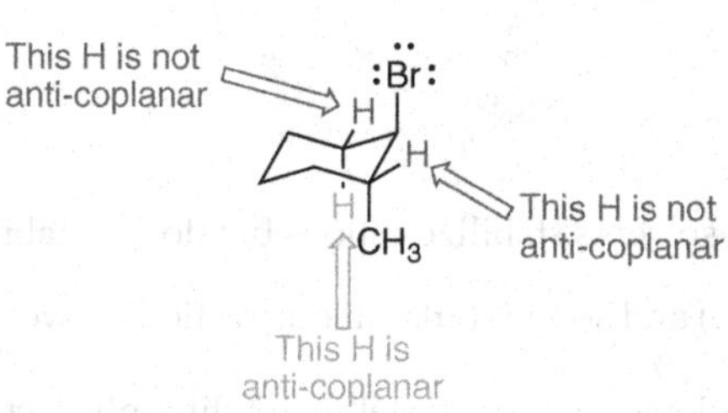

Figure 14.19 Example non-Zaitsev alkene produced by an E2 elimination reaction with a cyclic alkyl halide.

E2 STEREOCHEMISTRY

The E2 elimination does require a specific arrangement of H–C–C–LG atoms and bonds to occur (Figure 14.18). Specifically, H–C–C–LG must all lie in the same plane (that is be coplanar) and H and LG must be anti (that is have 180° dihedral).

To accommodate this stereoselectivity requirement, the E2 reaction will occasionally produce the non-Zaitsev alkene (Figure 14.19).
Rational:

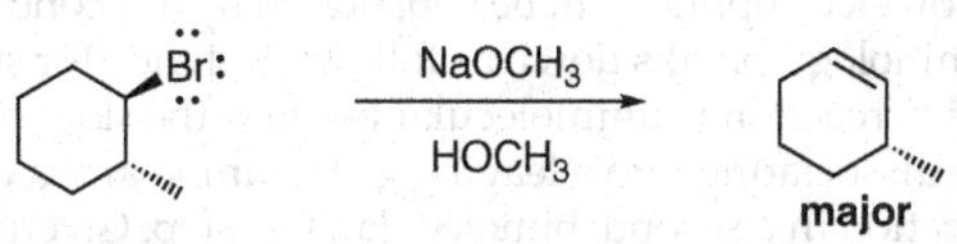

In acyclic systems with high degrees of freedom, deprotonation to form the most substituted alkene is not usually an issue. Only in rings with limited degrees of freedom does the feasibility of different deprotonations need to be considered.

Problem 14.5 Predict the major product(s) of the following E2 eliminations.

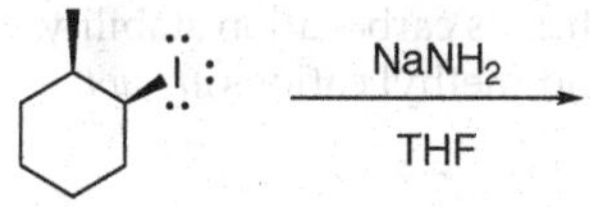

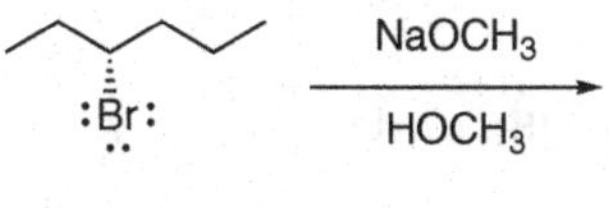

SOLVOLYSIS

To understand the effect of different solvents in their potential impact on reactions, consider the dissolution of sodium bromide. The ionic species sodium bromide will not dissolve in non-polar solvents. Neither the sodium cation nor the bromide anion is stabilized (Figure 14.20) by a

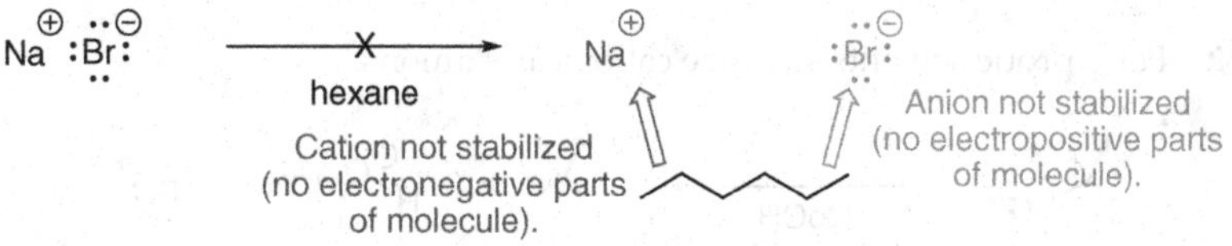

Figure 14.20 Non-polar solvents do not stabilize cations nor anions.

149

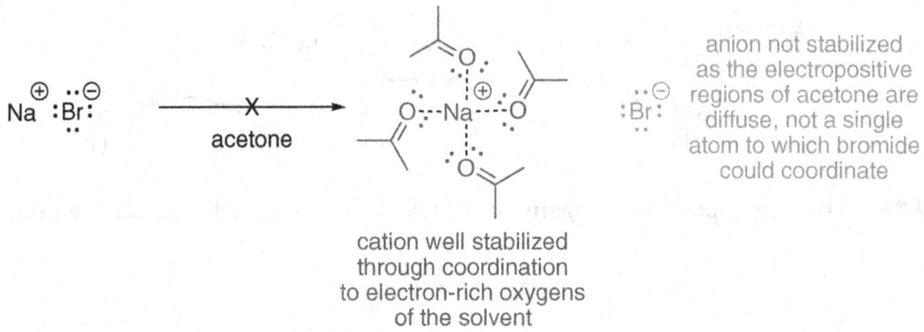

Na⁺ :Br:⁻ →(X) acetone →

anion not stabilized as the electropositive regions of acetone are diffuse, not a single atom to which bromide could coordinate

cation well stabilized through coordination to electron-rich oxygens of the solvent

Figure 14.21   Polar aprotic solvents stabilize cations but do not stabilize anions.

non-polar solvent (like hexane) and so the coulombic attraction between opposite charges keeps the ions together as a solid.

In polar aprotic solvents, cations can be well stabilized through electrostatic attraction between the cation and electronegative portion of the molecule (Figure 14.21). The anion is not well stabilized for most polar aprotic solvents, like acetone, and so typically the unshielded anion strongly attracts the cation. Thus, sodium bromide does not dissolve in most polar aprotic solvents either (DMF and DMSO can dissolve salts to a moderate extent).

Sodium bromide does dissolve well in polar protic solvents: water (94 g/mL at 25 °C), methanol (17 g/mL at 25 °C), ethanol (2 g/mL at 25 °C), and formic acid (19 g/mL at 25 °C). This is because both the anion and cation are stabilized through coordination with the solvent (oxygen-rich atoms to the cation and hydrogen bonds to the anion, Figure 14.22). The coordination with the solvent effectively shields the ions and coulombic attraction is overcome.

For highly substituted carbon electrophiles, the rate of S_N2 can be slowed (2°) or effectively 0 (3°). Secondary and tertiary carbon electrophiles can be substituted by a second mechanistic pathway: the S_N1 reaction. Again, the terminology breaks down as follows: S stands for substitution, $_N$ for nucleophilic, 1 for unimolecular. This reaction is unimolecular because the slow step of the reaction involves only the carbon electrophile dissociating into a leaving group anion and a carbocation. The nucleophile then attacks the carbocation in a second, bimolecular, fast step. Given the reaction requires the formation of a cation and anion, the reaction requires a polar protic solvent to be employed.

The mechanism of the S_N1 reaction, as outlined in Figure 14.24, is a multi-step reaction, and the mechanism relies upon the same ion stabilization that dissolving sodium bromide requires.[7] First, the carbon electrophile dissociates into a carbocation and leaves group anion. Then the nucleophile attacks the carbocation. Depending upon the nucleophile, a third step is deprotonation (Figure 14.24).

Because carbocation intermediates form there are two concerns. The first is carbocation stability. Tertiary and secondary carbocations *are* likely to form, while primary and methyl cations *are not*

Na⁺ :Br:⁻ → methanol →

Cation well stabilized through coordination to electron-rich oxygens of the solvent.

Anion well stabilized through hydrogen bonds to the solvent.

Figure 14.22   Polar protic solvents stabilize cations and anions.

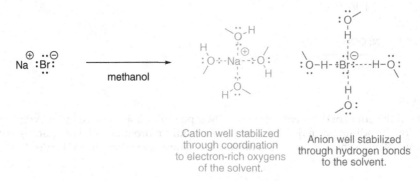

Figure 14.23 S_N1 solvolysis reaction of *tert*-butyl bromide in methanol.

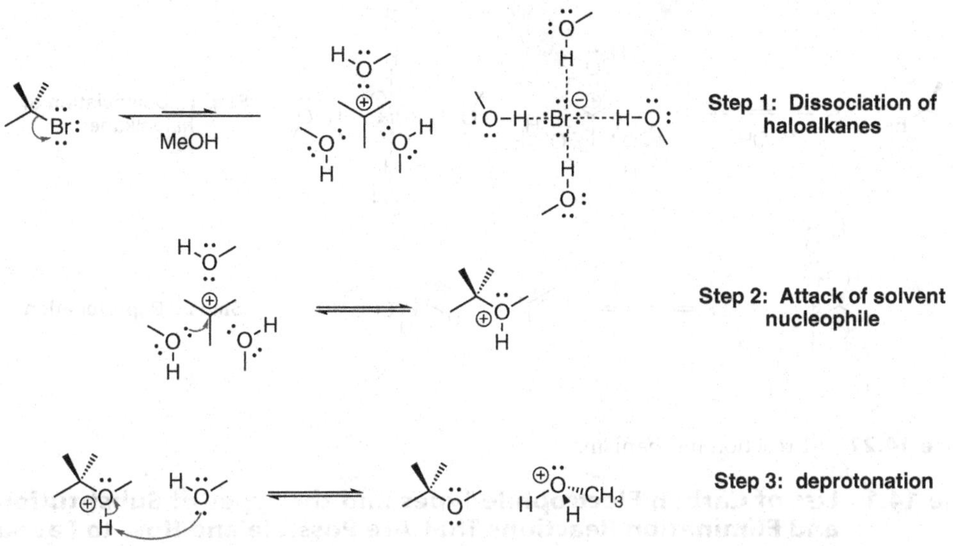

Step 1: Dissociation of haloalkanes

Step 2: Attack of solvent nucleophile

Step 3: deprotonation

Figure 14.24 S_N1 reaction mechanism.

likely to form. As such, S_N1 reactions are possible for tertiary and secondary carbon electrophiles, but not for methyl or primary carbon electrophiles. The second concern is the potential for carbocation rearrangements.

Problem 14.6 Propose a mechanism for the following S_N1 reaction (using the reaction in Figure 14.24 as a template).

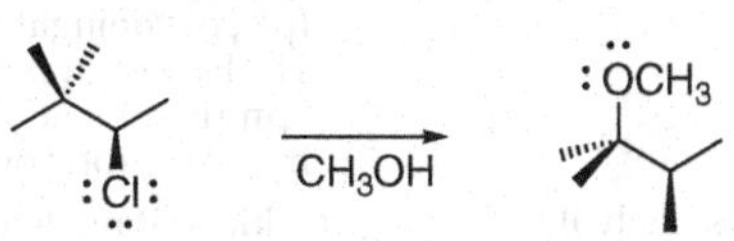

Mechanism:

S_N1 STEREOCHEMISTRY

In addition, because carbocations are planar, sp^2-hybridized carbons, the incoming nucleophile can attack either side of the carbocation. When a chirality centre is created, therefore, both R and S products are produced in equal amounts (Figure 14.25).

UNIMOLECULAR SUBSTITUTION AND ELIMINATION COMPETITION

The competing elimination is the E1 reaction, Figure 14.26. The E1 is like the S_N1 reaction mechanistically. The E1 is always produced in small amounts during S_N1 reactions and is typically just the minor product of S_N1 reactions. E1 is only favoured when the reaction is heated, and the E1-alkene product is distilled off (this drives the equilibrium towards the E1 product by shifting the equilibrium).

The E1 reaction proceeds mechanistically in a similar fashion to the S_N1 (Figure 14.27). First, the carbon electrophile dissociates into a carbocation and a leaving group anion. The difference is that the nucleophile then removes a β-proton. The E1 nomenclature denotes that the reaction is an

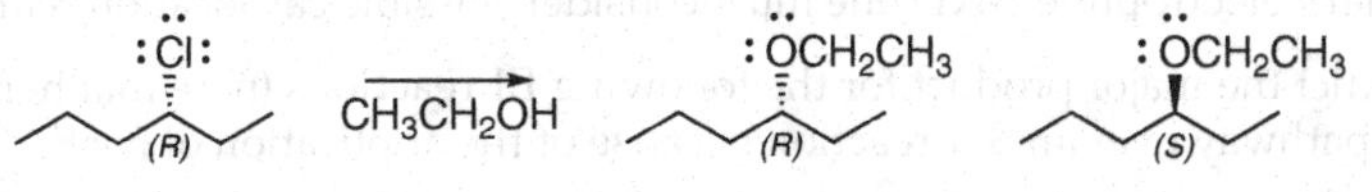

Figure 14.25 Racemization of a stereocenter during an S_N1 reaction.

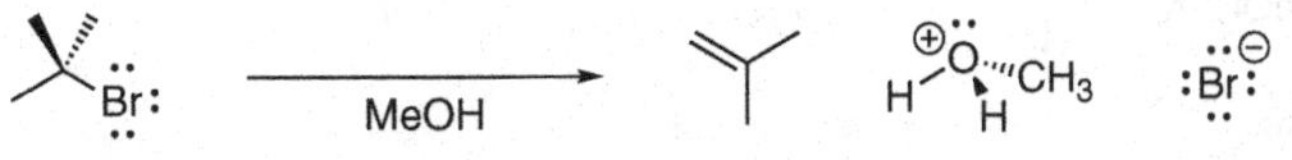

Figure 14.26 E2 elimination of *tert*-butyl bromide to yield 2-methylpropene.

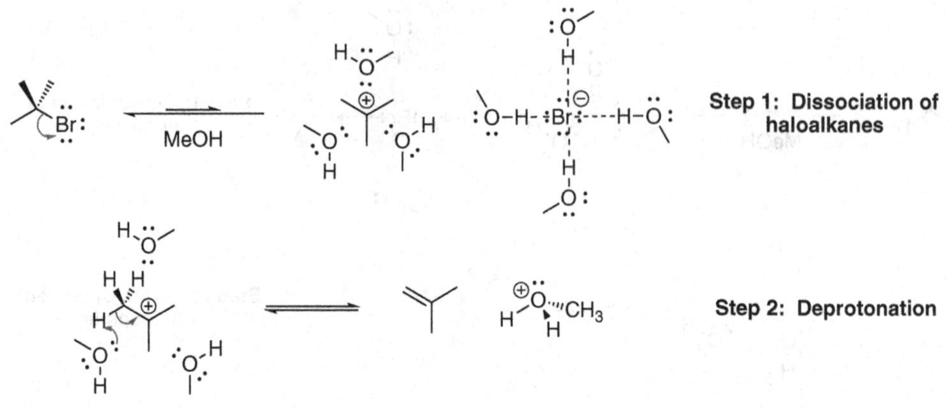

Step 1: Dissociation of haloalkanes

Step 2: Deprotonation

Figure 14.27 E1 reaction mechanism.

Table 14.1 List of Carbon Electrophile Types and the Types of Substitution and Elimination Reactions That Are Possible and How to Favour Them

	Methyl	Primary	Secondary	Tertiary
	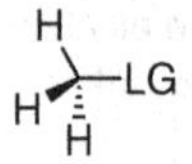	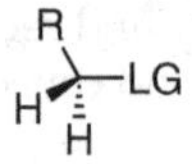	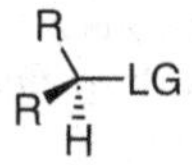	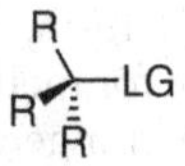
S_N2	Only S_N2 is possible	S_N2 is most likely reaction	S_N2 is most likely with weakly basic nucleophiles (pK_a of conjugate acid < 11). Fastest in polar aprotic solvents, slower in polar protic ones.	S_N2 not possible (under any normal circumstances).
E2	E2 not possible	E2 becomes likely if nucleophile is bulky, strong base or if R has steric bulk and the nucleophile is strongly basic	E2 likely if nucleophile is strongly basic (pK_a of conjugate acid > 11).	E2 most likely when a strongly basic nucleophile is used.
S_N1	S_N1 not possible (carbocation too unstable to form)	S_N1 not possible (carbocation too unstable to form)	S_N1 likely with weak/non-basic nucleophile and polar, protic solvents.	S_N1 likely with weak/non-basic nucleophile and polar, protic solvents.
E1	E1 not possible (carbocation too unstable to form)	E1 not possible (carbocation too unstable to form)	E1 likely only if heat applied and no good nucleophile present in polar protic solvent.	E1 likely only if heat applied and no good nucleophile present in polar protic solvent.

elimination (E) and unimolecular (1, that is the slow step involves only one molecule, the carbon electrophile). Unlike the E2, the E1 reaction has no stereoselectivity restrictions and will always produce the most substituted alkene product. Also, since a carbocation is produced, the E1 is never possible for primary electrophile AND one must consider possible carbocation rearrangements.

Problem 14.7 Predict the major product for the following E1 reactions (note that here the E1 is the predominant pathway over an S_N1 reaction because of the application of heat).

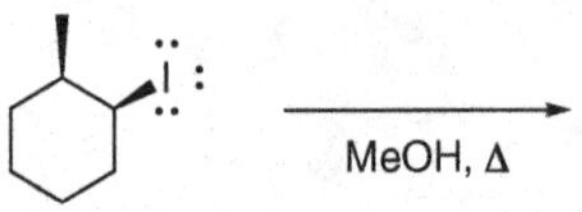

MeOH, Δ

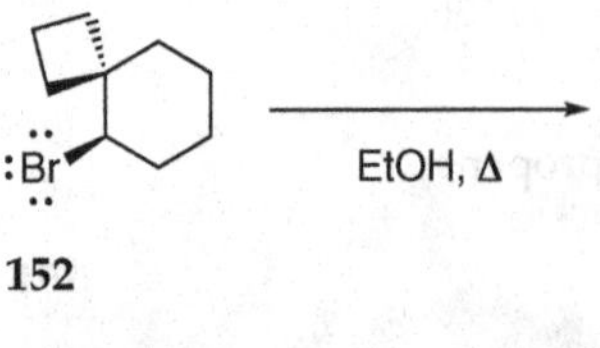

EtOH, Δ

IN COMPETITION: S$_N$2, S$_N$1, E2, E1

The best course of action in deciding what type of reaction is most likely to occur for a given carbon electrophile is to first distinguish what type of carbon electrophile is reacting. Table 14.1 breaks down all the considerations that one needs to consider in deciding the type of reaction that is likely to occur.

GENERAL PRACTICE PROBLEMS

Problem 14.8 Predict the major product(s) and rationalize the outcome (that after you have predicted a particular pathway explain what factors led you to propose that as the most likely pathway).

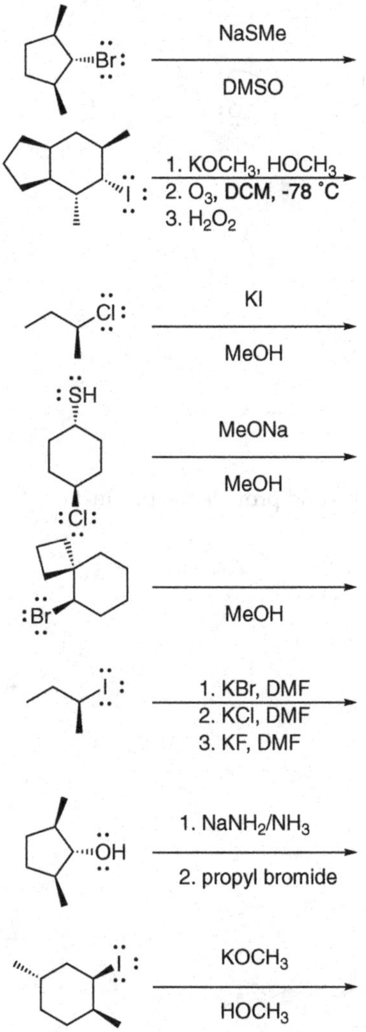

NaSMe

DMSO

1. KOCH$_3$, HOCH$_3$
2. O$_3$, DCM, -78 °C
3. H$_2$O$_2$

KI

MeOH

MeONa

MeOH

MeOH

1. KBr, DMF
2. KCl, DMF
3. KF, DMF

1. NaNH$_2$/NH$_3$

2. propyl bromide

KOCH$_3$

HOCH$_3$

Problem 14.9 For each chemical transformation give the appropriate reagents and conditions for each step. Fill in any intermediate or intermediary products in the boxes provided.

Problem 14.10 Look at the following examples of synthetic examples and provide the product or reagents necessary in each case.

NOTES

1. Swain, C.G.; Scott, C.B. Quantitative correlation of relative rates. Comparison of hydroxide ion with other nucleophilic reagents toward alkyl halides, esters, epoxides, and acyl halides.[1] *J. Am. Chem. Soc.*, **1953**, 75 (1), 141–147. DOI: 10.1021/ja01097a041

2. The rates of reaction are assembled from the following references. The exact rates are for specific, not general, reactions.

   i)   DeTar, D.F. Effect of alkyl groups on the rate of S_N2 reactions. *J. Org. Chem.*, **1980**, *45* (25), 5174–5176. DOI: 10.1021/jo01313a030

   ii) Okamoto, K.; Nitta, I.; Imoto, T.; Shingu, H. Kinetic studies of bimolecular nucleophilic substitution. II. Structural effects of alkyl halides on the rate of S_N2 reactions – a reinvestigation of the linear free-energy relationships for the structural variation of alkyl groups. *Bull. Chem. Soc. Japan*, **1967**, *40* (8), 1905–1908. DOI: 10.1246/bcsj.40.1905

   iii) Pines, E.; Fleming, G. Proton transfer in mixed water-organic solvent solutions: correlation between rate, equilibrium constant, and the proton free energy of transfer. *J. Phys. Chem.*, **1991**, *95* (25), 10448–10457. DOI: 10.1021/j100178a036

3. Walden, P. Über die gegenseitige Umwandlung optischer Antipoden. *Ber. Dtsch. Chem. Ges.*, **1896**, *29* (1), 133–138. DOI: 10.1002/cber.18960290127

4. Note that when an alkoxide is used in an S_N2 reaction, this is called a Williamson ether synthesis after Alexander Williamson. Williamson, A. XLV. Theory of ætherification. *London, Edinburgh, Dublin Philos. Mag. J. Sci.*, **1850**, *37* (251), 350–356. DOI: 10.1080/14786445008646627

5. Dhar, M.L.; Hughes, E.D.; Ingold, C.K.; Mandour, A.M.M.; Maw, G.A.; Woolf, L.I. 426. Mechanism of Elimination Reactions. Part XVI. Constitutional Influences in Elimination. A General Discussion. *J. Chem. Soc.*, **1948**, 2093–2119. DOI: 10.1039/JR9480002093

6. Saytzeff, A. Zur Kenntniss der Reihenfolge der Analgerung und Ausscheidung der Jodwasserstoffelement in organischen Verbindungen. *Liebigs Ann.*, **1875**, *179*, 296–301. DOI: 10.1002/jlac.18751790304

7. Dostrovsky, I.; Hughes, E.D.; Ingold, C.K. 50. Mechanism of Substitution at a Saturated Carbon Atom. Part XXXII. The Role of Steric Hinderance, Magnitude of Steric Effects, Range of Occurrence of Steric and polar Effects and Place of the Wanger Rearrangement in Nucleophilic Substitution and Elimination. *J. Chem. Soc.*, **1946**, 173–194. DOI: 10.1039/JR9460000173

15 Alcohols

Alcohols are molecules that contain a hydroxyl (–OH) group. The simplest example of an alcohol is methanol.

Alcohols can react as both electrophiles and nucleophiles. The hydrogen atom attached to the oxygen atom is very electrophilic (blue in the electrostatic potential map [Figure 15.1]) and alcohols are moderately acidic (typical pK_a values range from 16 to 18). In addition, the alcohol oxygen atom has two lone pairs and can react as a Brønsted–Lowry and Lewis base (nucleophile).

ALCOHOL NOMENCLATURE

In terms of priority ranking, alcohol functional groups take precedence over alkenes and alkynes. Alcohols are given the suffix -ol in IUPAC nomenclature. Alcohols are below carbonyl functional groups, in priority. In these compounds, alcohols are named as substituents (hydroxy). Alcohols are classified as methyl, primary (1°), secondary (2°), or tertiary (3°) based on the substitution at the carbon bearing the hydroxide (HO⁻) group (Figure 15.2).

HYDROGEN BONDING AND PHYSICAL PROPERTIES

Alcohols are the first functional group that will be considered that can engage in hydrogen bonding. Hydrogen bonding is the interaction of a hydrogen atom attached to an O, N, or F atom with the lone pair of an O, N or F atom (Figure 15.3).

This leads to higher intermolecular forces between alcohols compared to isomeric molecules without –OH groups, which leads to significantly higher melting and boiling points.

SYNTHESIS OF ALCOHOLS

Alcohols can be synthesized by several different methods (as seen in previous chapters):

Acid-catalyzed hydration of an alkene

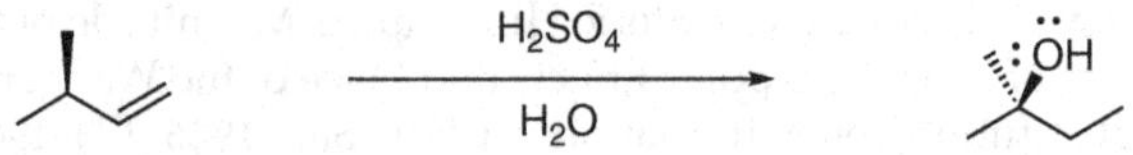

Oxymercuration reduction of an alkene

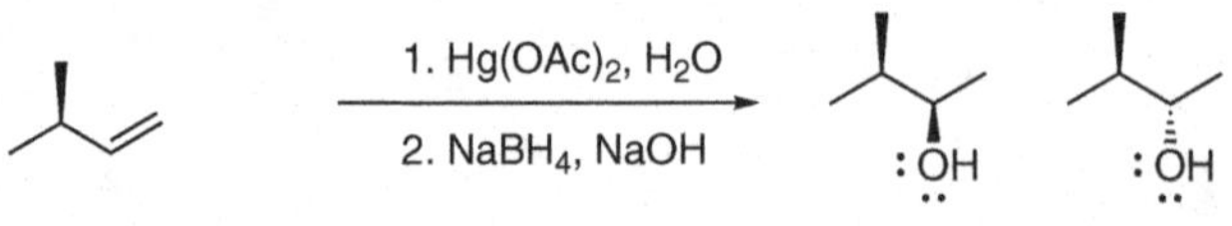

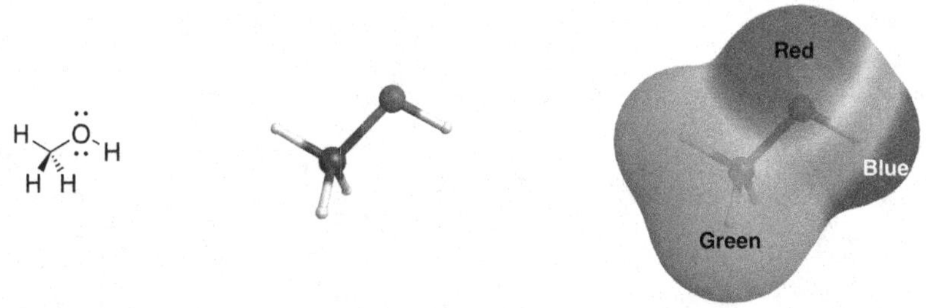

Figure 15.1 Structure (*left*), model (*middle*) and electrostatic potential map (*right*) of methanol (red corresponds to high electron density and blue corresponds to low electron density).

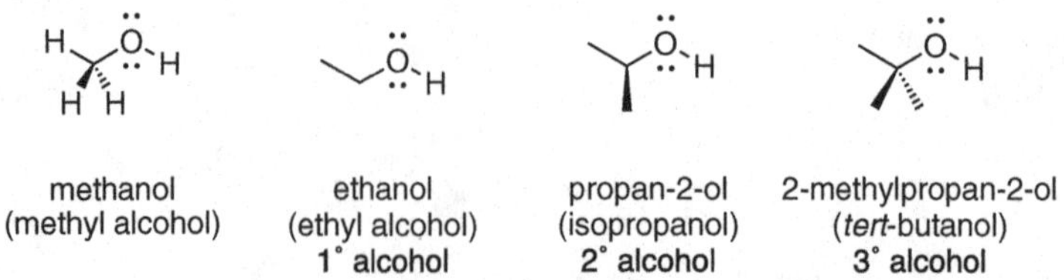

| methanol (methyl alcohol) | ethanol (ethyl alcohol) 1° alcohol | propan-2-ol (isopropanol) 2° alcohol | 2-methylpropan-2-ol (*tert*-butanol) 3° alcohol |

Figure 15.2 Examples of different alcohol types with their IUPAC names and common names in parentheses.

DOI: 10.1201/9781003479352-15

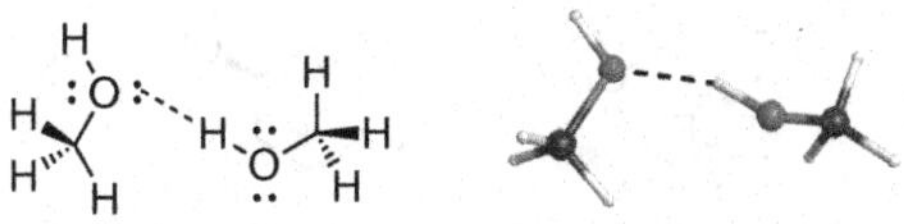

Figure 15.3 Model of a hydrogen bond between methanol molecules.

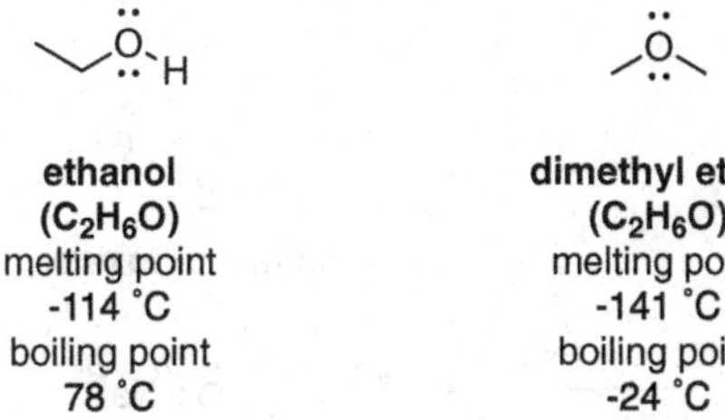
ethanol
(C_2H_6O)
melting point
-114 °C
boiling point
78 °C
dimethyl ether
(C_2H_6O)
melting point
-141 °C
boiling point
-24 °C

Figure 15.4 Examples of alkoxide formation using different alkali metals.

Hydroboration oxidation of an alkene

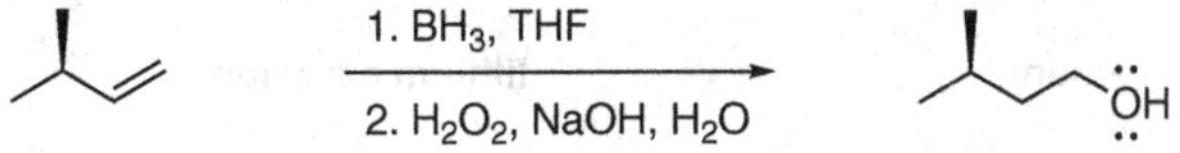
1. BH3, THF
2. H2O2, NaOH, H2O
OH

S_N2 substitution of primary alkyl halides with hydroxide as a nucleophile

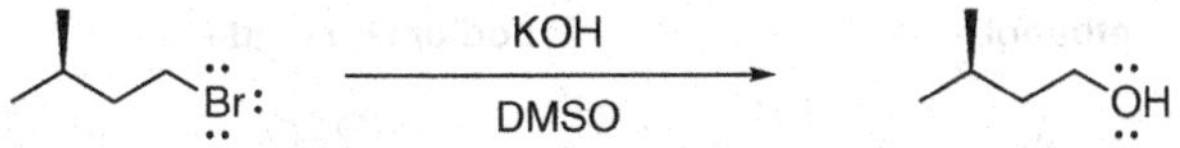
Br:
KOH
DMSO
OH

As shall be seen in Chapters 16, 19, 20, and 21, alcohols can also be prepared by opening epoxides and by reducing carbonyl-containing compounds.

ALCOHOLS AS ELECTROPHILES AND ALCOHOL ACIDITY

Strongly basic nucleophiles can attack alcohols, which act as electrophiles or Brønsted–Lowry acids. This produces an alkoxide, which is a strong base and good (much better than the starting alcohol) nucleophile.

Problem 15.1 Draw the appropriate arrows to show the deprotonations below.

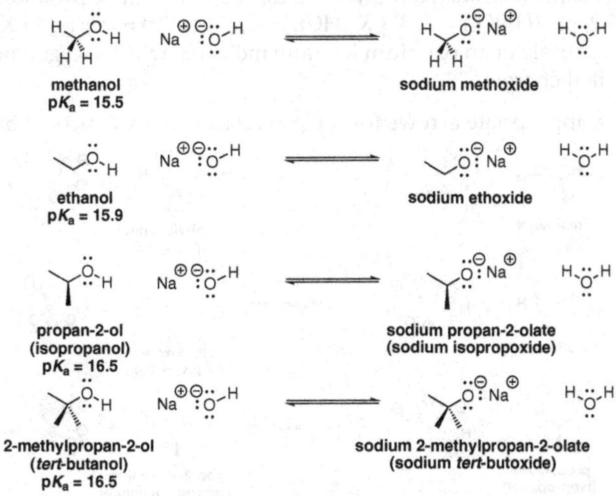
methanol
pKa = 15.5
sodium methoxide
ethanol
pKa = 15.9
sodium ethoxide
propan-2-ol
(isopropanol)
pKa = 16.5
sodium propan-2-olate
(sodium isopropoxide)
2-methylpropan-2-ol
(tert-butanol)
pKa = 16.5
sodium 2-methylpropan-2-olate
(sodium tert-butoxide)

While there is a marginal effect on pK_a due to substitution at the α-carbon, the acidity of alcohols can be greatly affected by the inductive effect and by conjugation (Figure 15.5).

Alkoxides can also be prepared through reaction of the appropriate alcohol with alkali metals Li, Na, and K (Figure 15.6). Rubidium (Rb) and caesium (Cs) could also be used, but the reaction of

pK$_a$ = 15.9   pK$_a$ = 12.5   pK$_a$ = 18   pK$_a$ = 10

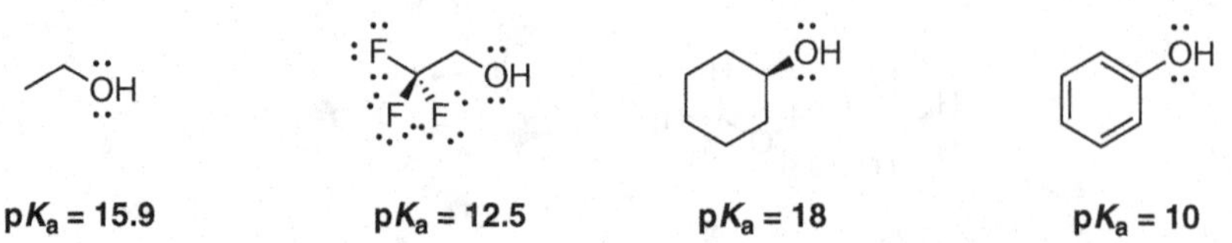

Figure 15.5 Examples of how the inductive effect (*left*) and conjugation (*right*) can affect alcohol acidity.

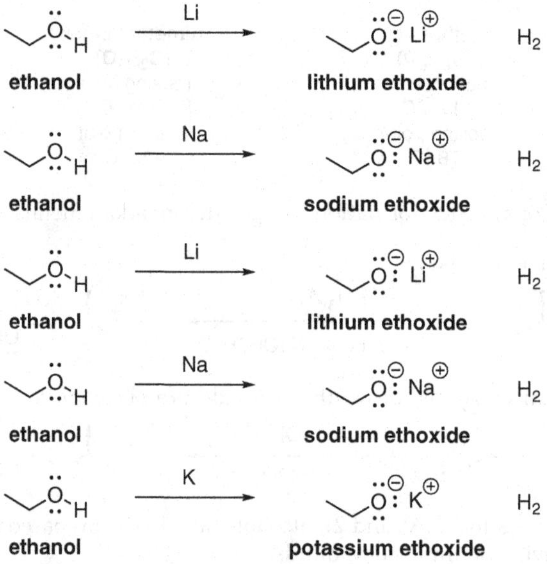

Figure 15.6 Examples of alkoxide formation using different alkali metals.

alcohols with Rb or Cs is so vigorous as to be explosive. The mechanism for this reaction involves single-electron transfer reduction (by the metal) of the alcohol and is beyond the scope of this chapter.

Alcohols are also reasonable Brønsted–Lowry bases. Alcohols can be protonated by any mineral acid (pK_a(H$_2$SO$_4$) = –5.2, pK_a(HNO$_3$) = –1.4, pK_a(HCl) = –7, pK_a(HBr) = –9), and pK_a(HI) = –10). A protonated alcohol is an example of an oxonium ion (any molecule where oxygen makes three bonds and has a positive formal charge).

Problem 15.2 Draw the appropriate arrows for the protonation of each alcohol by sulfuric acid.

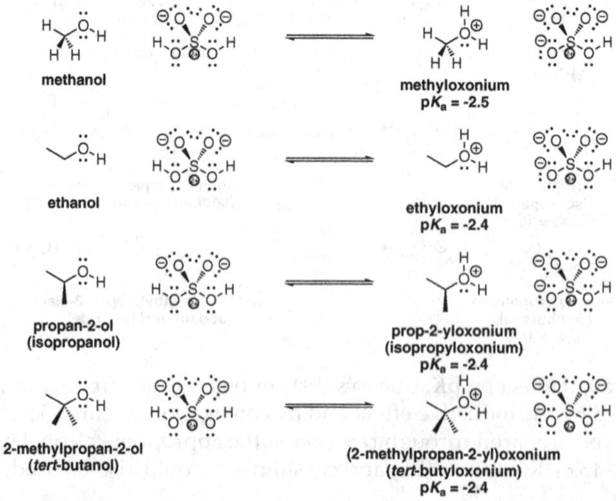

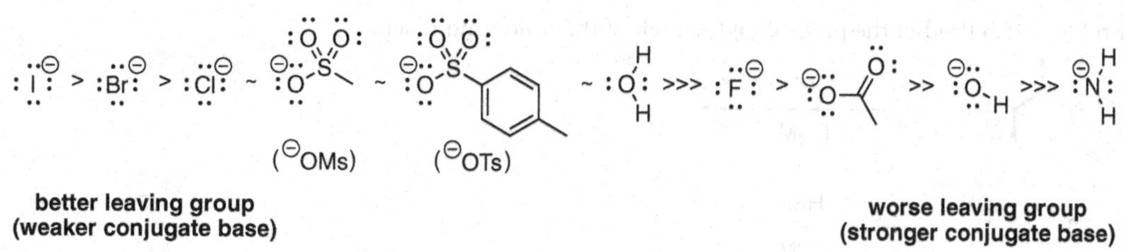

Figure 15.7 Order of leaving group ability.

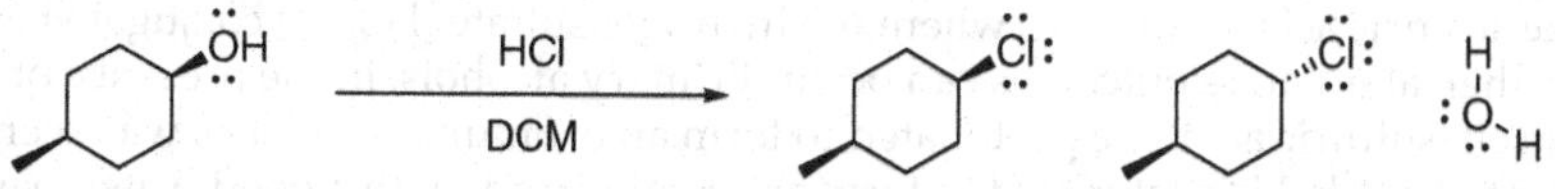

Figure 15.8 Dehydration of 3,3-dimethylbutan-2-ol with phosphoric acid.

Neutral alcohols are terrible electrophiles for $S_N2/E2/S_N1/E1$ chemistry, as the potential leaving group (hydroxide, $^-$OH) is a bad leaving group. Recall that leaving group ability and basicity track together (Figure 15.7).

ALCOHOL SUBSTITUTION WITH MINERAL ACIDS

However, after protonation, the resulting oxonium ion has a good (H_2O) leaving group. The protonation of an alcohol, therefore, renders the resulting oxonium ion reactive towards $S_N2/E2/S_N1/E1$ reactions.

Problem 15.3 Consider the following reaction of butanol with hydrogen chloride.

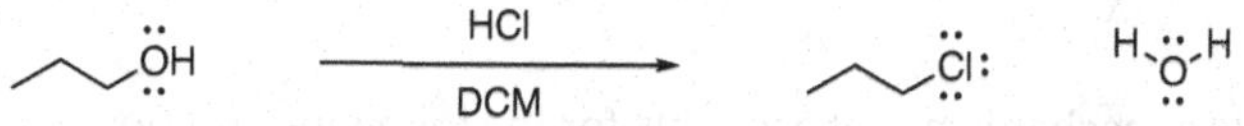

Draw the appropriate arrows to complete the mechanism.

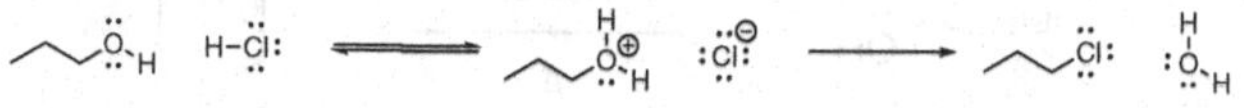

The first step of the mechanism is the same Brønsted–Lowry acid–base reaction as was seen in Problem 15.3. The resulting oxonium ion is then susceptible to an S_N2 reaction with the chloride conjugate base nucleophile.[1] This type of reaction is generalizable for hydrogen bromide (HBr) and hydrogen iodide (HI). Hydrogen fluoride (HF) cannot be utilized in these types of reactions as it is not a strong enough acid (pK_a is 3.2) to significantly protonate the alcohol.

Secondary and tertiary alcohols can also undergo substitution reactions with the hydrogen halides (HX); however, because the carbocation that would form from the loss of water would form a reasonable carbocation, the reaction proceeds predominantly via an S_N1 reaction mechanism (this can be seen in the stereo-random product mixture that results) rather than an S_N2 mechanism.

Problem 15.4 Draw the appropriate arrows to complete the mechanism.

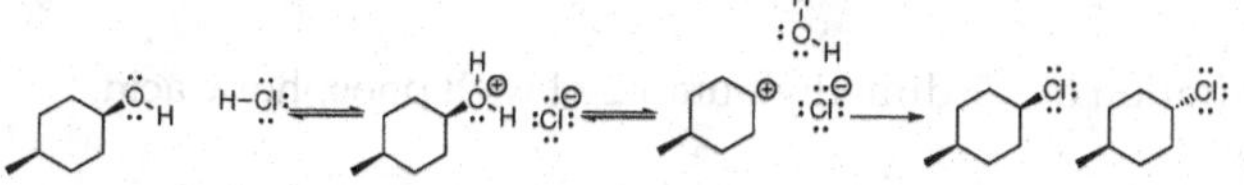

In the above reactions, hydrogen halide (HX) was utilized to form the oxonium ion. The conjugate base in each case (chloride, bromide, or iodide) is a good nucleophile and can displace water in a substitution reaction (S_N1 or S_N2 depending on the type of alcohol).

Problem 15.5 Predict the product(s) for each of the following reactions.

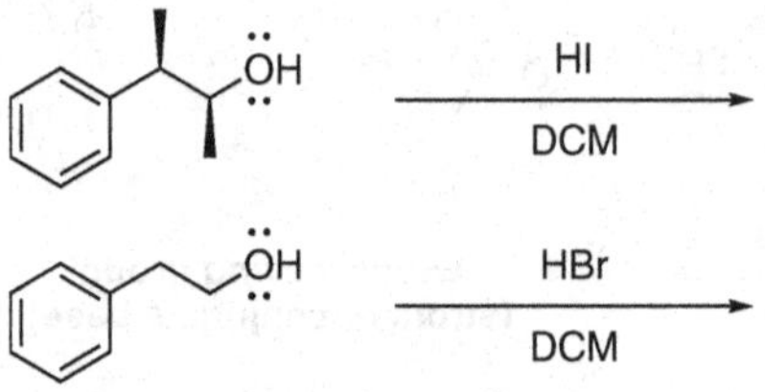

DEHYDRATION OF PRIMARY ALCOHOLS – CONDENSATION

If an acid like sulfuric acid is utilized, where the hydrogensulfate (HSO_4^-) conjugate base is non-nucleophilic, then alternative outcomes can occur. Primary alcohols, in the presence of neat, only the reagent itself, sulfuric acid, are protonated to form an oxonium ion, just as was seen above. One equivalent of alcohol is protonated to form an oxonium ion (with a good water leaving group) and then a second equivalent of alcohol acts as a nucleophile.[2]

Problem 15.6 Draw the appropriate electron-pushing arrows to complete the mechanism for the reaction of ethanol with sulfuric acid.

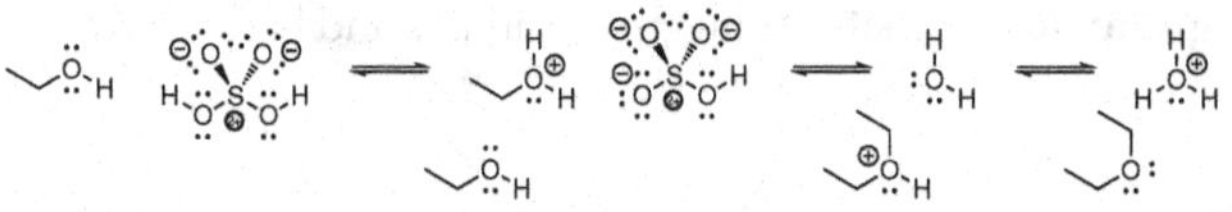

Problem 15.7 Predict the product(s) for each of the following reactions.

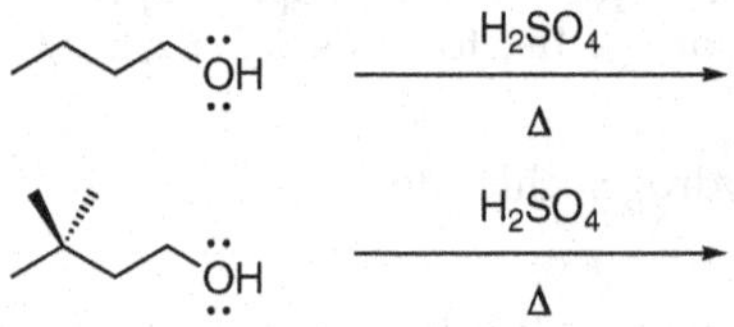

Problem 15.8 Provide a mechanism that accounts for the following reaction.

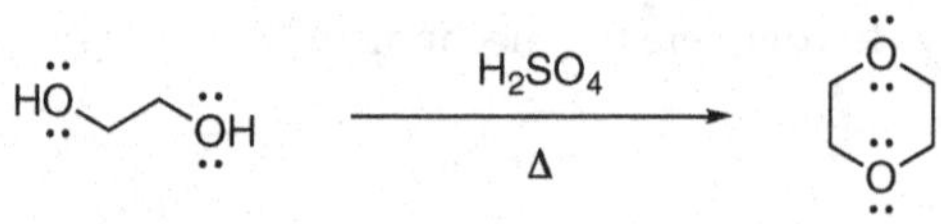

DEHYDRATION OF SECONDARY AND TERTIARY ALCOHOLS – ALKENE SYNTHESIS

Unlike primary alcohols, secondary and tertiary alcohols do not condense to form ethers; rather, secondary, and tertiary alcohols undergo dehydration reactions to produce alkenes (Figure 15.9).

In the reaction in Figure 15.9,[3] the major alkene product is a Zaitsev product that shows a skeletal rearrangement, which implies the formation of a carbocation intermediate.

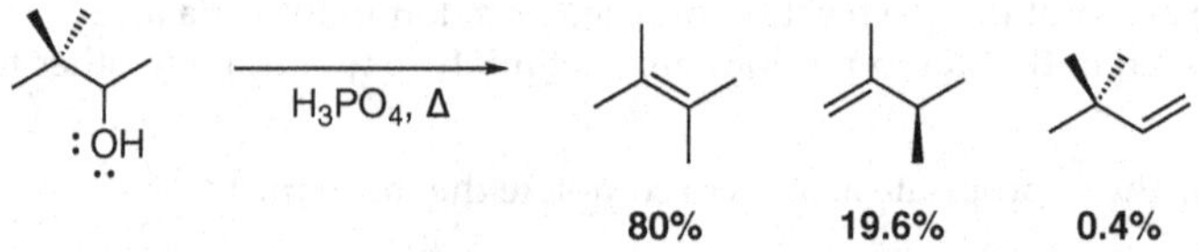

Figure 15.9 Dehydration of 3,3-dimethylbutan-2-ol with phosphoric acid.

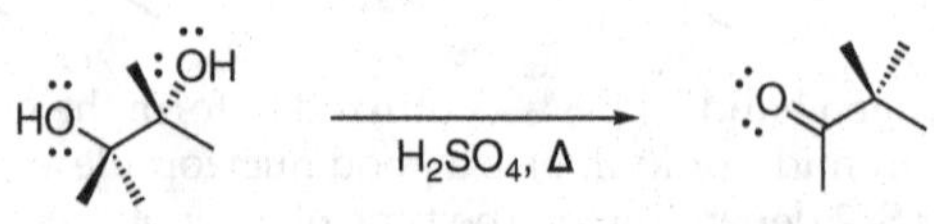

Figure 15.10 Pinacol rearrangement of 2,3-dimethylbutan-2,3-diol.

Problem 15.9 Utilizing what has been seen above for reactions, provide an electron-pushing mechanism that accounts for the major dehydration product in Figure 15.9.

DEHYDRATION OF VICINAL DIOLS – PINACOL REARRANGEMENT

Vicinal diols (glycols, prepared from alkenes reacting with osmium tetroxide or with potassium permanganate) react with concentrated sulfuric acid to produce a carbonyl product by what is known as the pinacol rearrangement (Figure 15.10).[4]

Problem 15.10 Draw the appropriate arrows to complete the pinacol rearrangement mechanism.

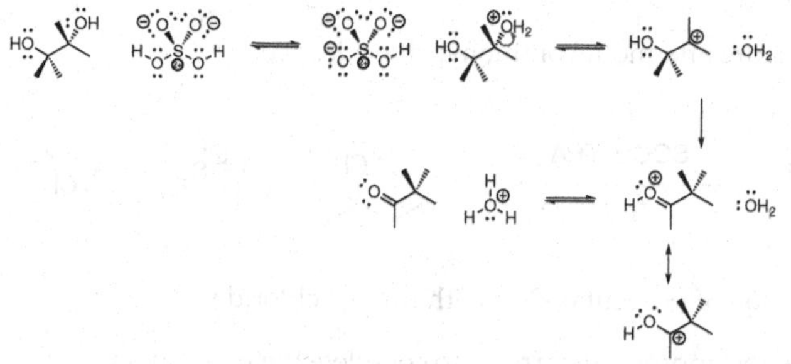

In the above pinacol rearrangement mechanism, the reaction proceeds as has already been seen, by protonation of one of the hydroxyl groups. The resulting oxonium ion can then lose water to form a carbocation (in asymmetrically substituted glycols the protonation and loss of water always proceeds via the most stable carbocation intermediate). At this point, the carbocation rearranges to form the more stable, delocalization-stabilized cation. Deprotonation then gives the carbonyl final product.

Problem 15.11 Predict the product(s) for each of the following reactions.

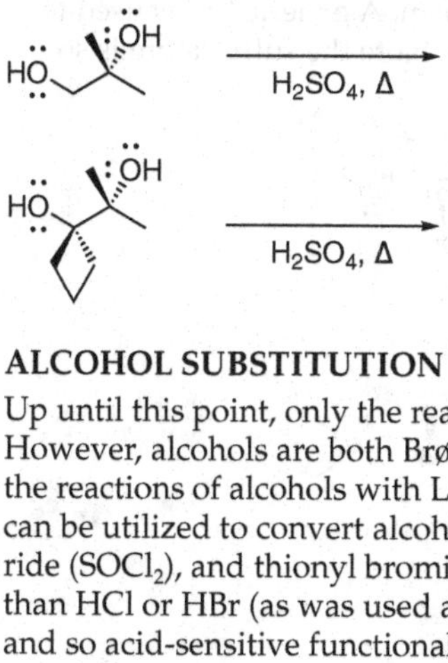

ALCOHOL SUBSTITUTION WITH LEWIS ACIDS

Up until this point, only the reactions of alcohols with Brønsted–Lowry acids have been considered. However, alcohols are both Brønsted–Lowry bases and Lewis bases. Attention will now be given to the reactions of alcohols with Lewis-acidic reagents. First, to be considered are those reagents that can be utilized to convert alcohols into alkyl halides: phosphorus tribromide (PBr_3), thionyl chloride ($SOCl_2$), and thionyl bromide ($SOBr_2$). The advantage of utilizing PBr_3, $SOCl_2$, or $SOBr_2$ rather than HCl or HBr (as was used above) is that PBr_3, $SOCl_2$, and $SOBr_2$ are not Brønsted–Lowry acids and so acid-sensitive functional groups – namely alkenes and alkynes – will not be affected.

Primary and secondary alcohols can be converted into primary and secondary bromoalkanes with phosphorus tribromide (Figure 15.11).[5] The substitution proceeds by an S_N2 mechanism. Note the inversion of stereochemistry (Figure 15.12).

Figure 15.11 Structure and model of phosphorus tribromide.

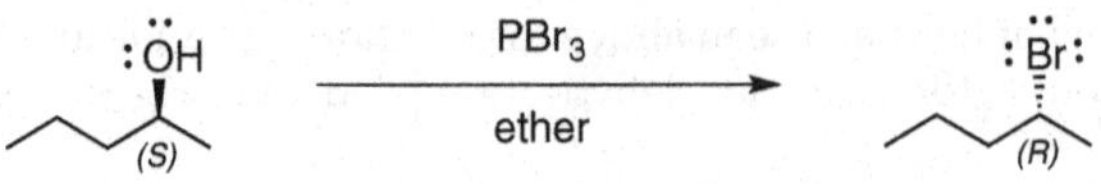

Figure 15.12 Reaction of (S)-pentan-2-ol with phosphorus tribromide.

Figure 15.13 Structure and model of thionyl chloride.

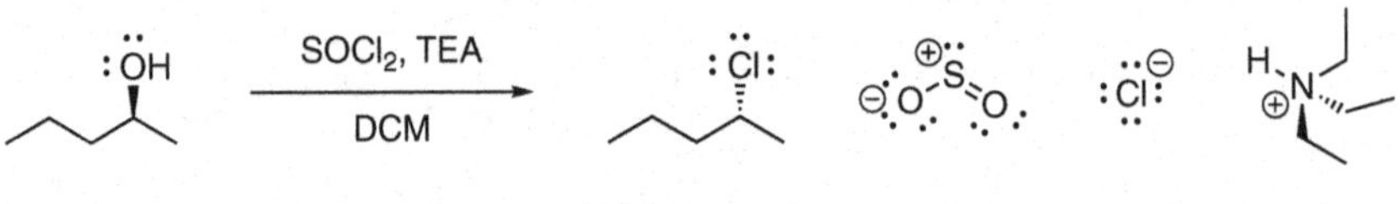

Figure 15.14 Reaction of (S)-pentan-2-ol with thionyl chloride.

Problem 15.12 Draw the appropriate arrows to complete the mechanism.

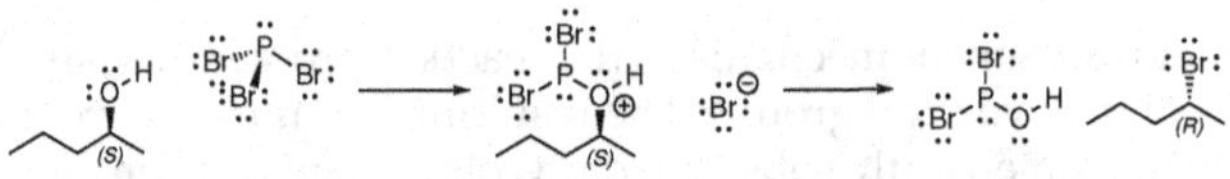

Primary and secondary alcohols can be converted into primary and secondary haloalkanes with thionyl chloride (Figure 15.13) and thionyl bromide.[6] The substitution proceeds by an S_N2 mechanism. Note the inversion of stereochemistry (Figure 15.14).

Problem 15.13 Draw the appropriate arrows to complete the mechanism. A generic "X" is used to stand for both Cl and Br as both proceed by the same mechanism. Note the sulfur atom is so electrophilic that the weak neutral alcohol nucleophile can attack.

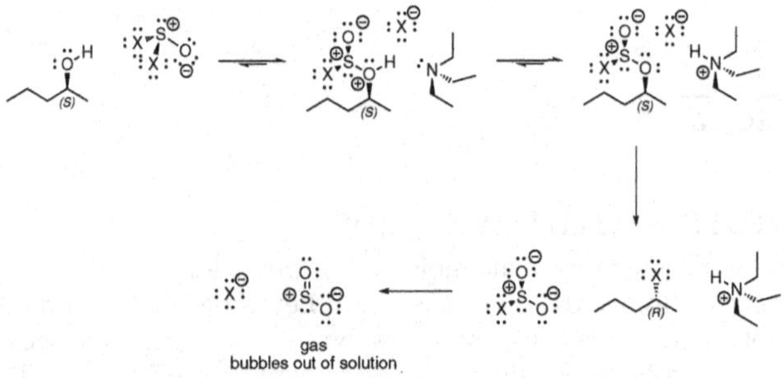

gas
bubbles out of solution

Problem 15.14 Predict the product(s) of each reaction.

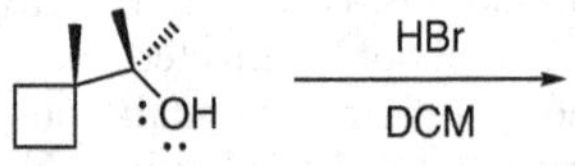

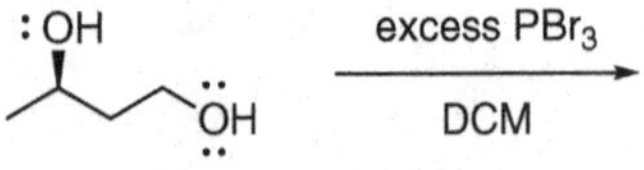

SULFONATE ESTERS

Alcohols also react with sulfonyl and sulfonic (Table 15. 1) Lewis acids – methanesulfonyl chloride (mesyl chloride, MsCl), p-toluenesulfonyl chloride (tosyl chloride, TsCl), and trifluoromethanesulfonic anhydride (triflic anhydride, Tf$_2$O) – to produce sulfonate esters.[7]

Table 15.1 Names, Abbreviations, Structures, and Models of Reagents Used to Make Sulfonate Esters

IUPAC name	Trivial name	Abbreviation	Skeletal structure	Computational model
methanesulfonyl chloride	mesyl chloride	MsCl	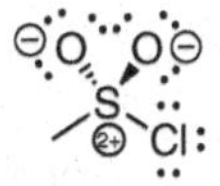	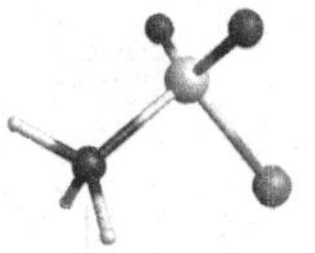
p-toluenesulfonyl chloride	tosyl chloride	TsCl	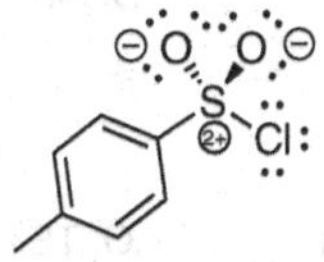	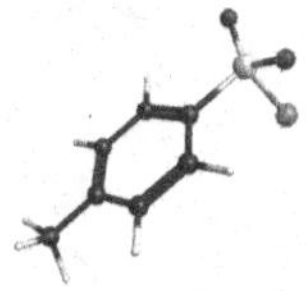
trifluoromethanesulfonic anhydride	triflic anhydride	Tf₂O	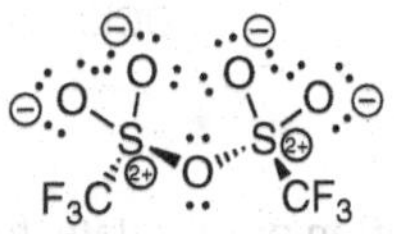	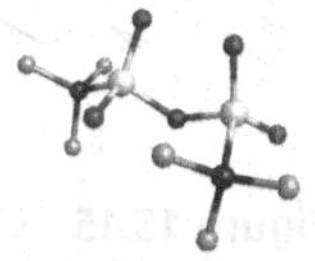

The reaction with sulfonyl chlorides and sulfonic anhydrides converts alcohols into sulfonate esters (Figure 15.15): mesylates (OMs), tosylates (OTs), and triflates (OTf). Sulfonate esters are excellent leaving groups, all are conjugate bases of strong, sulfonic acids. Therefore, sulfonate esters are great substrates for substitution (S_N1/S_N2) and elimination (E1/E2) reactions. Sulfonate esters can be used interchangeably.

Problem 15.15 Draw the appropriate arrows for the mechanism of sulfonate ester formation, an S_N2 reaction followed by deprotonation.[8] Note this is a generic reaction R stands for CH_3, p-toluene, or CF_3 and X stands for Cl or OTf.

Figure 15.16 shows an example of the types of reactions that sulfonate esters can be used for. While only mesylates are shown undergoing reactions in Figure 15.16, the sulfonate esters can all be used more or less interchangeably.

Problem 15.16 Provide the product(s) for the following reaction.

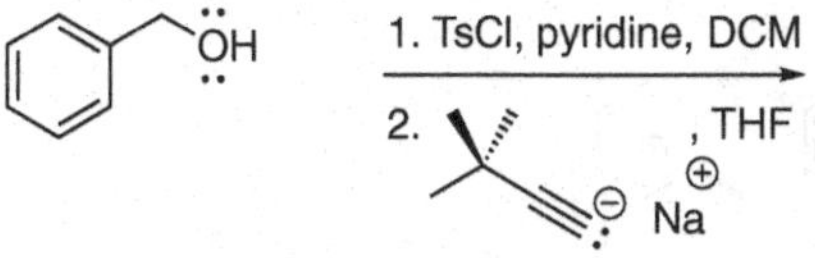

1. TsCl, pyridine, DCM

2. , THF

Problem 15.17 Provide the reagent(s) necessary to affect the following transformation.

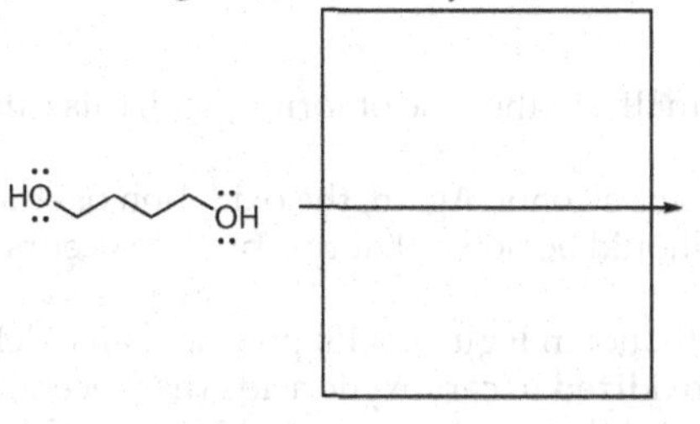

ALCOHOL OXIDATION

The final class of reactions that will be considered for alcohols are oxidation reactions. Consider the full oxidation of methane (Figure 15.17).

The first type of oxidation in Figure 15.17, from an alkane to an alcohol, is beyond the scope of this chapter but is of intense interest in current research. Methanol can then be oxidized to form-aldehyde. Notice that this oxidation involves the breaking of one C–H bond and the formation of a C–O bond. Formaldehyde can then be oxidized to formic acid (through a subsequent conversion

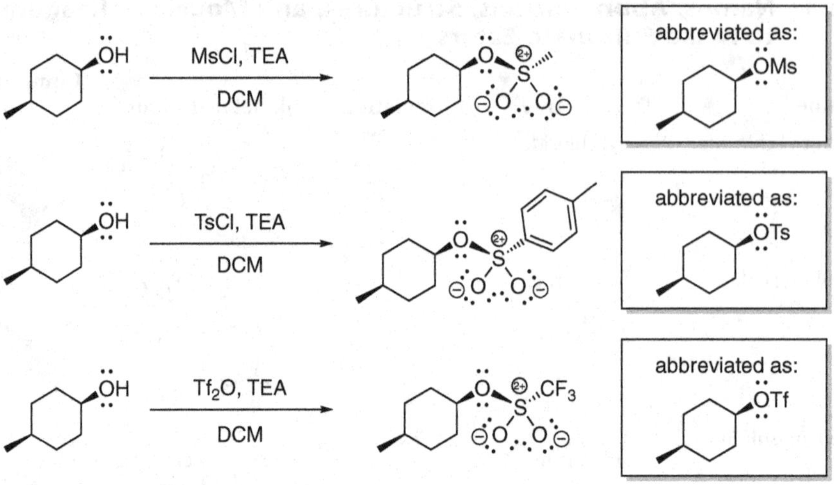

Figure 15.15 Creation of mesylate, tosylate, and triflate esters from *cis*-4-methylcyclohexanol.

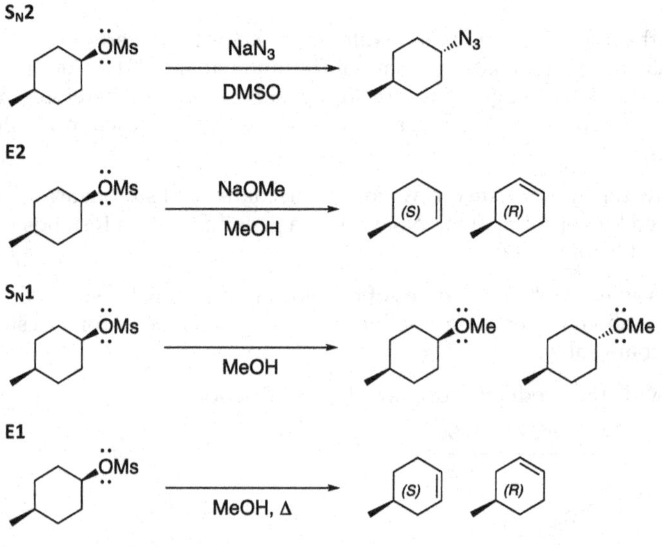

Figure 15.16 Reactions of mesylates.

of another C–H bond to a C–O bond). Finally, in the case of formic acid, this can be completely oxidized to carbon dioxide.

Now consider the oxidation of a primary alcohol. Again, the oxidation of ethane to ethanol is beyond the scope of this chapter but it should be noted that this process occurs *in vivo* (cytochrome P450 can oxidize alkanes to alcohols).

The oxidation of a primary alcohol (ethanol in Figure 15.18) produces an aldehyde (acetaldehyde in this case). Aldehydes can be further oxidized to carboxylic acids (here acetic acid). This is possible because of the presence of a C–H bond that can be converted into a C–O bond.

Consider the case of secondary alcohol. The oxidation of propane is again beyond the scope of this course. However, it should be noted that secondary alcohols (here isopropanol) are oxidized to ketones (here acetone). No further oxidation is possible in this case as there are no additional C–H bonds to convert to C–O bonds.

Finally, for tertiary alcohols, there is no further oxidation possible as the alcoholic carbon has no C–H bonds that can be converted into C–O bonds.

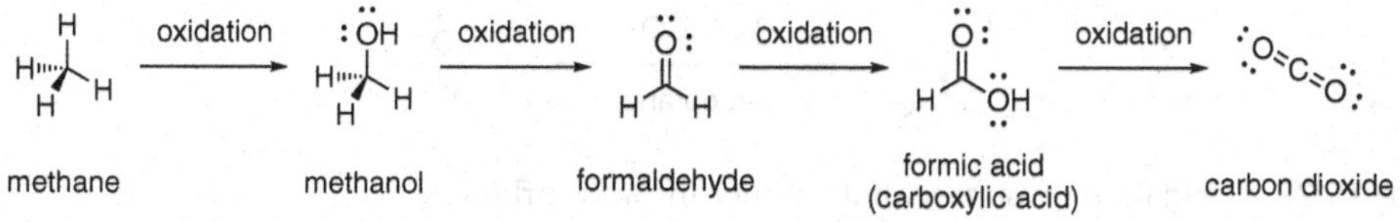

Figure 15.17 Stepwise oxidation of methane to carbon dioxide.

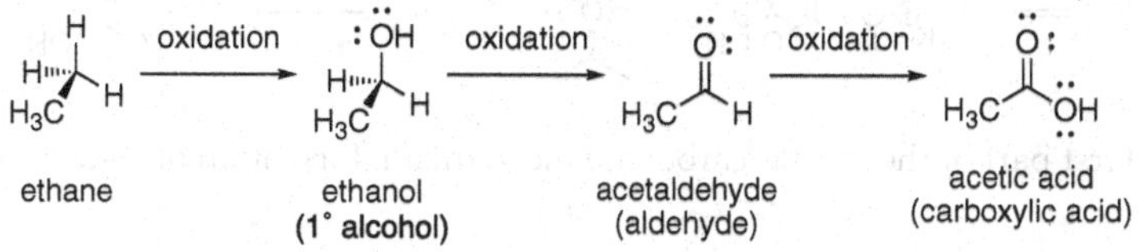

Figure 15.18 Stepwise oxidation of ethane to a primary alcohol and ultimately to acetic acid.

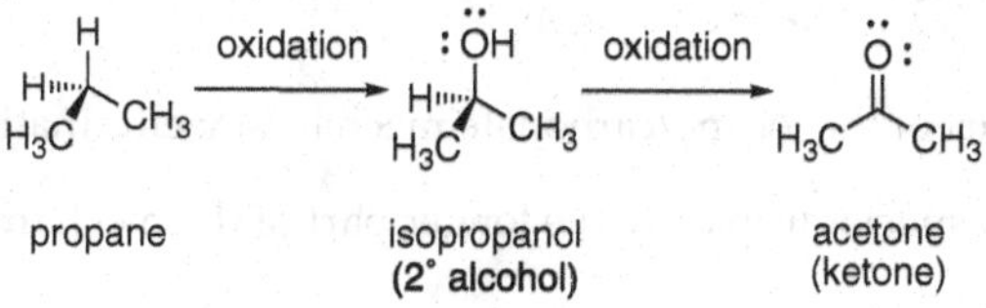

Figure 15.19 Stepwise oxidation of ethane to a primary alcohol and ultimately to acetic acid.

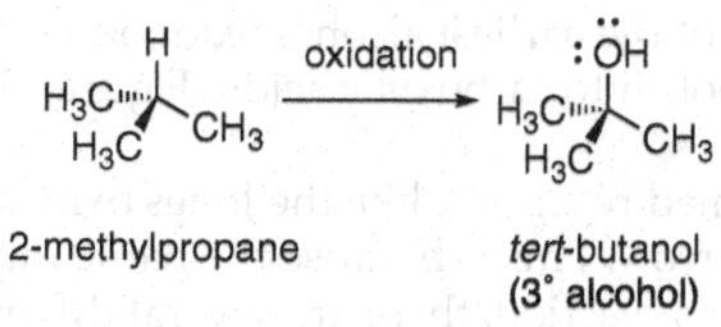

Figure 15.20 Stepwise oxidation of ethane to a primary alcohol and ultimately to acetic acid.

Alcohols can be oxidized by several different reagents. In general, oxidation of an alcohol is accomplished by a mechanism like an E2 elimination (Figure 15.21).

The specific base and leaving group (LG) that are employed vary depending upon the specific reaction. Also, the exact mechanism whereby an alcohol (R–OH) is converted into the R–O–LG species depends on the reagents used.

First, the oxidation of alcohols with iodine and potassium carbonate will be considered as it highlights several trends that shall be seen in more complicated reactions, and it reviews the chemistry seen in previous chapters.[9] This reaction is also a newer, green oxidation method that uses relatively benign reagents and has high atom economy. Other oxidation methods shown in this chapter are either classic oxidation methods or widely used because they are more functional group tolerant.

Let us consider the mechanism for the iodine oxidation reaction. In the first part of the mechanism, the alcohol nucleophile attacks (via an S_N2 mechanism) the iodine. Subsequent deprotonation by the K_2CO_3 gives the R–O–LG species.

The oxidation then occurs by E2 elimination to give the ketone product (Figure 15.24).

While the reaction in Figure 15.22 is not commonly used (several unwanted side reactions are possible), it does highlight the common mechanistic tropes for every oxidation: part one is the

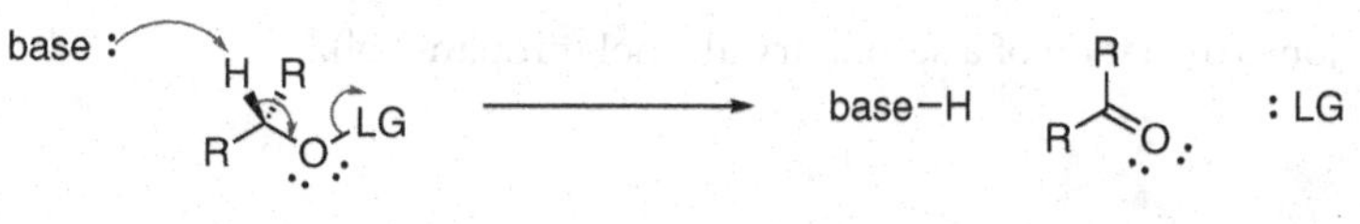

Figure 15.21 Common elimination step in the oxidation reactions to be considered.

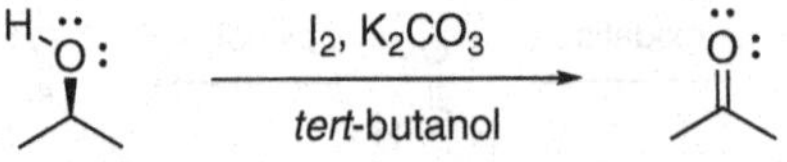

Figure 15.22 Oxidation of isopropanol with iodine and carbonate.

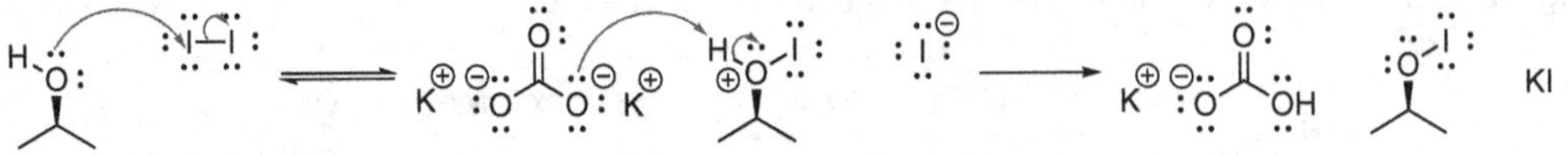

Figure 15.23 First part of the iodine/carbonate mechanism: formation of R–O–LG.

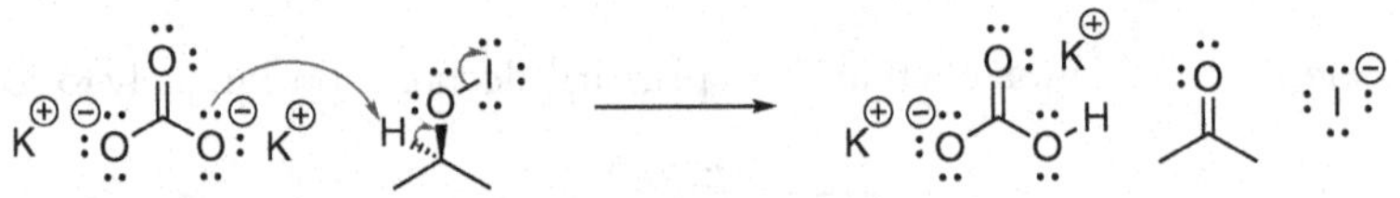

Figure 15.24 Second part of the iodine/carbonate mechanism: elimination.

formation of the R–O–LG species (typically the longer part of the mechanism) and part two is E2 elimination.

Jones Oxidation

Now attention will be given to some of the more common oxidation reactions. First, is the Jones oxidation reaction, which is one of the earliest alcohol oxidation methods.[10] Jones oxidation reactions will convert primary alcohols into carboxylic acids (Figure 15.25) and secondary alcohols into ketones (Figure 15.26).

It should be noted that for named reactions, like the Jones oxidation, it is common to just list the name of the reaction above the arrow. This is because the name implies a specific set of reagents/conditions. In the case of the Jones reaction, there are several different possible sets of reagents that all give rise to a common, active reagent (Figure 15.27).

In terms of being able to recognize each of these as a Jones oxidation, it is important to be able to note, solely, that you have some form of chromium(VI) oxide in sulfuric acid. For the mechanism of the Jones oxidation, only the active, oxidizing agent, chromic acid (H_2CO_4), will be shown (Figure 15.28) as it is present under all conditions listed in Figure 15.27.

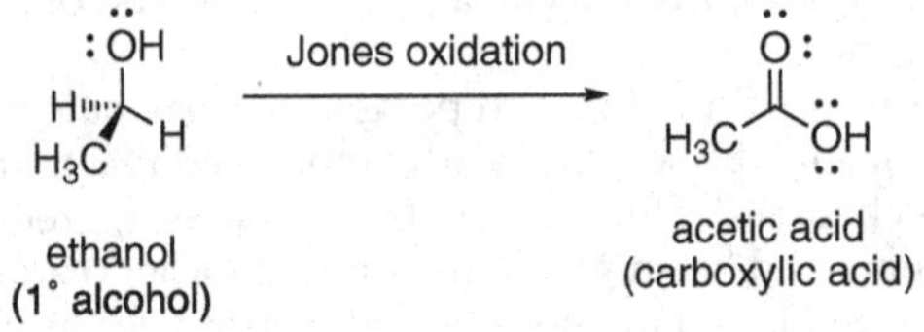

Figure 15.25 Jones oxidation of a primary alcohol (ethanol).

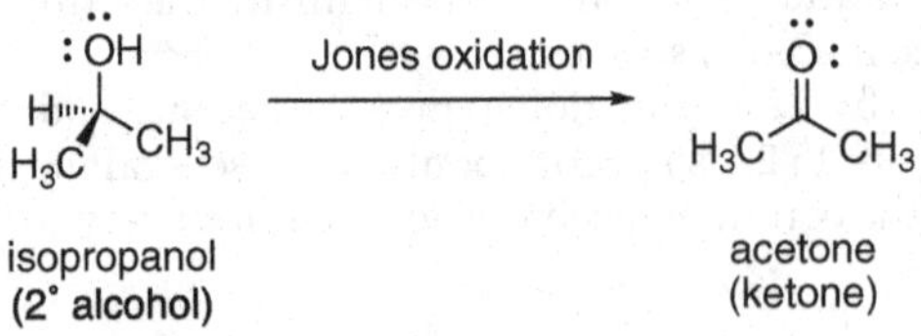

Figure 15.26 Jones oxidation of a secondary alcohol (propan-2-ol).

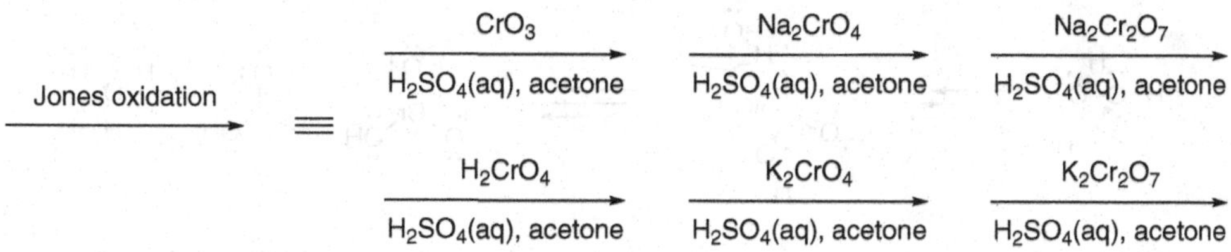

Figure 15.27 All of the different reagent combinations that are possible in a Jones oxidation reaction.

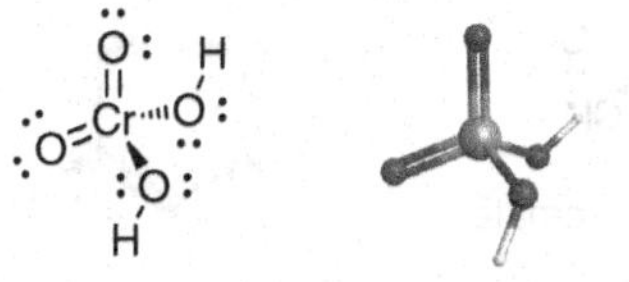

Figure 15.28 Structure and model of the active oxidizing agent in a Jones oxidation – chromic acid.

Problem 15.18 Draw the arrows to complete the Jones oxidation mechanism.

Part I: formation of R–O–LG

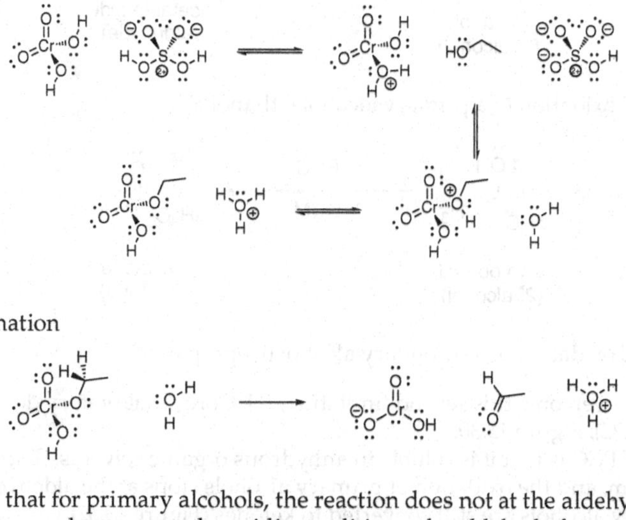

Part II: E2 elimination

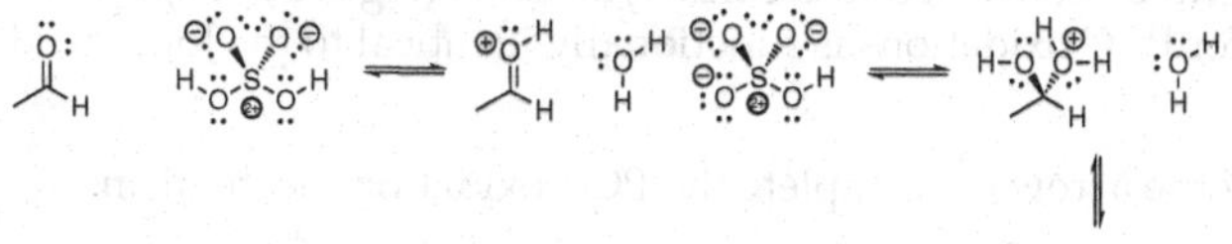

Problem 15.19 Note that for primary alcohols, the reaction does not at the aldehyde product but continues. This occurs because, under acidic conditions, the aldehyde forms a hydrate, draw the arrows to complete the hydrate mechanism.

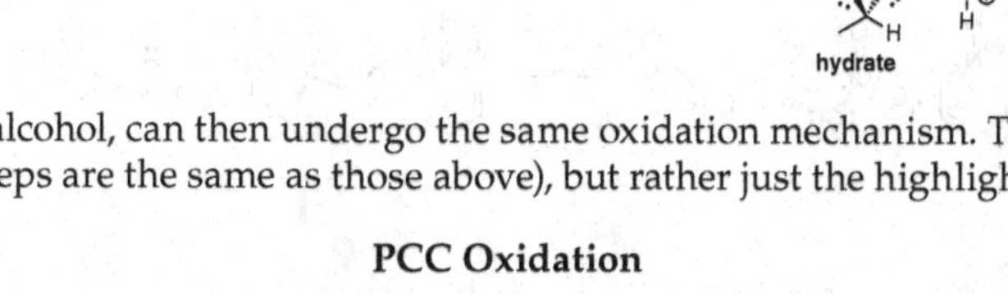

hydrate

The hydrate, as an alcohol, can then undergo the same oxidation mechanism. The total mechanism is not shown (the steps are the same as those above), but rather just the highlights (Figure 15.29).

PCC Oxidation

Although the Jones oxidation reaction is facile and cheap, it suffers from several disadvantages. First, chromium(VI) is carcinogenic. Second, Jones oxidation always converts primary alcohols into

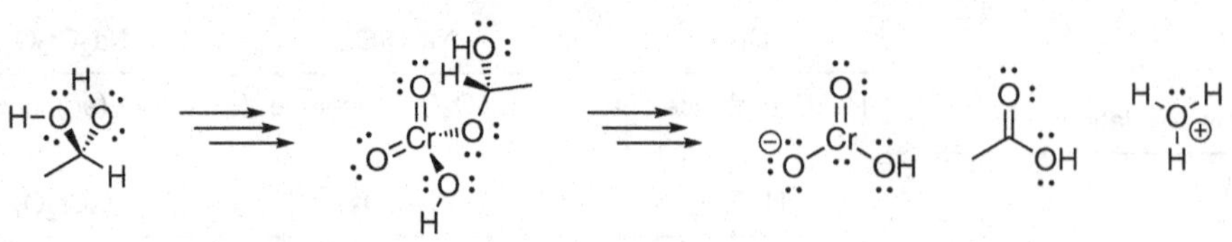

Figure 15.29 Abbreviated sequence showing the conversion of a geminal diol hydrate to a carboxylic acid.

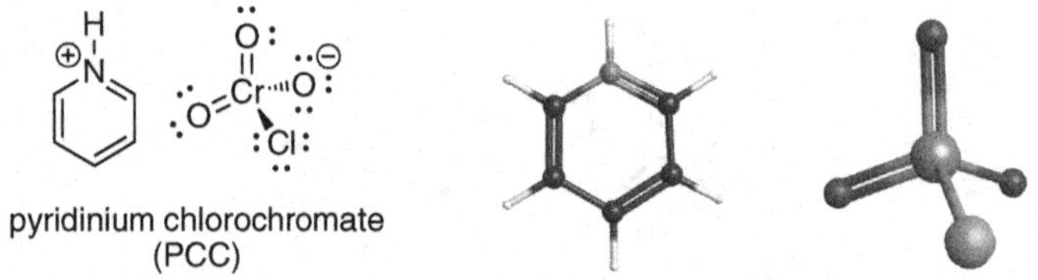

pyridinium chlorochromate
(PCC)

Figure 15.30 Structure and model of the active oxidizing agent in a PCC oxidation – pyridinium chlorochromate (PCC).

ethanol
(1° alcohol)

acetaldehyde
(aldehyde)

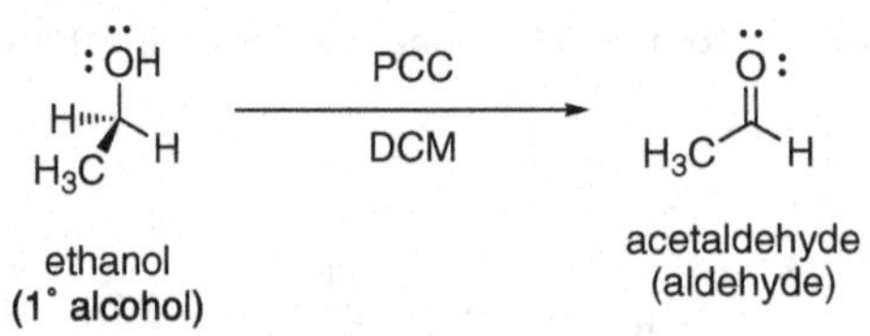

Figure 15.31 PCC oxidation of a primary alcohol (ethanol).

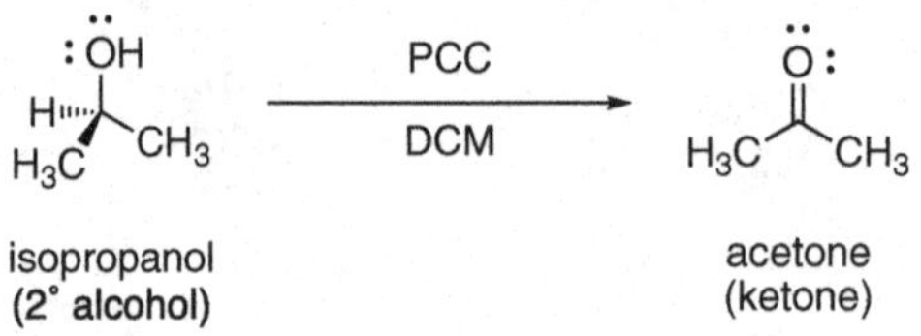

isopropanol
(2° alcohol)

acetone
(ketone)

Figure 15.32 PCC oxidation of a secondary alcohol (isopropanol).

carboxylic acids. To overcome this second limitation, E.J. Corey's laboratory developed pyridinium chlorochromate (PCC), Figure 15.30.[11]

The advantage of PCC is that it is soluble in anhydrous organic solvents. Since there is no water, hydrates do not form, and the oxidation of primary alcohols stops at the aldehyde product (Figure 15.31) and secondary alcohols are still converted to ketones (Figure 15.32).

Note, that the absence of water is important. If PCC is used with water as the solvent, then primary alcohols will still be converted into carboxylic acids (Figure 15.33).

The mechanism for PCC oxidations is functionally identical to the Jones oxidation mechanism seen above.

Problem 15.20 Draw the arrows to complete the PCC oxidation mechanism.

Part I: formation of R–O–LG

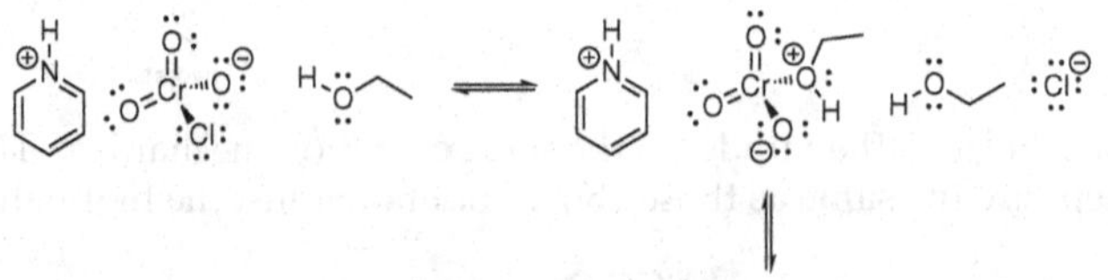

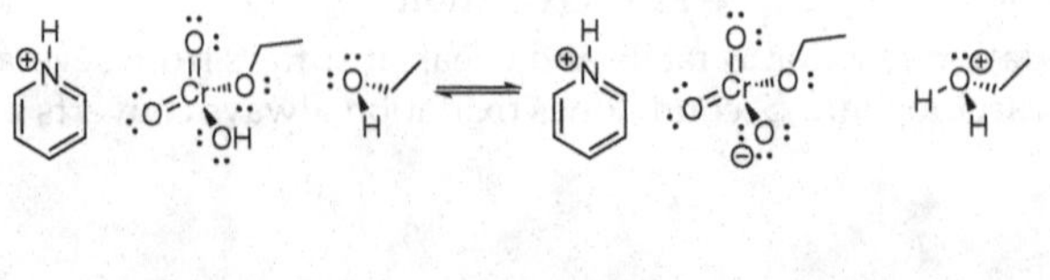

Part II: E2 elimination

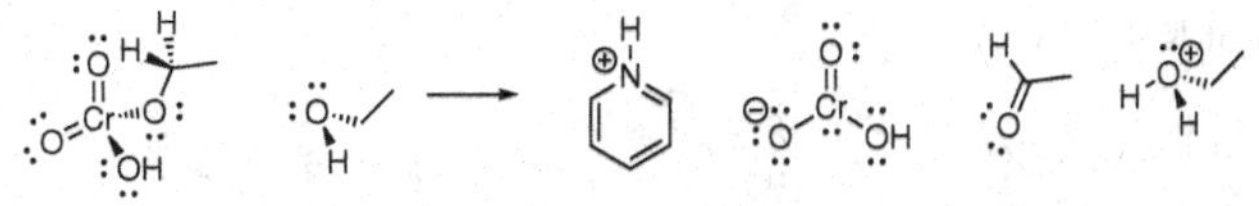

Swern Oxidation

Pyridinium chlorochromate is useful for preparing aldehydes and ketones from primary and secondary alcohols; however, PCC still utilizes a toxic and carcinogenic chromium(VI) reagent. Safer, less toxic reagents have been developed to produce aldehydes and ketones. The first to be studied will be the Swern oxidation.[12] The Swern oxidation uses oxalyl chloride ((COCl)$_2$), dimethyl sulfoxide (DMSO), triethylamine (TEA), and DCM solvent (Figure 15.34).

The Swern oxidation is used to convert primary alcohols into aldehydes (Figure 15.35) and secondary alcohols into ketones (Figure 15.36).

The reaction mechanism for the Swern oxidation is longer than the previous oxidation mechanisms, and there is a slight twist to the elimination step.

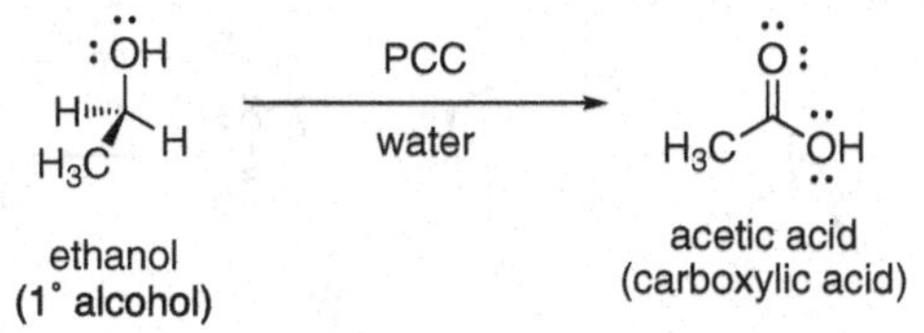

Figure 15.33 PCC oxidation of a primary alcohol (ethanol) with water present.

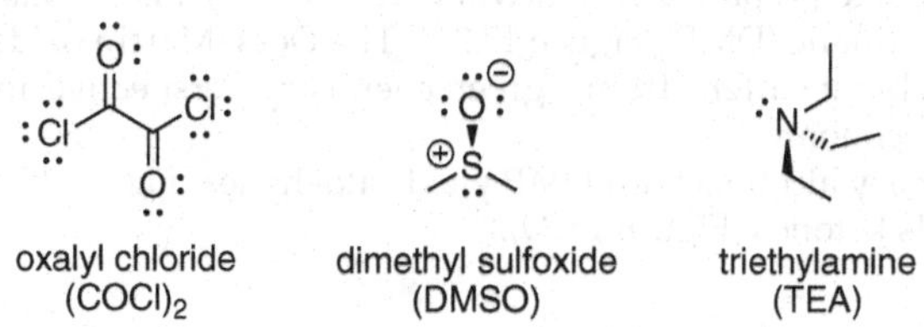

Figure 15.34 Reagents used in Swern oxidation: oxalyl chloride, dimethyl sulfoxide, and triethylamine.

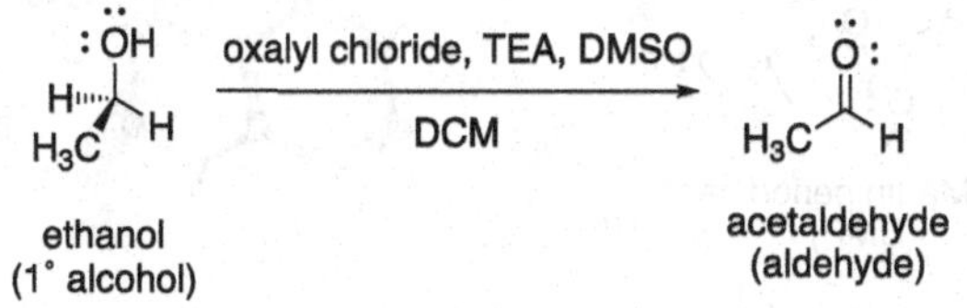

Figure 15.35 Swern oxidation of a primary alcohol (ethanol).

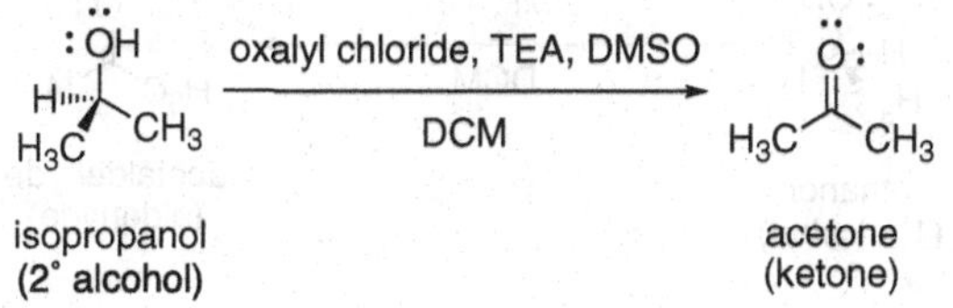

Figure 15.36 Swern oxidation of a secondary alcohol (isopropanol).

Problem 15.21 Draw the arrows to complete the Swern oxidation mechanism.

Part I: formation of R–O–LG

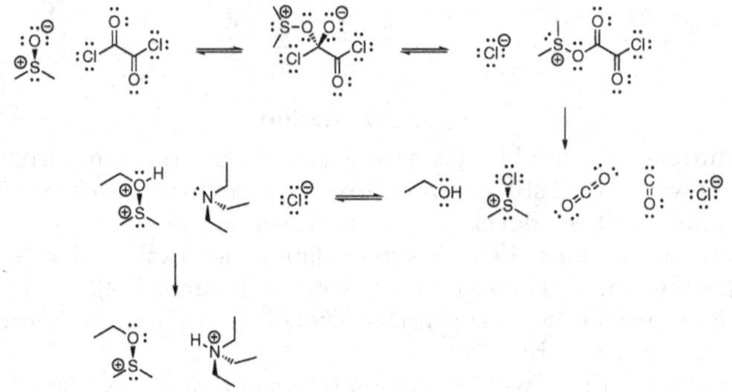

Part II: Elimination (note this is not an E2 elimination and is one of the very few cases of intra-molecular proton transfer)

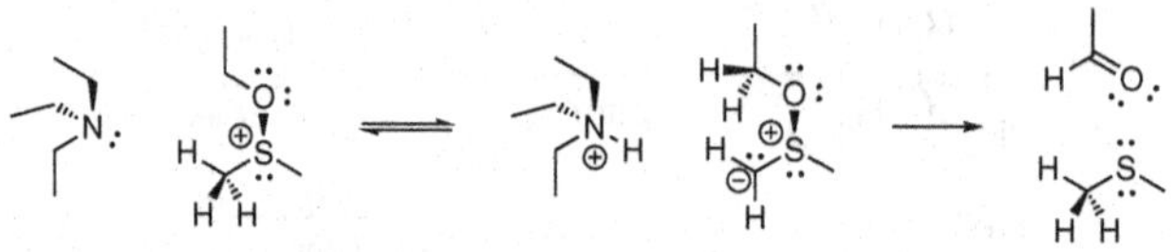

DMP Oxidation

While the Swern oxidation reaction has proven to be broadly useful, the reaction is noxiously smelly. Another oxidation to prepare aldehydes and ketones is the Dess-Martin oxidation, which uses Dess–Martin periodinane (DMP), Figure 15.37.[13] The Dess–Martin oxidation is particularly mild with broad utility, but it suffers, from a green chemistry perspective, in terms of having a particularly poor atom economy.

The oxidation of primary alcohols with DMP yields aldehydes (Figure 15.38) and oxidation of a secondary alcohol yields ketones (Figure 15.39).

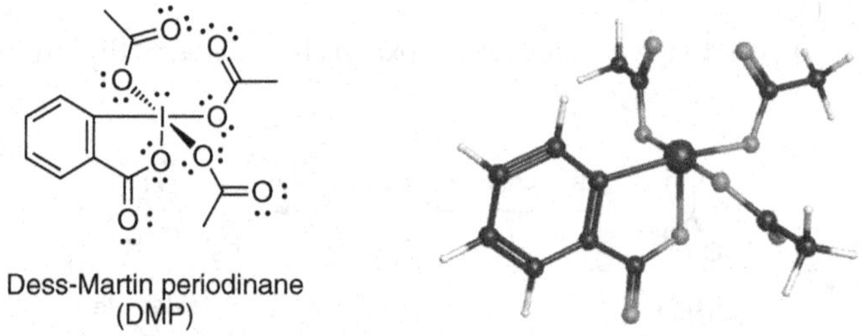

Dess-Martin periodinane
(DMP)

Figure 15.37   Structure and model of the active oxidizing agent Dess-Martin periodinane (DMP).

ethanol
(1° alcohol)

acetaldehyde
(aldehyde)

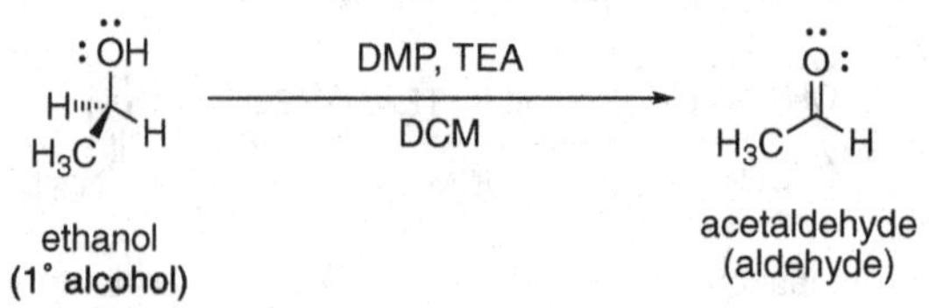

Figure 15.38   DMP oxidation of a primary alcohol (ethanol).

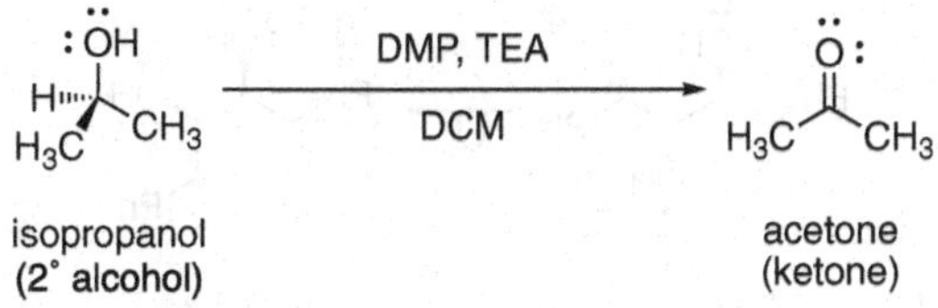

Figure 15.39 DMP oxidation of a secondary alcohol (isopropanol).

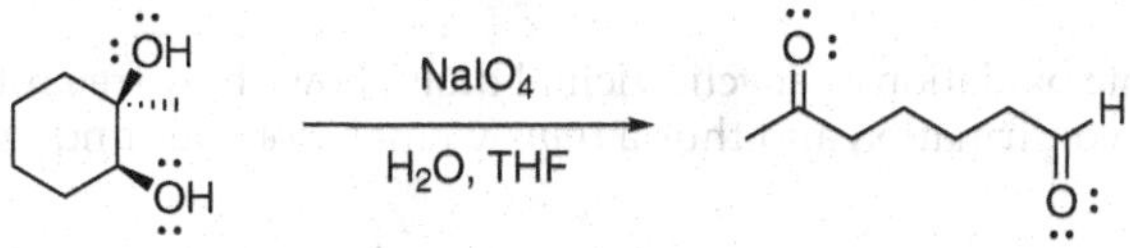

Figure 15.40 Periodate oxidation of *cis*-vicinal diol: 1-methylcyclohexa-1,2-diol.

Problem 15.22 Draw the arrows to complete the Dess-Martin oxidation mechanism.

Part I: formation of R–O–LG

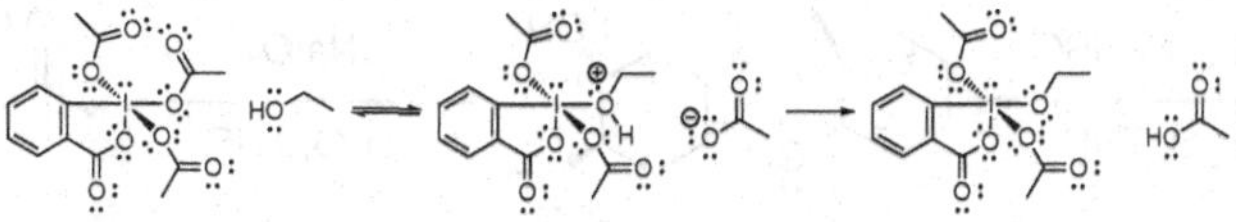

Part II: E2 elimination

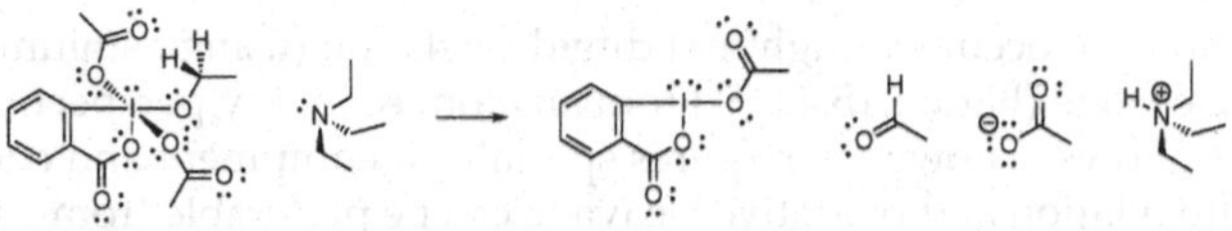

OXIDATIVE CLEAVAGE

The last oxidation to be considered for this chapter is the oxidation of vicinal diols (glycols). *Cis*-vicinal diols undergo oxidative cleavage with periodic acid to give aldehyde or ketone products (Figure 15.40).[14] It should be noted that periodic acid comes in two, forms meta-periodic acid (HIO_4) and orthoperiodic acid (H_5IO_6) which are in equilibrium with each other. As such it is common to see HIO_4 and H_5IO_6 written as the reagent for these reactions. It is also common to employ the sodium salt sodium periodate ($NaIO_4$) as the reagent for this reaction. The mechanism for this reaction is beyond the scope of this chapter.

This reaction is strictly limited to *cis*-vicinal diols. The same conditions applied to *trans*-vicinal diols give no reaction (Figure 15.41).

While the stringent requirement for a *cis* arrangement is of concern for cyclic diols, this is typically not a serious concern for acyclic diols as the carbon-carbon skeleton can rotate to accommodate the *cis*-geometric requirement (Figure 15.42).

This oxidative cleavage of vicinal diols by periodate should recall the previous oxidative cleavage reaction, ozonolysis (Figure 15.43).

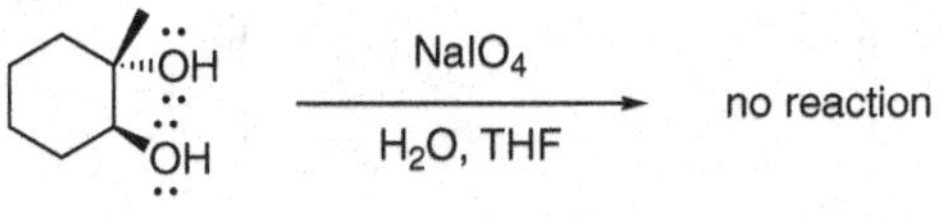

Figure 15.41 Lack of reaction when *trans*-vicinal diols are subjected to periodate oxidation conditions.

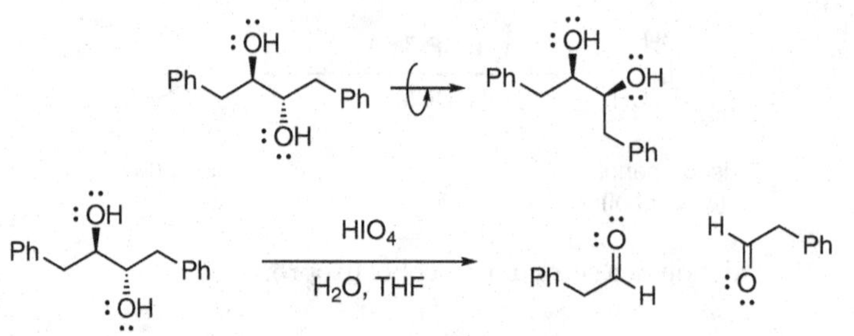

Figure 15.42 Periodate oxidation of acyclic vicinal diols shows how free rotation allows alcohols to adopt the necessary confirmation and thus a *trans*-vicinal diol does undergo reaction.

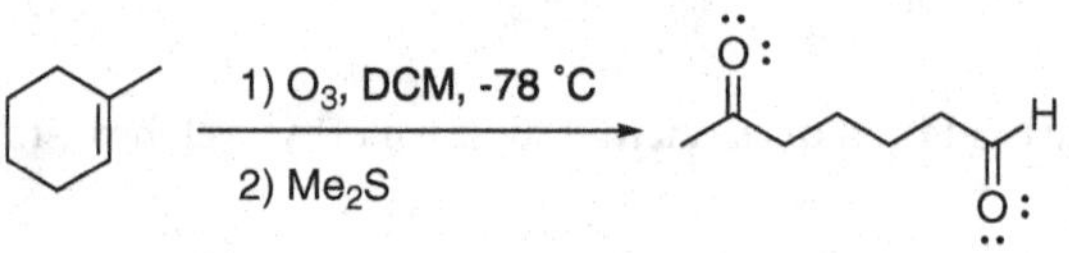

Figure 15.43 Ozonolysis of 1-methylcyclohexene with a reductive workup.

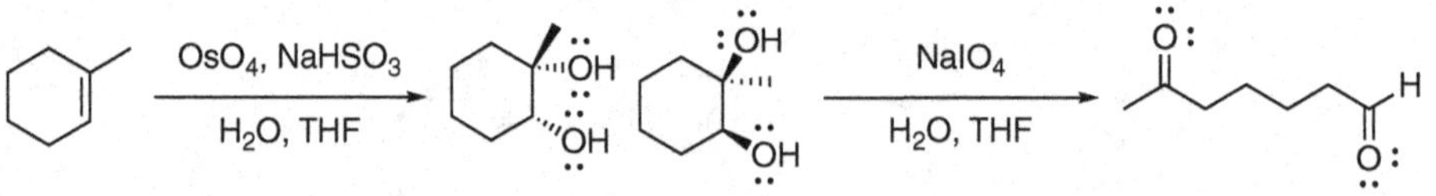

Figure 15.44 Dihydroxylation of 1-methylcyclohexene followed by periodate oxidation.

The same net reaction can occur through first dihydroxylation (using osmium tetroxide) followed by periodate cleavage (Figure 15.44).[15] From an atom economy perspective, ozonolysis is the best methodology. Ozonolysis, however, requires specialized equipment and ozone is hazardous to work with. Dihydroxylation and oxidative cleavage can be preferable from an operational ease standpoint.

GENERAL PRACTICE PROBLEMS

Problem 15.23 Provide the reagent(s) necessary to affect each transformation from a) to j).

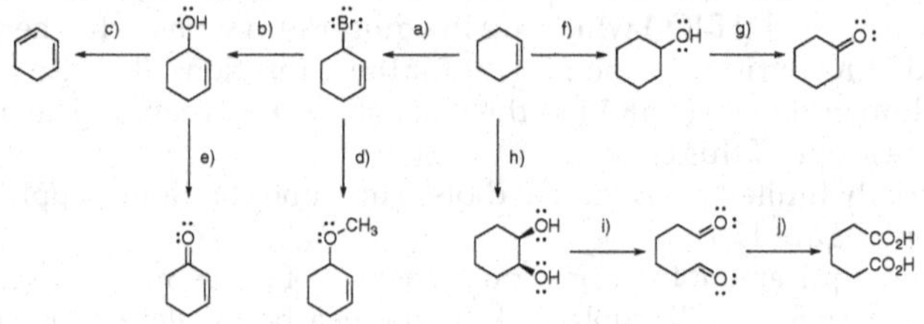

a.	b.
c.	d.
e.	f.
g.	h.
i.	j.

Problem 15.24 Predict the product(s).

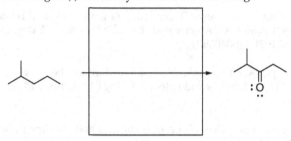

Problem 15.25 Provide the reagent(s) necessary to affect the following transformation and explain why HCl cannot be used for this reaction.

Problem 15.26 Provide the reagent(s) necessary to affect the following.

NOTES

1. i) Kilpi, S.; Puranen, U.H. Reaktionsgeschwindigkeit und Aktivität der Chlorwasserstoffsäure in Wasser-Äthylalkoholgemischen. *Z. Phys. Chem.*, **1929**, *145A* (1), 109–136. DOI: 10.1515/zpch-1929-14510

   ii) Kilpi, S. Einwirkung von Chlorwasserstoff auf Propylalkohol, *Z. Phys. Chem.*, **1933**, *166A* (1), 285–304. DOI: 10.1515/zpch-1933-16629

   iii) Hinshelwood, C.N. 132. The reaction of hydrogen chloride with methyl alcohol. *J. Chem. Soc.*, **1935**, 599–601. DOI: 10.1039/JR9350000599

2. The first such example of this reaction was reported by Valerius Cordus in 1540. He also demonstrated the resulting ether's anesthetic efficacy on his chickens. Bause, G.S. Cordus' Synthesis of Ether. *Anesthesiology*, **2009**, *111* (4), 804. DOI: 10.1097/01.anes.0000361292.45205.5e

3. Taber, R.L; Grantham, G.D.; Champion, W.C. The dehydration of 3,3-dimethyl-2-butanol. *J. Chem. Educ.*, **1969**, *46* (12), 849–850. DOI: 10.1021/ed046p849

4. Fittig, R. 41. Ueber einige Derivate des Acetons. *Liebigs Ann.*, **1860**, *114* (1), 54–63. DOI: 10.1002/jlac.18601140107

5. Reynolds, R.B.; Adkins, H. The Relationship of the Constitution of Certain Alkyl Halides to the Formation of Nitroparaffins and Alkyl Nitrites. *J. Am. Chem. Soc.*, **1929**, *51* (1), 279–287. DOI: 10.1021/ja01376a037

6. Carius, L. Ueber die Chloride des Schwefels und deren Derivate. *Liebigs Ann.*, **1859**, *111* (1), 93–113. DOI: 10.1002/jlac.18591110111

7. Otto, R.; Gruber, O.V. Untersuchungen aus dem chemischen Laboratorium zu Greifswald. 35) Ueber toluolschweflige Säure. *Liebigs Ann.*, **1867**, *142* (1), 92–102. DOI: 10.1002/jlac.18671420107

8. Gordon, I.M.; Maskill, H.; Ruasse, M.-F. Sulphonyl Transfer Reactions. *Chem. Soc. Rev.*, **1989**, *18*, 123–151. DOI: 10.1039/CS9891800123

9. Mori, N.; Togo, H. Facile oxidative conversion of alcohols to esters using molecular iodine. *Tetrahedron*, **2005**, *61* (24), 5915–5925. DOI: 10.1016/j.tet.2005.03.097

10. Bowden, K.; Heilbron, I.M.; Jones, E.R.H.; Weedon, B.C.L. 13. Researches on Acetylenic Compounds. Part I. The Preparation of Acetylenic Ketones by Oxidation of Acetylenic Carbinols and Glycols. *J. Chem. Soc.*, **1946**, 39–45. DOI: 10.1039/JR9460000039

11. Corey, E.J.; Suggs, J.W. Pyridinium chlorochromate. An efficient reagent for oxidation of primary and secondary alcohols to carbonyl compounds. *Tetrahedron Lett.*, **1975**, *16* (31), 2647–2650. DOI: 10.1016/S0040-4039(00)75204-X

12. i) Omura, K.; Swern, D. Oxidation of alcohols by "activated" dimethyl sulfoxide. A preparative, steric, and mechanistic study. *Tetrahedron*, **1978**, *34* (11), 1651–1660. DOI: 10.1016/0040-4020(78)80197-5

ii.) Mancuso, A.J.; Huang, S.-L.; Swern, D. Oxidation of long-chain and related alcohols to carbonyls by dimethyl sulfoxide "activated" by oxalyl chloride. *J. Org. Chem.*, **1978**, *43* (12), 2480–2482. DOI: 10.1021/jo00406a041

13. Dess, D.B.; Martin, J.C. Readily accessible 12-I-5 oxidant for the conversion of primary and secondary alcohols to aldehydes and ketones. *J. Org. Chem.*, **1983**, *48* (22), 4155–4156. DOI: 10.1021/jo00170a070

14. Malaprade, L. Action of polyalcohols on periodic acid and alkaline periodate. *Bull. Soc. Chim. Fr.*, **1934**, *3* (1), 833–852.

15. Pappo, R.; Allen Jr., D.S.; Lemieux, R.U. and Johnson, W.S. Notes – Osmium Tetroxide-Catalyzed Periodate Oxidation of Olefinic Bonds. *J. Org. Chem.*, **1956**, *21* (4), 478–479. DOI: 10.1021/jo01110a606

16 Ethers and Epoxides

Ethers are molecules that contain an oxygen connected to two, non-hydrogen groups (R–O–R). The simplest example of an ether is dimethyl ether (Figure 16.1).

Ethers are unreactive towards several different reagents: bases/nucleophiles (notice the lack of blue, electrophilic sites on the electrostatic potential map, Figure 16.1) and oxidizing agents. For this reason, ethers are routinely used as solvents for organic reactions and, as we will see below, protecting groups.

ETHER NOMENCLATURE

The trivial name of ethers comes from naming the alkyl groups as substituents followed by the term ether, for example, dimethyl ether for CH_3OCH_3. When the groups are not identical, asymmetric ethers, then the substituent names are listed alphabetically, as in ethyl methyl ether for $CH_3CH_2OCH_3$.

In IUPAC nomenclature, ethers are named alkanes with alkoxy substituents. So dimethyl ether (CH_3OCH_3) is properly named methoxymethane. When the substituents are asymmetric, the larger substituent is named the parent chain, for example, $CH_3CH_2OCH_3$ would be methoxyethane.

SYNTHESIS OF ETHERS

Ethers can be synthesized by several different methods:

Acid-catalyzed addition of alcohols to an alkene

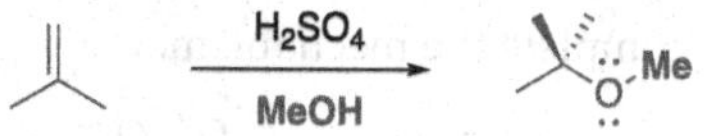

Oxymercuration reduction of an alkene

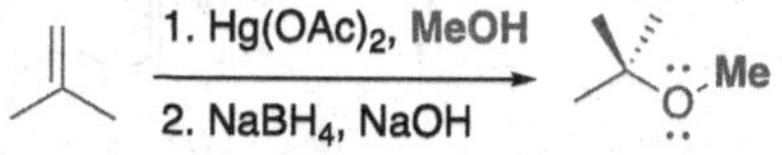

William ether synthesis (S_N2 substitution with alkoxide nucleophiles)

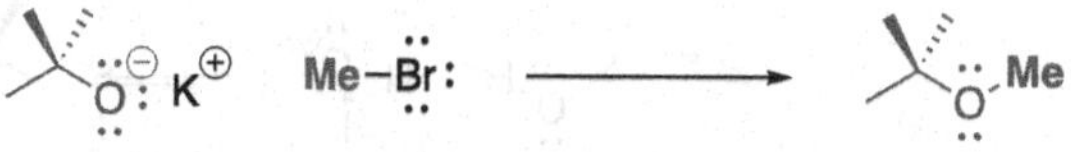

Solvolysis (S_N1 substitution with alcohol nucleophiles)

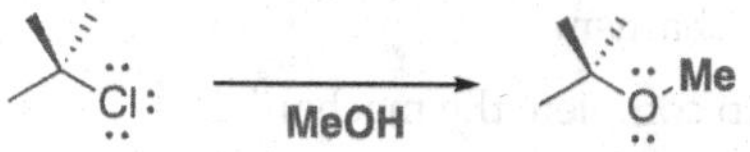

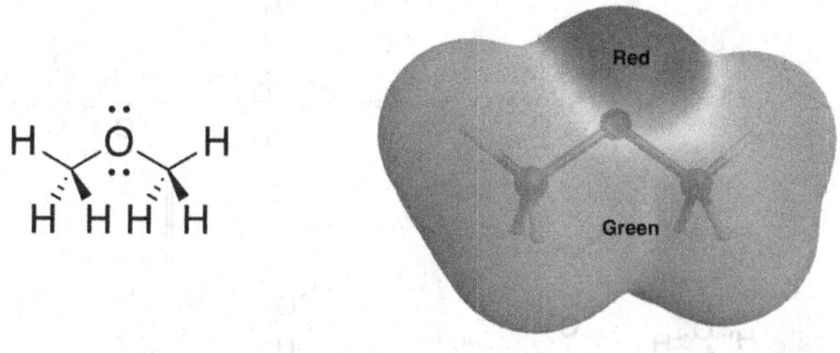

Figure 16.1 Structure (*left*) of dimethyl ether and electrostatic potential map (*right*) of dimethyl ether (red corresponds to high electron density and blue corresponds to low electron density).

DOI: 10.1201/9781003479352-16

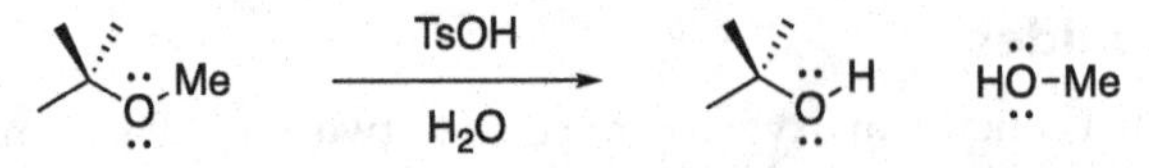

Figure 16.2 Acid-catalyzed hydrolysis of methyl *tert*-butyl ether (MTBE).

Acid-catalyzed dehydration of primary alcohols

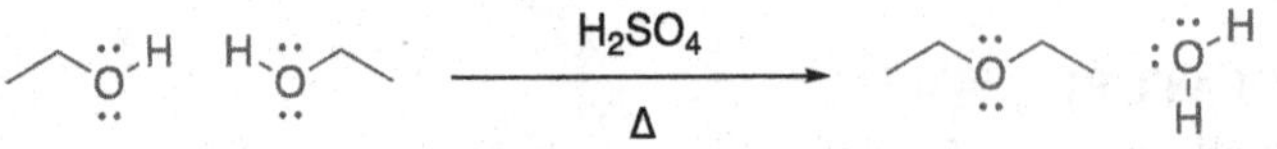

ETHER CLEAVAGE

Ethers, like alcohols, are nucleophilic (see the red, high-electron-density area on the electrostatic potential map, Figure 16.1). However, ethers are weak nucleophiles that require very strongly acidic electrophiles. The ether oxygen can be protonated under strongly acidic conditions to create a good leaving group.

There are two common reactions employing acidic conditions with ethers. The first is acid-catalyzed hydrolysis of ethers, which produces two alcohols (Figure 16.2).[1]

The mechanism for ether hydrolysis differs depending on the proclivity of the molecule to form stable carbocations. If a stable carbocation can be formed – tertiary or secondary – then the mechanism proceeds via an S_N1 mechanism.

Problem 16.1 Draw the arrows to complete the mechanism.

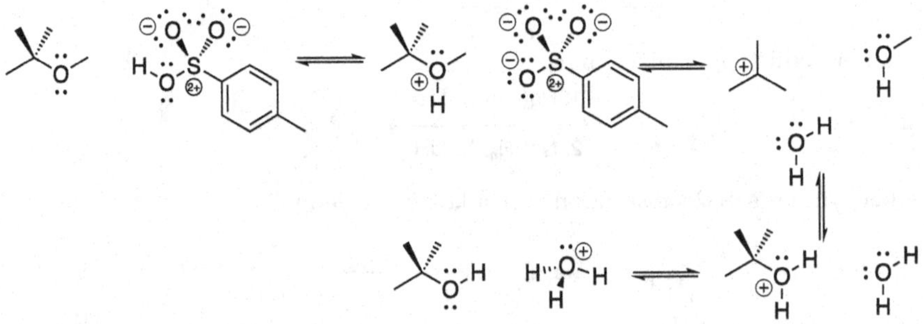

In contrast, if the loss of an alcohol would produce an unstable, primary, carbocation, then the reaction proceeds via an S_N2 mechanism.

Problem 16.2 Draw the arrows to complete the mechanism.

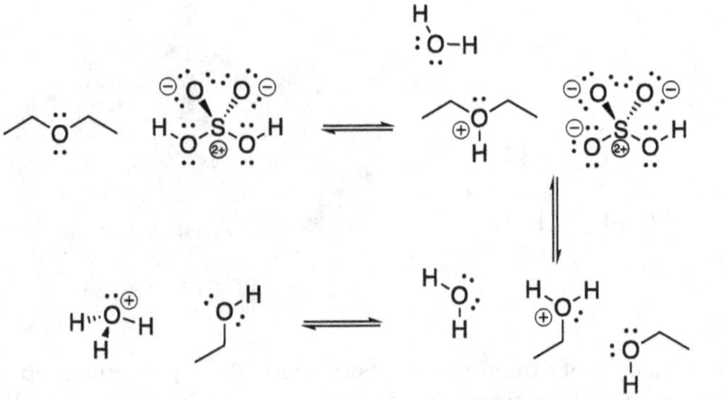

The second acidic ether cleavage uses HX acids: HCl, HBr, and HI. These reactions proceed similarly to the hydrolysis reactions but produce alkyl halides and alcohol products (Figure 16.3).[2]

Note that the mechanism for ether cleavage again differs depending on the proclivity of the molecule to form stable carbocations. If a stable carbocation can be formed – tertiary or secondary – then the mechanism proceeds via an S_N1 mechanism.

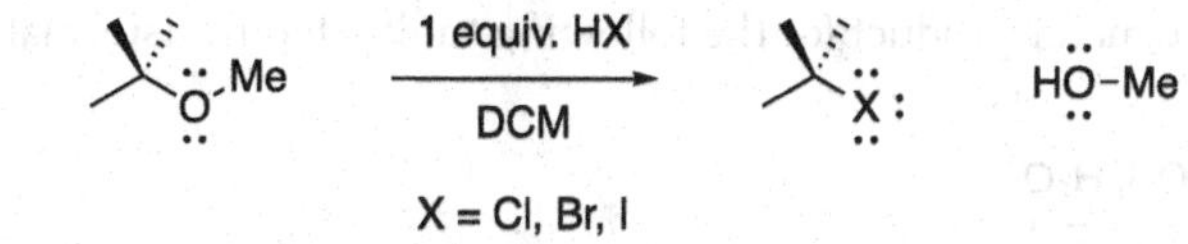

Figure 16.3 Acidic cleavage of MTBE with one equivalent of HX.

Problem 16.3 Draw the arrows to complete the mechanism.

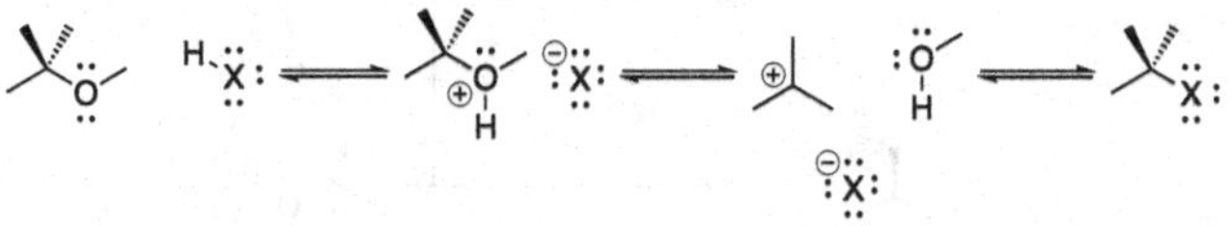

Again, if the loss of an alcohol would produce an unstable, primary, carbocation, then the reaction proceeds via an S_N2 mechanism.

Problem 16.4 Draw the arrows to complete the mechanism.

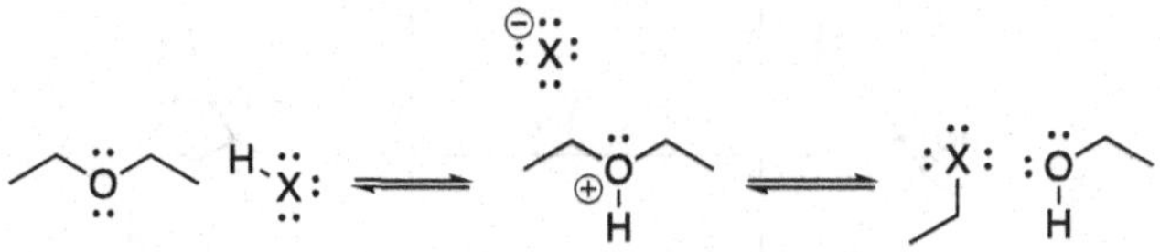

With excess HX (two equivalents or more) the ultimate products are two haloalkanes (Figure 16.4).[3] The mechanism proceeds the same as in Problem 16.4 to produce a haloalkane and an alcohol. The alcohol is then converted into a haloalkane as was seen previously.

Problem 16.5 Predict the product(s) for each of the following reactions.

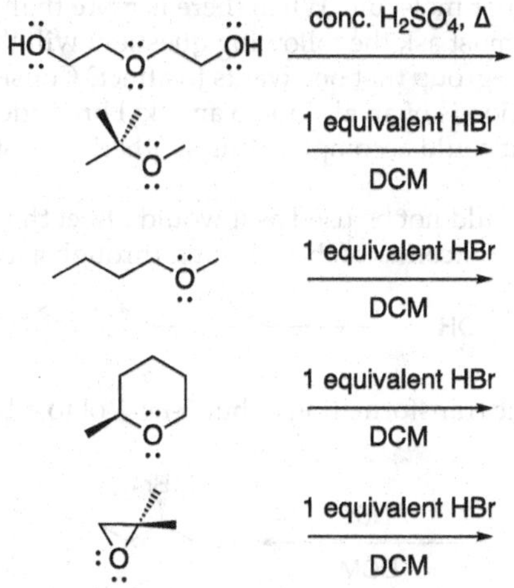

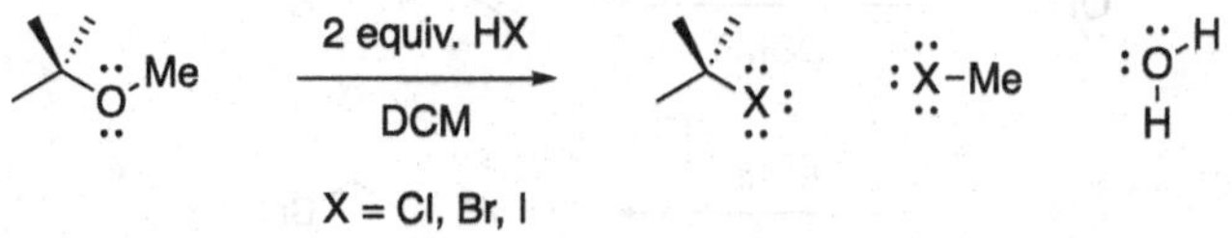

Figure 16.4 HX acid cleavage of MTBE with two equivalents of HX.

177

Problem 16.6 Predict the major product for the following multi-step transformation.

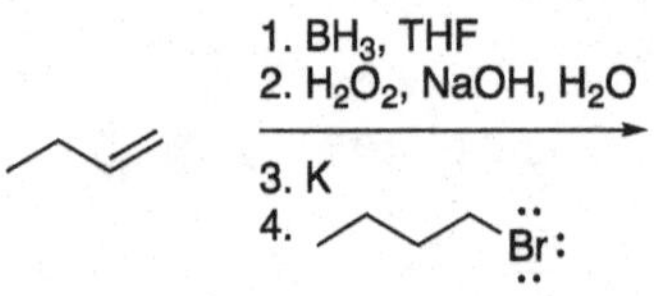

1. BH$_3$, THF
2. H$_2$O$_2$, NaOH, H$_2$O

3. K
4. (structure: ⌃⌃⌃Br:)

Problem 16.7 Provide the reagent(s) necessary to affect the following transformations.

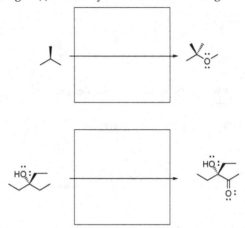

ETHERS AS PROTECTING GROUPS

Because ethers are only reactive under such limited (acidic) conditions, ethers are often employed as protecting groups. In organic synthesis, it is often the case that a molecule will have multiple functional groups on a particular molecule. When there is more than one functional group on a particular molecule, then one must ask the following question: will the reaction being carried out affect more than the functional group that one wants to affect? Consider the following example, where the desired transformation is of an alcohol to an alkyl bromide (Figure 16.5).

The available conditions that could accomplish this synthetic transformation are seen in Figure 16.6.

As can be seen, HBr/DCM could not be used as it would affect the alkene moiety. In this instance, therefore, the desired outcome can be achieved through a judicious choice of reagents.

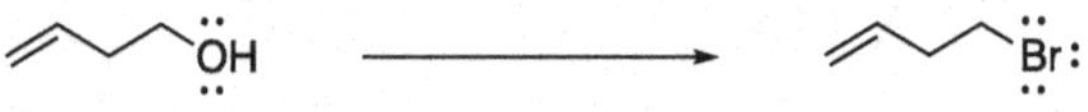

Figure 16.5 Desired synthetic transformation of but-3-en-1-ol to 4-bromobut-1-ene.

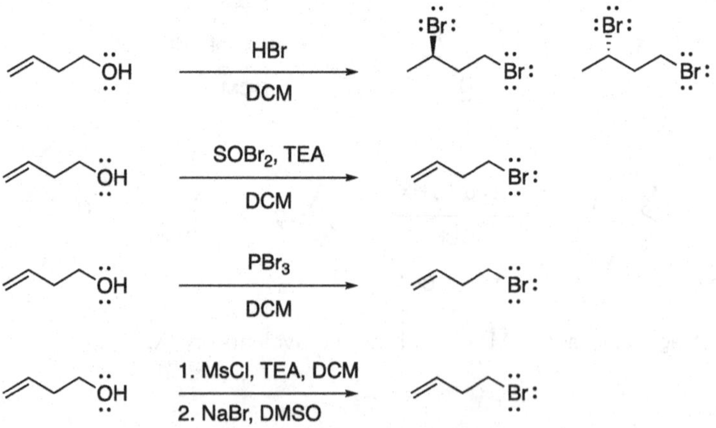

Figure 16.6 Possible bromination reactions of but-3-en-1-ol.

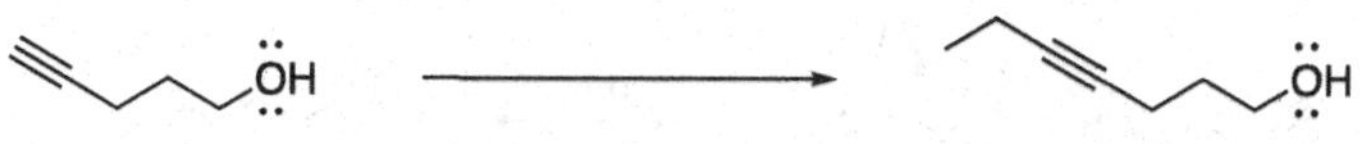

Figure 16.7 Desired synthetic transformation of pent-4-yn-1-ol to hept-4-yn-1-ol.

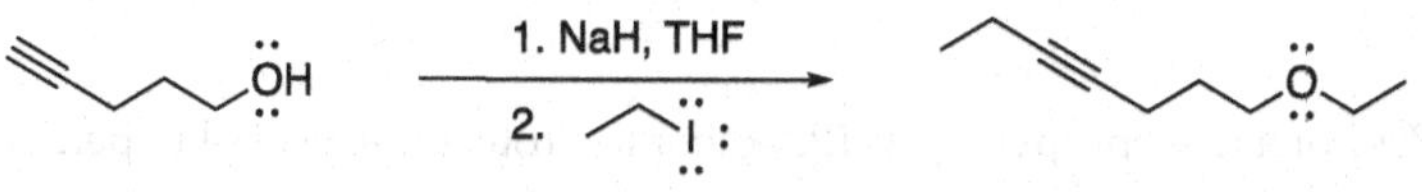

Figure 16.8 Possible substitution reactions of pent-4-yn-1-ol.

Now consider the following transformation, where the desired outcome is substitution of a terminal alkyne (Figure 16.7).

In this instance, there is only one possible set of conditions (Figure 16.8), which will lead to both alkyne substitution and Williamson ether synthesis, recall that alkyne pK_a values are about 25 and alcohol pK_a values are about 18 and so there is no way to avoid formation of the alkoxide by judicious choice of reagents.

To get only substitution of the alkyne, new synthetic methodologies would have to be created. One of the main avenues of research in modern organic chemistry is the development of new methodologies that selectively lead to only desired products in an atom-economical way. In the absence of a new methodology, however, another way is to protect the alcohol (Figure 16.9) with a protecting group (PG is a general symbol). The desired transformation is carried out and the protecting group is removed, or in common parlance we say that the desired functional group is deprotected.

A very common protecting group for alcohols, there are others that will not be considered here, are silyl ethers ($R-O-SiR_3$).[4] There are several, commonly employed silyl ether protecting groups, each with their own abbreviation. Despite their diversity in appearance, all silyl ethers are created through a common set of conditions (R_3Si-Cl with TEA in DCM), Figures 16.10 through 16.13.

The mechanism for formation of a silyl ether is unique among mechanisms seen to date. First, a pentavalent silicon intermediate is formed, then deprotonation followed by loss of chloride yields the final silyl ether.

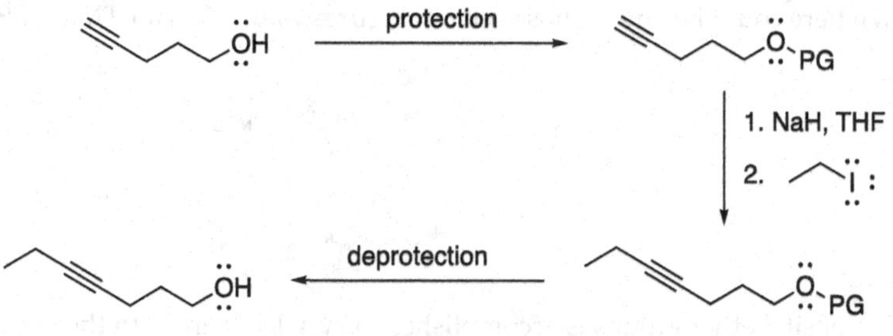

Figure 16.9 Use of a protecting group (PG) to carry out the desired alkyne substitution.

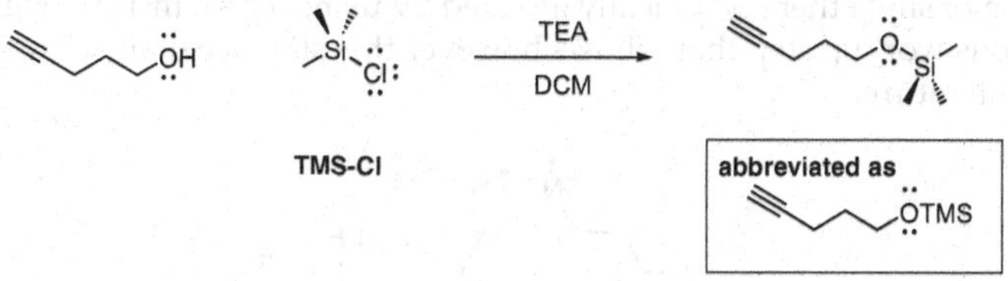

Figure 16.10 Use of a trimethylsilyl (TMS) ether to protect the alcohol in pent-4-yn-1-ol.

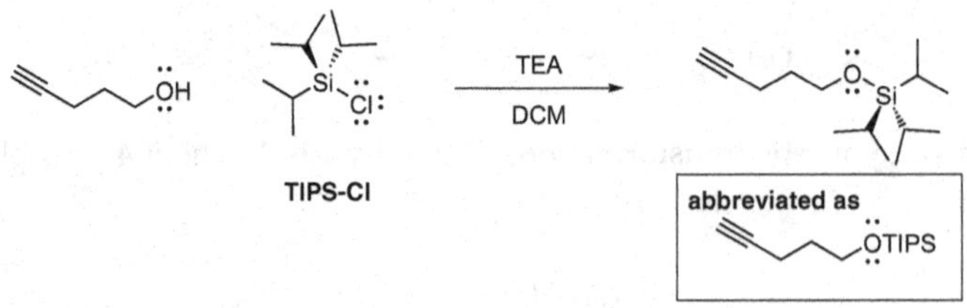

Figure 16.11 Use of a triisopropylsilyl (TIPS) ether to protect the alcohol in pent-4-yn-1-ol.

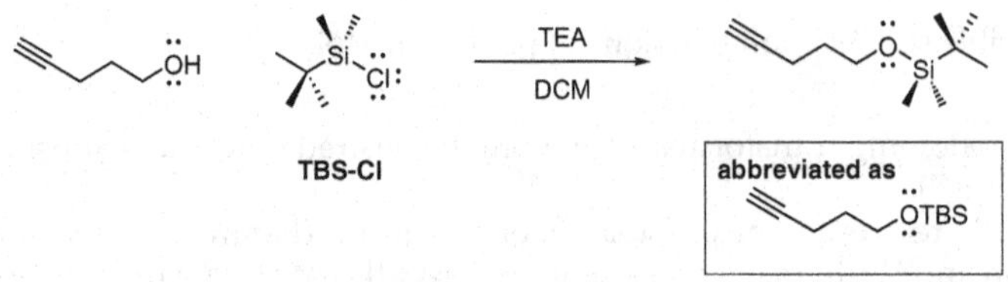

Figure 16.12 Use of a tert-butyldimethylsilyl (TBS) ether to protect the alcohol in pent-4-yn-1-ol.

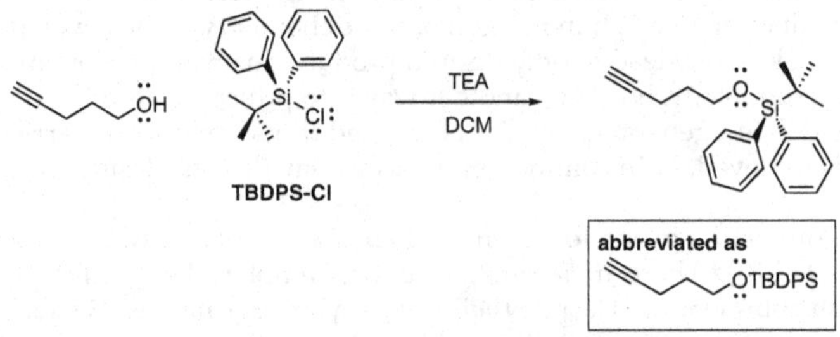

Figure 16.13 Use of a *tert*-butyldiphenylsilyl (TBDPS) ether to protect the alcohol in pent-4-yn-1-ol.

Problem 16.8 Draw appropriate electron-pushing arrows to complete the mechanism (note, the R groups shown here could be any of those seen in Figures 16.10–16.13 for TMS, TES, TIPS, TBS, or TBDPS).

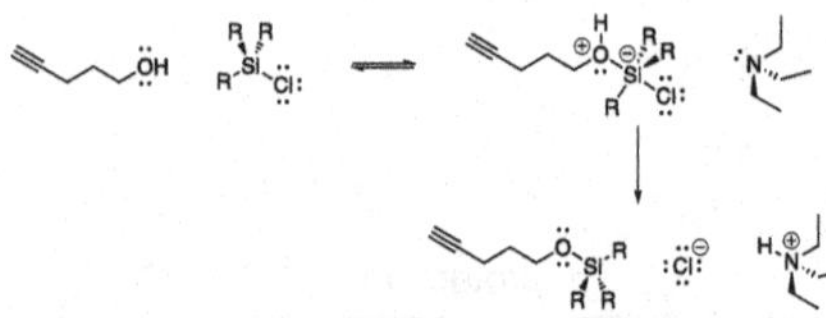

The deprotection of silyl ether groups is accomplished very selectively with the use of fluoride (F–). While HF can be used, the most common source of fluoride for deprotections is tetrabutylammonium fluoride (TBAF), Figure 16.14.

The deprotection of silyl ethers is typically affected by using TBAF in THF (Figure 16.15). There is always an aqueous workup step that follows however this step is usually omitted in textbooks and the chemical literature.

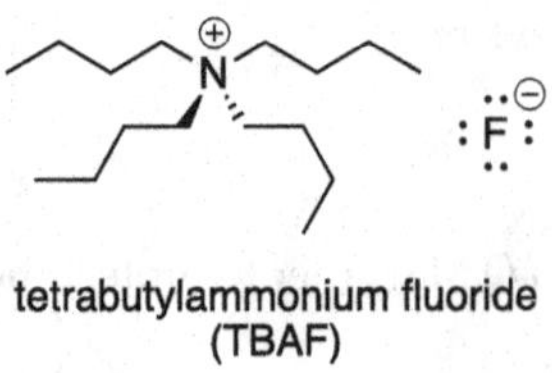

tetrabutylammonium fluoride
(TBAF)

Figure 16.14 Tetrabutylammonium fluoride (TBAF), a common fluoride ion source.

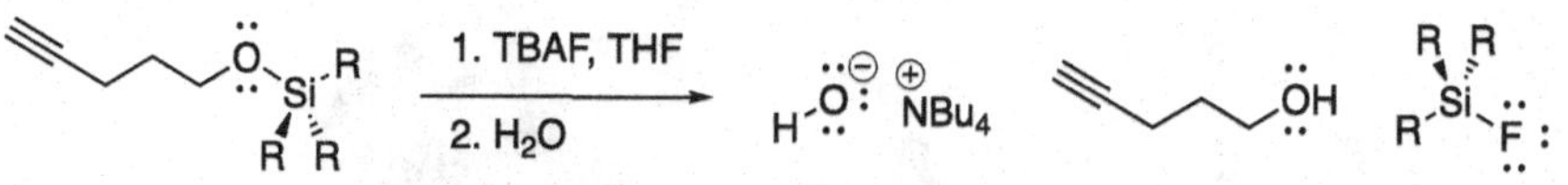

Figure 16.15 Silyl ether deprotection reaction using TBAF in THF with aqueous workup.

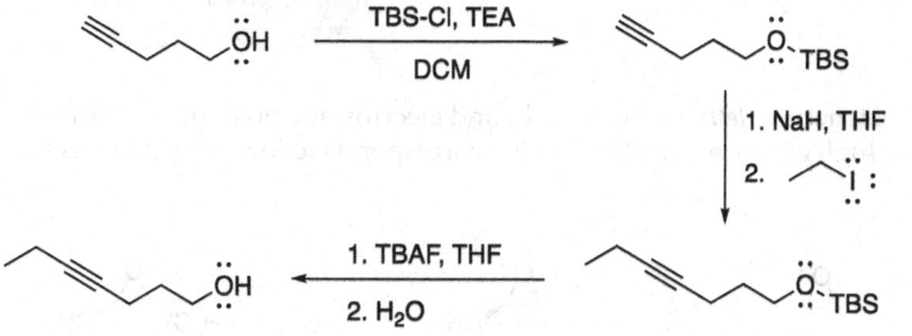

Figure 16.16 Successful synthesis of hept-4-yn-1-ol with a protecting group (TBS).

The mechanism for deprotection of a silyl ether is effectively the reverse of the protection mechanism: first, a pentavalent silicon intermediate is formed, then loss of the alkoxide and protonation give the alcohol product.

Problem 16.9 Draw appropriate electron-pushing arrows to complete the mechanism (note, the R groups shown here could be any of those seen in Figures 16.10–16.13 for TMS, TES, TIPS, TBS, or TBDPS).

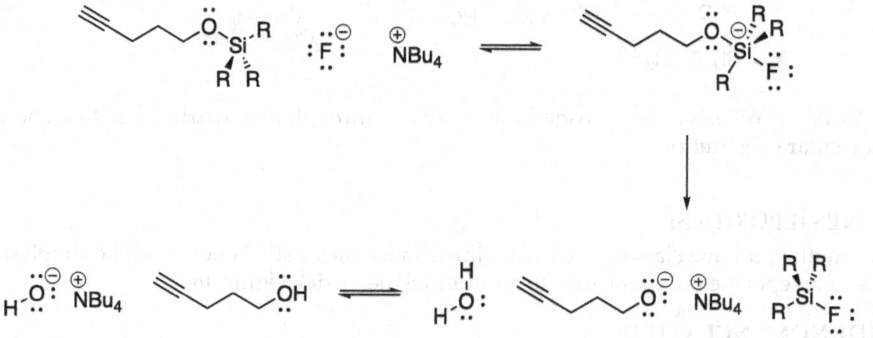

Overall, then, the desired synthetic transformation could be affected by utilizing the following steps:

Problem 16.10 For each of the following transformations, indicate which of the following could be executed with a judicious choice of reagents, which would require a protecting group to carry out, and which are impossible to affect with current knowledge of reagents and protecting groups. Then provide the appropriate steps necessary to carry out each of the following.

181

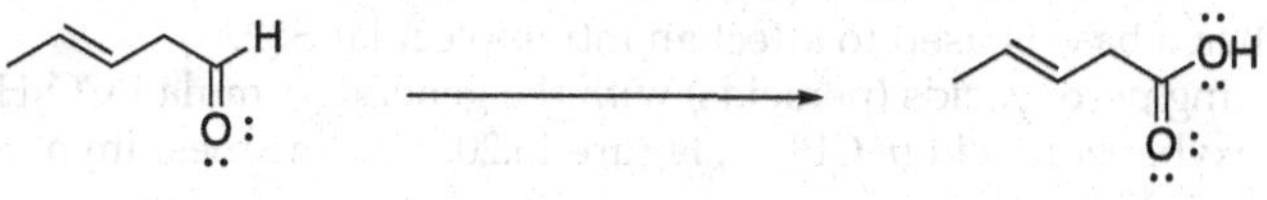

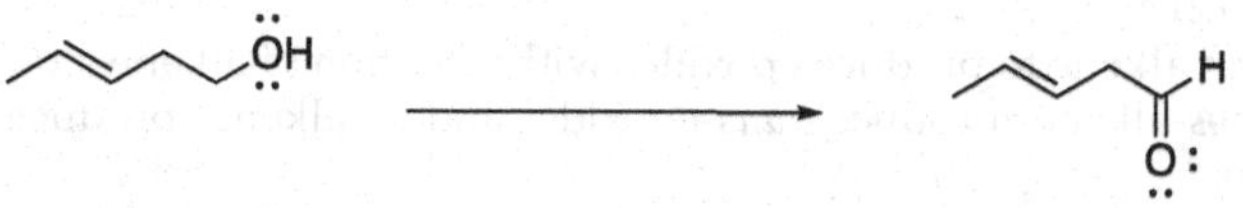

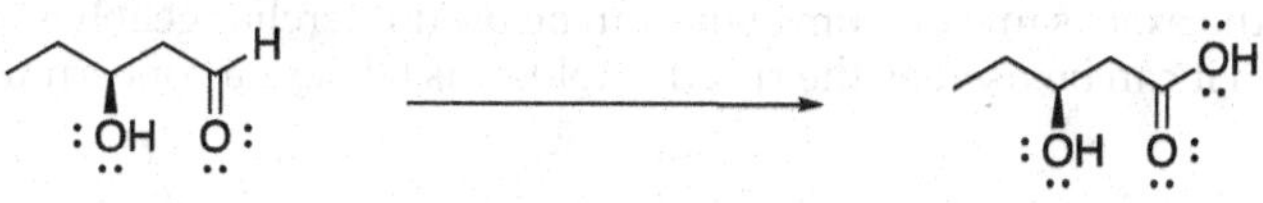

Figure 16.17 Structure (*left*), model (*middle*) and electrostatic potential map (*right*) of oxirane (red corresponds to high electron density and blue corresponds to low electron density).

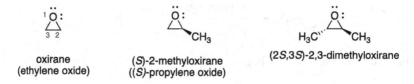

oxirane
(ethylene oxide)

(*S*)-2-methyloxirane
((*S*)-propylene oxide)

(2*S*,3*S*)-2,3-dimethyloxirane

Figure 16.18 IUPAC numbering and nomenclature of some simple epoxides. Trivial names are given in parentheses.

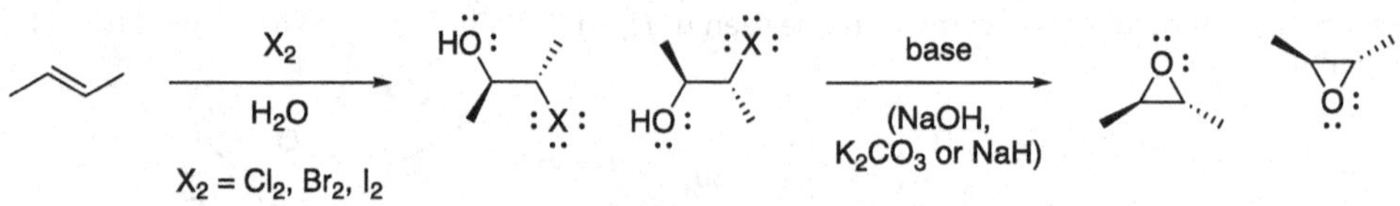

Figure 16.19 Synthesis of an epoxide from an alkene through halohydrin formation and then intramolecular substitution.

OXIRANES (EPOXIDES)

There is another, unique class of ether-containing molecules, called epoxides. The simplest example of an epoxide is oxirane (trivial name ethylene oxide), Figure 16.17.

EPOXIDE NOMENCLATURE

Since the simplest epoxide is oxirane, in the IUPAC nomenclature system, epoxides are named as derivatives of oxirane (Figure 16.18). are then named using the usual rules for substitution (note in this system, oxygen is position one and the carbon atoms are positions two and three in the ring).

SYNTHESIS OF EPOXIDES

Epoxides can be synthesized by several different methods. The first, which is more atom-economical, is through an extension of previous chemistry (Figure 16.19). Namely, an alkene is first converted into a halohydrin and then a base is used to affect an intramolecular S_N2.[5]

Epoxides are also synthesized using peroxyacids (peracids) with the general formula RCO_3H.[6] The most used are *meta*-chloroperoxybenzoic acid (*m*CPBA), Figure 16.20,[7] and magnesium monoperoxyphthalate (MMPP), Figure 16.21.[8]

Peracids like *m*CPBA react with an alkene to produce epoxides with retention of alkene stereochemistry, which means that *trans*-alkenes produce *trans*-epoxides and *cis*-alkenes produce *cis*-epoxides (Figure 16.22).

MMPP produces (Figure 16.23) the exact same outcome (and can be used interchangeably with *m*CPBA). MMPP is preferable when it can be used as the risk of explosions (always a concern with peracids) is negligible.

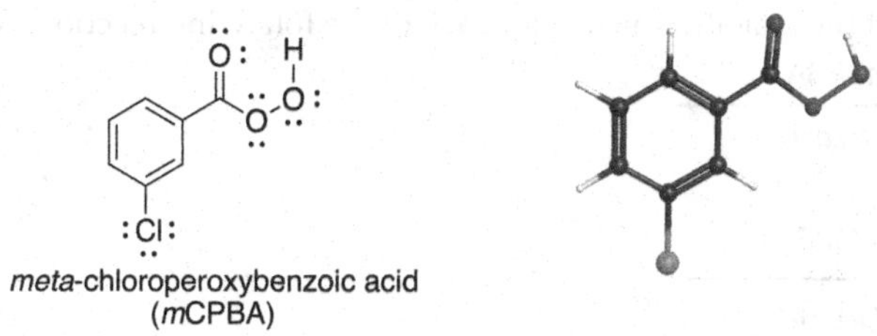

Figure 16.20 Structure (*left*) and 3D model (*right*) of *meta*-chloroperoxybenzoic acid (*m*CPBA).

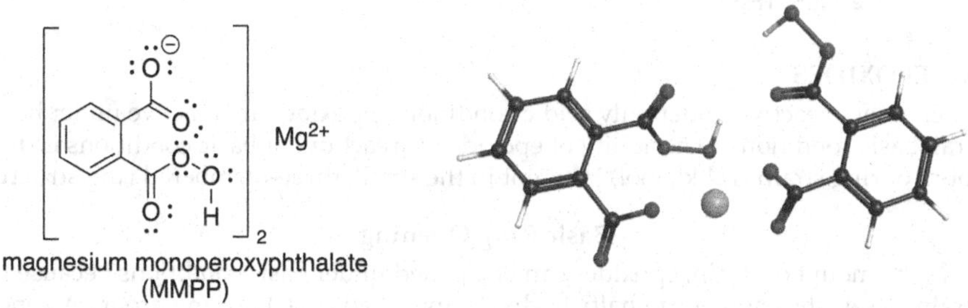

Figure 16.21 Structure (*left*) and 3D model (*right*) of magnesium monoperoxyphthalate (MMPP).

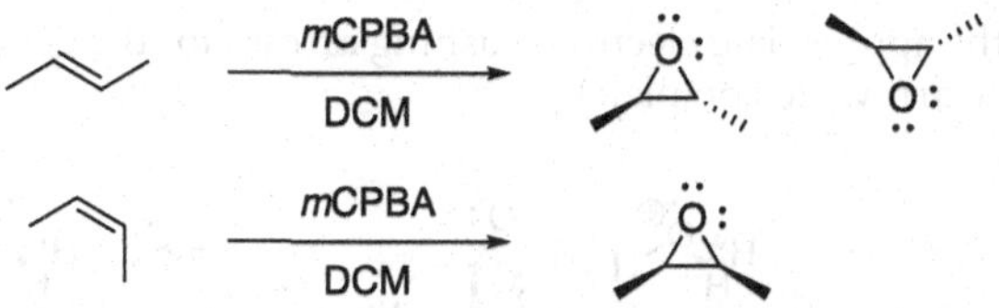

Figure 16.22 Epoxide formation (with retention of *cis*- and *trans*-stereochemistry) with *m*CPBA.

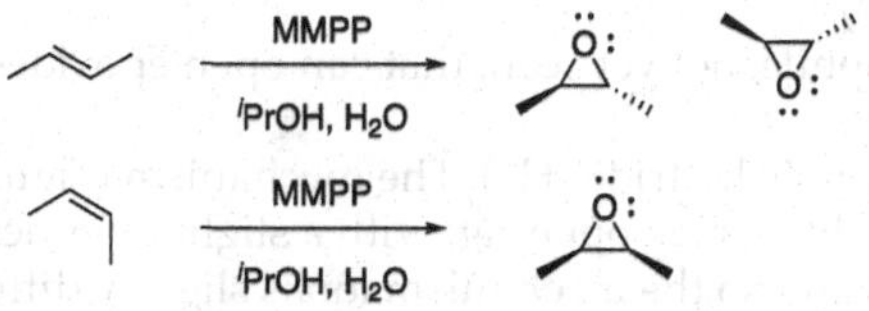

Figure 16.23 Epoxide formation (with retention of *cis*- and *trans*-stereochemistry) with MMPP.

Problem 16.11 The mechanism of peracid epoxidation is a single, concerted step that transfers the oxygen atom from the peracid to the alkene. Draw all the electron-pushing arrows necessary to accomplish this transformation.

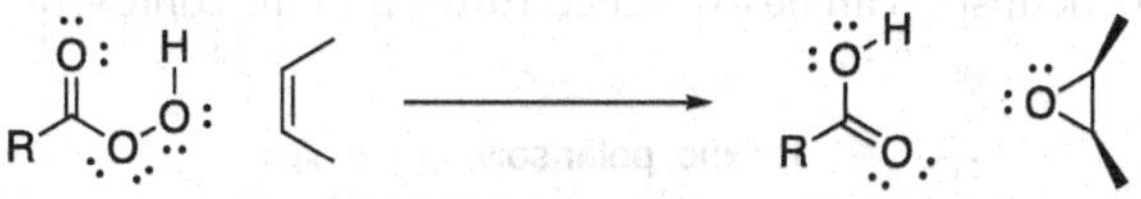

Problem 16.12 Predict the major product(s) for each of the following reactions.

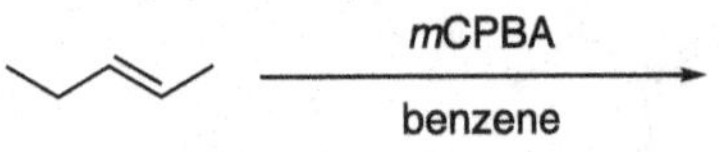

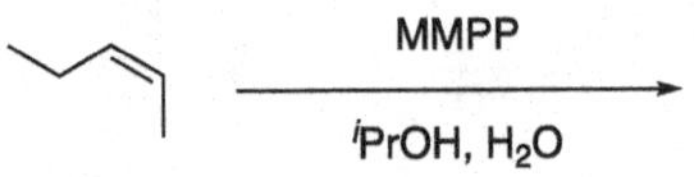

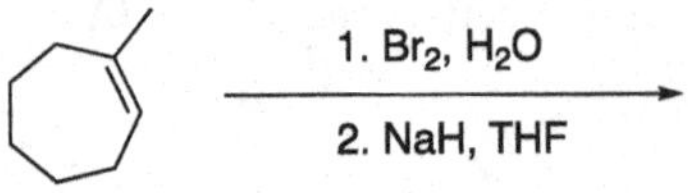

OPENING EPOXIDES

Unlike other ethers, reactive under only acidic conditions, epoxides are reactive under both acidic and neutral/basic conditions.[9] The ability of epoxides to react under basic conditions is due to the high amount of ring strain (114 kJ/mol) inherent in the small, three-membered ring structure.

Basic Ring Opening

Given the high amount of strain, epoxides can be opened under basic conditions because the relief of ring strain offsets the energetic penalty of displacing a bad (RO–) leaving group. A general neutral/basic reaction is seen in Figure 16.24.

Notice that the nucleophile attacks the least hindered side of the epoxide (the least substituted side), opening the ring. In neutral/basic ring-opening reactions of an epoxide proceed by a normal S_N2 mechanism, in which steric repulsion dictates the outcome.

Problem 16.13 Provide the appropriate electron-pushing arrows for the general neutral/basic mechanism (draw the arrows to complete).

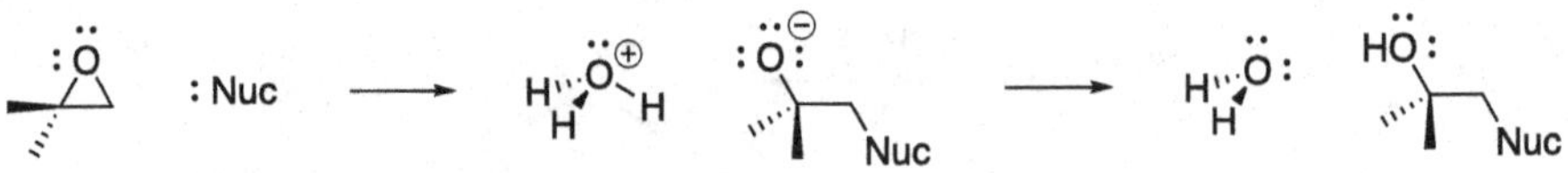

There are several neutral, basic nucleophiles that will work for epoxide opening that have been seen previously (Figure 16.25).

There is also a new nucleophile, not yet seen, that can open epoxides, lithium aluminium hydride (LAH),[10] Figure 16.26.

LAH is a source of nucleophilic hydride (H–). The mechanism (Figure 16.27) is very similar to that already seen for neutral/basic ring opening, with a slight modification. The hydride is not free but connected to aluminium and so the mechanism looks slightly different than those shown in Problem 16.13 (also note that the fate of the aluminium is ignored here for clarity).

Acidic Ring Opening

As ethers, epoxides are also reactive under acidic conditions. What is of particular interest, is that under acidic-ring-opening conditions the final product is the opposite of the regiochemistry (Figure 16.28) seen in the basic ring-opening case.

The reversal in regiochemistry can be explained fully on in the context of the mechanism.

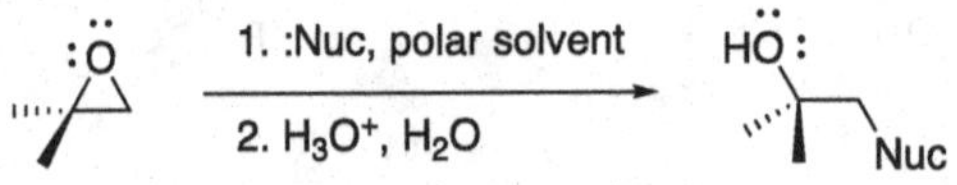

Figure 16.24 Generic neutral/basic ring-opening reaction of an epoxide.

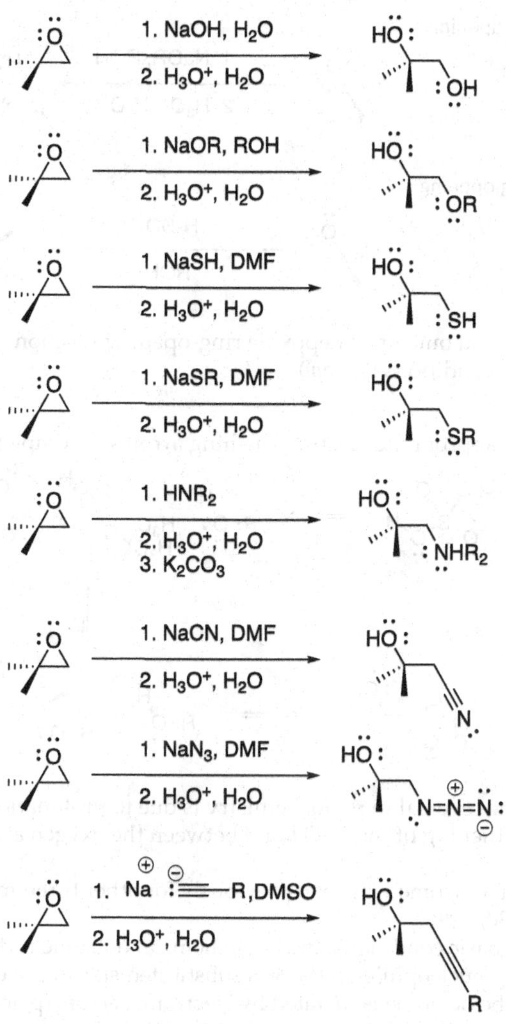

Figure 16.25 Examples of neutral/basic ring-opening reactions involving 2,2-dimethyloxirane.

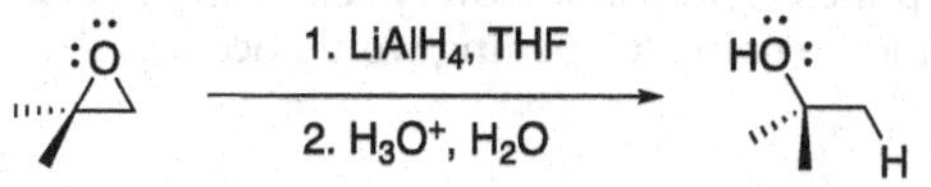

Figure 16.26 Ring-opening reaction of 2,2-dimethyloxirane with LiAlH₄.

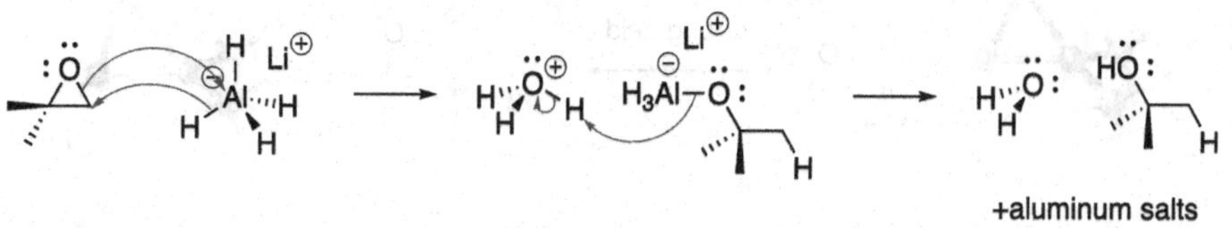

Figure 16.27 Mechanism of LAH epoxide opening.

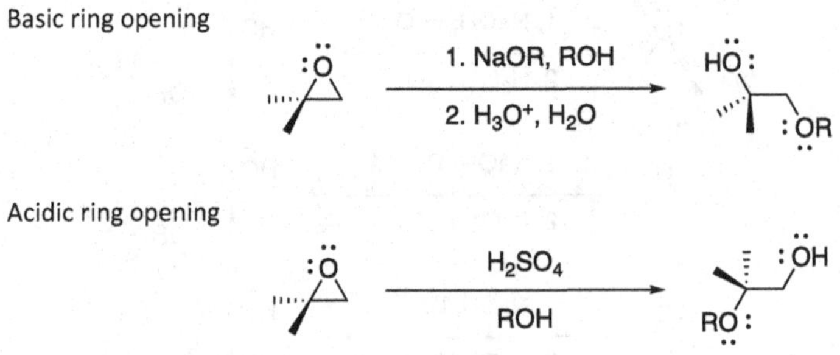

Basic ring opening

$$\text{1. NaOR, ROH} \quad \text{2. } H_3O^+, H_2O$$

Acidic ring opening

$$\text{H}_2\text{SO}_4 \quad \text{ROH}$$

Figure 16.28 Regiochemical outcome of epoxide ring-opening reaction under neutral/basic conditions (*top*) and acidic conditions (*bottom*)

Problem 16.14 Provide the appropriate electron-pushing arrows to complete the mechanism.

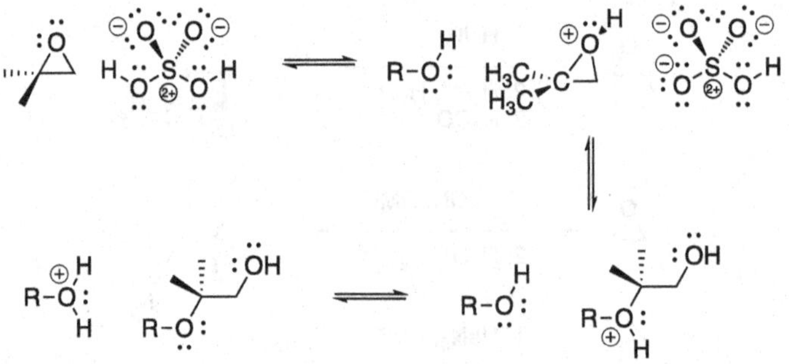

As seen in Figure 16.29, the reversal in regiochemistry is due to protonation of the epoxide oxygen atom, which leads to lengthening of the C–O bond between the oxygen atom and the more substituted carbon atom.

The same regiochemical outcome was seen previously for other three-membered-ring, cationic intermediates (Figure 16.30).

Altogether then, under basic conditions, the regiochemical outcome is dictated by steric repulsion (placing the incoming nucleophile on the less substituted side of the epoxide). In contrast, under acidic conditions, the outcome is dictated by electronic factors (placing the incoming nucleophile on the more substituted side of the epoxide).

There are two, common Brønsted–Lowry conditions for opening epoxides to make β–alkoxylalcohols/vicinal diols (depending on the identity of the solvent) and β–haloalcohols (Figure 16.31).

For strongly basic nucleophiles, a Brønsted–Lowry acid could not be used (as the acid would protonate the nucleophile and no epoxide opening would occur).

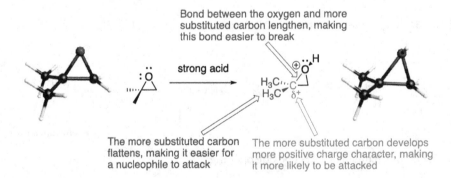

Bond between the oxygen and more substituted carbon lengthen, making this bond easier to break

strong acid

The more substituted carbon flattens, making it easier for a nucleophile to attack

The more substituted carbon develops more positive charge character, making it more likely to be attacked

Figure 16.29 Effect of protonation on epoxide ring structure.

Oxymercuration (mercurinium ion)

Halogenation (halonium ion, here a bromonium ion is shown)

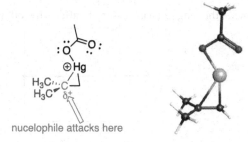

Figure 16.30 Structures and 3D models of a three-membered ring, cationic intermediates.

Figure 16.31 Acidic epoxide opening of 2,2-dimethyloxirane with sulfuric acid and a protic solvent (*top*) and with HX in DCM (*bottom*)

Problem 16.15 Predict the major product(s) for each of the following reactions.

1. *m*CPBA, benzene

2. KN$_3$, DMF

1. *m*CPBA, benzene

2. H$_2$SO$_4$, H$_2$O
3. conc. H$_2$SO$_4$, Δ

1. *m*CPBA, benzene

2. H$_2$SO$_4$, H$_2$O
3. (COCl)$_2$, DMSO,
   TEA, DCM

GENERAL PRACTICE PROBLEMS

Problem 16.16 Provide the reagent(s) necessary to affect the following transformation.

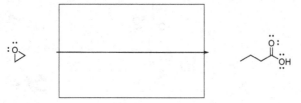

Problem 16.17 Provide the reagents to effect each transformation a)–i).

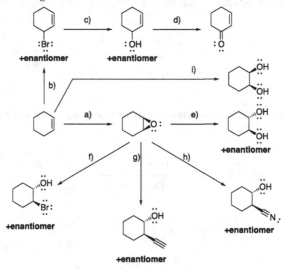

NOTES

1. i) Smith, L.; Wode, G.; Widhe, T. Kinetik der Wasseraddition einiger Oxidoverbindungen. Eine neue Wasserstoffionenkatalyse. *Z. Phys. Chem.*, **1927**, *130U* (1), 154–166. DOI: 10.1515/zpch-1927-13020

   ii) Skrabal, R. Zur Hydrolysegeschwindigkeit der Äther. *Z. physik. Chem.*, **1939**, *185A* (1), 81–96. DOI: 10.1515/zpch-1939-18507

   iii) Burwell Jr., R.L. The Cleavage of Ethers. *Chem. Rev.*, **1954**, *54* (4), 615–685. DOI: 10.1021/cr60170a003

2. i) Butlerow, A. Ueber die Aethylmilchsäure. *Liebig Ann.*, **1861**, *118* (3), 325–330. DOI: 10.1002/jlac.18611180311

   ii) Saytzeff, C. Ueber Paraoxybenoësäure, Zerstzungsproduct der Anissäure durch Jodwasserstoffsäure. *Liebig Ann.*, **1863**, *127* (2), 129–137. DOI: 10.1002/jlac.18631270202

   iii) Graebe, C. II. Untersuchungen über die Oxysäuren der aromatischen Reihe. *Liebig Ann.*, **1866**, *139* (2), 134–149. DOI: 10.1002/jlac.18661390203

3. Stone, H.; Shechter, H. A new method for the preparation of organic iodides. *J. Org. Chem.*, **1950**, *15* (3), 491–495. DOI: 10.1021/jo01149a008

4. Corey, E.J.; Venkateswarlu, A. Protection of hydroxyl groups as tert-butyldimethylsilyl derivatives. *J. Am. Chem. Soc.*, **1972**, *94* (17), 6190–6191. DOI: 10.1021/ja00772a043

5. Winstein, S.; Lucas, H.J. Retention of Configuration in the Reaction of the 3-Bromo-2-butanols with Hydrogen Bromide. *J. Am. Chem. Soc.*, **1939**, *61* (6), 1576–1581. DOI: 10.1021/ja01875a070

6. Prileschajew, N. Oxydation ungesättigter Verbindungen mittels organischer Superoxyde. *Ber. Dtsch. Chem. Ges.*, **1909**, *42* (4), 4811–4815. DOI: 10.1002/cber.190904204100

7. Gipstein, E.; Nichik, F.; Offenbach, J.A. *m*-Chloroperbenzoic acid as a reagent for the determination of unsaturation in natural and cyclized rubber. *Anal. Chim. Acta.*, **1968**, *43*, 129–131. DOI: 10.1016/s0003-2670(00)89187-4

8. Brougham, P.; Cooper, M.S.; Cummerson, D.A.; Heaney, H.; Thompson, N. Oxidation Reactions using Magnesium Monoperphthalate: A comparison with *m*-Chloroperoxybenzoic acid. *Synthesis*, **1987**, *1987* (11), 1015–1017. DOI: 10.1055/s-1987-28153

9. Burwell Jr., R.L. The Cleavage of Ethers. *Chem. Rev.*, **1954**, *54* (4), 615–685. DOI: 10.1021/cr60170a003

10. Per IUPAC rules, this should properly be named aluminium lithium hydride.

17 Organometallic Chemistry

Magnesium, Lithium, Copper, and Carbenes

Until now, attention has been paid to organic molecules with functional groups (halogens and alcohols) that cause the carbon to which they are attached to become electrophilic (as can be seen on the electrostatic potential maps (Figure 17.1), the carbon atoms appear blue to blue-green, which indicates low-electron density).

In this chapter, attention will turn to organometallic compounds (those with a carbon–metal bond). Since carbon is more electronegative than almost all metals, organometallic compounds contain electron-rich (orange to red in electrostatic potential maps in Figure 17.2) carbon atoms. Since organometallic carbon atoms are electron-rich, these are some of the few classes of molecules with nucleophilic carbon atoms. Two, early types of organometallic reagents are organomagnesium compounds (Grignard reagents, RMgBr)[1] and organolithium (RLi)[2] compounds.

A later type of organometallic reagent (Figure 17.3) is the diorganocopper(I) lithium compounds (Gilman reagents R$_2$CuLi).[3]

ORGANOMETALLIC REAGENT AGGREGATION

It should be noted here that the way in which organometallic reagents are depicted is somewhat misleading. It is most common that organometallic reagents are depicted as single, discrete molecules (Figure 17.4).

These reagents exist as aggregates (dimers, trimers, tetramers, octomers, etc.) in solution (Figure 17.5).[4] This aggregation behaviour is often ignored, but it will occasionally be brought up to rationalize certain phenomena.

ORGANOMETALLIC REAGENT SYNTHESIS

To synthesize the organometallic reagents (Figure 17.6), the reactions start with some sort of brominated hydrocarbon (bromoalkane, bromoalkene, or bromobenzene).

The Gilman reagents are made (Figure 17.7) by reacting an organolithium reagent with a copper(I) salt, usually: CuCl, CuBr, CuI, or CuCN.

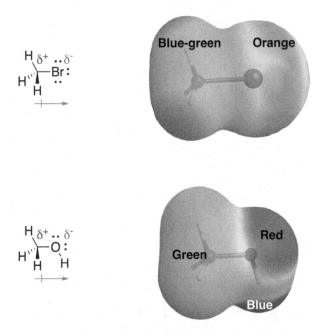

Figure 17.1 Structure and electrostatic potential maps for methyl bromide (*top*) and methanol (*bottom*).

DOI: 10.1201/9781003479352-17

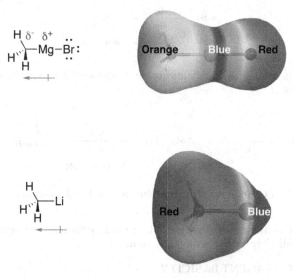

Figure 17.2 Structure and electrostatic potential maps for methylmagnesium bromide (*top*) and methyl lithium (*bottom*).

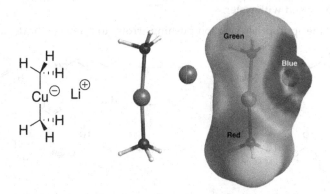

Figure 17.3 Structure, 3D model, and electrostatic potential map for dimethylcopper(I) lithium.

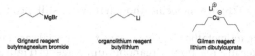

Grignard reagent
butylmagnesium bromide

organolithium reagent
butyllithium

Gilman reagent
lithium dibutylcuprate

Figure 17.4 Simplified skeletal structure representations of Grignard, organolithium, and Gilman reagents.

Grignard reagent
butylmagnesium bromide

organolithium reagent
butyllithium

Gilman reagent
lithium dibutylcuprate

Figure 17.5 More realistic skeletal structure representations of Grignard, organolithium, and Gilman reagents as they occur in solution.

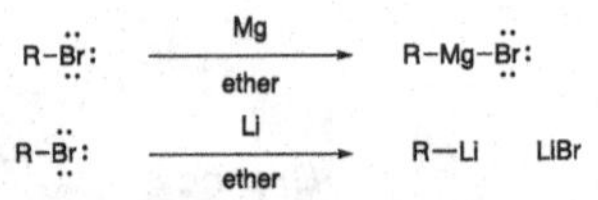

Figure 17.6 Synthesis of Grignard reagents (*top*) and organolithium reagents (*bottom*) from brominated hydrocarbons.

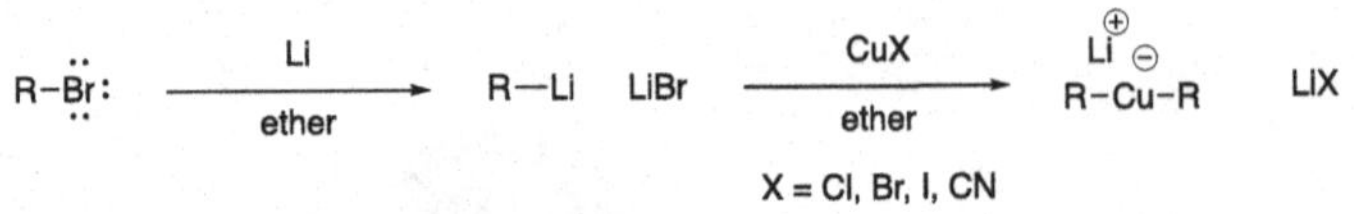

Figure 17.7 Synthesis of Gilman reagents from brominated hydrocarbons.

Note that the mechanism for the formation of Grignard, organolithium, and Gilman reagents is beyond the scope of this chapter and will not be shown here.

ORGANOMETALLIC REAGENT BASICITY

Organometallic reagents are strongly basic, nucleophilic reagents. Because of their high basicity (organometallic reagents are considered as analogues of carbanions) Grignard reagents and organolithium reagents are incompatible with all acidic functional groups (carboxylic acids, thiols, phenols, and alcohols) and with water.

Problem 17.1 Draw the appropriate electron-pushing arrows to complete the deprotonations.

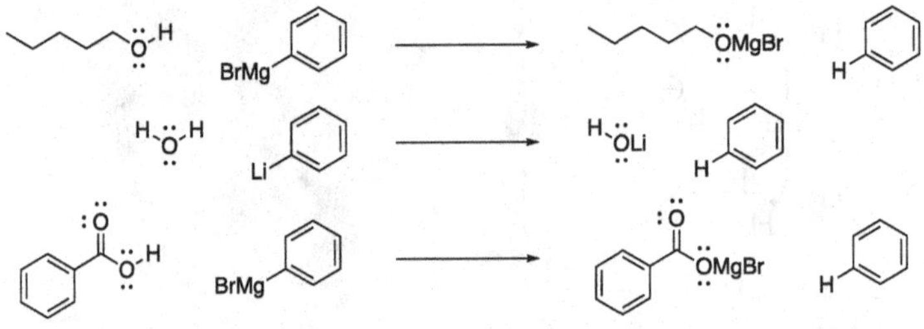

Gilman reagents are "softer" nucleophilic species than either Grignard reagents or organolithium reagents. That is, they are less basic (copper has a much greater electronegativity (1.90 versus 2.55 for carbon) than either lithium (0.98) or magnesium (1.31)). Gilman reagents can tolerate alcohol functional groups (Figure 17.8) in a molecule (though they are still incompatible with carboxylic acids and with water).

The high basicity of organometallic reagents means that special care needs to be taken (no acidic functional groups or water in a reaction). Organolithium reagents (Figure 17.9) are, in fact, most often used as strong bases, especially the butyllithium (BuLi) reagents.

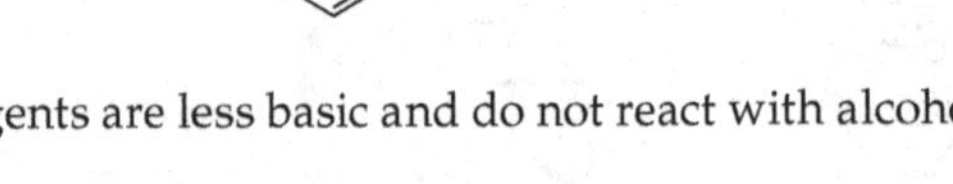

Figure 17.8 Gilman reagents are less basic and do not react with alcohol functional groups.

n-butyllithium
(nBuLi)

sec-butyllithium
(sBuLi)

tert-butyllithium
(tBuLi)

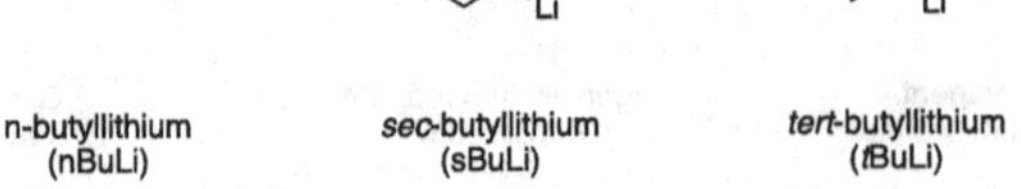

Figure 17.9 Common organolithium reagents, isomers of butyllithium.

Problem 17.2 Draw the appropriate electron-pushing arrows to complete the following deprotonation reaction.

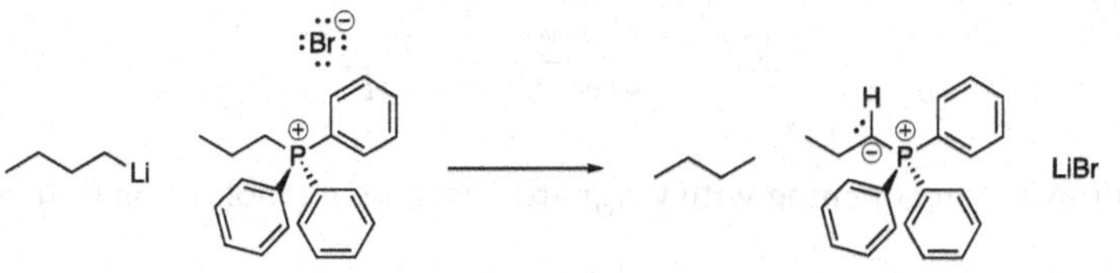

Problem 17.3 Predict the major products for each of the following deprotonation reactions.

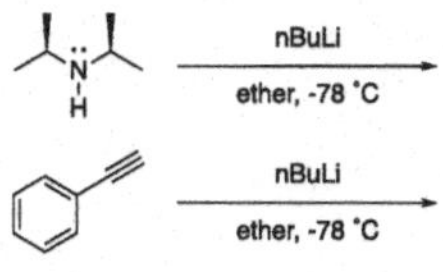

ORGANOMETALLIC REAGENTS AS NUCLEOPHILES

As is true for almost every strong base, Grignard reagents, organolithium reagents, and Gilman reagents are also strong nucleophiles. Organometallic reagents are commonly employed to open epoxides (Figures 17.10 and 17.11).

Problem 17.4 Draw the appropriate electron-pushing arrows for epoxide opening by an organometallic reagent (R–M here is used as a generic stand-in as Grignard, organolithium, and Gilman reagent).

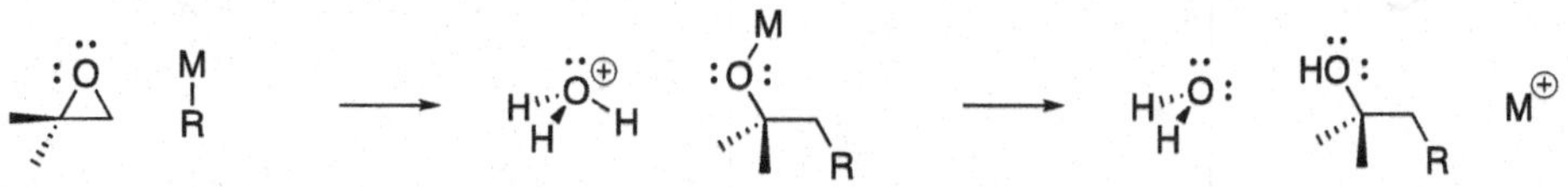

In the next chapters, attention will be paid to how these organometallic nucleophiles react with carbonyl-containing molecules.

Problem 17.5 Predict the major product(s) for each of the following reactions.

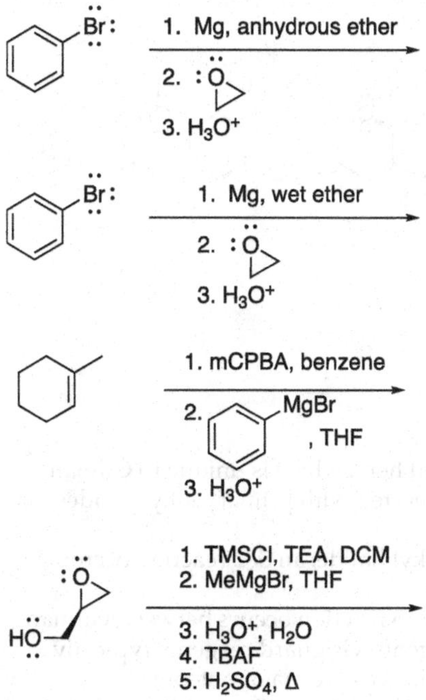

1. Mg, anhydrous ether
2. :O (epoxide)
3. H_3O^+

1. Mg, wet ether
2. :O (epoxide)
3. H_3O^+

1. mCPBA, benzene
2. (phenyl)MgBr, THF
3. H_3O^+

1. TMSCl, TEA, DCM
2. MeMgBr, THF
3. H_3O^+, H_2O
4. TBAF
5. H_2SO_4, Δ

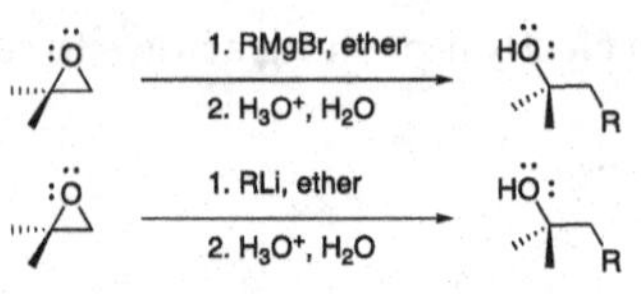

Figure 17.10 Epoxide ring opening with Grignard (*top*) and organolithium (*bottom*) reagents.

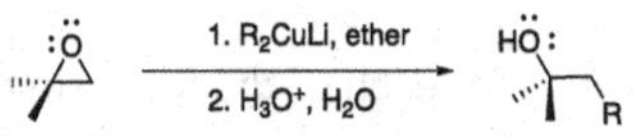

Figure 17.11 Epoxide ring opening with a Gilman reagent.

Problem 17.6 Provide the reagent(s) necessary to affect the following transformations.

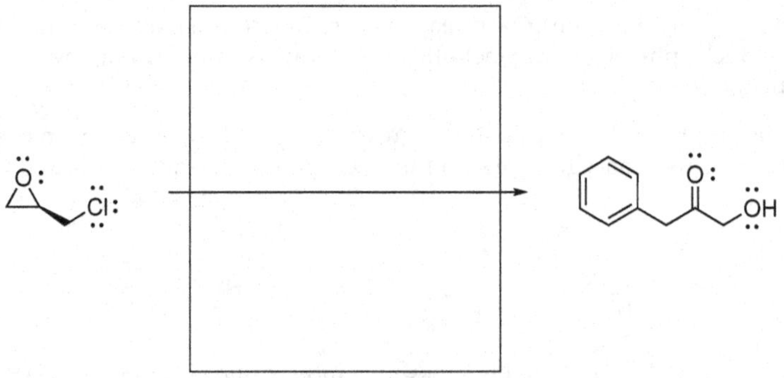

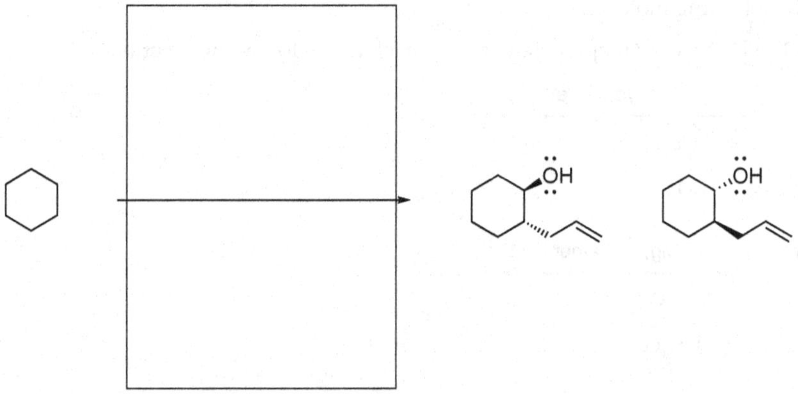

GILMAN REAGENT HALOGEN SUBSTITUTION

A final type of organometallic reactivity that will be considered here, which is unique to Gilman reagents, is halogen substitution.[5] When a Gilman reagent is reacted with primary alkyl halides, an S_N2 reaction occurs (Figure 17.12).

And when a Gilman reagent is reacted with a secondary alkyl halide, an S_N2 reaction occurs (Figure 17.13).

Tertiary alkyl halides do not undergo S_N2 reactions, and so no reaction occurs between Gilman reagents and tertiary alkyl halides. In contrast to Gilman reagents, Grignard reagents typically produce E2 elimination products (Figure 17.14) with alkyl halides as the major outcome.[6]

Moreover, organolithium reagents engage in halogen-lithium exchange (Figure 17.15) rather than in S_N2 reactions with alkyl halides.[6]

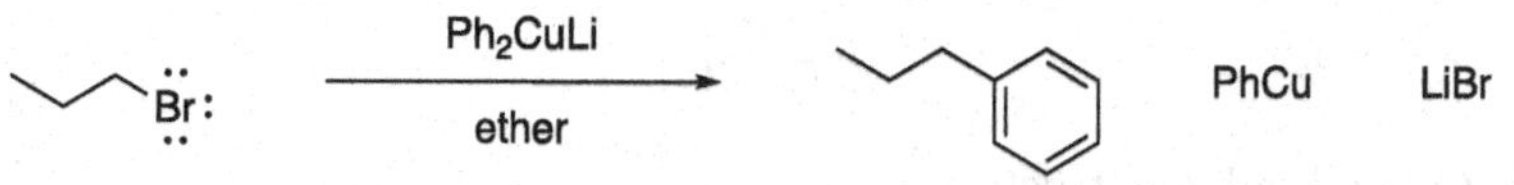

Figure 17.12 S$_N$2 substitution of a primary alkyl halide with diphenylcopper(I) lithium.

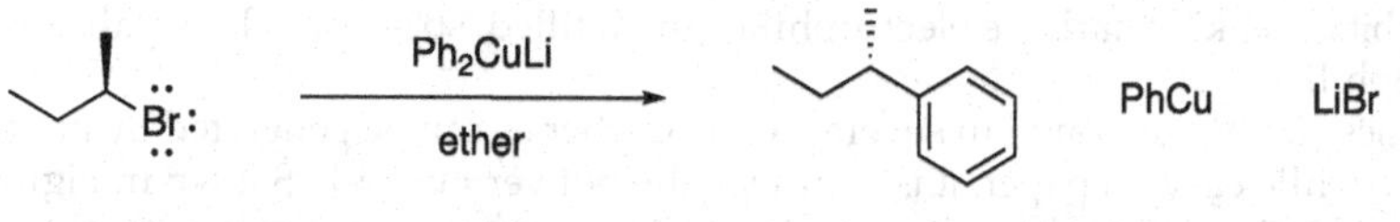

Figure 17.13 S$_N$2 substitution of a secondary alkyl halide with diphenylcopper(I) lithium.

Figure 17.14 E2 elimination of a secondary alkyl halide with phenylmagnesium bromide.

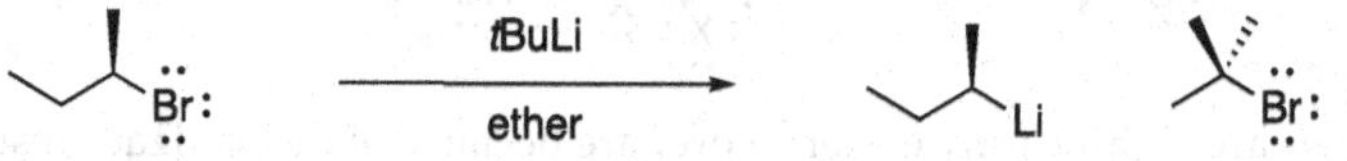

Figure 17.15 Lithium-halogen exchange of a secondary alkyl halide with *tert*-butyl lithium.

Problem 17.7 Predict the major product(s) for each of the following reactions.

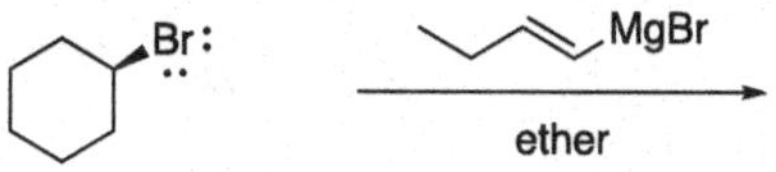

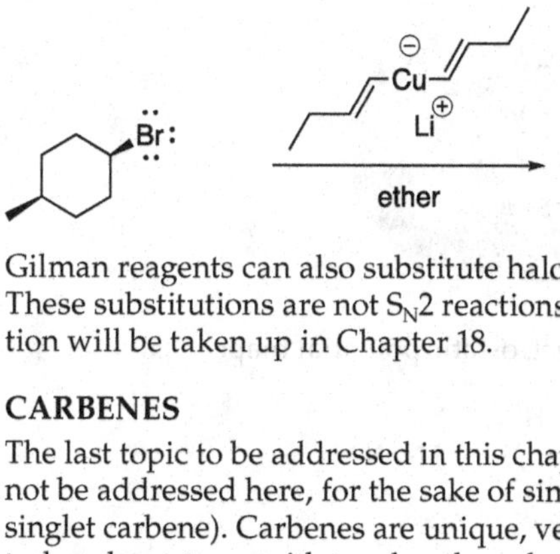

Gilman reagents can also substitute halogen atoms on alkenes and on benzene rings (Figure 17.16). These substitutions are not S$_N$2 reactions, and the details of haloalkene and halobenzene substitution will be taken up in Chapter 18.

CARBENES

The last topic to be addressed in this chapter involves singlet carbenes (since triplet carbenes will not be addressed here, for the sake of simplicity from here on the use of the term carbene will imply singlet carbene). Carbenes are unique, very reactive species. Carbenes are defined as formally neutral, carbon atoms with two bonds and a lone pair (Figure 17.17).[7]

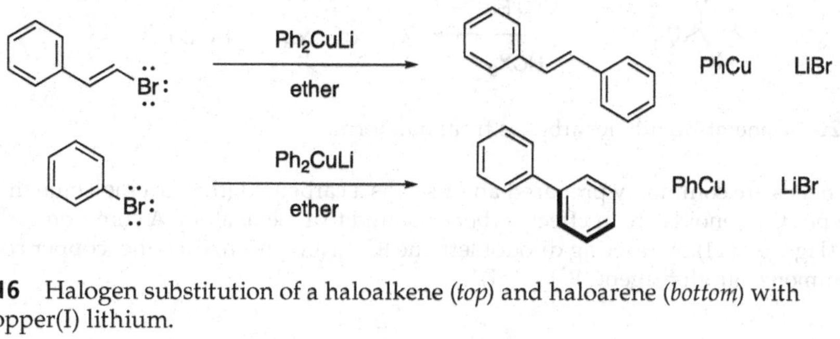

Figure 17.16 Halogen substitution of a haloalkene (*top*) and haloarene (*bottom*) with diphenylcopper(I) lithium.

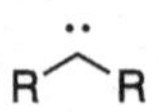

$$R \overset{\cdot\cdot}{\frown} R$$

Figure 17.17 General carbene structure.

The electronic structure of a carbene (Figure 17.18) consists of a sp²-hybridized carbon with an unfilled p-orbital (making carbene electrophilic) and a filled, sp2-orbital lone pair (making carbenes nucleophilic).

Free carbenes can be generated in several ways. Carbenes can be generated by photolysis of diazoalkanes (while easy on paper, it is synthetically not very useful). Shown in Figure 17.19 is the preparation of the simplest carbene methylidene, CH_2, from diazomethane (CH_2N_2).

It is significantly easier (Figure 17.20) to generate dihalocarbenes from chloroform ($HCCl_3$), bromoform ($HCBr_3$), and iodoform (HCI_3) reacting with potassium *tert*-butoxide.[8]

Problem 17.8 Draw the appropriate arrows to complete the mechanism for dihalocarbene formation.

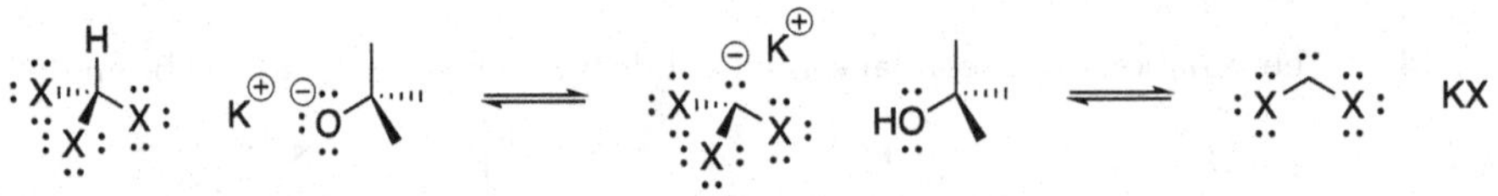

The dihalocarbenes are significantly easier to prepare because of delocalization stabilization of the hypovalent carbon atom.

Problem 17.9 Draw all contributing structures for dichlorocarbene.

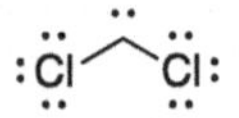

$$:\overset{\cdot\cdot}{\underset{\cdot\cdot}{Cl}} \overset{\cdot\cdot}{\frown} \overset{\cdot\cdot}{\underset{\cdot\cdot}{Cl}}:$$

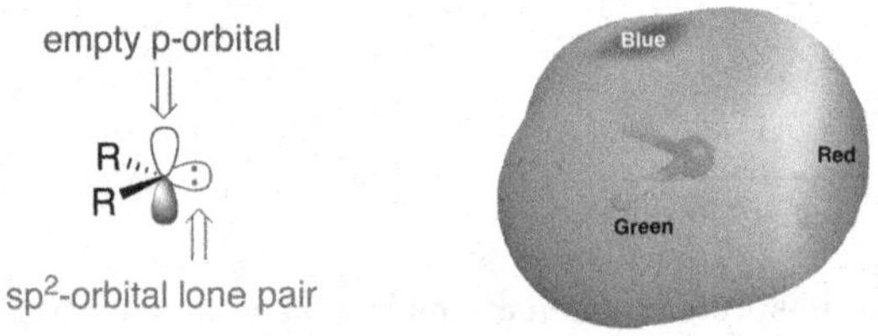

Figure 17.18 General carbene electronic structure and electrostatic potential map.

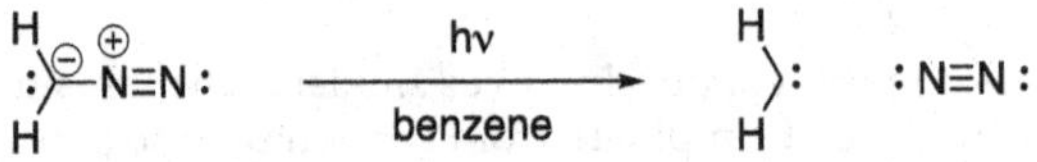

Figure 17.19 Generation of methylidene from diazomethane.

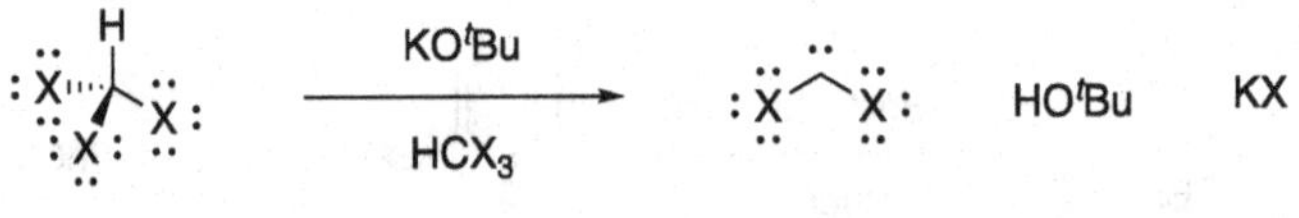

Figure 17.20 Generating dihalocarbenes from haloform.

Finally, carbenes are commonly prepared and used as a carbenoid (that is a molecule that reacts like a carbene). Carbenoids are in effect carbenes bound to a metal atom. A common carbenoid can be formed (Figure 17.21) by reacting diiodomethane (CH_2I_2) with Zn/Cu (zinc–copper couple) to give the Simmons–Smith reagent (ICH_2ZnI).[9]

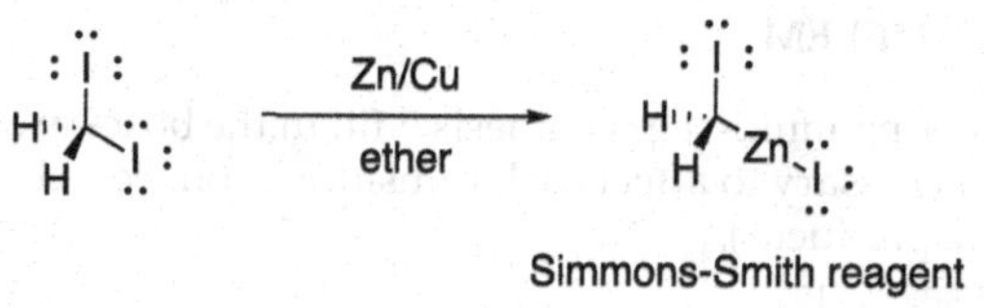

Figure 17.21 Formation of the Simmons–Smith reagent from diiodomethane.

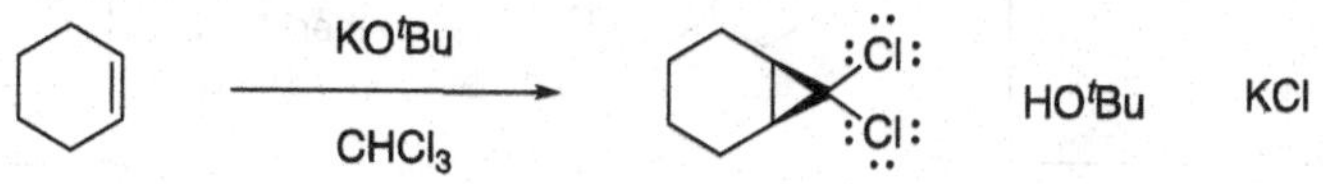

Figure 17.22 Cyclopropanation of cyclohexene with dichlorocarbene.

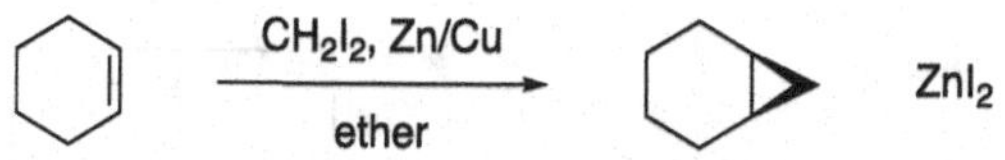

Figure 17.23 Cyclopropanation of cyclohexene via the Simmons–Smith reaction.

CYCLOPROPANATION

While there are numerous reactions in which carbenes can be employed, for this chapter only the reaction of carbenes to prepare cyclopropanes will be considered. Dichlorocarbene, generated in situ, will produce a dichlorocyclopropane ring (Figure 17.22).[10]

Problem 17.10 Draw the appropriate electron-pushing arrows to complete the formation of cyclopropane from dichlorocarbene.

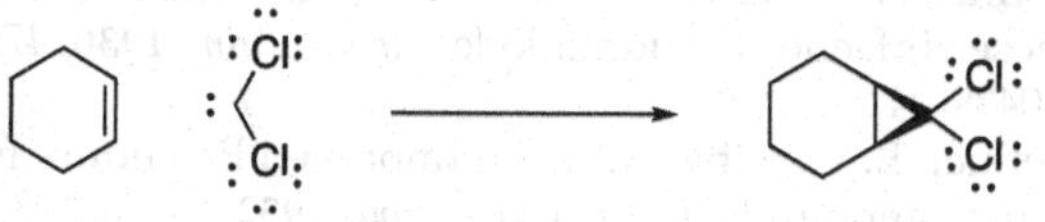

The Simmons–Smith reaction produces a cyclopropane ring with no halogen atoms on the ring (Figure 17.23).

Problem 17.11 Draw the appropriate electron-pushing arrows to complete the formation of cyclopropane from the Simmons–Smith reagent.

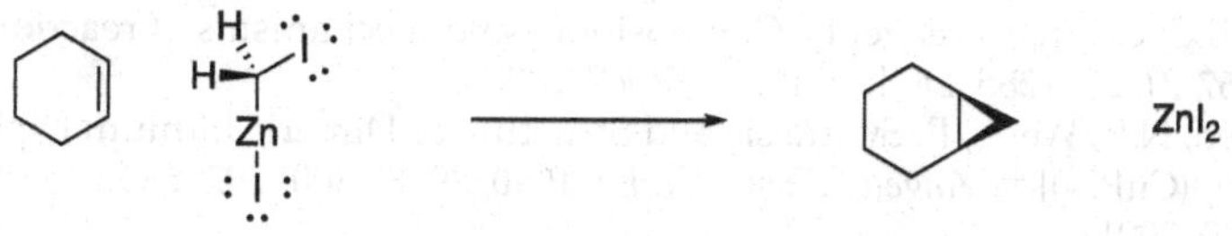

Problem 17.12 Predict the major product(s) for each of the following transformations.

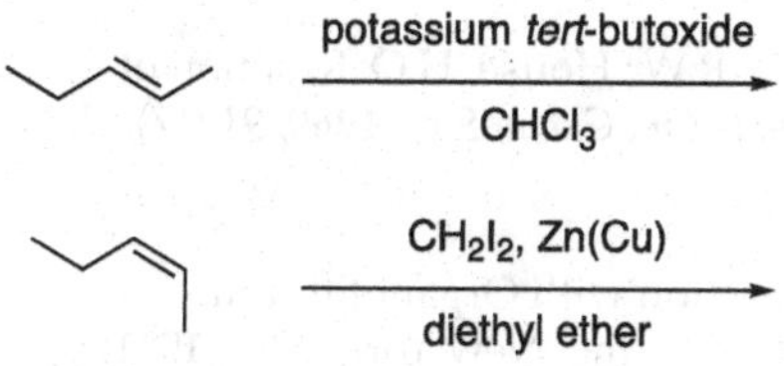

GENERAL PRACTICE PROBLEMS

Problem 17.13 For the following multi-step synthesis,[11] fill in the boxes around an arrow with the appropriate reagent(s) necessary to affect each transformation. For a box at the end of an arrow provide the appropriate product(s).

NOTES

1. Grignard, V. Sur quelques Nouvelles combinaisons organometalliques du magnesium et leur application à des syntheses d'alcools et d'hydrocarbures. *C. R. Acad. Sci.*, **1900**, *130*, 1322–1324.

2. i) Ziegler, K.; Colonius, H. Untersuchungen über alkali-organische Verbindungen. V. Eine bequeme Synthese einfacher Lithiumalkyle. *Liebigs Ann.*, **1930**, *479* (1), 135–149. DOI: 10.1002/jlac.19304790111
   ii) Gilman, H.; Zoellner, E.A.; Selby, W.M. An Improved Procedure for the Preparation of Organolithium Compounds. *J. Am. Chem. Soc.*, **1932**, *54* (5), 1957–1962. DOI: 10.1021/ja01344a033

3. Gilman, H.; Jones, R.G.; Woods, L.A. The Preparation of Methylcopper and some Observations on the Decomposition of Organocopper Compounds *J. Org. Chem.*, **1952**, *17* (12), 1630–1634. DOI: 10.1021/jo50012a009

4. i) Ashby, E.C. Grignard reagents. Compositions and mechanisms of reaction. *Q. Rev. Chem. Soc.*, **1967**, *21*, 259–285. DOI: 10.1039/QR9672100259
   ii) Lorenzen, N.P.; Weiss, E. Synthesis and Structure of Dimeric Lithium Diphenylcuprate: [{Li(OEt)$_2$}(CuPh$_2$)]$_2$†‡ *Angew. Chem. Int. Ed.*,**1990**, *29* (3), 300–302. DOI: 10.1002/anie.199003001
   iii) Reich, H.J. Role of Organolithium Aggregates and Mixed Aggregates in Organolithium Mechanisms. *Chem. Rev.*, **2013**, *113* (9), 7130–7178. DOI: 10.1021/cr400187u

5. Whitesides, G.M.; Fischer Jr., W.F.; San Filippo Jr., J.; Bashe, R.W.; House, H.O. Reaction of lithium dialkyl- and diarylcuprates with organic halides. *J. Am. Chem. Soc.*, **1969**, *91* (17), 4871–4882. DOI: 10.1021/ja01045a049

6. Jones, R.J.; Gilman, H. Halogen-Metal Interconversion Reaction with Organolithium Compounds. In *Organic Reactions* Vol. VI. John Wiley and Sons, Inc. New York, N.Y., 1951, pp. 339–356. DOI: 10.1002/0471264180.or006.07

7. Buchner, E.; Feldmann, L. Diazoessigester und Toluol. *Ber. Dtsch. Chem. Ges.*, **1903**, *36* (3), 3509–3517. DOI: 10.1002/cber.190303603139

8. Geuther, A. Ueber die Zersetzung des Chloroforms durch alkoholische Kalilösung. *Liebigs Ann.*, **1862**, *123* (1), 121–122. DOI: 10.1002/jlac.18621230109

9. i) Simmons, H.E.; Smith, R.D. A New Synthesis of Cyclopropanes from Olefins. *J. Am. Chem. Soc.*, **1958**, *80* (19), 5323–5324. DOI: 10.1021/ja01552a080
   ii) Simmons, H.E.; Smith, R.D. A New Synthesis of Cyclopropanes[1]. *J. Am. Chem. Soc.*, **1959**, *81* (16), 4256–4264. DOI: 10.1021/ja01525a036

10. Doering, W. von E.; Hoffmann, A.K. The Addition of Dichlorocarbene to Olefins. *J. Am. Chem. Soc.*, **1954**, *76* (23), 6162–6165. DOI: 10.1021/ja01652a087

11. Holub, N.; Neidhöfer, J.; Blechert, S. Total Synthesis of (+)-*trans*-195A. *Org. Lett*, **2005**, *7* (7), 1227–1229. DOI: 10.1021/ol0474610

18 Organometallic Chemistry

Palladium

In the previous chapter, attention was given to Grignard, organolithium, and Gilman reagents. While these reagents are well-established components of synthesis, they suffer from poor atom economy (producing a stoichiometric amount of metal and halide waste). Modern organic chemistry relies heavily upon catalytic, cross-coupling chemistry, which uses a substoichiometric amount of metal and reduces waste.

In this chapter, an understanding of palladium cross-coupling chemistry will be built from previous examples and emphasis will be placed upon general reactivity trends and product prediction. Note that cross-coupling chemistry, which lies at the intersection of inorganic and organometallic chemistry, does not tend to use electron-pushing arrows. As such, electron-pushing arrows will not be used in this chapter.

COMMON ORGANOMETALLIC MECHANISTIC STEPS
Oxidative Addition

During the previous discussion of Grignard reaction chemistry, the fine details of the reaction were ignored. It is, however, instructive to consider them here. The first step of the Grignard reaction (Figure 18.1) is the addition of a magnesium atom to the C–X bond.

Considering the Figure 18.1 reaction, the oxidation number of magnesium goes from 0 to +2 (which can be indicated in the reaction itself, Figure 18.2). As such, the addition of a metal atom, like magnesium, into a C–X bond is called oxidative addition[1] and is a common step that will be seen throughout this chapter.

Problem 18.1 Predict the product for the following oxidative addition steps.

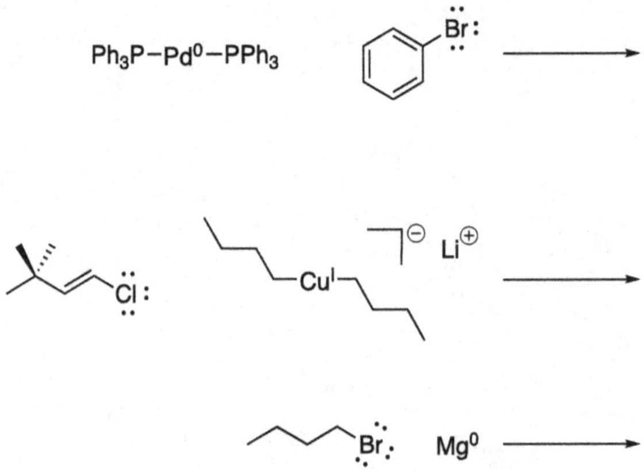

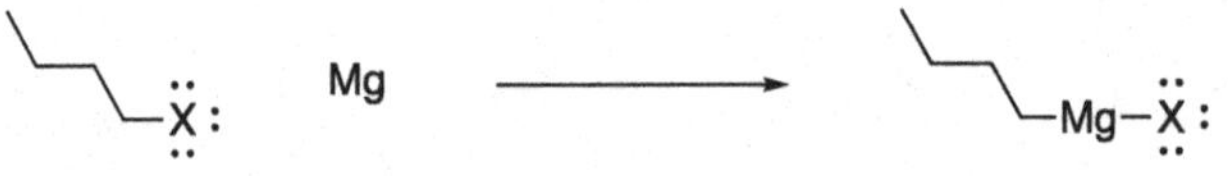

Figure 18.1 Insertion of a magnesium atom into a C–X bond.

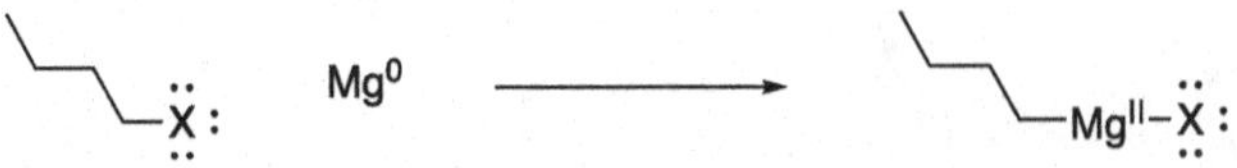

Figure 18.2 Insertion of a magnesium atom into a C–X bond showing the oxidation of the magnesium atom.

DOI: 10.1201/9781003479352-18

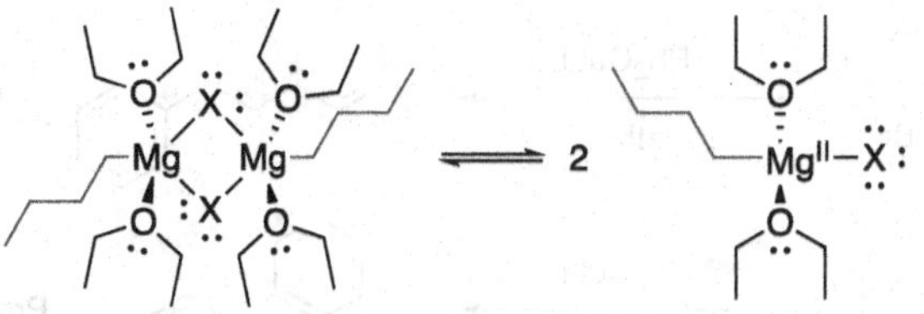

Figure 18.3 Solution equilibrium of a Grignard reagent in diethyl ether.

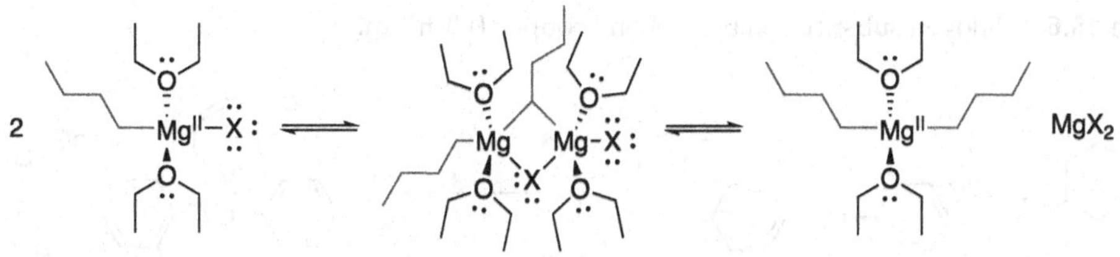

Figure 18.4 Schlenk equilibrium of a Grignard reagent in diethyl ether.

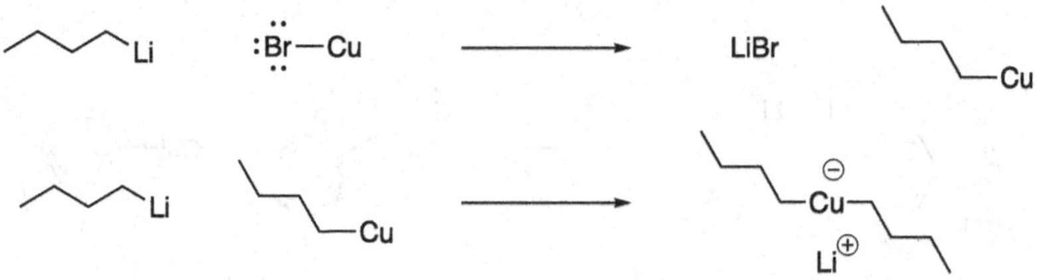

Figure 18.5 Transmetalation of a butyl group from lithium to copper(I).

Transmetalation

Recall that the Grignard reagent can aggregate (Figure 18.3) in solution (ether is both a common solvent and it acts as a ligand (Lewis base that binds to a metal) for magnesium).

While the aggregation can give symmetrical dimers like that seen in Figure 18.3, the dimerization can also give asymmetric dimers. The asymmetric dimer can then give rise to two new species (Figure 18.4) MgX_2 and MgR_2 (this is known as the Schlenk equilibrium[2]).

The key process to notice in the Schlenk equilibrium is the transfer of a carbon group from one metal to another. This process is known as transmetalation,[3] another common trope in cross-coupling chemistry. Here the transmetalation step (butyl transfer from one magnesium atom to another) occurs between two magnesium atoms.

It is far more common for transmetalation to occur where a less electronegative metal or metalloid transfers a R group to a more electronegative metal. While not discussed as a transmetalation step, the formation of a Gilman reagent (Figure 18.5) occurs when two organolithiums (lithium has a Pauling electronegativity value of 0.98) transfer their R groups from the lithium atom to a copper atom (copper has a Pauling electronegativity value of 1.90).

Problem 18.2 Predict the products for the following transmetalation steps.

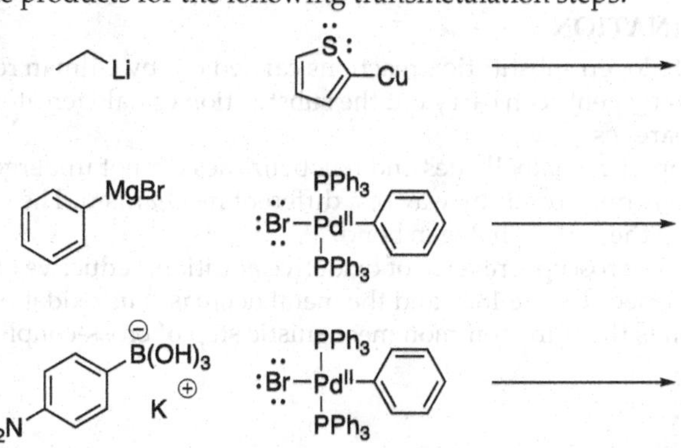

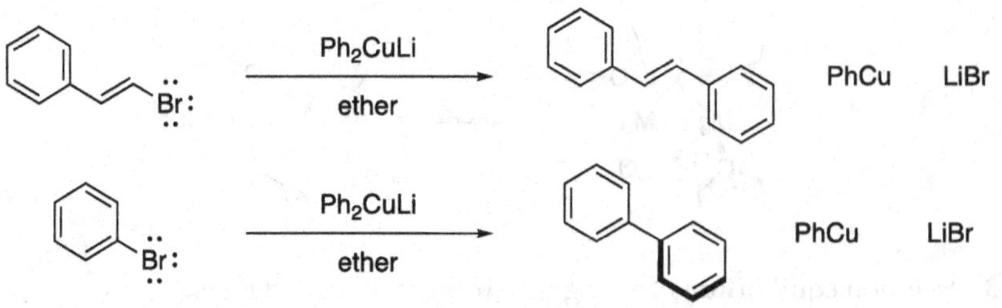

Figure 18.6 Halogen substitution by diphenylcopper(I) lithium.

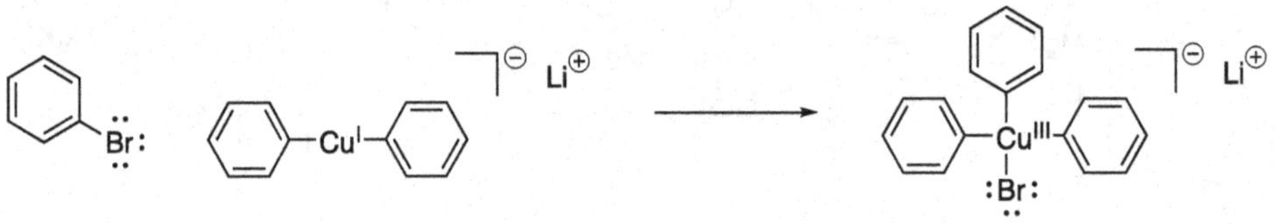

Figure 18.7 Oxidative addition of bromobenzene to lithium diphenylcuprate.

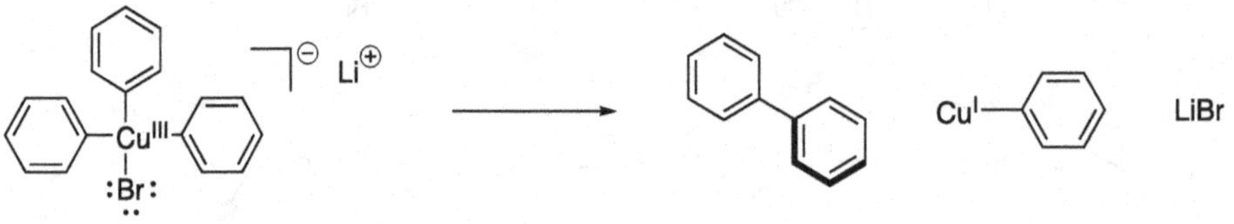

Figure 18.8 Reductive elimination to create biphenyl.

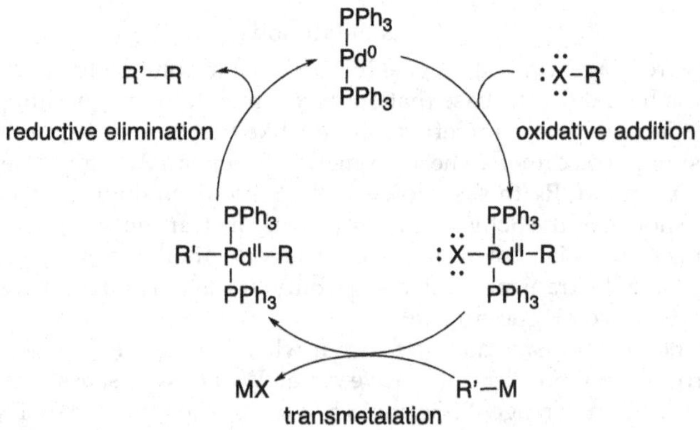

Figure 18.9 General mechanism for palladium(0)-catalyzed cross-coupling reactions.

REDUCTIVE ELIMINATION

Finally, consider the halogen substitution reactions carried out by Gilman reagents. As was seen in Chapter 17, Gilman reagents can carry out the substitution of halogen atoms (Figure 18.6) on haloalkenes and haloarenes.

As was seen in Chapter 14, haloalkenes and halobenzenes cannot undergo S_N1 nor S_N2 reactions. As such, these reactions must occur by way of a different mechanism. The first step (Figure 18.7) is oxidative addition into the carbon-halogen bond.

The new step is the microscopic reverse of oxidative addition: reductive elimination.[4] Here a new C–C bond is formed (Figure 18.8) and the metal decreases its oxidation number by two. Reductive elimination is the third common mechanistic step of cross-coupling reactions.

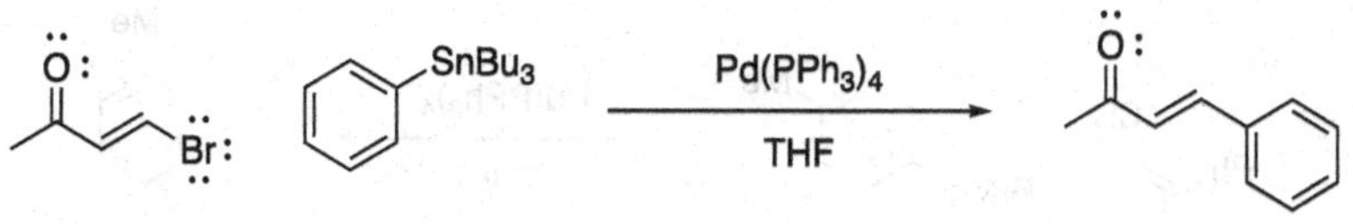

Figure 18.10 Example Stille coupling reaction.

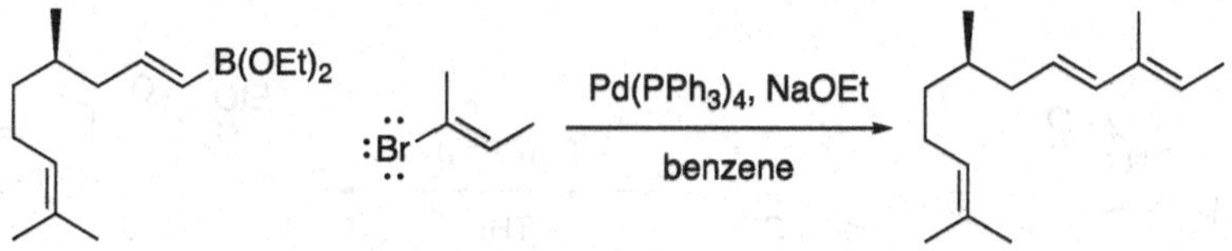

Figure 18.11 Example Suzuki coupling reaction.

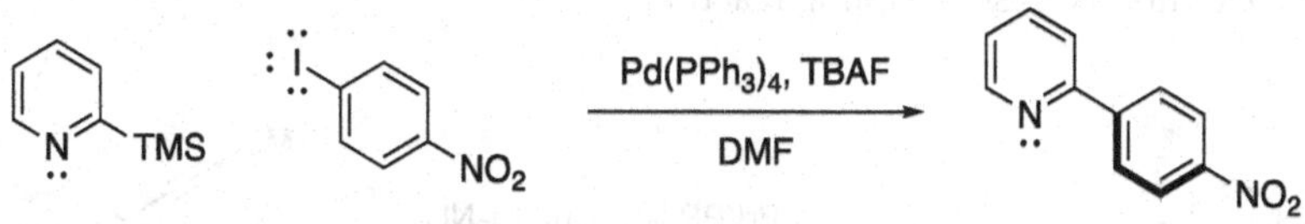

Figure 18.12 Example Hiyama coupling reaction

Problem 18.3 Predict the products for the following reductive elimination steps (note sp carbon atoms move faster than sp^2 carbon atoms which move faster than sp^3 carbon atoms).

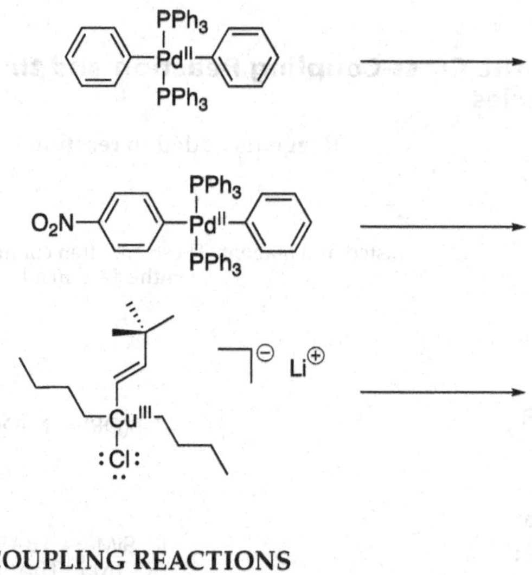

PALLADIUM CROSS-COUPLING REACTIONS

Palladium cross-coupling reactions consist of three separate steps: oxidative addition, transmetalation, and reductive elimination. The most significant variation among all the cross-coupling reactions is what the transmetalation step is. In this chapter, a limited number of cross-coupling reactions will be considered: Stille,[5] Suzuki,[6] Hiyama,[7] Kumada,[8] Sonogashira,[9] and Negishi.[10] The reaction mechanism for all these reactions is generally represented by a catalytic cycle, a general example shown in Figure 18.9.

The only significant difference for most of the palladium(0) cross-coupling reactions is the nature of R′–M (Table 18.1).

In summary, palladium cross-coupling reactions provide a unique means to construct C–C bonds. Table 18.1 shows example reactions for each of the named cross-coupling reactions considered in this chapter. The mechanism for each follows the catalytic cycle seen in Figure 18.9.

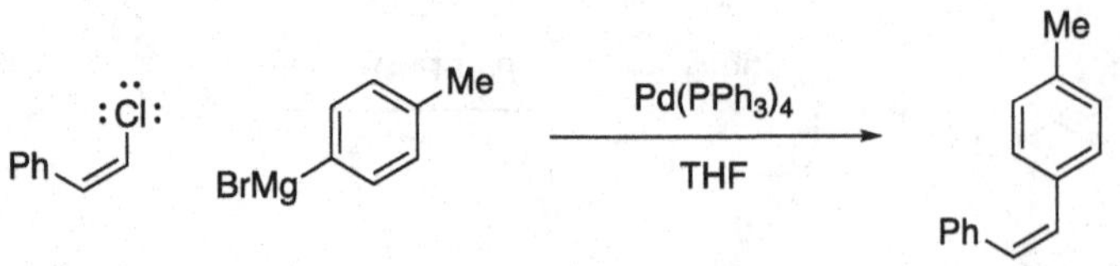

Figure 18.13 Example Kumada coupling reaction.

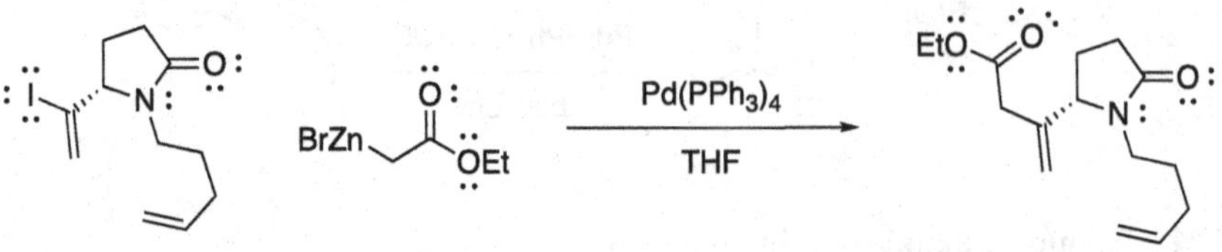

Figure 18.14 Example Negishi coupling reaction.

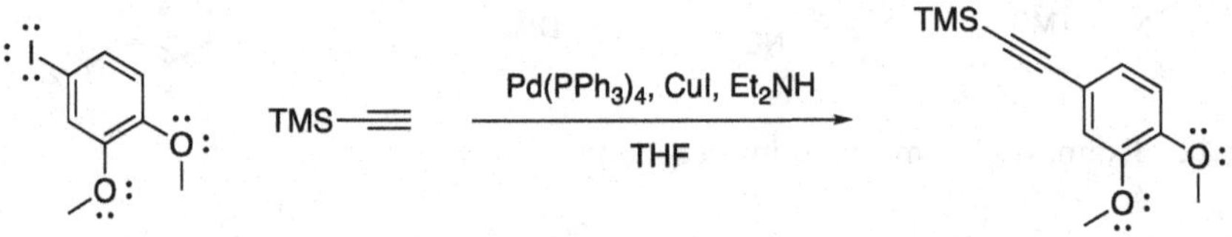

Figure 18.15 Example Sonogashira coupling reaction.

Table 18.1 Name of the Cross-Coupling Reaction and the Identity of the R′–M Species

Coupling	R′–M	Reagents added to reaction to generate R′–M
Stille	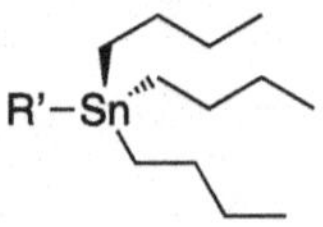 (R'-SnBu₃)	Listed as a reagent. These are often commercially available or are synthesized in advance.
Suzuki	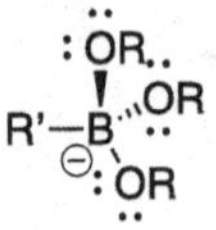	R′–B(OR)₂ + NaOR
Hiyama	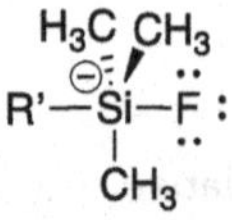	R′–SiMe₃ + TBAF (R′–TMS +TBAF)
Kumada	R′–MgBr	Usually listed as a reagent. In practice, these are made in a separate flask and added to the reaction mixture.
Negishi	R′–ZnBr	Usually listed as a reagent. In practice, these are made in a separate flask and added to the reaction mixture.
Sonogashira	R′–≡–Cu	R′–≡, CuX, R₂NH

GENERAL PRACTICE PROBLEMS

Problem 18.4 Utilizing the information from above predict the major product for each of the following reactions.

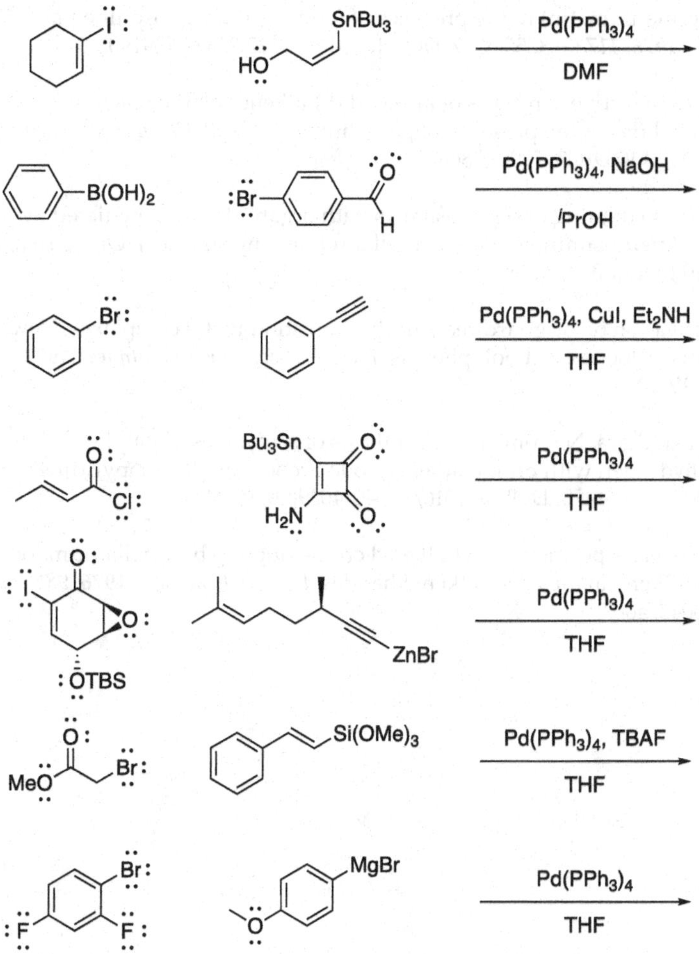

NOTES

1. i) Vaska, L.; DiLuzio, J.W. Hydrido Complexes of Iridium. *J. Am. Chem. Soc.*, **1961**, *83* (3), 756–756. DOI: 10.1021/ja01464a059

   ii) Collman, J.P. Patterns of Organometallic Reactions Related to Homogenous Catalysis. *Acc. Chem. Res.*, **1968**, *1*, 136–143. DOI: 10.1021/ar50005a002

   iii) Labinger, J.A. Tutorial on oxidative addition. *Organometallics*, **2015**, *34* (20), 4784–4795. DOI: 10.1021/acs.organomet.5b00565

2. Schlenk, W.; Schlenk, W. Jr. Über die Konstitution der Grignardschen Magnesiumverbindungen. *Ber. Dtsch. Chem. Ges.*, **1929**, *62* (4), 920–924. DOI: 10.1002/cber.19290620422

3. Negishi, E. Selective Carbon-Carbon Bond Formation *via* Transition Metal Catalysis: Is Nickel or Palladium Better than Copper? in *Aspects of Mechanism and Organometallic Chemistry*, Ed.: J. H. Brewster. Pleanum Press, New York, **1978**, 285–317. DOI: 10.1007/978-1-4684-3393-7_12

4. i) Davidson, P.J.; Lappert, M.F.; Pearce, R. Stable homoleptic metal alkyls. *Acc. Chem. Res.*, **1974**, *7*, 209–217. DOI: 10.1021/ar50079a001

ii) Kochi, J.K. Electron-transfer mechanisms for organometallic intermediates in catalytic reactions. *Acc. Chem. Res.*, **1974**, *7*, 351–360. DOI: 10.1021/ar50082a006

5. Azarian, D.; Dua, S.S.; Eaborn, C.; Walton, D.R.M. Reactions of organic halides with R_3MMR_3 compounds (M = silicon, germanium, tin) in the presence of tetrakis(triarylphosphine)palladium. *J. Organomet. Chem.*, **1976**, *117* (3), C55–C57. DOI: 10.1016/S0022-328X(00)91902-8

6. Miyaura, N.; Suzuki, A. Stereoselective synthesis of arylated (*E*)-alkenes by the reaction of alk-1-enylboranes with aryl halides in the presence of palladium catalyst. *J. Chem. Soc., Chem. Commun.*, **1979**, 866–867. DOI: 10.1039/C39790000866

7. Hatanaka, Y.; Hiyama, T. Cross-coupling of organosilanes with organic halides mediated by a palladium catalyst and tris(diethylamino)sulfonium difluorotrimethylsilicate. *J. Org. Chem.*, **1988**, *53* (4), 918–920. DOI: 10.1021/jo00239a056

8. Corriu, R.J.P.; Masse, J.P. Activation of Grignard reagents by transition-metal complexes. New and simple synthesis of trans-stilbenes and polyphenyls. *J. Chem. Soc., Chem. Commun.*, **1972**, 144a. DOI: 10.1039/C3972000144A

9. Sonogashira, K.; Tohda, Y.; Hagihara, N. Convenient synthesis of acetylenes. Catalytic substitutions of acetylenic hydrogen with bromo alkenes, iodo arenes, and bromopyridines. *Tetrahedron Lett.*, **1975**, *16* (50), 4467–4470. DOI: 10.1016/S0040-4039(00)91094-3

10. Baba, S.; Negishi, E. A novel stereospecific alkenyl-alkenyl cross-coupling by a palladium- or nickel-catalyzed reaction of alkenylalanes with alkenyl halides. *J. Am. Chem. Soc.*, **1976**, *98*, 6729–6731. DOI: 10.1021/ja00437a067

19 Aldehydes and Ketones

Aldehydes and ketones are molecules that contain a carbonyl group (C=O). Aldehydes have at least one hydrogen attached to the carbonyl carbon (abbreviated as RCHO) and ketones have two carbon substituents attached to the carbonyl carbon. An example of an aldehyde and a ketone are the isomers propanal (Figure 19.1) and acetone (Figure 19.2).

Aldehydes and ketones can react as both electrophiles and nucleophiles. The carbon attached to the oxygen is highly electrophilic (deep blue in the electrostatic potential maps in Figures 19.1 and 19.2) and the neighbouring hydrogens are also relatively acidic (typical pK_a values range from 16 to 20, a fact that will be investigated more thoroughly in Chapter 21). In addition, the carbonyl oxygen also has two lone pairs and can react as a Brønsted–Lowry base and as a Lewis base (nucleophile).

NOMENCLATURE OF ALDEHYDES AND KETONES

In terms of priority, aldehyde and ketone functional groups take precedence over alkenes, alkynes, and alcohols. In IUPAC nomenclature, aldehydes are named with the suffix -al and ketones are named, with an appropriate locant number, with the suffix -one (Figure 19.3).

PHYSICAL PROPERTIES OF ALDEHYDES AND KETONES

Aldehydes and ketones are inherently very polar molecules that, due to the carbonyl group, have a strong dipole moment. While aldehydes and ketones have a higher dipole moment than the analogous alcohol, the alcohol typically has a higher boiling point due to hydrogen bonding (Figure 19.4).

SYNTHESIS OF ALDEHYDES

Recall that aldehydes and ketones can be synthesized by several different methods (as seen in previous chapters). Aldehydes can be synthesized by:

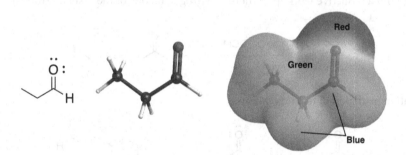

Figure 19.1 Structure (*left*), 3D model (*middle*), and electrostatic potential map (*right*) of an aldehyde: propanal (red corresponds to high electron density and blue corresponds to low electron density).

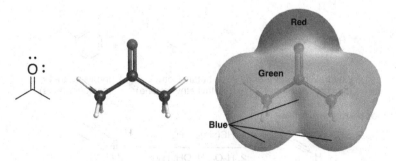

Figure 19.2 Structure (*left*), 3D model (*middle*), and electrostatic potential map (*right*) of a ketone: acetone (red corresponds to high electron density and blue corresponds to low electron density).

DOI: 10.1201/9781003479352-19

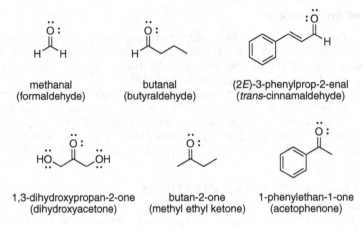

| methanal (formaldehyde) | butanal (butyraldehyde) | (2*E*)-3-phenylprop-2-enal (*trans*-cinnamaldehyde) |

| 1,3-dihydroxypropan-2-one (dihydroxyacetone) | butan-2-one (methyl ethyl ketone) | 1-phenylethan-1-one (acetophenone) |

Figure 19.3 Example aldehydes (*top*) and ketones (*bottom*) showing IUPAC names (and common names).

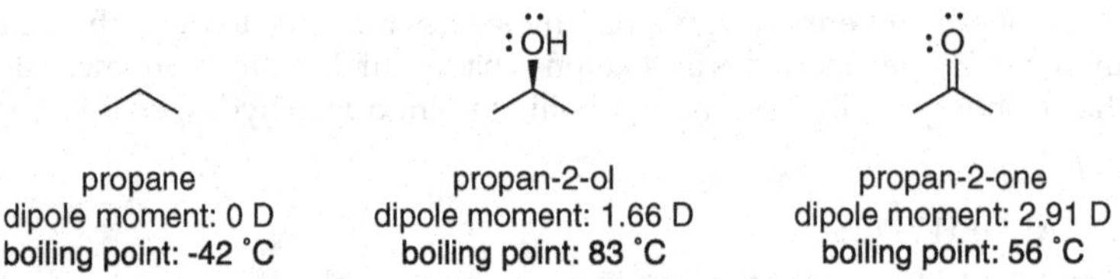

propane	propan-2-ol	propan-2-one
dipole moment: 0 D	dipole moment: 1.66 D	dipole moment: 2.91 D
boiling point: -42 °C	boiling point: 83 °C	boiling point: 56 °C

Figure 19.4 Physical properties of propane, propan-2-ol, and propan-2-one.

Ozonolysis with a reductive workup (if there a hydrogen on one of the alkene carbons)

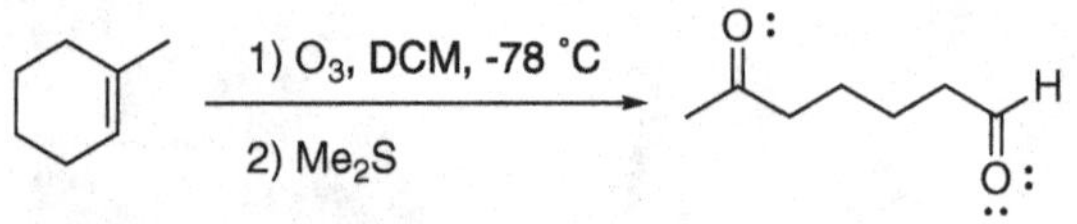

$$\text{1) O}_3\text{, DCM, -78 °C} \qquad \text{2) Me}_2\text{S}$$

Hydroboration oxidation of terminal alkynes

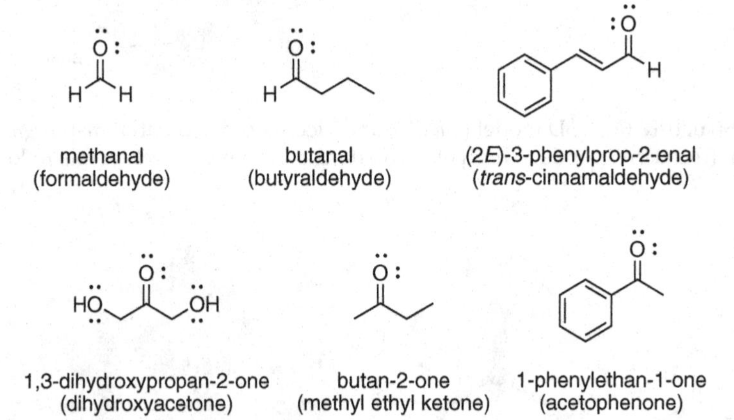

| methanal (formaldehyde) | butanal (butyraldehyde) | (2*E*)-3-phenylprop-2-enal (*trans*-cinnamaldehyde) |

| 1,3-dihydroxypropan-2-one (dihydroxyacetone) | butan-2-one (methyl ethyl ketone) | 1-phenylethan-1-one (acetophenone) |

PCC oxidation of primary alcohols

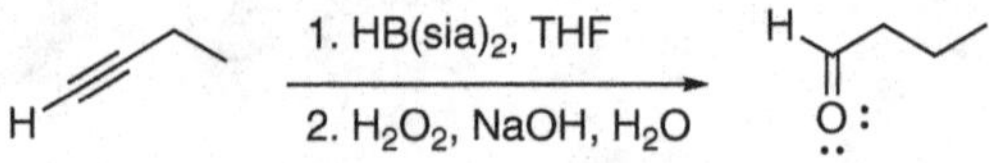

$$\text{1. HB(sia)}_2\text{, THF} \qquad \text{2. H}_2\text{O}_2\text{, NaOH, H}_2\text{O}$$

Swern oxidation of primary alcohols

ethanol
(1° alcohol)

acetaldehyde
(aldehyde)

PCC

DCM

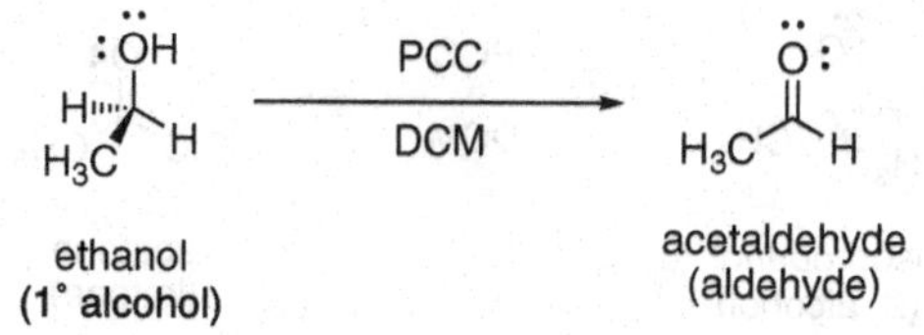

Dess-Martin oxidation of primary alcohols

ethanol
(1° alcohol)

acetaldehyde
(aldehyde)

PCC

DCM

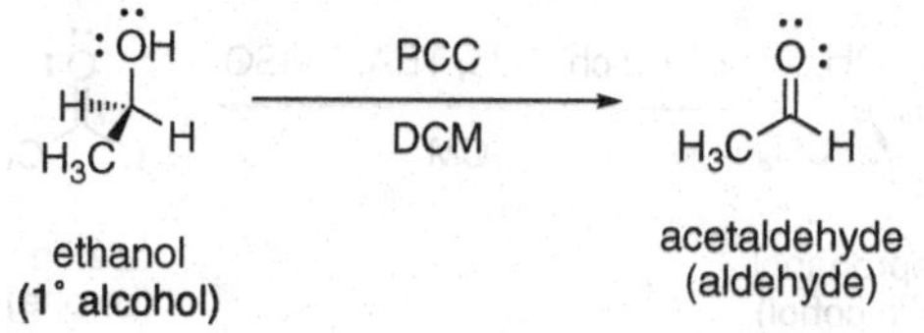

Periodic acid cleavage of diols (if one of the diol alcohols is secondary)

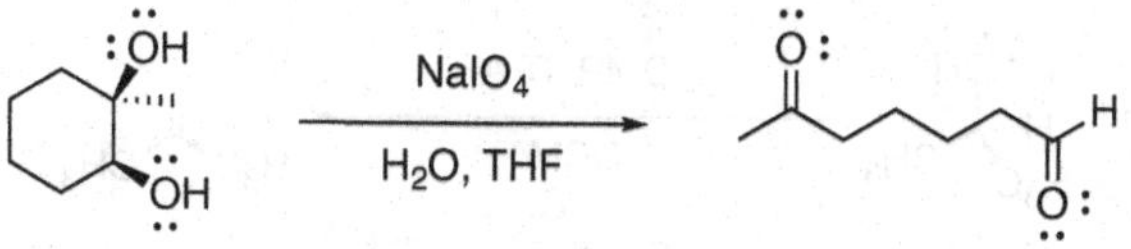

$NaIO_4$

H_2O, THF

SYNTHESIS OF KETONES

Meanwhile, ketones can be synthesized by similar methods, though there is a broader range of options because ketones, unlike aldehydes, cannot be oxidized further:

Ozonolysis (ketones are guaranteed whenever an alkene has two carbon substituents)

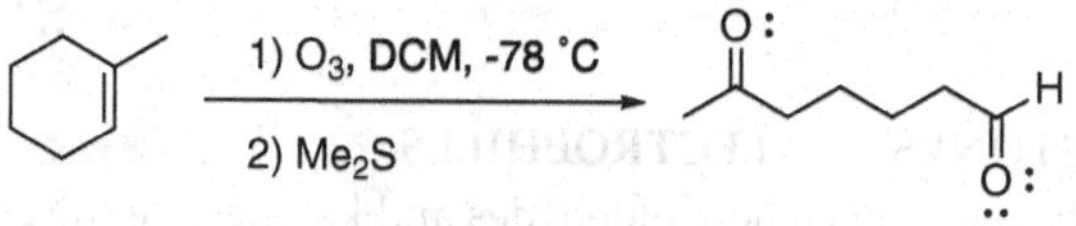

1) O_3, DCM, -78 °C

2) Me_2S

Acidic hydration of alkynes

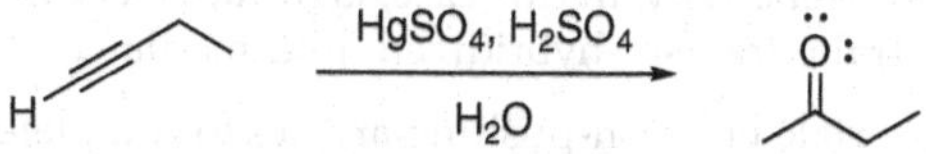

$HgSO_4$, H_2SO_4

H_2O

Pinacol rearrangement

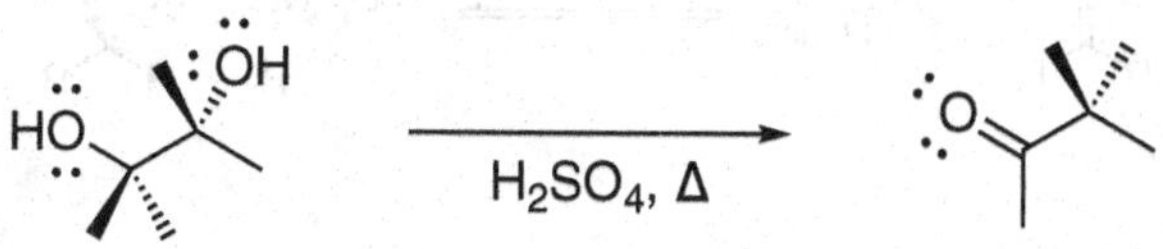

H_2SO_4, Δ

Jones oxidation of secondary alcohols

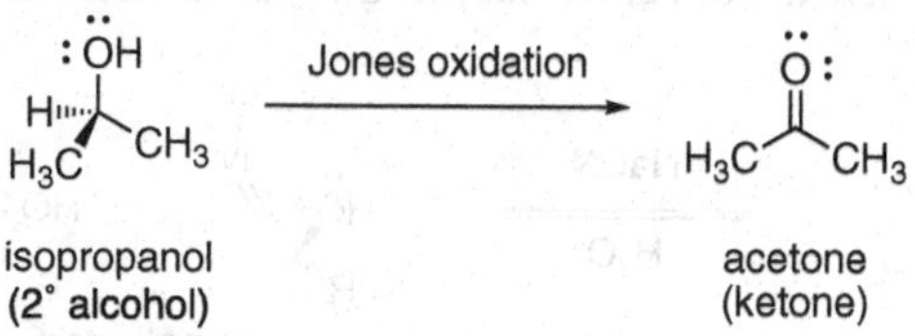

isopropanol
(2° alcohol)

Jones oxidation

acetone
(ketone)

209

PCC oxidation of secondary alcohols

isopropanol
(2° alcohol)

acetone
(ketone)

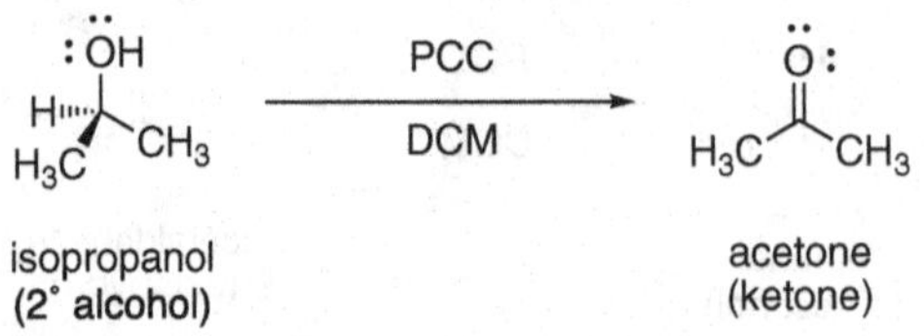

Swern oxidation of secondary alcohols

isopropanol
(2° alcohol)

acetone
(ketone)

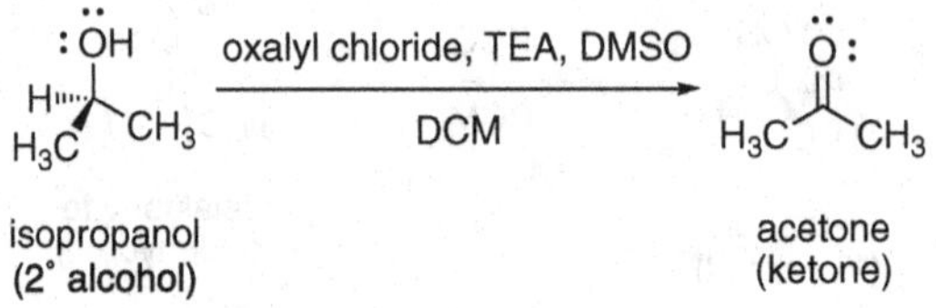

Dess-Martin oxidation of secondary alcohols

isopropanol
(2° alcohol)

acetone
(ketone)

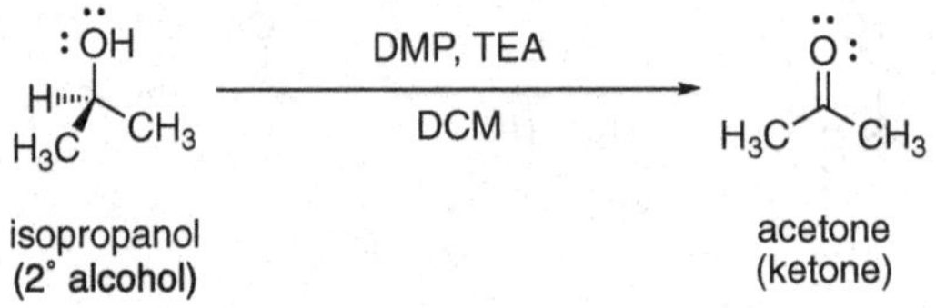

Periodic acid cleavage of diols (if one of the diol carbons is a tertiary alcohol)

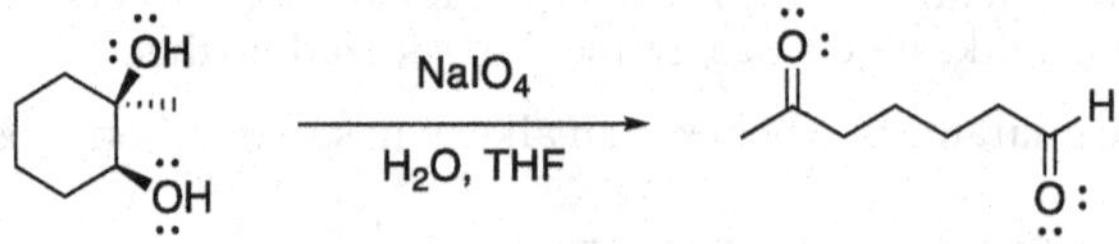

ALDEHYDES AND KETONES AS ELECTROPHILES

First, we will consider the reactions where aldehydes and ketones are the electrophile. A wide variety of nucleophiles can attack the highly electrophilic carbonyl carbon of aldehydes and ketones. The general pattern of reactivity, in this case, is the attack of the nucleophile on the carbonyl carbon to produce a tetrahedral (sp^3-hybridized) intermediate.

Problem 19.1 Draw the appropriate electron-pushing arrows to complete a generic mechanism.

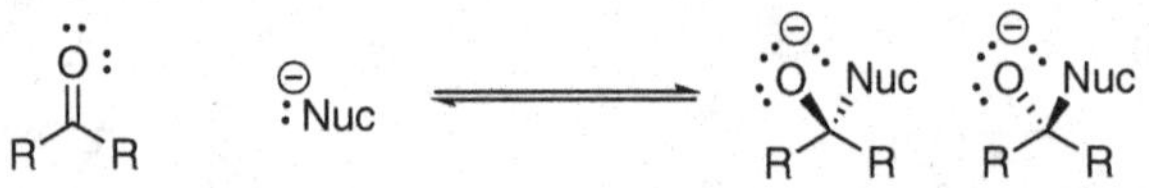

Carbon Nucleophiles

One of the simplest carbon nucleophiles is cyanide (CN$^-$), which upon reaction with an aldehyde or ketone produces a cyanohydrin (note that the potential creation of a chirality centre as the planar sp^2 carbon can be attacked from either face leading to different stereochemical outcomes, Figure 19.5).[1]

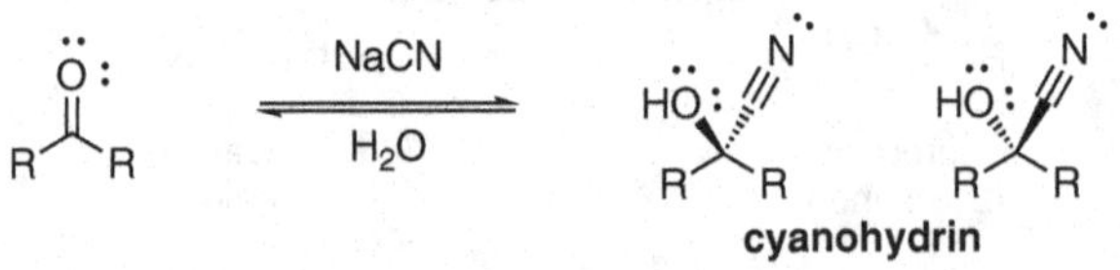

cyanohydrin

Figure 19.5 Generic reaction showing the synthesis of a cyanohydrin from a carbonyl and cyanide.

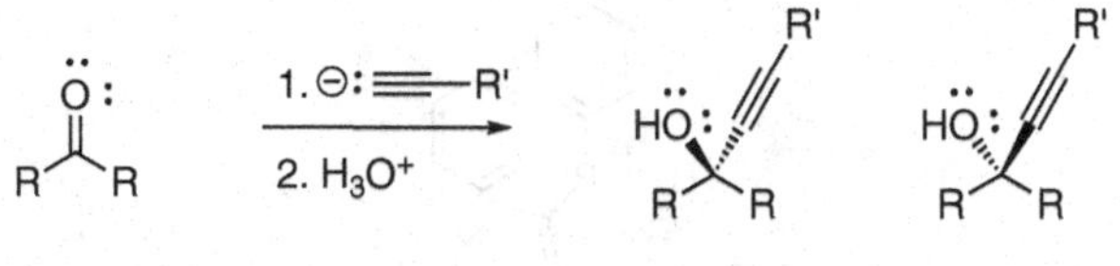

Figure 19.6 Generic Favorskii reaction showing the addition of an acetylide to a carbonyl.

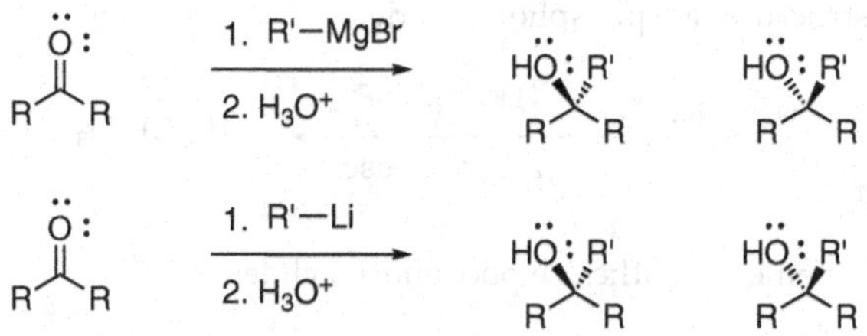

Figure 19.7 Generic reactions showing the addition of a Grignard reagent (*top*) and an organo-lithium reagent (*bottom*) to a carbonyl.

Problem 19.2 Draw appropriate electron-pushing arrows to complete the mechanism.

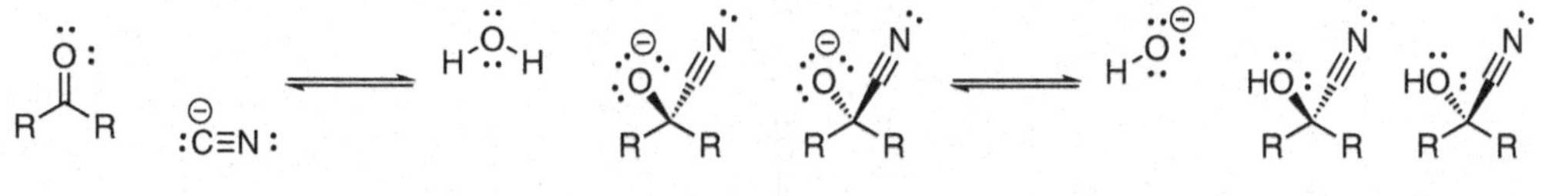

cyanohydrin

Structurally analogous to cyanide are the acetylide carbon nucleophiles. These can be generated by the deprotonation of terminal alkynes and then added to the aldehyde or ketone. From this reaction, the Favorskii reaction (Figure 19.6),[2] tertiary propargylic alcohols are produced.

Problem 19.3 Draw appropriate electron-pushing arrows to complete the mechanism.

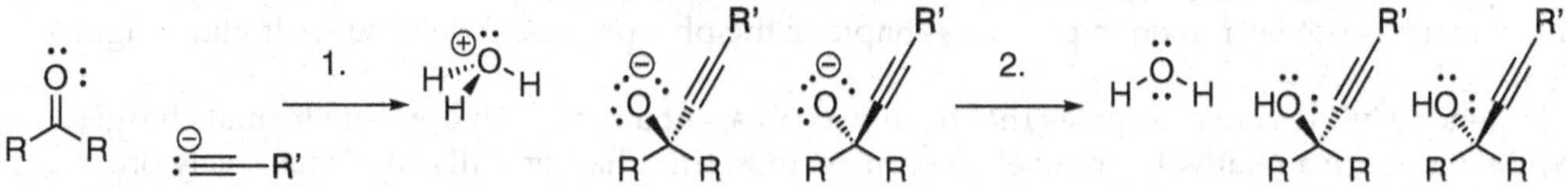

Finally, one can also use Grignard and organolithium reagents as carbon nucleophiles to add to aldehydes and ketones (Figure 19.7).

Problem 19.4 Draw appropriate electron-pushing arrows to complete the mechanisms of Grignard reagents and organolithium reagents reacting with carbonyls. Note in reference to Chapter 17 and the discussion of organometallic molecules aggregating in solution, these mechanisms shown below are an oversimplification)

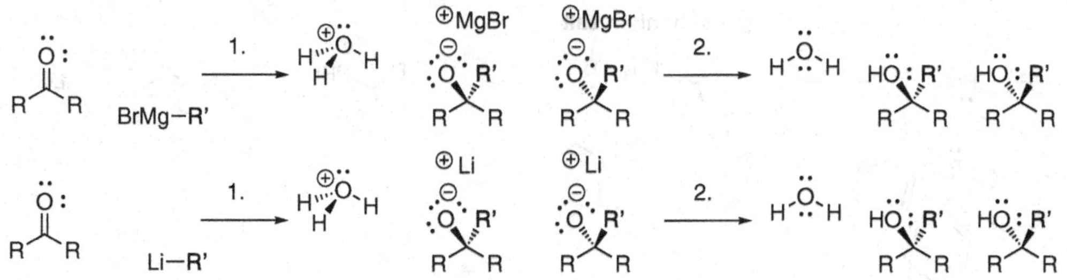

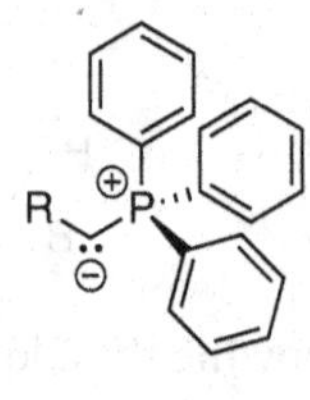

Figure 19.8 Generic structure of a phosphorus ylide.

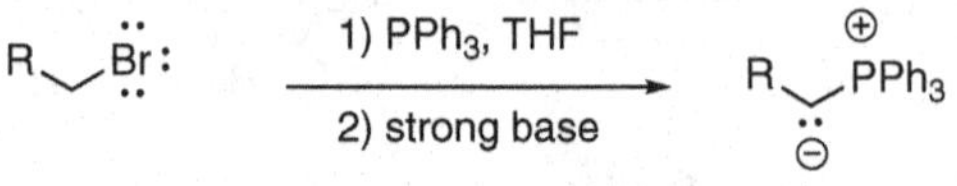

Figure 19.9 General scheme to synthesize phosphorus ylides.

Problem 19.5 Draw the structure of the product formed by treating each compound with propyl magnesium bromide followed by an aqueous HCl workup.

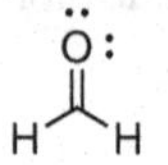

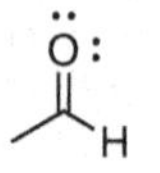

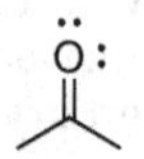

One of the most important carbon nucleophiles that is commonly used with aldehydes and ketones is one that has not been seen in previous chapters: the phosphorus ylide (also spelt ylid) (Figure 19.8).

In general, ylides are molecules that have two, adjacent atoms with opposite formal charges. While ylides are a relatively broad class of molecules, this chapter will only focus solely on phosphorus ylides, which contain a negatively charged carbon atom next to a formally positive phosphorus atom. These phosphorus ylides can be prepared in two steps (Figure 19.9): first, an alkyl halide undergoes an S_N2 reaction with triphenyl phosphine (PPh_3) and then deprotonation by nBuLi (or NaH or $NaNH_2$).

Problem 19.6 Draw appropriate electron-pushing arrows to complete the ylide formation mechanism. Note that because this process relies on an S_N2 reaction, primary alkyl halides are favoured for ylide formation.

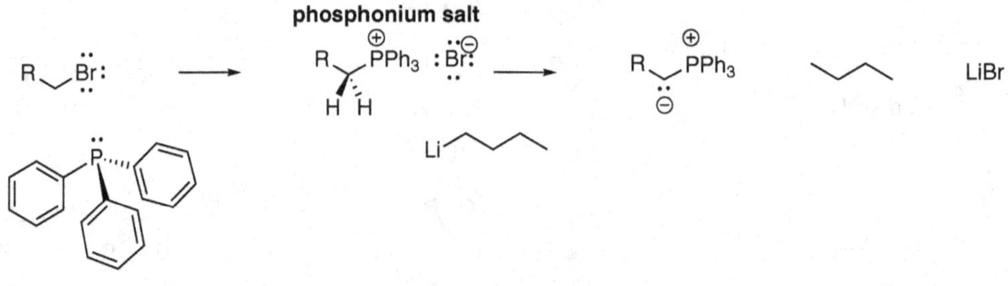

phosphonium salt

There are two types of ylides that can be prepared (Figure 19.10): unstabilized ylides (where R, in Figure 19.8, contains no conjugated electron-withdrawing group (C=O, C≡N, or NO_2) and stabilized ylides (where R, in Figure 19.8, contains a conjugated electron-withdrawing group, C=O, C≡N, or NO_2).

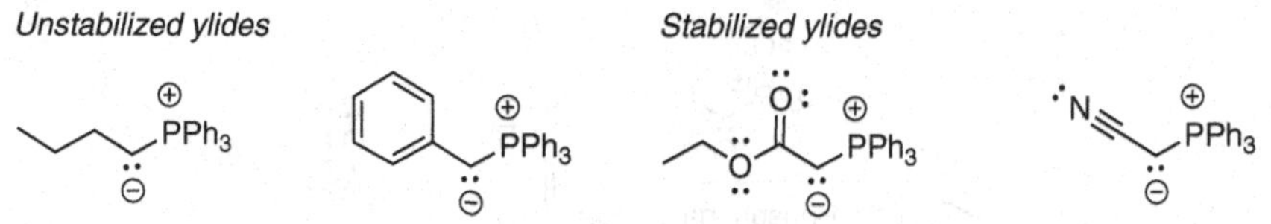

Figure 19.10 Examples of unstabilized ylides (*left*) and stabilized ylides (*right*).

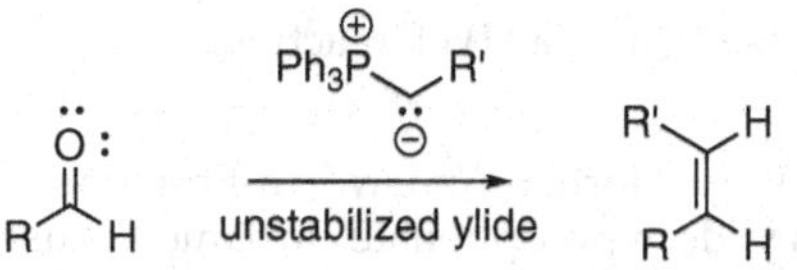

Figure 19.11 Synthesis of a Z-alkene from an unstabilized ylide.

Ylides are used, in the Wittig reaction,[3] as carbon nucleophiles to produce alkenes from aldehydes and ketones. When unstabilized ylides react with carbonyls, they produce Z-alkenes as the major product (Figure 19.11). This is because the lower-energy pathway (lower-energy transition state) when the ylide attacks the carbonyl is the formation of a *cis*-oxaphosphetane intermediate, and thence the Z-alkene. Unstabilized ylides react irreversibly with the carbonyl and so the kinetic product (that which is formed fastest) is what is produced, namely the Z-alkene.

Problem 19.7 Draw electron-pushing arrows to complete the mechanism.

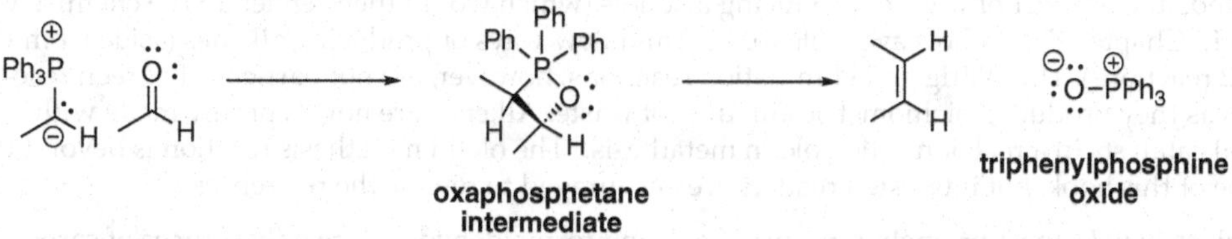

When stabilized ylides react with carbonyls, they produce E-alkenes as the major product (Figure 19.12). This is because the stabilized ylides react reversibly (notice the first step is an equilibrium step) with the carbonyl and so produce the most stable *trans*-oxaphosphetane intermediate will be produced and thence the E-alkene.

Problem 19.8 Draw electron-pushing arrows to complete the mechanism.

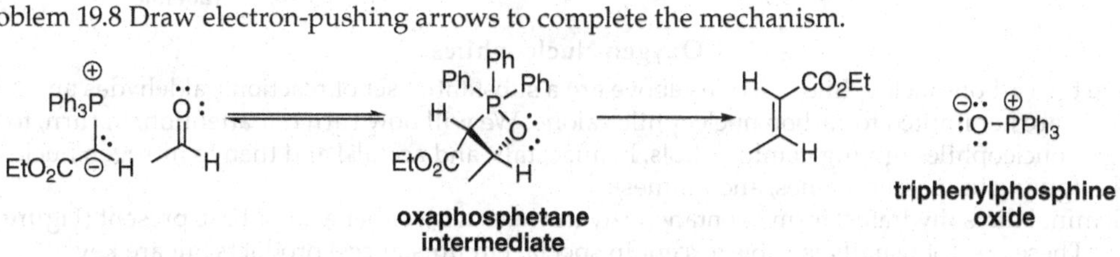

If the outcome of a reaction is dependent upon the type of reactant, then a question arises: how could one produce an E-alkene from an unstabilized ylide? Chemists are rarely satisfied with the limitations named above because the synthesis of a defined, specific product requires the ability to tailor the synthetic outcome to the synthetic needs of the chemist. The way to create E-alkenes

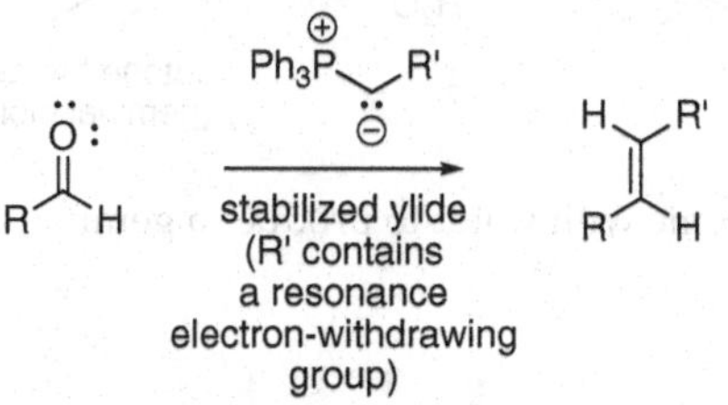

Figure 19.12 Synthesis of an E-alkene from a stabilized ylide.

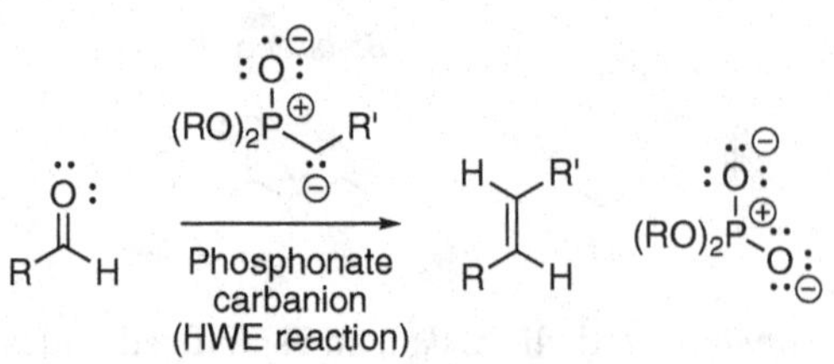

Figure 19.13 Synthesis of an *E*-alkene via HWE reaction.

from even unstabilized ylides is the Horner-Wadsworth-Emmons (HWE) reaction (Figure 19.13).[4] Instead of using a phosphorus ylide, a phosphonate carbanion is used. The phosphonate structure (with three inductive, electron-withdrawing oxygen atoms) stabilizes the negative charge, thus regardless of the R group attached to the carbanion an *E*-alkene will be produced.

Problem 19.9 Draw electron-pushing arrows to complete the mechanism.

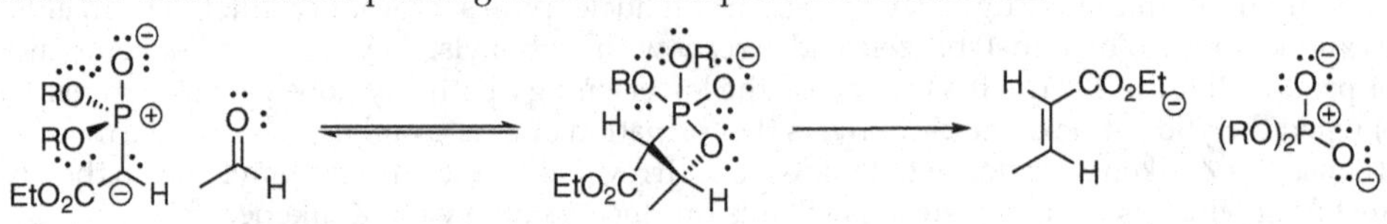

As can be seen above, the use of carbon nucleophiles with aldehydes and ketones dramatically expands the number of ways of producing alcohols (which would then render all the chemistry seen in Chapter 15 Alcohols available for use) and new ways of producing alkenes (aside from E1 or E2 reactions). The Wittig and elimination reactions, however, are not particularly green reactions as they produce stoichiometric amounts of waste. Alkenes are now typically made with metal catalysts in a reaction called olefin metathesis.[5] The olefin metathesis reaction is beyond the scope of this book, but interested readers are encouraged to review the references.

Problem 19.10 Using 1-bromobutane, propanal, and formaldehyde as the only sources of carbon, provide a synthesis (employing any other reagents) of 2-ethylhexanol.

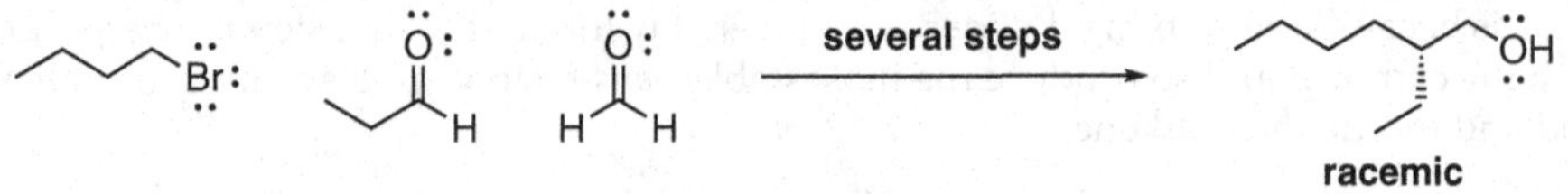

racemic

Oxygen Nucleophiles

While the carbon-nucleophile reactions above are a substantial set of reactions, aldehydes and ketones are not limited to carbon nucleophiles alone. We will now turn our attention, in turn, to oxygen nucleophiles (giving geminal diols, hemiacetals, and acetals) and then to nitrogen nucleophiles (giving imines, enamines, and amines).

Geminal diols (hydrates) form spontaneously in water with either acid or base present (Figure 19.14). These are not usually isolable (except in special circumstances) products but are key intermediates in other reactions (recall the geminal diol intermediate in the Jones oxidation mechanism).

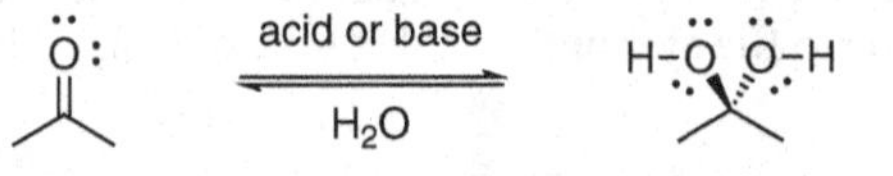

acetone hydrate
(geminal diol)

Figure 19.14 Reaction of acetone with water to produce a geminal diol.

Problem 19.11 Draw electron-pushing arrows to complete the acid-mediated mechanism. Note that any strong, aqueous acid (H_2SO_4, HCl, HBr, HI) could be used.

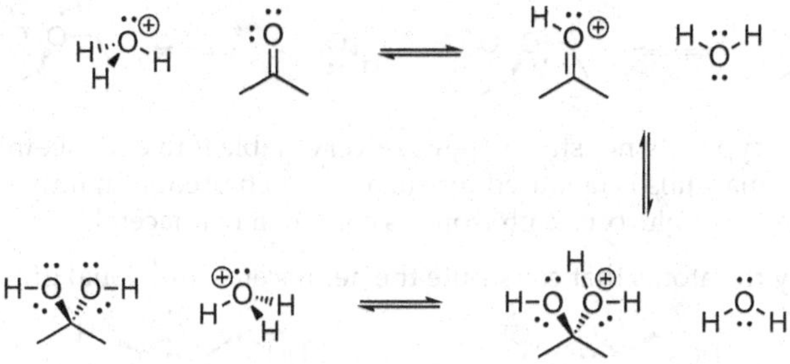

Problem 19.12 Draw electron-pushing arrows to complete the base-mediated mechanism.

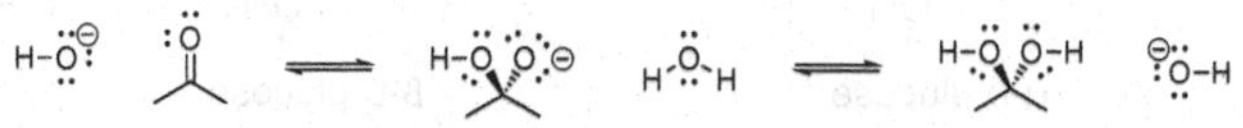

Problem 19.13 Under aqueous, acidic conditions carbonyls exist in equilibrium with their hydrate. Typically, for cyclohexanone, the hydrate is higher in energy and the carbonyl is the dominant species at equilibrium. For cyclopropanone, however, the hydrate is lower in energy and the major species at equilibrium. Provide a rationale for this difference in behaviour for cyclopropanone.

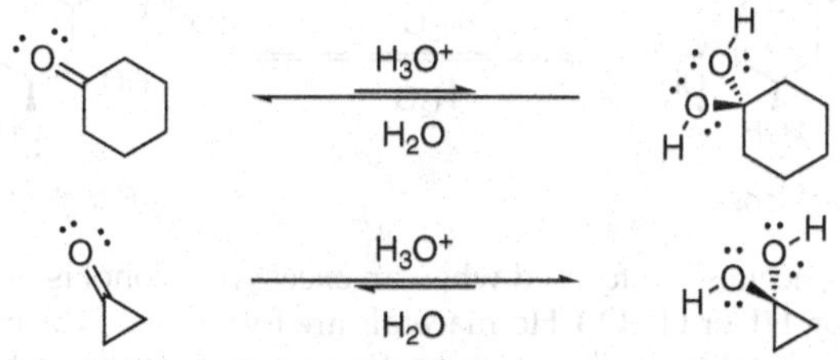

If an alcohol solvent (ROH) is used instead of water, then a hemiacetal is formed spontaneously with either acid or base present (Figure 19.15). These are not usually isolable products unless a five- or six-membered ring is formed. Hemiacetals are also intermediates on the way to forming acetals, which are stable and very useful (Figure 19.16).

Problem 19.14 Draw electron-pushing arrows to complete the acid-mediated mechanism. Note that HCl is shown as the acid, but any strong acid (H_2SO_4, HCl, HBr, HI) could be used, and ethanol is shown as the alcohol but almost any alcohol could be used.

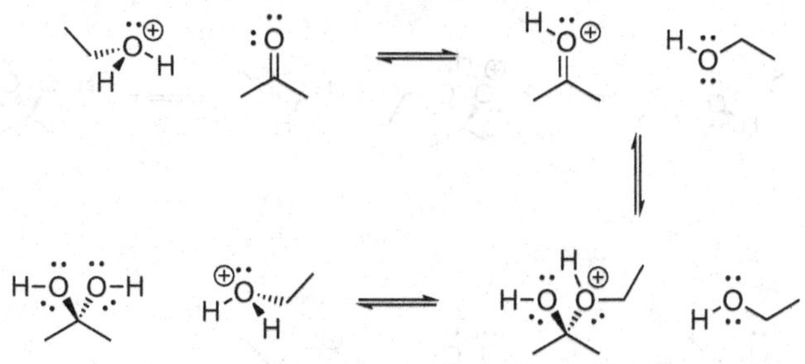

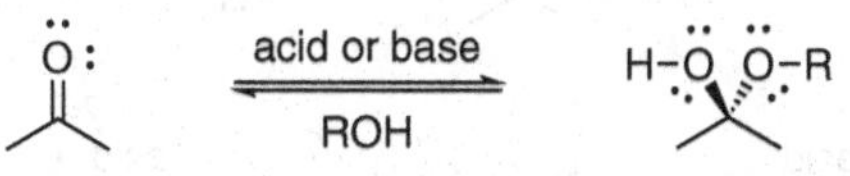

Figure 19.15 Reaction of acetone with an alcohol to produce a hemiacetal.

215

Problem 19.15 Draw electron-pushing arrows to complete the base-mediated mechanism. Note that experimentally, basic conditions are rarely used to form hemiacetals.

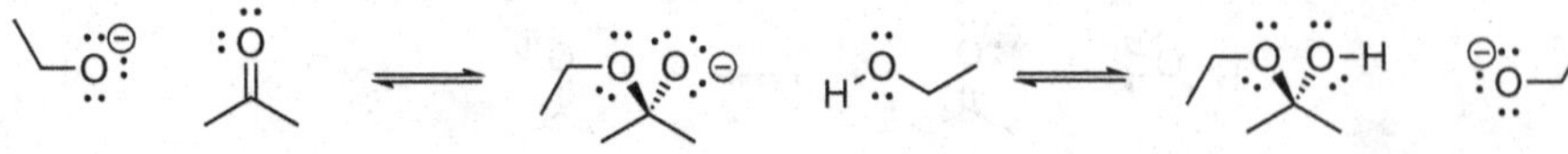

While hemiacetals are typically not stable, they are very stable if five- or six-membered rings can form. Formation of hemiacetals is favoured for sugars, which predominantly exist in this form. Consider the two isomers of glucose. Both isomers contain a hemiacetal.

Problem 19.16 Identify the atoms that constitute the hemiacetal functional group in each molecule.

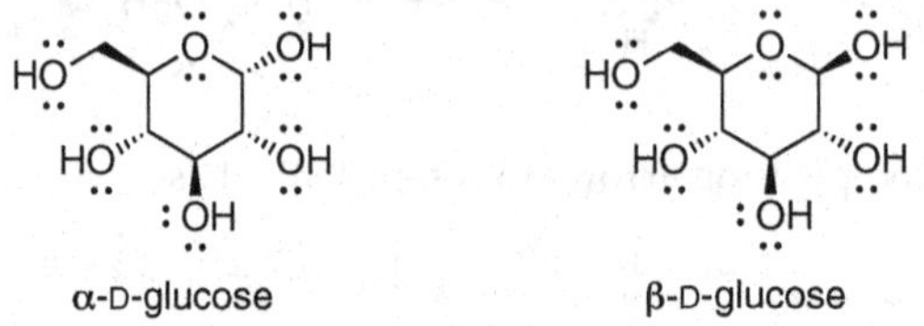

α-D-glucose   β-D-glucose

These two isomers (called anomers) can interconvert in an aqueous solution (this is known as mutarotation).

Problem 19.17 Provide an arrow-pushing mechanism to account for the isomerization of α-D-glucose into β-D-glucose.

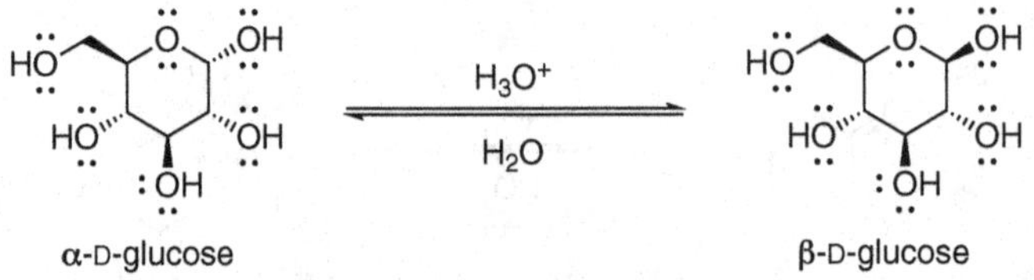

H_3O^+

H_2O

α-D-glucose   β-D-glucose

In contrast to hemiacetals, acetals are formed when an excess of alcohol is used with a strong acid (most commonly HCl, HBr, HI, or H_2SO_4). Hemiacetals are formed (see the mechanism in Problem 19.14), but these, typically, unstable species keep reacting to produce the acetal product (Figure 19.16).

Problem 19.18 Draw electron-pushing arrows to complete the mechanism.

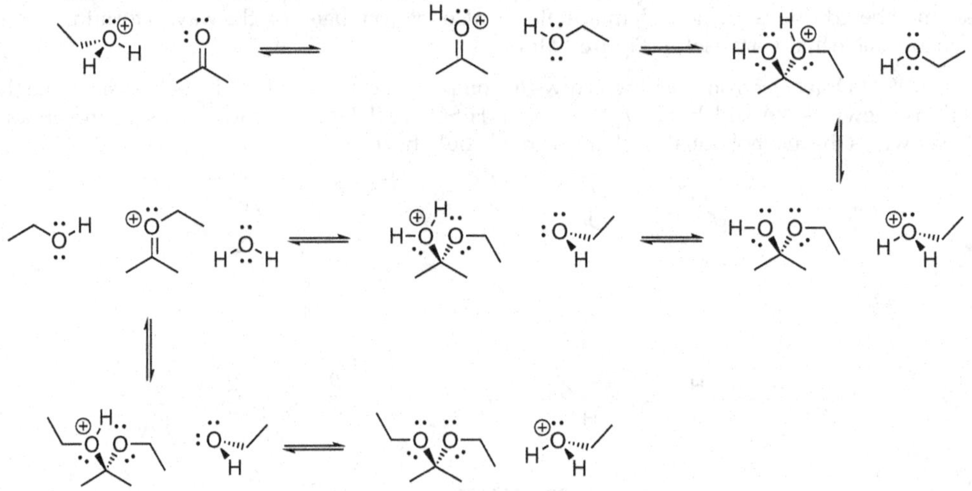

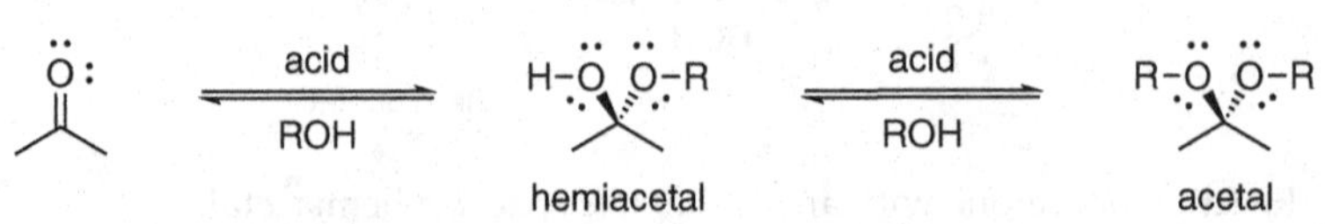

acid / ROH

hemiacetal

acid / ROH

acetal

Figure 19.16 Reaction of acetone with alcohol and acid to produce an acetal.

216

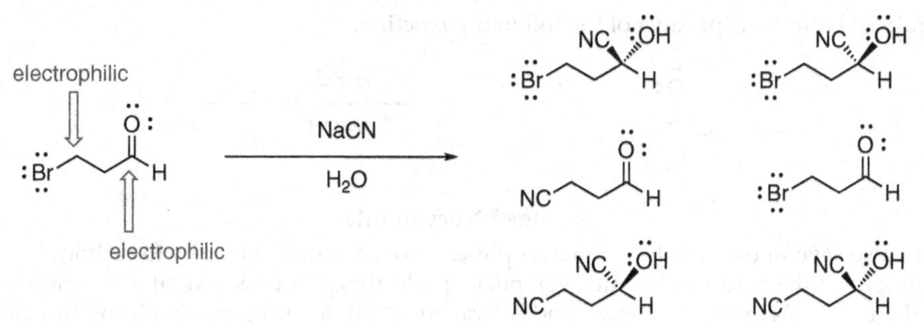

Figure 19.17 Possible outcomes of the reaction of cyanide with 3-bromopropanal.

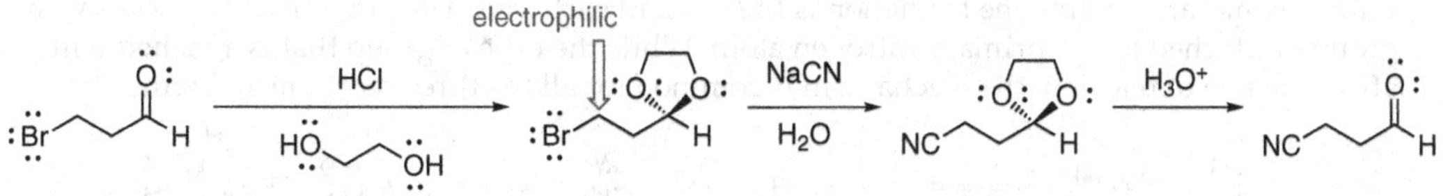

Figure 19.18 Use of an acetal protecting group to favour S_N2 substitution.

Problem 19.19 Provide an electron-pushing mechanism that accounts for the formation of the major product below.

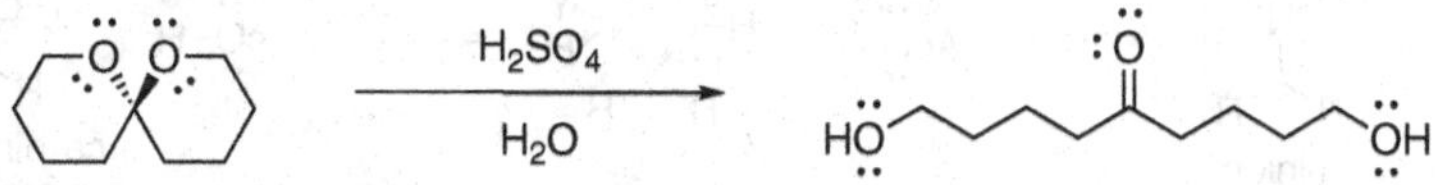

Acetals are quite stable to several conditions (except aqueous, acidic conditions) and are therefore employed (like silyl ethers for alcohols) as protecting groups for carbonyls. Consider the potential synthetic step in Figure 19.17. Because we have two electrophilic sites, there are five possible products that could form (which is likely to leave unreacted starting material left over too).

Given the susceptibility of the carbonyl carbon and the alkyl bromide carbon to nucleophilic attack, the reaction in Figure 19.17 produces a mess of products. Acetal protection of the carbonyl would leave the resulting alkyl bromide carbon as the only electrophilic site likely to be attacked (Figure 19.18). The aldehyde can then be deprotected through hydrolysis.

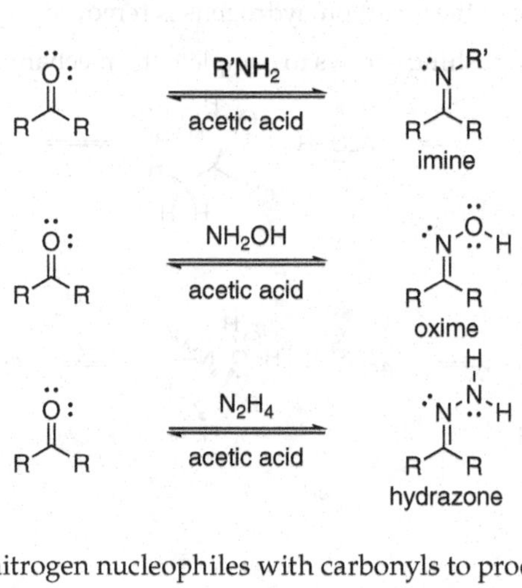

Figure 19.19 Reaction of nitrogen nucleophiles with carbonyls to produce imines, oximes, and hydrazones.

Problem 19.20 Predict the product of the following reaction.

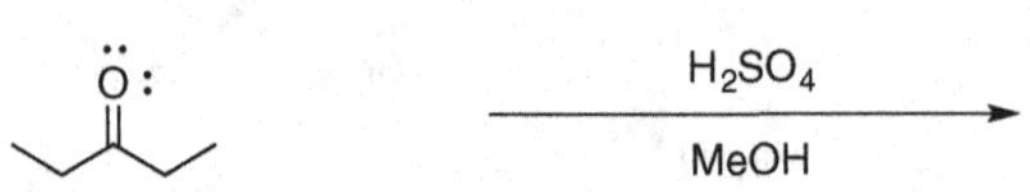

$$\text{H}_2\text{SO}_4 \quad / \quad \text{MeOH}$$

Nitrogen Nucleophiles

There are also several useful nitrogen nucleophiles also that can add readily to carbonyls. Depending upon the nature of the nitrogen nucleophile, this produces several different functional groups (Figure 19.19): imines,[6] oximes,[7] and hydrazones.[8] These nitrogen-containing functional groups are also important in biology and in biochemistry.

Problem 19.21 Draw electron-pushing arrows to complete the mechanism. Note the mechanism for imine, oxime, and hydrazone formation is the same in each case. The point of difference is what group is attached to our primary nitrogen atom. While the type of group that is attached will affect the rate of reaction, the mechanism is common for all the three reactions in Figure 19.19).

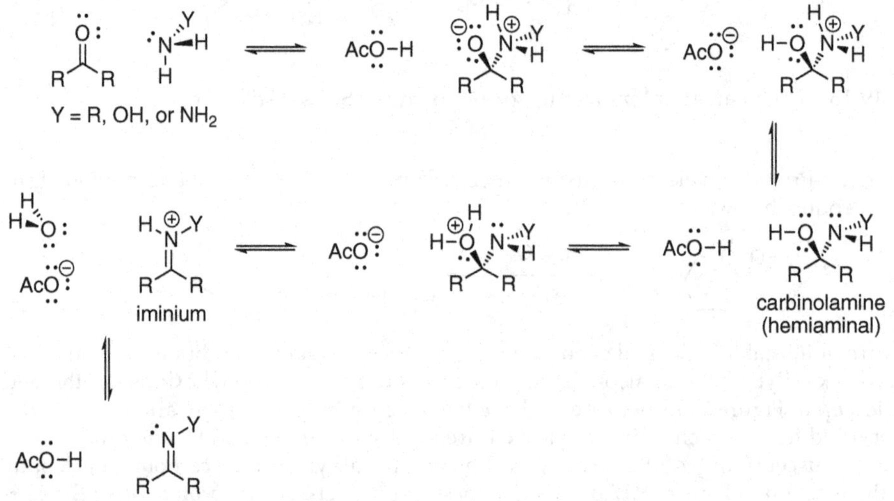

Y = R, OH, or NH$_2$

iminium

carbinolamine (hemiaminal)

For a secondary nitrogen, an enamine is produced (Figure 19.20).[9] This difference in outcome is due entirely to the fact that once the key iminium ion intermediate forms, there is no acidic proton on the nitrogen. As such, one of the α-carbon hydrogens is removed.

Problem 19.22 Draw electron-pushing arrows to complete the mechanism.

iminium

carbinolamine (hemiaminal)

enamine

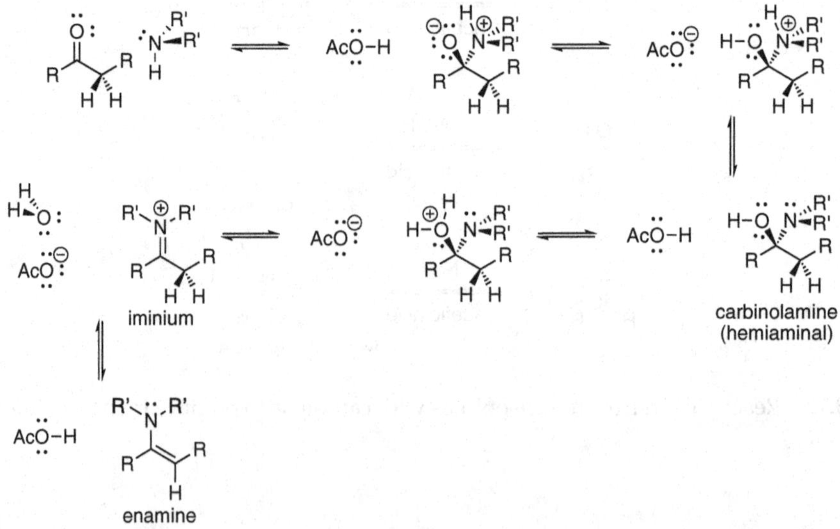

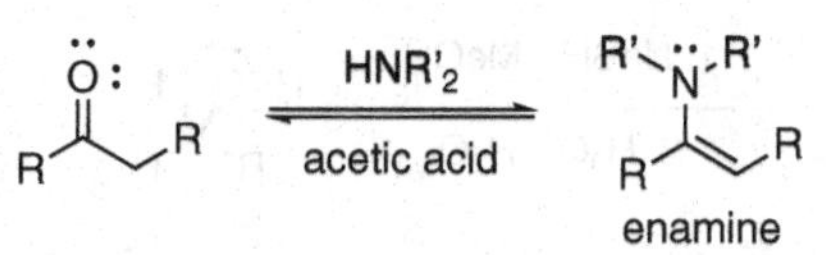

Figure 19.20 Addition of a secondary amine to a carbonyl to produce an enamine.

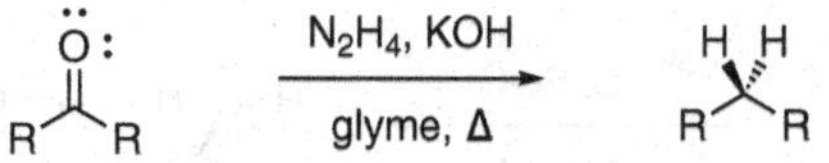

Figure 19.21 Wolff–Kischner reduction of a carbonyl group to an alkane.

Problem 19.23 Provide an electron-pushing mechanism that explains the formation of the major product of this reaction. Be sure to show all appropriate arrows, lone-pair electrons, formal charges, intermediates, and stereochemistry.

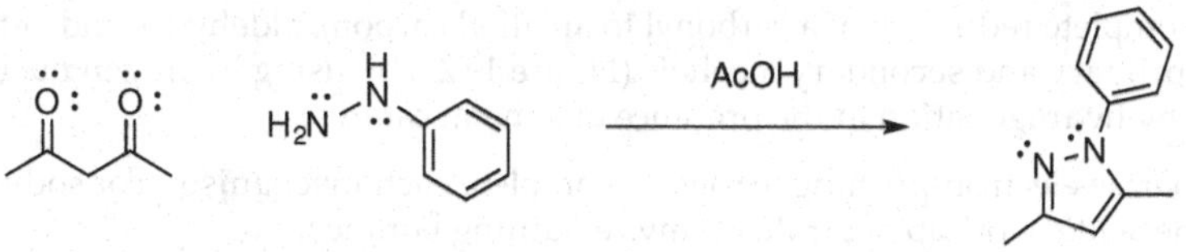

Problem 19.24 Identify the nitrogen species that will produce an enamine when reacted with acetone.

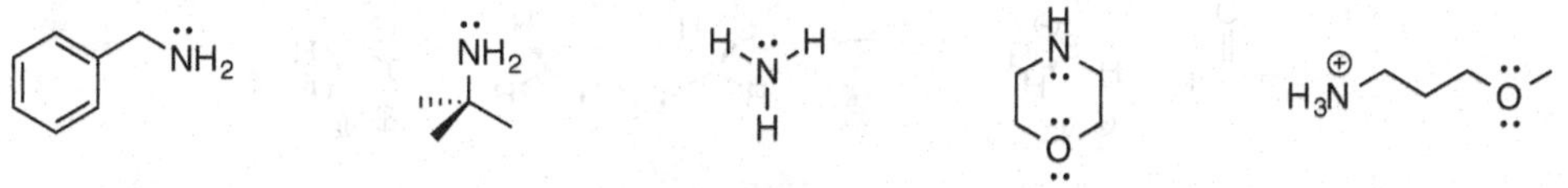

Aldehyde and Ketone Reduction

Hydrazones serve a unique purpose in aldehyde and ketone chemistry. Namely, under highly basic conditions, the aldehyde or ketone can be fully reduced to an alkane (Figure 19.21) in a reaction called the Wolff–Kischner reduction.[10]

Problem 19.25 Draw electron-pushing arrows to complete the mechanism.

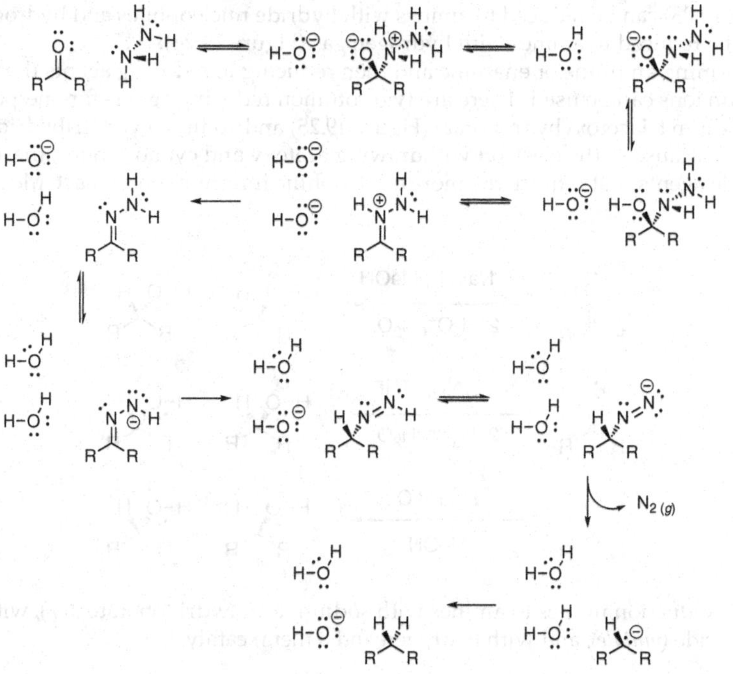

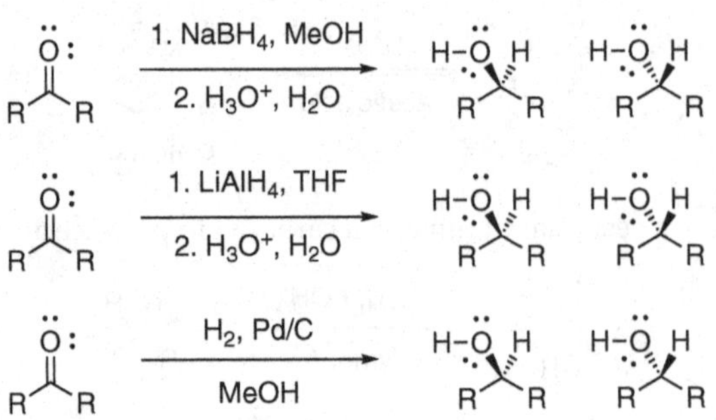

Figure 19.22 Reduction of a carbonyl to an alcohol with sodium tetrahydridoborate (*top*), with lithium aluminium hydride (*middle*), and with hydrogen and a metal catalyst.

While the Wolff–Kischner reaction is an extreme reduction (both in terms of the required conditions and the complete reduction of a carbonyl to an alkyl carbon), aldehydes and ketones can be reduced to primary and secondary alcohols (Figure 19.22) by using hydrogen nucleophiles (hydrides)[11] or by hydrogenation in the presence of a metal catalyst.

Problem 19.26 Draw electron-pushing arrows to complete each mechanism. For sodium tetrahydridoborate, the acidic workup neutralizes any remaining boranes.

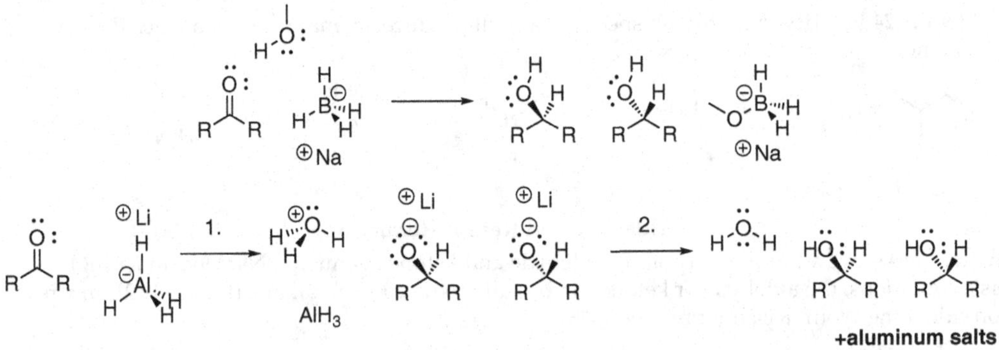

Imines (Figure 19.23) can be reduced to amines with hydride nucleophiles and hydrogen gas. Enamines can be reduced to amines with hydrogen gas (Figure 19.24).

Instead of forming an imine or enamine and then reducing it, reducing agents that are selective for iminium ions can be used. There are two common reducing agents for one-pot, reductive aminations: sodium triacetoxyhydroborate (Figure 19.25) and sodium cyanotrihydridoborate (Figure 19.26).[12] Because of the electron-withdrawing acetoxy and cyano groups, these are less reactive hydride agents that require the more electrophilic iminium ion to react. Mechanistically,

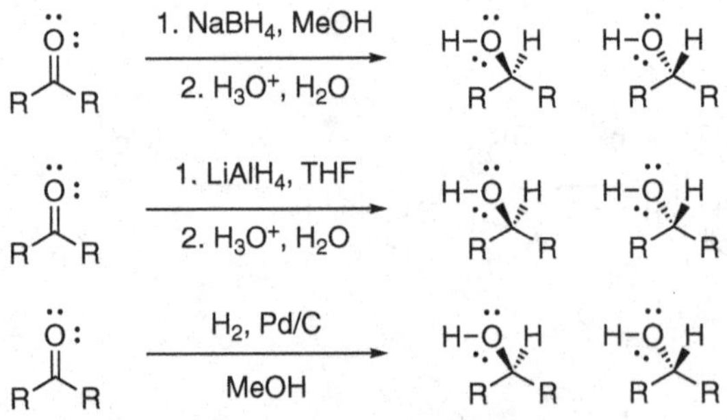

Figure 19.23 Reduction imines to amines with sodium tetrahydridoborate (*top*), with lithium aluminium hydride (*middle*), and with hydrogen and a metal catalyst.

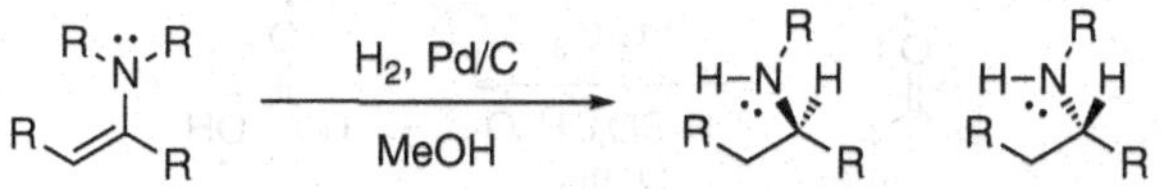

Figure 19.24 Reduction enamines to amines with hydrogen and a metal catalyst.

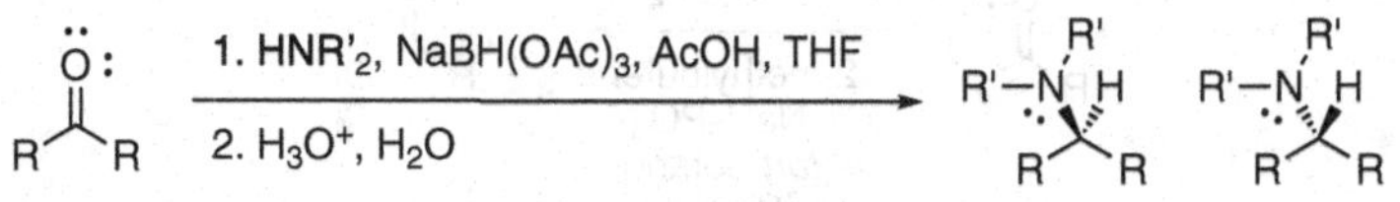

Figure 19.25 Reductive amination of carbonyls with amines (almost any amine will work so long as it has a single hydrogen atom on the nitrogen) and sodium triacetoxyhydroborate.

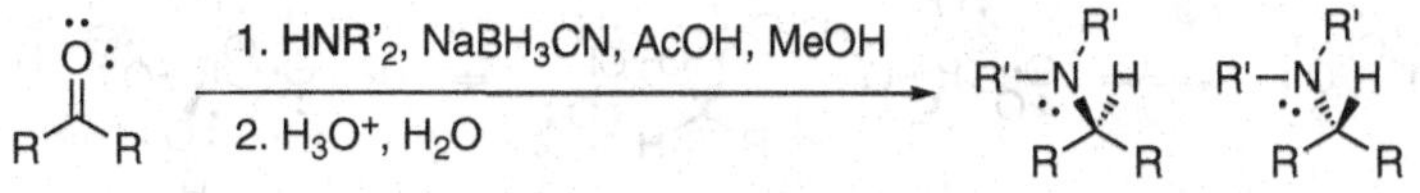

Figure 19.26 Reductive amination of carbonyls with amines (almost any amine will work so long as it has a single hydrogen atom on the nitrogen) and sodium cyanotrihydridoborate.

these proceed with the mechanisms seen on page 158 except that once the iminium is formed it is reduced by the hydride source.

Problem 19.27 Predict the products for each of the following reactions.

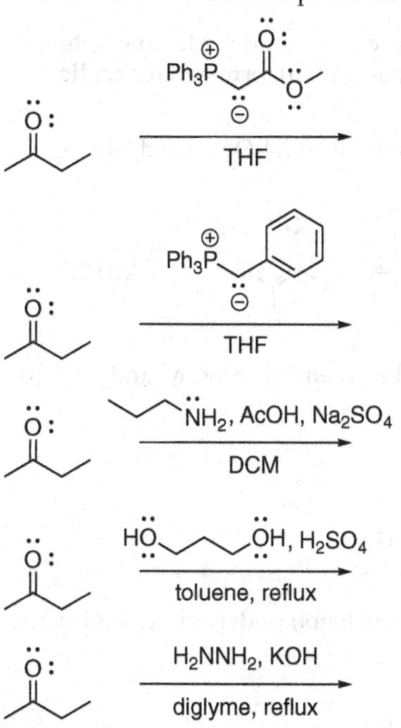

Aldehyde Oxidation

It is worth adding at this point that while ketones cannot be readily oxidized, aldehydes can be readily oxidized to carboxylic acids under different conditions.

Jones Oxidation (this mechanism is detailed in the Chapter 15)

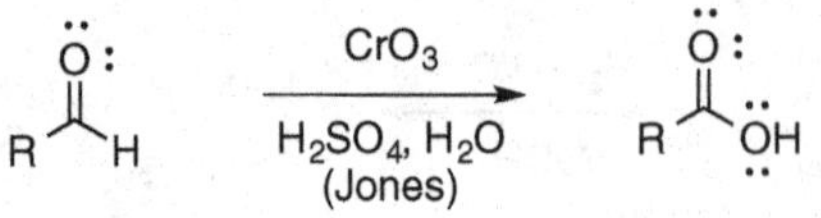

Pinnick Oxidation[13]

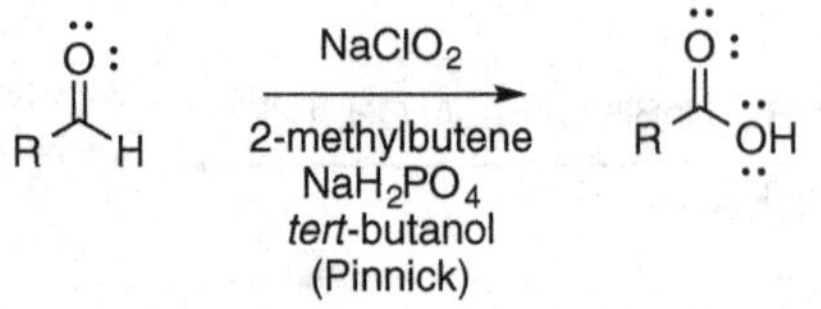

Problem 19.28 Draw electron-pushing arrows to complete the mechanism.

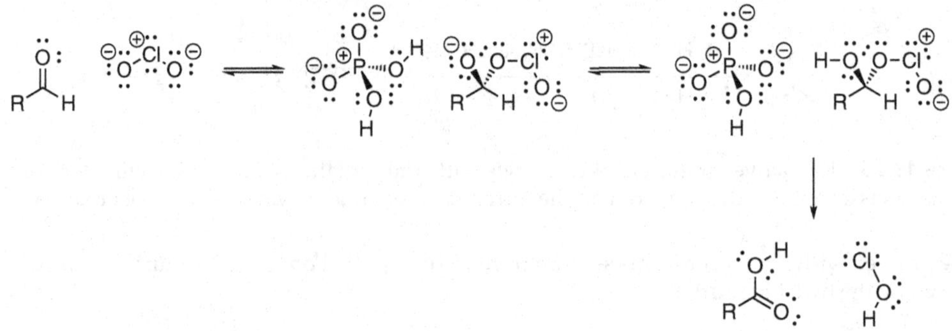

Reactions at α-Carbons

Finally, we will briefly consider the electrophilicity or acidity of α-carbons, aldehydes and ketones that can interconvert between the carbonyl (keto) and enol (alkene-alcohol) forms under acidic conditions (Figure 19.27).

Problem 19.29 Draw electron-pushing arrows to complete the mechanism (H₃O⁺ is used as a generic acid).

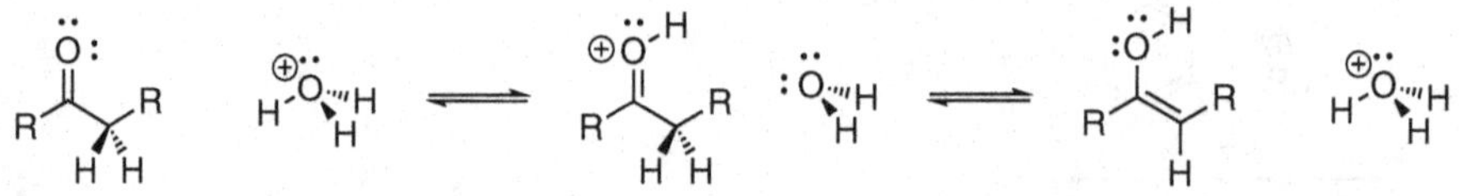

Under basic conditions, aldehydes and ketones can interconvert between the carbonyl and enolate forms (Figure 19.28).

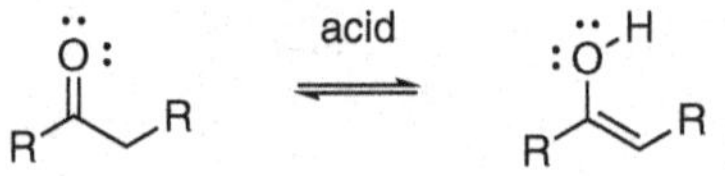

Figure 19.27 Isomerization of carbonyl (keto) to an alkene-alcohol (enol) under acidic conditions.

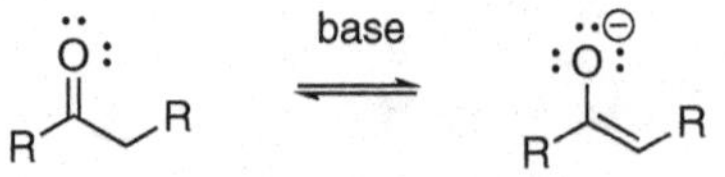

Figure 19.28 Conversion of a carbonyl to an enolate under basic conditions.

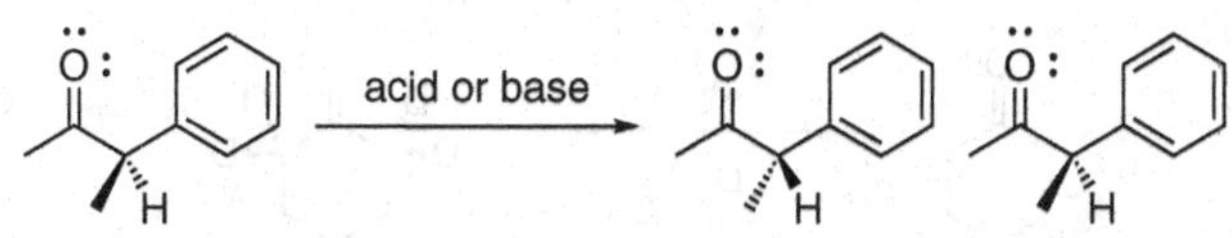

Figure 19.29 Racemization of α-carbon stereochemistry under acidic or basic conditions.

Problem 19.30 Draw electron-pushing arrows to complete the mechanism (hydroxide is used as a generic base).

Conversion of a carbonyl into an enol or an enolate converts a sp^3-hybridized, tetrahedral α-carbon into a sp^2-hybridized, trigonal planar atom. Any stereochemistry at the α-carbon position will therefore be racemized by the addition of acid/base (Figure 19.29).

Problem 19.31 Draw electron-pushing arrows to complete the acid-mediated mechanism.

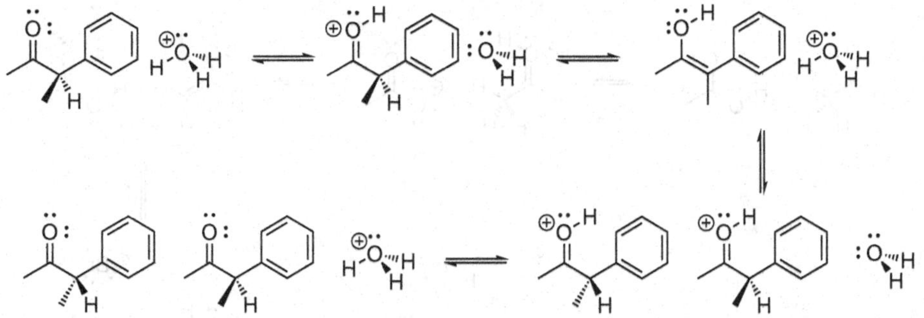

Problem 19.32 Draw electron-pushing arrows to complete the base-mediated mechanism.

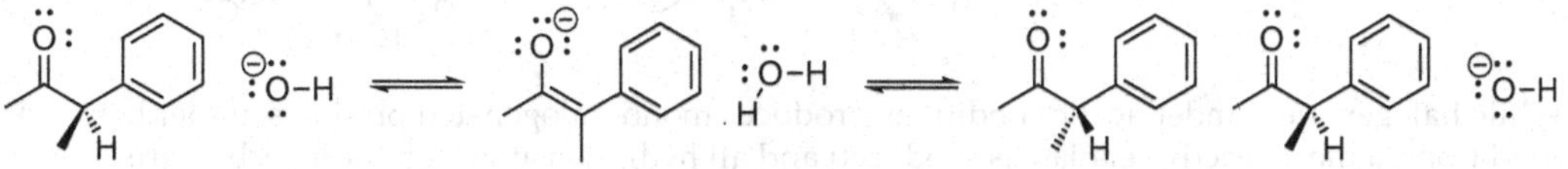

Problem 19.33 The two diastereomeric aldehydes **Y** and **Z** shown below interconvert in acid. Draw a reasonable arrow-pushing mechanism for the conversion of **Y** to **Z**.

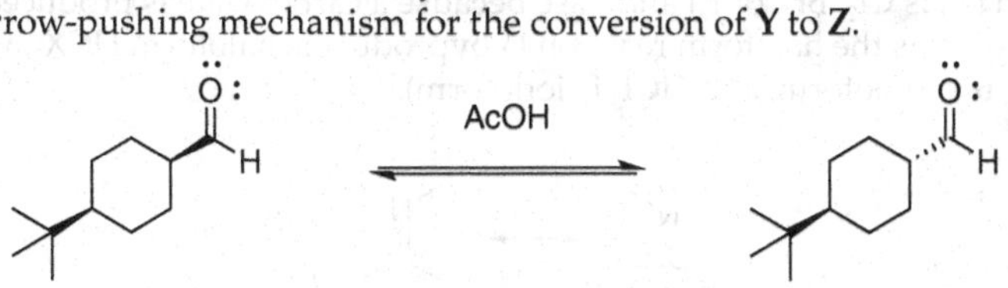

Problem 19.34 What do you predict the concentrations of **Y** and **Z** to be at equilibrium? (That is, do you expect that [**Y**] > [**Z**], [**Y**] < [**Z**], or [**Y**] = [**Z**]?). Use a clear, 3D drawing to rationalize your prediction.

Because of the increased acidity of the α-carbon, several reactions can occur involving the enol/enolate form of an aldehyde/ketone, including deuterium exchange, racemization, and halogenation. Deuterium exchange replaces all hydrogen atoms on α-carbons with deuterium atoms when reacted in D_2O with base (Figure 19.30).

Problem 19.35 Using the mechanism in Problem 19.32, propose a reasonable mechanism to account for the mechanism in Figure 19.30.

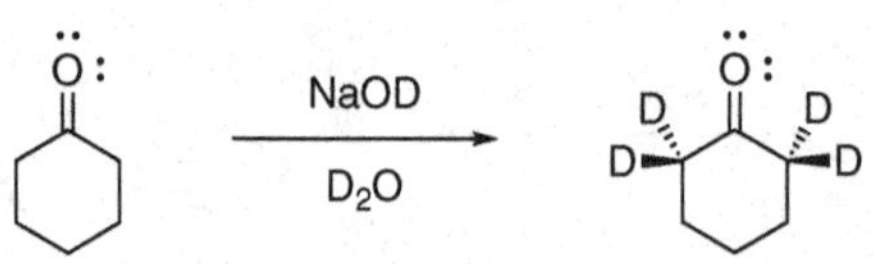

Figure 19.30 Replacement of α-carbon hydrogen atoms with deuterium.

Enamines, enols, and enolates are common nucleophiles in organic chemistry. When considering the contributing structure in Figure 19.31, each is nucleophilic at both the α-carbon atom and the heteroatom. Normally, however, the only observed products are those that result from nucleophilic attack by the α-carbon atom. We will see our first reactions where enols and enolates act as nucleophiles below (α-halogenation) and then this type of chemistry will be considered more extensively in Chapter 22.

The α-carbon can be mono-halogenated by adding acid (typically acetic acid) to generate an enol and X_2 (X_2 is Cl_2, Br_2, or I_2) (Figure 19.32). As we saw in the electrophilic addition to alkenes, the added X_2 is the electrophile is attacked by our nucleophile (here the enol).

Problem 19.36 Draw electron-pushing arrows to complete the mechanism.

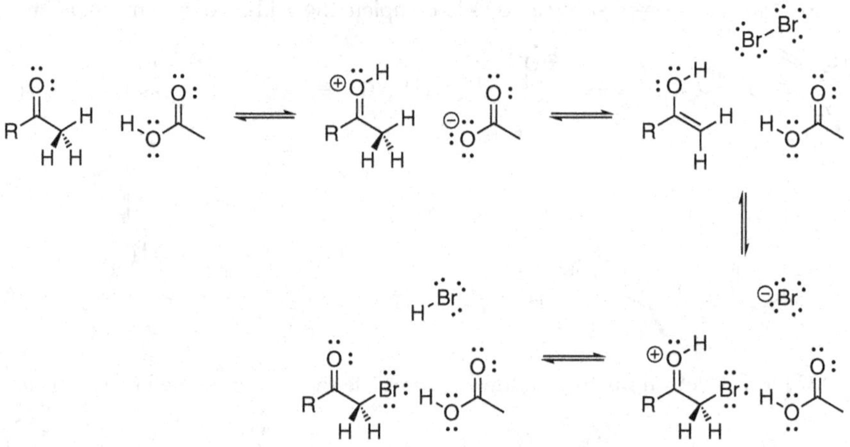

While halogenation under acidic conditions produces mono-halogenated products, under basic conditions, a more reactive enolate is produced and all hydrogen atoms on the α-carbon are replaced with halogen atoms. This is even more dramatic when a methyl ketone is exposed to a mixture of halogen (X_2 is Cl_2, Br_2, or I_2) and base because a carboxylate is produced (Figure 19.33). This reaction is known as the haloform reaction (a byproduct is haloform HCX_3, where $HCCl_3$ is chloroform, $HCBr_3$ is bromoform, and HCI_3 is iodoform).[14]

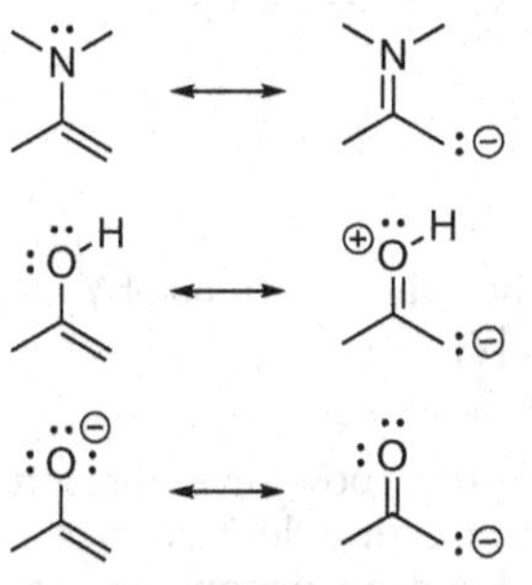

Figure 19.31 Contributing structures of an enamine (*top*), enol (*middle*), and enolate (*bottom*).

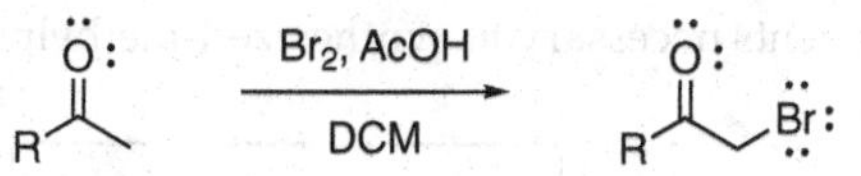

Figure 19.32 Mono halogenation of a carbonyl under acidic conditions.

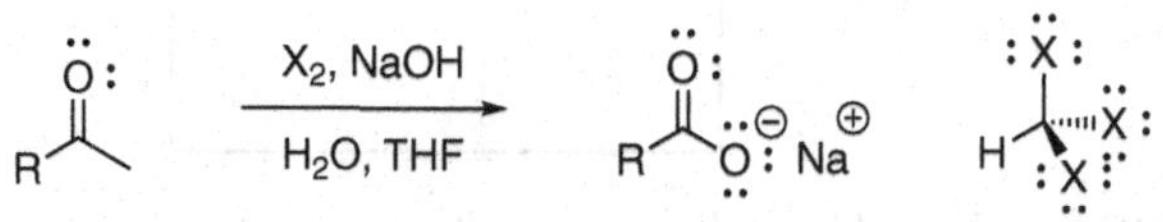

Figure 19.33 Conversion of a methyl ketone to a carboxylate via the haloform reaction.

Problem 19.37 Draw electron-pushing arrows to complete the mechanism.

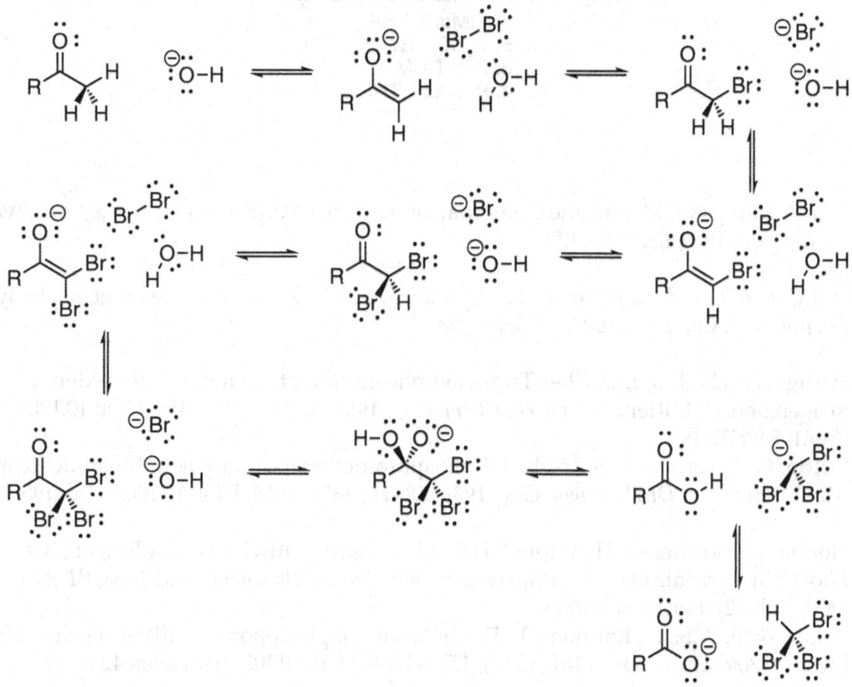

GENERAL PRACTICE PROBLEMS

Problem 19.38 Provide the necessary reagents to prepare *N,N*-dimethylcyclohexaneamine from cyclohexene.

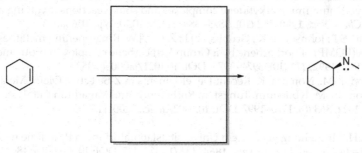

Problem 19.39 Provide the reagents necessary to synthesize 1-methylpiperidine from pentane-1,5-diol.

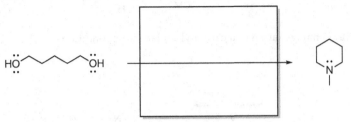

Problem 19.40. Predict the product of the following multi-step synthesis.

1. Hg(OAc)$_2$, H$_2$O
2. NaBH$_4$, NaOH, H$_2$O
3. (COCl)$_2$, DMSO, TEA, DCM

4. EtMgBr, THF
5. H$_2$SO$_4$, Δ
6. HBr, DCM
7. Ph$_2$CuLi, THF

NOTES

1. Gautier, A.; Simpson, M. Sur une Combinaison Directe d'Aldéhyde et d'Acide Cyanhydrique. *C. R. Acad. Sci.*, **1867**, *65*, 414–417.

2. Favorsky, A. Action de la potasse caustique sur les mélanges des cétones avec le phénylacé-tyléne. *Bull. Soc. Chim. Fr.*, **1907**, *2*, 1087–1088.

3. i) Wittig, G.; Schöllkopf, U. Über Triphenyl-phosphin-methylene as olefinbildende Reagenzien. (I. Mitteil. *Ber. Dtsch. Chem. Ges.*, **1954**, *87* (9), 1318–1330. DOI: 10.1002/cber.19540870919
   ii) Wittig, G.; Haag, W. Über Triphenyl-phosphinmethylene as an olefinbildende Reagenzien. (II. Mitteil.[1]) *Ber. Dtsch. Chem. Ges.*, **1955**, *88* (11), 1654–1666. DOI: 10.1002/cber.19550881110

4. i) Horner, L.; Hoffmann, H.; Wippel, H.G. Phosphororganische Verbindungen, XII. Phosphinoxyde als Olefinierungsreagenzien. *Ber. Dtsch. Chem. Ges.*, **1958**, *91* (1), 61–63. DOI: 10.1002/cber.19580910113
   ii) Wadsworth, W.S., Jr.; Emmons, W.D. The utility of phosphonate carbanions in olefin synthesis. *J. Am. Chem. Soc.*, **1961**, *83* (7), 1733–1738. DOI: 10.1021/ja01468a042

5. i)   Hérisson, J.-L.; Chauvin, Y. Catalyse de transformation des oléfines par les complexes du tungstène. II. Télomérisation des oléfines cycliques en presence d'oléfines acycliques. *Makromol. Chem.*, **1971**, *141* (1), 161–176. DOI: 10.1002/macp.1971.021410112
   ii) Schrock, R.R. Murdzek, J.S.; Bazan, G.C.; Robbins, J.; DiMare, M.; O'Regan, M. Synthesis of molybdenum imido alkylidene complexes and some reactions involving acyclic olefins. *J. Am. Chem. Soc.*, **1990**, *112* (10), 3875–3886. DOI: 10.1021/ja00166a023
   iii) Nguyen, S.T.; Johnson, L.K.; Grubbs, R.H.; Ziller, J.W. Ring-opening metathesis polymerization (ROMP) of norbornene by a Group VIII carbene complex in protic media. *J. Am. Chem. Soc.*, **1992**, *114* (10), 3974–3975. DOI: 10.1021/ja00036a053
   iv) Dawood, K.M.; Nomura, K. Recent Developments in Z-Selective Olefin Metathesis Reactions by Molybdenum, Tungsten, Ruthenium, and Vanadium Catalyst. *Adv. Synth. Catal.*, **2021**, *363* (8), 1970–1997. DOI: 10.1002/adsc.202001117

6. i) Schiff, H. Mittheilungen aus dem Universitätslaboratorium in Pisa: Eine neue Reihe organischer Basen. *Liebigs Ann.*, **1864**, *131* (1), 112–117. DOI: 10.1002/jlac.18641310112
   ii) Ladenburg, A. Ueber die Imine. [Vorläufige Mittheilung.] *Ber. Dtsch. Chem. Ges.*, **1883**, *16* (1), 1149–1152. DOI: 10.1002/cber.188301601259

7. Meyer, V.; Janny, A. Ueber die Einwirkung von Hydroxylamin auf Aceton. *Ber. Dtsch. Chem. Ges.*, **1882**, *15* (1), 1324–1326. DOI: 10.1002/cber.188201501285

8. Curtius, T.; Thun, K. Einwirkung von Hydrazinhydrat auf Monoketone und Orthodiketone. *J. Prakt. Chem.*, **1891**, *44*, 161–186. DOI: 10.1002/prac.18910440121

9. i) Wittig, G.; Blumenthal, H. Über die Einwirkung von Ammoniak und Ammoniak-Derivaten auf *o*-Acetylacetophenole. *Ber. Dtsch. Chem. Ges.*, **1927**, *60* (5), 1085–1094. DOI: 10.1002/cber.19270600515
   ii) Mannich, C.; Davidsen, H. Über einfache Enamine mit tertiär gebundenem Stickstoff. *Ber. Dtsch. Chem. Ges.*, **1936**, *69* (9), 2106–2112. DOI: 10.1002/cber.19360690921

10. i) Kishner, N. Каталитическое разложение алкилиденгидразинов, как метод получениыа углеводородов. [Kataliticheskoe razlozhenie alkilidengidrazinov, kak metod polucheniya uglevodorodov]. *J. Russ. Phys. Chem. Soc.*, **1911**, *43*, 582–595.
   ii) Wolff, L. Chemischen Institut der Universität Jena: Methode zum Ersatz des Sauerstoffatoms der Ketone und Aldehyde dur Wasserstoff. [Erste Abhandlung.]. *Liebigs Ann.*, **1912**, *394* (1), 86–108. DOI: 10.1002/jlac.19123940107

11. i) Chaikin, S.W.; Brown, W.G. Reduction of Aldehydes, Ketones and Acid Chlorides by Sodium Borohydride. *J. Am. Chem. Soc*, **1949**, *71* (1), 122–125. DOI: 10.1021/ja01169a033
   ii) Nystrom, R.F. and Brown, W.G. Reduction of Organic Compounds by Lithium Aluminium Hydride. I. Aldehydes, Ketones, Esters, Acid Chlorides and Acid Anhydrides. *J. Am. Chem. Soc.*, **1947**, *69* (5), 1197–1199. DOI: 10.1021/ja01197a060

12. i) Gribble, G.W.; Lord, P.D.; Skotnicki, J.; Dietz, S.E.; Eaton, J.T.; Johnson, J. Reactions of sodium borohydride in acidic media. I. Reduction of indoles and alkylation of aromatic amines with carboxylic acids. *J. Am. Chem. Soc.*, **1974**, *96* (25), 7812–7814. DOI: 10.1021/ja00832a035
   ii) Borch, R.F.; Bernstein, M.D.; Durst, H.D. Cyanohydridoborate anion as a selective reducing agent. *J. Am. Chem. Soc.*, **1971**, *93* (12), 2897–2904. DOI: 10.1021/ja00741a013

13. Lindgren, B.O.; Nilsson, T. Preparation of Carboxylic Acids from Aldehydes (Including Hydroxylated Benzaldehydes) by Oxidation with Chlorite. *Acta. Chem. Scand.*, **1973**, *27*, 888–890. DOI: 10.3891/acta.chem.scand.27-0888

14. i) Serullas, G.-S. Nouveau compose d'iode, d'hydrogène et de carbone ou hydriodure de carbone. *Ann. Chim. Phys.*, **1822**, *20*, 165–168.
   ii) von Liebig, J. Ueber die Verbindungen, welche durch die Einwirkung des Chlors auf alcohol, Aether, ölbildendes Gas und Essiggeist entstehen. *Ann. Phys.*, **1832**, *100* (2), 243–295. DOI: 10.1002/andp.18321000206

20 Carboxylic Acids

Carboxylic acids are molecules that contain a carboxyl group (COOH). An example of a carboxylic acid is acetic acid (Figure 20.1).

As can be seen in the electrostatic potential map (Figure 20.1), the carbonyl oxygen is very electron-rich and nucleophilic. In this chapter, we will see chemistry where the carbonyl oxygen acts as a nucleophile or a base. The hydroxyl oxygen is relatively non-nucleophilic. We can explain this by considering the contributing structures of acetic acid (Figure 20.2).

In considering the contributing structures (Figure 20.2), we can see that the carbonyl oxygen atom is rendered more electron-rich than a normal carbonyl oxygen atom through delocalization. Simultaneously, the hydroxyl oxygen is rendered less electron-rich than a normal hydroxyl oxygen. This difference in electron density can be seen in Figure 20.1 when one focuses on the oxygen atoms. The carbonyl oxygen is very red (very electron-rich), and the hydroxyl oxygen is yellow (only somewhat electron-rich).

NOMENCLATURE OF CARBOXYLIC ACIDS

In terms of priority, carboxylic acid functional groups take precedence over alkenes, alkynes, alcohols, aldehydes, and ketones. If an aldehyde or ketone is present, it is named as an oxo substituent. In IUPAC nomenclature, carboxylic acids are named with the suffix -oic acid. If deprotonated, the carboxylate is given the suffix -oate (Figure 20.3).

PHYSICAL PROPERTIES OF CARBOXYLIC ACIDS

The electrostatic potential map in Figure 20.1 also highlights the highly electrophilic (acidic) nature of the carboxylic hydrogen atom (deep blue). Carboxylic acids are aptly named because they are reasonably acidic (typical pK_a from 2 to 5). Many organic molecules, especially those containing nitrogen, will be protonated by carboxylic acids. The highly acidic hydrogen atom also contributes to the strong hydrogen-bonding ability of carboxylic acids, which tend to have higher

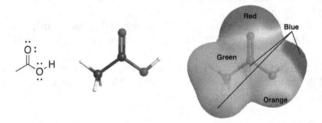

Figure 20.1 Structure (*left*), 3D model (*middle*), and electrostatic potential map (*right*) of an example carboxylic acid: acetic acid (red corresponds to high electron density and blue corresponds to low electron density).

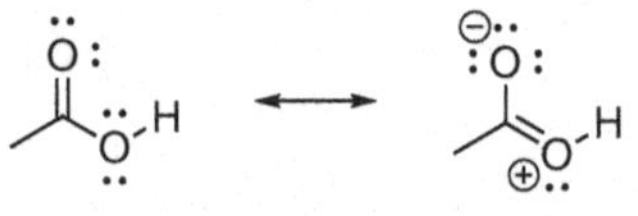

Figure 20.2 Contributing structures of acetic acid.

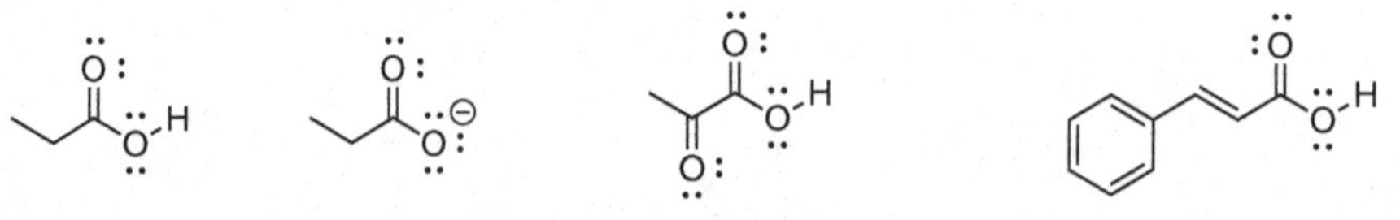

propanoic acid (propionic acid) | propanoate (propionate) | 2-oxopropanoic acid (pyruvic acid) | (2*E*)-3-phenylprop-2-enoic acid (*trans*-cinnamic acid)

Figure 20.3 Example carboxylic acids showing IUPAC names (and common names).

DOI: 10.1201/9781003479352-20

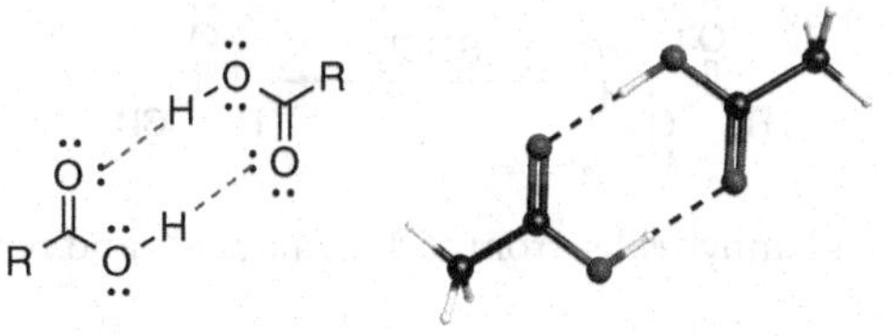

Figure 20.4 Structure (*left*) and 3D model (*right*) of carboxylic acids forming dimers.

water solubility than non-carboxylic acids of similar molecular weight. Carboxylic acids also have high melting and boiling points because of strong intermolecular hydrogen bonding (Figure 20.4) that leads to dimer formation.

SYNTHESIS OF CARBOXYLIC ACIDS

Carboxylic acids can be synthesized by several different methods:

Jones oxidation of primary alcohols

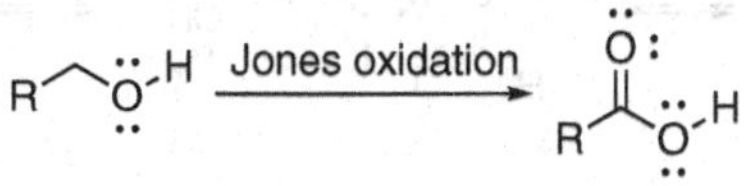

Jones oxidation of aldehydes

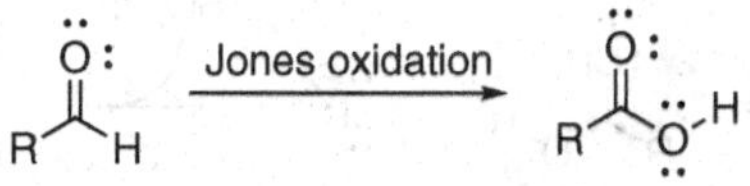

Pinnick oxidation of aldehydes

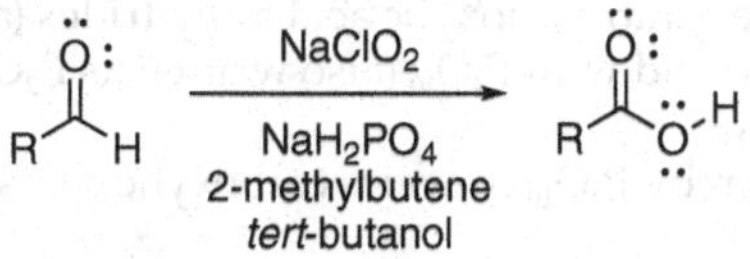

Addition of a Grignard reagent to dry ice

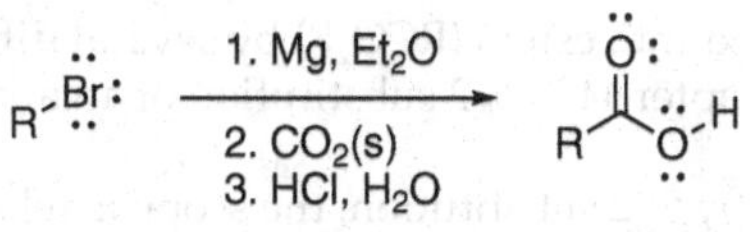

Problem 20.1 Draw the appropriate arrows to complete the mechanism.

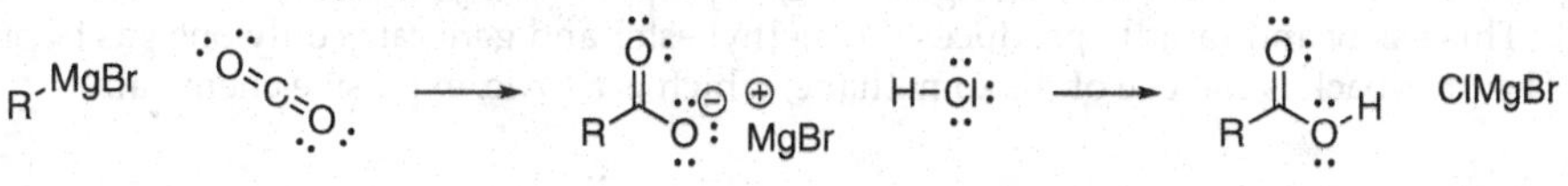

In Chapter 21, we will see that carboxylic acids can also be produced through hydrolysis of carboxylic acid derivatives, including acid chlorides, acid anhydrides, esters, and amides. Turning our attention to the reactions of carboxylic acids. We will first consider reactions to convert carboxylic acids into these various derivatives.

SYNTHESIS OF ACID CHLORIDES

Carboxylic acids can be converted into a much more reactive compound: acid chlorides. Acid chlorides replace the carboxylic –OH group with a –Cl (Figure 20.5).

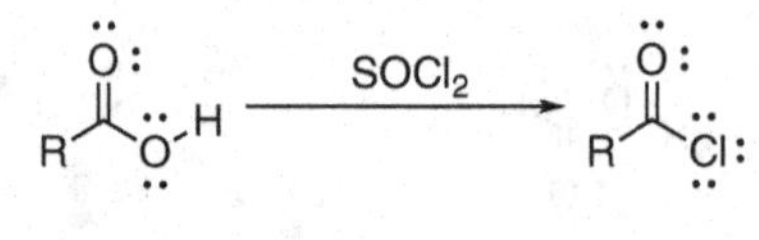

Figure 20.5 Synthesis of acid anhydrides from carboxylic acids and neat thionyl chloride.

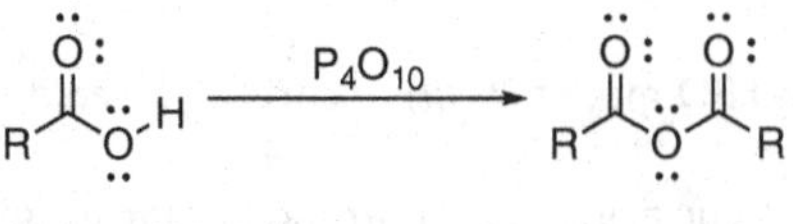

Figure 20.6 Synthesis of carboxylic acid anhydrides from carboxylic acids and P_4O_{10}.

This is accomplished using neat (no solvent present) thionyl chloride (a compound seen in Chapter 15 to convert alcohols to chlorides).[1] This reaction begins when the very electrophilic thionyl chloride is attacked by the carbonyl oxygen.

Problem 20.2 Draw appropriate electron-pushing arrows to complete the mechanism.

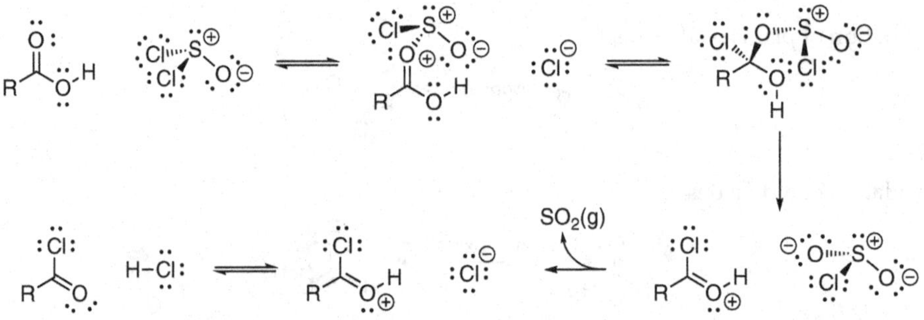

SYNTHESIS OF ACID ANHYDRIDES

Carboxylic acids can be converted into carboxylic acid anhydrides (anhydride meaning without water) by reacting the carboxylic acid with P_4O_{10} (also written as P_2O_5, Figure 20.6) or similar phosphorus-oxygen compounds.[2]

Note that the mechanism whereby P_4O_{10} converts carboxylic acids to anhydrides is beyond the scope of this chapter.

SYNTHESIS OF ESTERS

Carboxylic acids can be converted into esters (RCO_2R) by several different methods. One method that will be familiar from the chapter 14 is S_N2 substitution of an alkyl halide by a carboxylate (Figure 20.7).

While esters can be prepared by S_N2 substitution, the scope is relatively limited. Carboxylates are sluggish nucleophiles and so the alkyl halide must not be too sterically encumbered.

A highly atom-economical method (Figure 20.8) to prepare methyl esters is to use diazomethane (CH_2N_2).[3] This easily and rapidly produces the methyl ester and generates only one gas byproduct ($N_2(g)$). The drawback is the use of diazomethane, which is a toxic, explosive chemical.

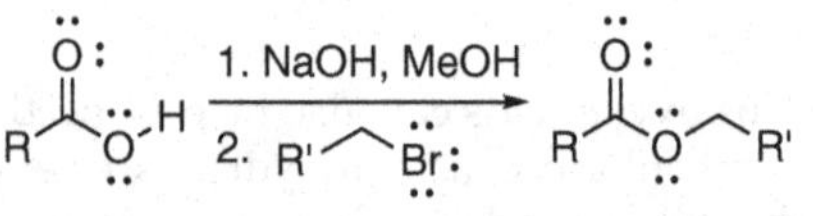

Figure 20.7 S_N2 synthesis of esters.

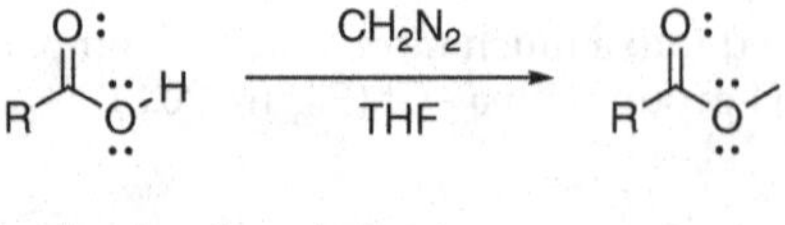

Figure 20.8 Diazomethane synthesis of methyl esters.

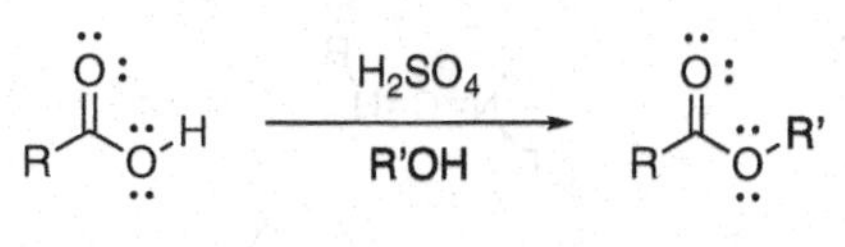

Figure 20.9 Fischer esterification reaction

Problem 20.3 Draw appropriate electron-pushing arrows to complete the mechanism.

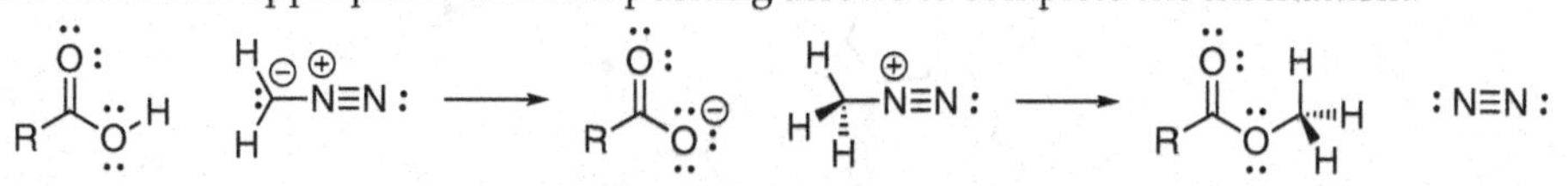

Finally, a more generalizable procedure is the Fischer esterification reaction.[4] In the Fischer esterification, the carboxylic acid combined with an alcohol in the presence of sulfuric acid (Figure 20.9). This reaction is highly generalizable and will work for almost any alcohol and carboxylic acid combination if they do not have other, acid-sensitive functional groups.

As can be seen in the mechanism in Problem 20.4, the entire process is in equilibrium. Therefore, we can control the outcome – ester formation or ester hydrolysis – through careful control of the reaction conditions and by shifting the equilibrium through added water or added alcohol.

Problem 20.4 Draw appropriate electron-pushing arrows to complete the mechanism.

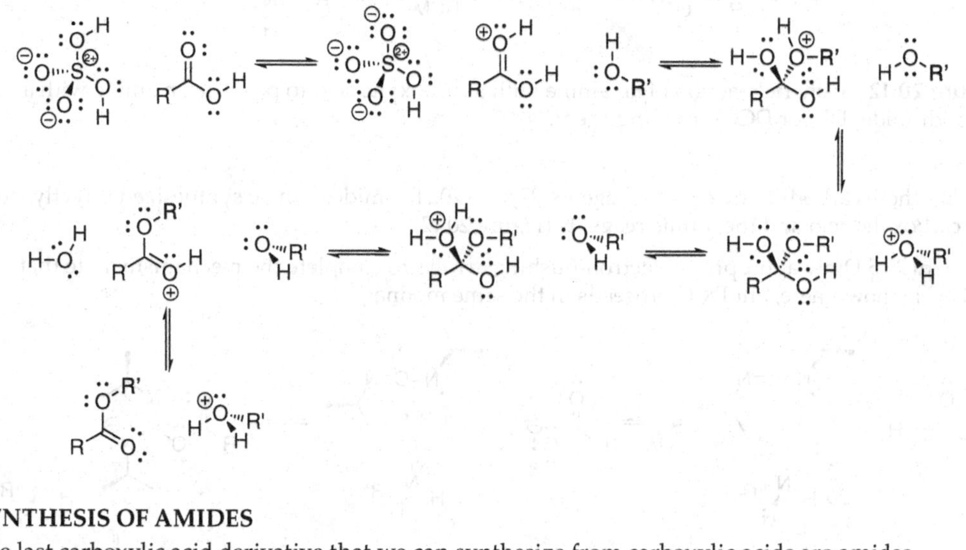

SYNTHESIS OF AMIDES

The last carboxylic acid derivative that we can synthesize from carboxylic acids are amides. Amides ubiquitous functional groups that appear in both organic and biochemical structures. Unlike esters considered above, amines cannot be synthesized directly from the acid and the amine. When an amine is present with a carboxylic acid, the dominant reaction is the acid-base reaction (Figure 20.10), which renders the amine non-nucleophilic (no lone pair electrons exist on the ammonium ion).

Moreover, the addition of an acid catalyst will also not help, as the acid would still protonate the amine and leave it unreactive. Therefore, to form an amide, a new reagent, a carbodiimide coupling reagent is used (Figure 20.11).[5] Note that carbodiimide coupling reagents could also be used (with an appropriate alcohol instead of an amine) to form esters from carboxylic acids.

Figure 20.10 Unproductive reaction of carboxylic acids with amines.

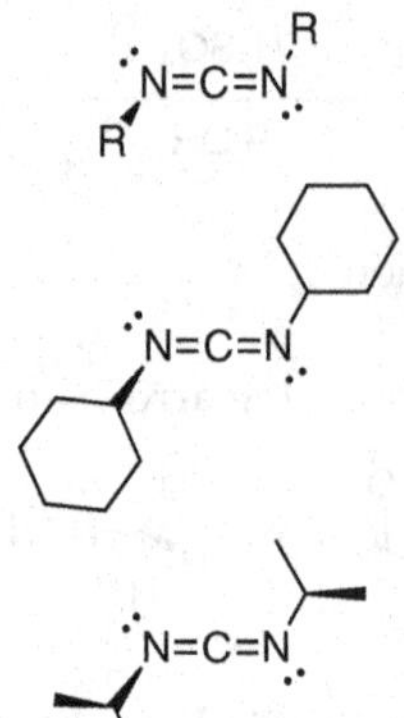

Figure 20.11 General structure of a diimide coupling reagent (*top*) and two common examples: N,N'-dicyclohexylcarbodiimide (DCC, *middle*), and N,N'-diisopropylcarbodiimide (DIC, *bottom*).

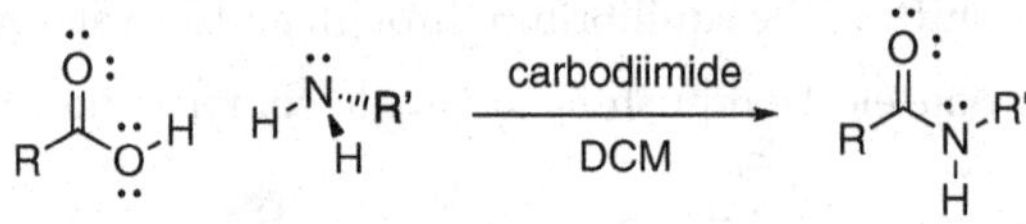

Figure 20.12 Generic reaction of an amine with a carboxylic acid to produce an amide with a carbodiimide (DIC or DCC) coupling agent.

With these carbodiimide coupling agents (Figure 20.11), amides can be synthesized directly from the carboxylic acid and the amine reagents (Figure 20.12).

Problem 20.5 Draw appropriate electron-pushing arrows to complete the mechanism (note that DIC is shown here, but DCC proceeds in the same manner).

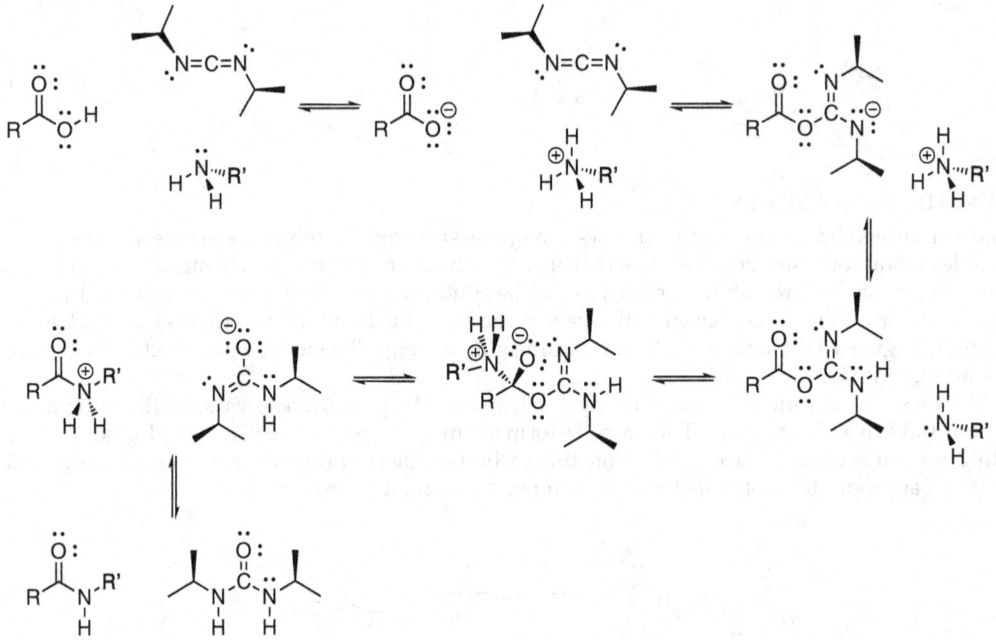

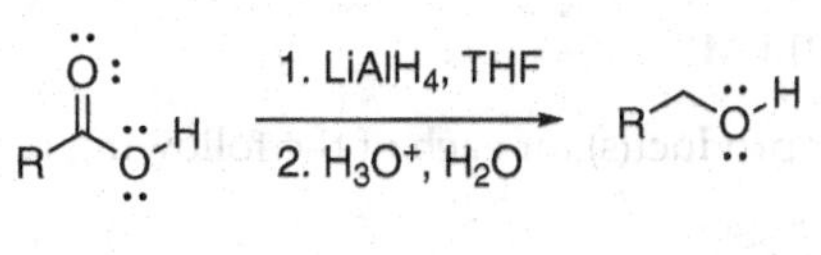

Figure 20.13   LAH reduction of carboxylic acids.

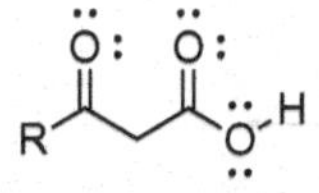

Figure 20.14   General structure of a β-ketocarboxylic acid (3-oxocarboxylic acid).

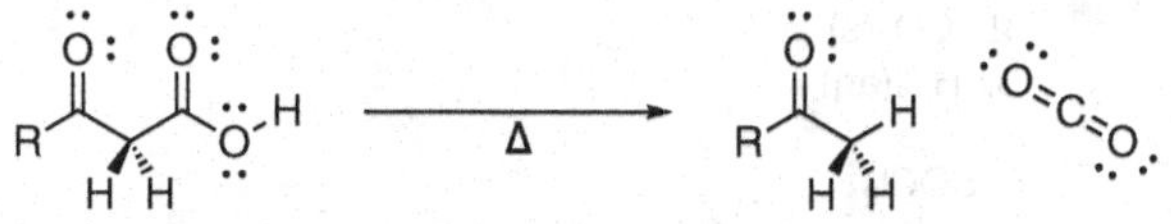

Figure 20.15   Decarboxylation of β-ketocarboxylic acid (3-oxocarboxylic acid).

Carboxylic acids can be reduced to primary alcohols (Figure 20.13) by lithium aluminium hydride (LiAlH$_4$).[6] Sodium tetrahydridoborate (NaBH$_4$) cannot be used for carboxylic acids as it is not a powerful enough reducing agent.

The mechanism for this reaction is challenging and beyond the scope of this chapter. The important point to be aware of is that LAH plays the same role as we saw previously, namely providing a hydride nucleophile to attack the carbonyl carbon.

The last topic to be considered is a special type of acid β-ketocarboxylic acids (Figure 20.14).

We are considering β-ketocarboxylic acids because they can undergo decarboxylation upon heating (Figure 20.15), which is a reaction that will be important in Chapter 21.

Problem 20.6 Draw appropriate electron-pushing arrows to complete the mechanism (note the β-ketocarboxylic acids undergo this reaction only after adopting the confirmation shown).

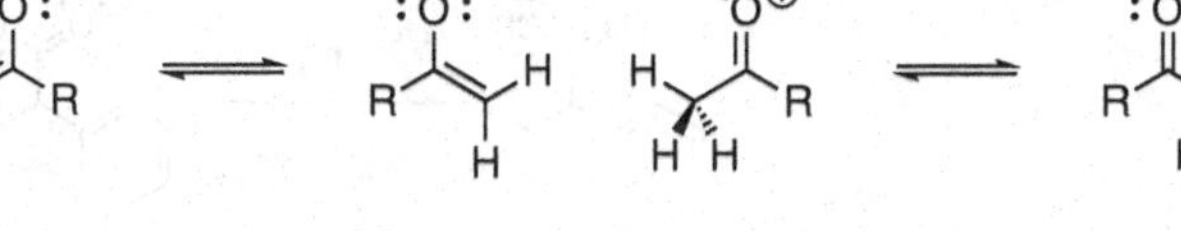

Problem 20.7 The resulting enol tautomerizes to the ketone by tautomerizing with another enol. Draw appropriate electron-pushing arrows to complete the mechanism.

GENERAL PRACTICE PROBLEMS

Problem 20.8 Predict the major product(s) for each of the following reactions.

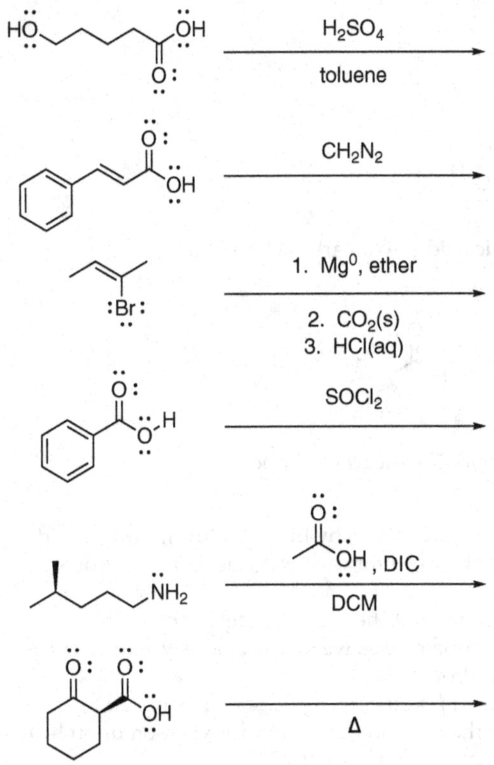

Problem 20.9 Provide the reagents necessary to affect each transformation from A to H.

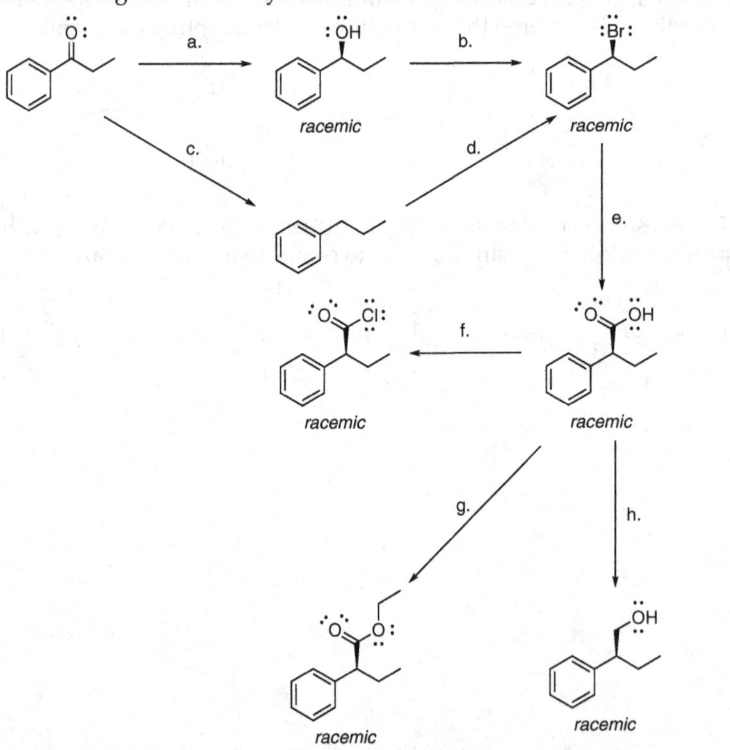

a.   b.

c.   d.

e.   f.

g.   h.

Problem 20.10 Predict the product of the following multi-step synthesis.

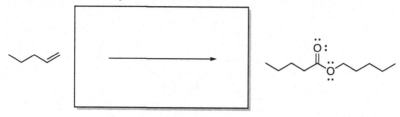

1. NBS, benzoyl peroxide, Δ
2. Mg, ether
3. CO_2 then aqueous workup

4. $LiAlH_4$, ether then H_3O^+
5. Dess-Martin periodinane
6. MeOH, H_2SO_4

Problem 20.11 Provide the reagents necessary to synthesize pentyl pentanoate (smells like apples) from with 1-pentene as the only source of carbon.

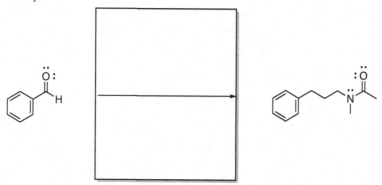

Problem 20.12 Provide the necessary reagents to prepare *N*-methyl-*N*-(3-phenylpropyl)acetamide from benzaldehyde.

NOTES

1. Carius, L. Ueber die Chloride des Schwefels. *Liebigs Ann.*, **1858**, *106* (3), 291–336. DOI: 10.1002/jlac.18581060309

2. Chiozza, L. Ueber wasserfreie organische Säuren. *Liebigs Ann.*, **1852**, *84* (1), 106–109. DOI: 10.1002/jlac.18520840111

3. Pechmann, H.v. Ueber Diazomethan. *Ber. Dtsch. Chem. Ges.*, **1895**, *28* (1), 855–861. DOI: 10.1002/cber.189502801189

4. Fischer, E.; Speier, A. Darstellung der Ester. *Ber. Dtsch. Chem. Ges.*, **1895**, *28* (3), 3252–3258. DOI: 10.1002/cber.189502803176

5. i) Schmidt, E.; Striewsky, W. Zur Kenntnis aliphatischer Carbodiimide (III. Mitteil.). *Ber. Dtsch. Chem. Ges.*, **1941**, *74* (7), 1285–1296. DOI: 10.1002/cber.19410740722

ii) Sheehan, J.C.; Hess, G.P. A New Method for Forming Peptide Bonds. *J. Am. Chem. Soc.*, **1955**, *77* (4), 1067–1068. DOI: 10.1021/ja01609a099

6. Nystrom, R.F. and Brown, W.G. Reduction of Organic Compounds by Lithium Aluminium Hydride. II. Carboxylic Acids. *J. Am. Chem. Soc.*, **1947**, *69* (10), 2548–2549. DOI: 10.1021/ja01202a082

21 Carboxylic Acid Derivatives

Carboxylic acid derivatives are molecules that can be synthesized from carboxylic acids and can be turned into carboxylic acids through hydrolysis. There are several carboxylic acid derivatives that will be considered in this chapter. The most reactive carboxylic acid derivatives are acid chlorides (Figure 21.1). In IUPAC nomenclature they are given the suffix -oyl chloride.

The second most reactive carboxylic acid derivatives are acid anhydrides (Figure 21.2). In IUPAC nomenclature they are named as carboxylic acids but given the suffix -oic anhydride instead of -oic acid.

One of the most common carboxylic acid derivatives are esters (Figure 21.3). Esters are substantially less reactive than acid chlorides or acid anhydrides. In IUPAC nomenclature the R group on the oxygen atom is named as a substituent (-yl ending) and the rest of the molecule gets the suffix –oate.

Amides (Figure 21.4) are common functional groups in organic chemistry and biochemistry (called peptides in biochemistry). One unique feature of amides is their increased acidity (pK_a near 17) over regular amines (pK_a near 38). Amides are the least reactive carboxylic acid derivative. In

Figure 21.1 Structure (*left*), 3D model (*middle*), and electrostatic potential map (*right*) of an example acid chloride: ethanoyl chloride (red corresponds to high electron density and blue corresponds to low electron density).

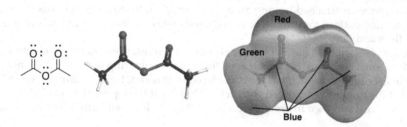

Figure 21.2 Structure (*left*), 3D model (*middle*), and electrostatic potential map (*right*) of an example anhydride: ethanoic anhydride (red corresponds to high electron density and blue corresponds to low electron density).

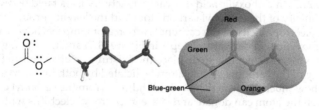

Figure 21.3 Structure (*left*), 3D model (*middle*), and electrostatic potential map (*right*) of an ester: methyl ethanoate (red corresponds to high electron density and blue corresponds to low electron density).

DOI: 10.1201/9781003479352-21

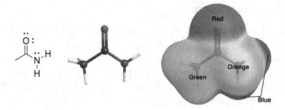

Figure 21.4 Structure (*left*), 3D model (*middle*), and electrostatic potential map (*right*) of an example amide: ethanamide (red corresponds to high electron density and blue corresponds to low electron density).

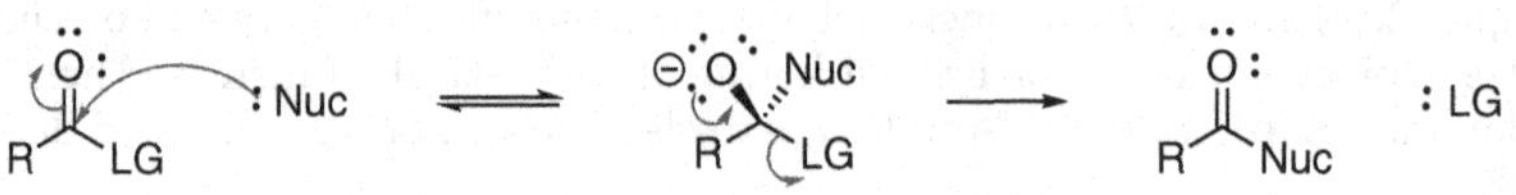

Figure 21.5 General mechanism of nucleophilic acyl substitution reactions: 1) nucleophilic attack to form a tetrahedral intermediate followed by 2) loss of leaving group to reform the carbonyl. LG could be –Cl, –O_2CR, –OR, or –NR_2. Claisen, L. Ueber die Einwirkung von Natriumalkylaten auf Benzaldehyd. *Ber. Dtsch. Chem. Ges.*, **1887**, *20* (1), 646–650. DOI: 10.1002/cber.188702001148

IUPAC nomenclature any substituent on the nitrogen atom is given the locant "*N-*" and the rest of the molecule gets the suffix -amide.

NUCLEOPHILIC ACYL SUBSTITUTION

In Chapter 19, we saw that carbonyls are an electrophilic site that nucleophiles can attack. For aldehydes and ketones, the nucleophilic attack was the end of the reaction. Carboxylic acid derivatives, however, have an attached leaving group (LG) attached to the carbonyl carbon. Therefore, the typical reaction pattern for carboxylic acid derivatives is nucleophilic acyl substitution (Figure 21.5) rather than addition.

There are a wide variety of nucleophilic acyl substitution reactions that are possible, but we will only consider a small subset of those reactions that entail converting one derivative into another. The complete list of reactions that will be considered in this chapter are summarized in Figure 21.6. Note that the graphic is arranged with acid chlorides at the top (as the most reactive carboxylic acid derivatives) and then in order of reactivity (high to low) with amides (the least reactive) at the bottom.[1]

Before looking at the types of reactions that each carboxylic acid derivative can undergo, let us first consider the overall reactivity trend (Figure 21.7). Our goal is to understand what factors contribute to this trend so that we may better understand the reactions that we consider in this chapter.

The trend (Figure 21.7) in carboxylic acid derivative reactivity is dictated by two complementary trends: the electrophilicity of the carbonyl carbon atom and the leaving group ability of the different derivatives. The electrophilicity of the carbonyl carbon atom impacts how readily the nucleophile attacks the carboxylic acid derivative (step 1 in Figure 21.5) and the leaving group ability impacts how easy it is for a nucleophile to displace the existing group (step 2 in Figure 21.5).

The electrophilicity of the carbonyl carbon atom is dictated by both inductive (how electronegative an atom is and how much electron density it withdraws from the carbonyl carbon atom) and conjugation (how well the atom can donate and share its p-orbital electrons with the carbon atom) effects. Figure 21.8 shows the relative electrophilicity of the different carboxylic acid derivative carbonyl carbon atoms.

Before we analyze Figure 21.8, we need to add a new layer to our understanding of conjugation: the quality of conjugated donation/withdrawal. We have, heretofore, treated all conjugated donation and withdrawal as the same; however, atoms that are in the same period on the periodic table will be better able to delocalize their electrons (their orbitals will overlap better due to closer sizes and energies) than atoms from different periods in the periodic table. This understanding will be necessary to understand Figure 21.8.

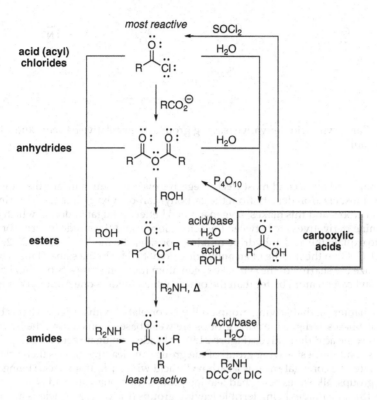

Figure 21.6 Nucleophilic acyl substitution reactions that will be considered in this chapter. The carboxylic acid derivatives are arranged in order of reactivity from high (acid chlorides) to low (amides).

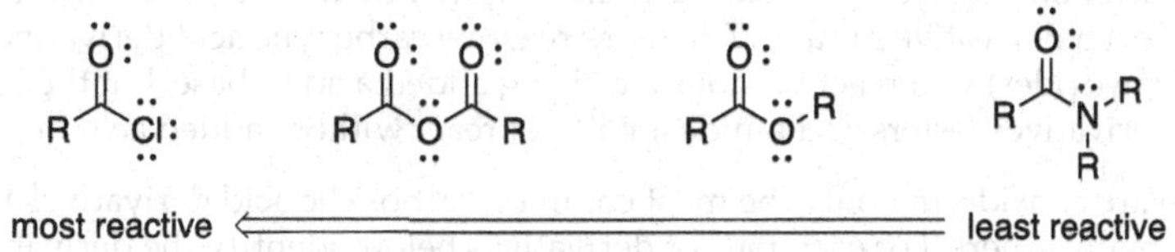

Figure 21.7 Nucleophilic acyl substitution reactivity trend.

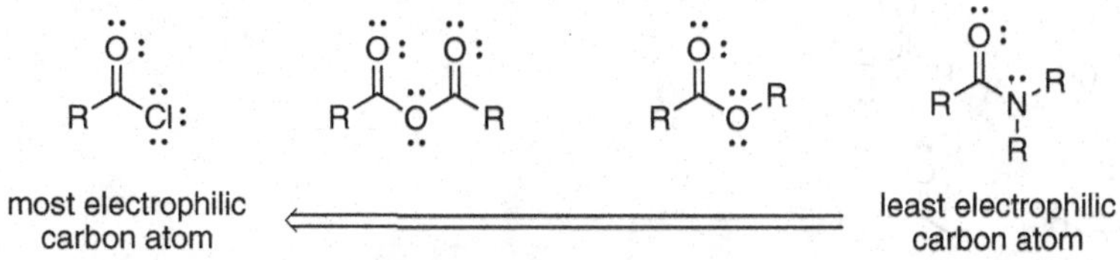

Figure 21.8 Relative electrophilicity of the carboxylic acid derivatives carbonyl carbon atoms.

Let us start by considering amides, the species with the least electrophilic carbonyl carbon atom, first. Nitrogen is the least electronegative of the different three atoms under consideration – Cl is 3.16, O is 3.44, and N is 3.04 – and so it withdraws the least amount of electron density from the carbonyl carbon atom. In terms of conjugated donation, nitrogen is not only in the same period as carbon, but also immediately adjacent to carbon and so it is an effective conjugated donor. On the other end of the scale, chlorine is more electronegative than nitrogen and so it withdraws more electron density from the carbonyl carbon atom. In addition, chlorine is in a different period than carbon and so it is a relatively poor conjugated donor. Together, these two factors make acid chloride carbonyl carbons very electrophilic. The significance of being in the same period becomes apparent when we consider the atoms in the middle. Oxygen is the most electronegative

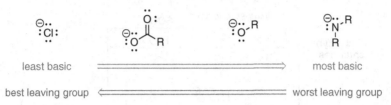

least basic ⟹ most basic

best leaving group ⟸ worst leaving group

Figure 21.9 Carboxylic acid derivative leaving groups, their relative basicity, and their relative leaving group ability.

atom of the group (and the second most electronegative overall), which means that it withdraws, inductively, the most electron density from the carbonyl carbon. Oxygen, unlike chlorine, is in the same period as carbon, and this makes oxygen a much better-conjugated donor, which is why the oxygen-containing derivatives are between the acid chloride and the amide in terms of electrophilicity. Finally, to explain the acid anhydride ranking, consider the model in Figure 21.2. The five atoms (O=C–O–C=O) in the middle of an anhydride do not all lie in the same plane. The molecule twists to minimize the sharing of the central oxygen atom (oxygen shares its p-orbital electrons with one carbonyl carbon atom better than the other), which renders one carbonyl carbon highly electrophilic.

Recall from Chapter 14, that leaving group ability is correlated with basicity. That is bad leaving groups are strong bases and good leaving groups are weak bases. Considering the leaving groups among the carboxylic acid derivatives (Figure 21.9), chloride (conjugate acid is HCl, $pK_a = -7$) is the weakest base and the best leaving group of the group. The leaving groups then trend in order with a carboxylate (the conjugate acid is a carboxylic acid with a pK_a from 2 to 5) being slightly worse leaving groups, alkoxides being bad leaving groups (the conjugate acid is an alcohol with a pK_a from 15 to 18), and amides being terrible leaving groups (the conjugate acid is an amine with a pK_a from 38 to 40). As such, the most reactive carboxylic acid derivative (acid chloride) has the best leaving group (chloride), and the least reactive carboxylic acid derivative (amide) has the worst leaving group (amide).

The significance of our analysis of Figure 21.8 and Figure 21.9 will be seen below as we consider the chemistry of each derivative in turn. The more reactive carboxylic acid derivatives (acid chlorides and acid anhydrides) will react without requiring added acid or base, but the less reactive carboxylic acid derivatives (esters and amides) will not react without added acid or base.

Problem 21.1 We are considering only the most common carboxylic acid derivatives in this chapter, but there are many others. For each pair of derivatives below, identify the derivative that is more reactive and explain your selection.

a. (a)

b. (b)

ACID CHLORIDE SUBSTITUTIONS

Now each carboxylic acid derivative and its reactions will be considered in turn. The acid chlorides, as the most reactive derivatives, have the greatest number of reactions they can undergo. The reactions, however, will all look remarkably similar, and no reaction will involve the addition of exogenous acid or base. As a general note on structure, the section for each derivative will look at its conversion into other derivatives first and then the section will end with hydrolysis to form carboxylic acids.

Acid chlorides can be converted into acid anhydrides (Figure 21.10) by reacting the acid chloride with a carboxylate salt.[2] Acid anhydrides can be prepared from the dehydration of a carboxylic acid with P_4O_{10}, but this will only produce symmetric (the two R groups are the same) anhydrides.

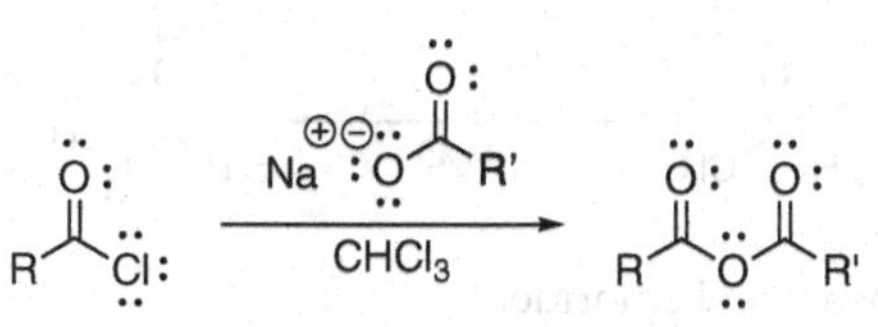

Figure 21.10   Conversion of an acid chloride to an acid anhydride.

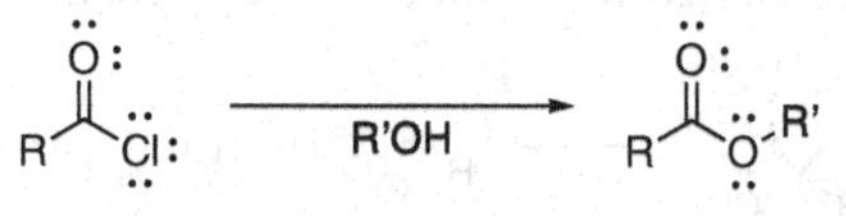

Figure 21.11   Conversion of an acid chloride to an ester (alcoholysis).

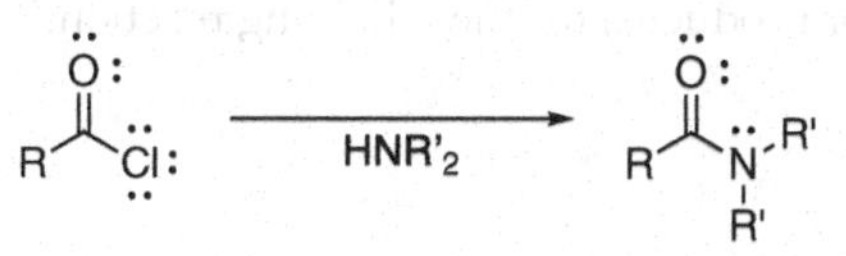

Figure 21.12   Conversion of an acid chloride to an amide (acylation of an amine).

Chemists have found useful roles for asymmetric or mixed anhydrides (the two R groups are different), and this is one way to prepare mixed anhydrides.

Problem 21.2 Draw the appropriate arrows to complete the mechanism.

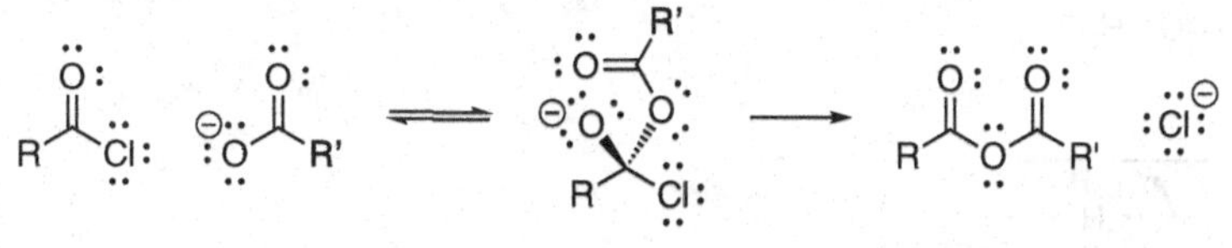

As we saw in Chapter 21, esters can be prepared from carboxylic acids through the Fischer esterification reaction. That reaction, however, requires strongly acidic conditions and it is an equilibrium process, which can limit overall yield. Esters can be prepared efficiently from acid chlorides by adding alcohol (usually both reagent and solvent) to the acid chloride (Figure 21.11).[3]

Problem 21.3 Draw the appropriate arrows to complete the mechanism.

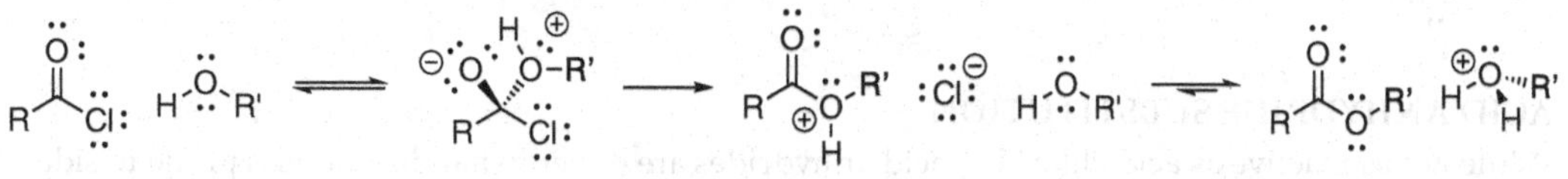

Amines will react with acid chlorides to produce amides (Figure 21.12).[4] Because amines are usually viewed as the more important component of the reaction, this is one of the reactions that is usually referred to as an acylation reaction (adding an acyl group). As with other reactions involving acid chlorides, the reagent (the amine) can play both the role of reactant and solvent.

Problem 21.4 Draw the appropriate arrows to complete the mechanism.

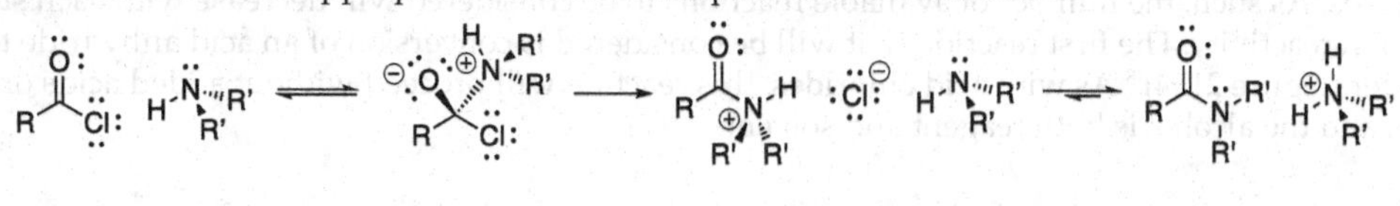

Finally, acid chlorides can be easily converted back to carboxylic acid by adding water (Figure 21.13). While facile, this reaction is typically an undesired one that can occur from acid chlorides being exposed to moisture in the air. As such, precautions are taken to protect the contents of an acid chloride bottle once it has been opened.

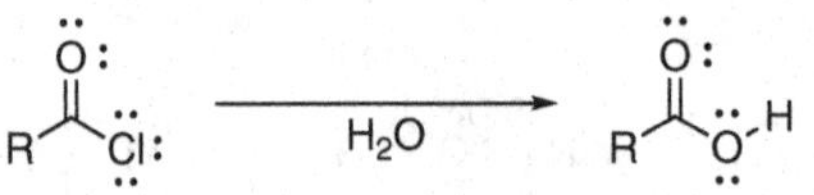

Figure 21.13 Hydrolysis of an acid chloride.

Problem 21.5 Draw the appropriate arrows to complete the mechanism.

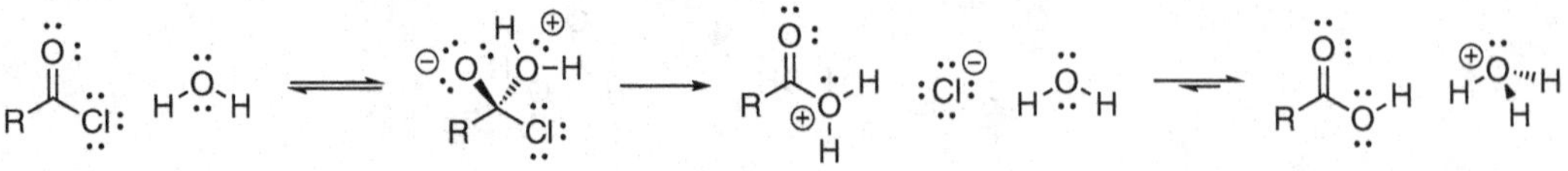

Problem 21.6 Predict the major product(s) for the following reactions.

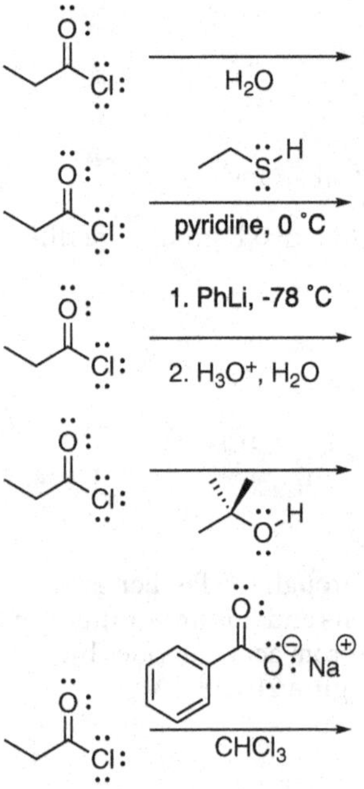

ACID ANHYDRIDE SUBSTITUTION

While not as reactive as acid chlorides, acid anhydrides are easier to handle and less prone to side reactions. As such acid anhydrides are typically more commonly employed than acid chlorides, even though the atom economy (the number of starting material atoms that end up in the final product) is lower (the carboxylate leaving group a substantially larger byproduct than chloride). For the following reactions, we are only considering the reactions of symmetrical anhydrides, not the reactions of mixed anhydrides mentioned above.

Acid anhydrides are less reactive than acid chlorides and cannot easily be converted into acid chlorides. As such, the number of available reactions to be considered will decrease with each step down in reactivity. The first reaction that will be considered is conversion of an acid anhydride to an ester (Figure 21.14).[5] As with acid chlorides, this reaction will proceed without added acids or bases and the alcohol is both reagent and solvent.

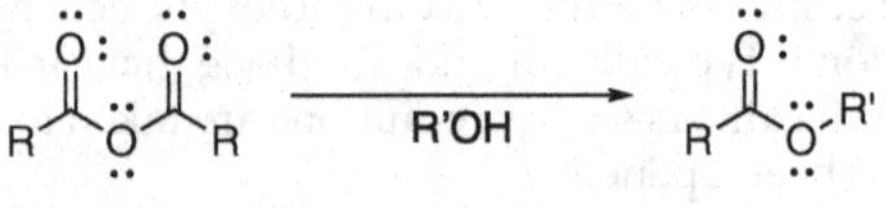

Figure 21.14 Conversion of an acid anhydride to an ester (alcoholysis).

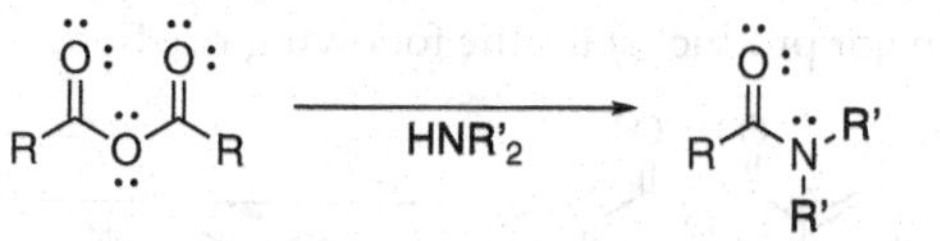

Figure 21.15 Conversion of an acid anhydride to an amine (acylation of an amine)

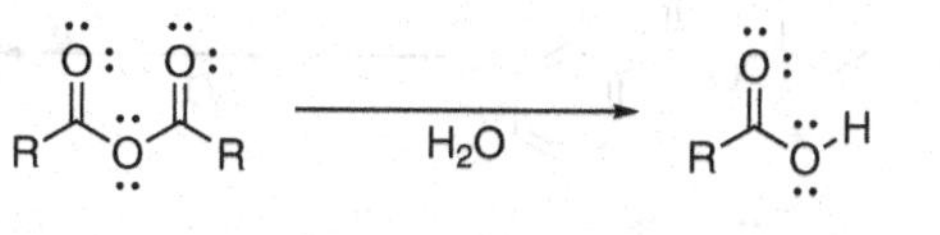

Figure 21.16 Hydrolysis of an acid anhydride.

Problem 21.7 Draw the appropriate arrows to complete the mechanism.

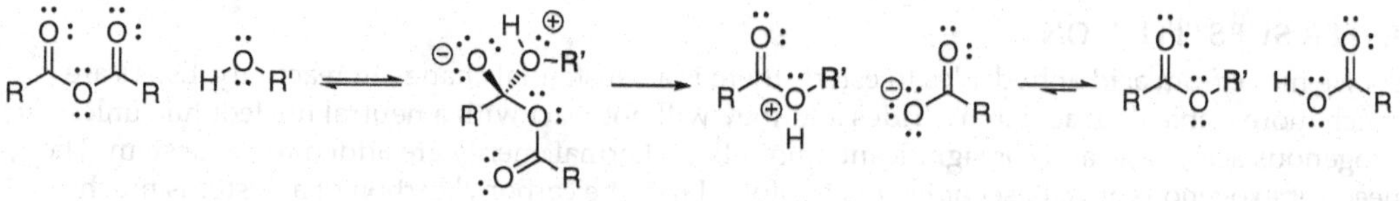

Similarly, acid anhydrides can be used to form amides (Figure 21.15), this is another example of what is referred to as an acylation reaction.[6] Acid anhydrides can be favoured for amine acylations, over acid chlorides, because of the increased stability (and reduced risk of over-acylation). The reaction proceeds without added acids or bases and the mechanism is very similar to that of the anhydride-alcohol reaction.

Problem 21.8 Draw the appropriate arrows to complete the mechanism.

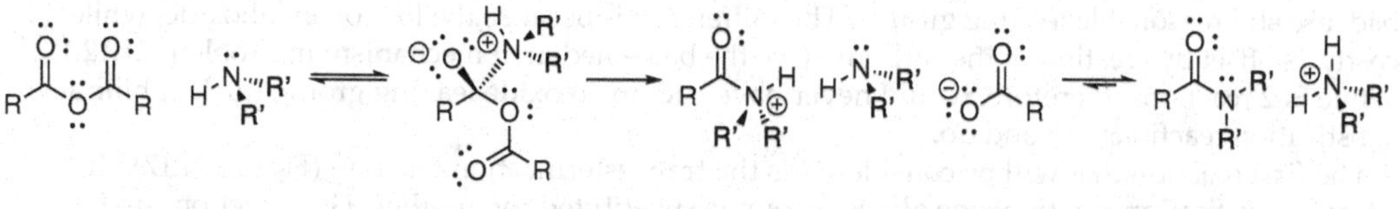

Finally, anhydrides can also be hydrolyzed into carboxylic acids (Figure 21.16).[7] The reaction is mechanistically identical to that of the reaction to produce esters from anhydrides (Figure 21.14). As with the hydrolysis of acid chlorides, it is usually an unwanted reaction that occurs when a bottle is opened to the air. As such, precautions are taken to protect the contents of an acid anhydride bottle once it has been opened.

Problem 21.9 Draw the appropriate arrows to complete the mechanism.

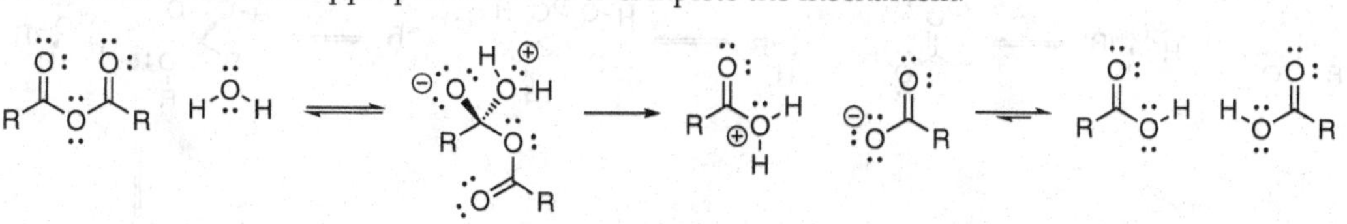

Problem 21.10 Predict the major product(s) for the following reactions.

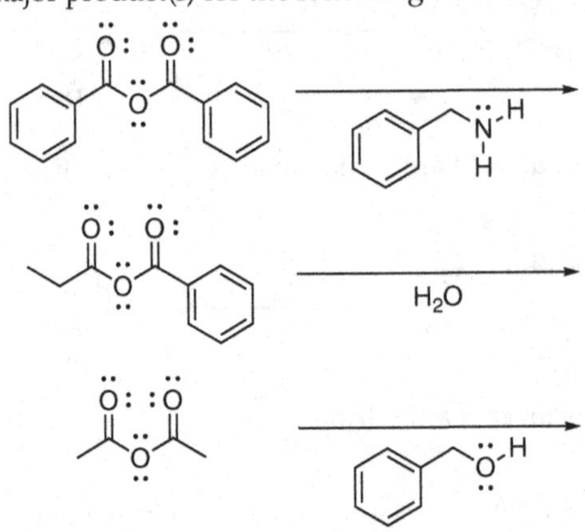

ESTER SUBSTITUTION

As we move from acid anhydrides to esters, there is a substantial change in reactivity. Esters are much more stable than acid anhydrides and they will not react with a neutral nucleophile unless exogenous acid, base, and/or significant amounts of thermal energy are added to the system. The need for exogenous acid, base, or heat is twofold. First, the carbonyl carbon of an ester is much less electrophilic. As such, weak nucleophiles, like alcohols and water, will not appreciably attack the carbonyl carbon. Added acid will make the carbonyl carbon more electrophilic and added base leads to the creation of stronger, anionic nucleophiles. Added heat helps to provide the necessary energy to overcome the higher activation energy for reactions of esters. Second, the leaving group of an ester (an alkoxide) is a substantially worse leaving group than we saw in the reactions of anhydrides (where carboxylate was the leaving group). Added acid can (as we saw in ether cleavage) facilitate the reaction by creating a good, alcohol leaving group. Added base creates an anion-rich environment where the loss of an alkoxide leaving group is less energetically costly. It should be noted, that for nucleophilic acyl substitution reactions, unlike S_N2 reactions, alkoxides, while bad, are still reasonable leaving groups. This difference is because the loss of an alkoxide, while costly, is offset by creation of the carbonyl (see the base-mediated mechanism in Problem 21.12.). While S_N2 reactions, therefore, would never have had an alkoxide leaving group, nucleophilic acyl substitution reactions can and do.

The first reaction that will be considered is the transesterification reaction (Figure 21.17).[8] In a transesterification reaction, one alkoxy group is substituted for another. This reaction can be advantageous, compared to using acid chlorides or acid anhydrides to synthesize esters, because esters are ubiquitous and quite stable. As was mentioned above, esters will not react with neutral nucleophiles like alcohols and so base or acid must be added to the reaction. Both mechanisms are shown in Problems 21.11 and 21.12.

Problem 21.11 Draw the appropriate arrows to complete the acid-mediated mechanism.

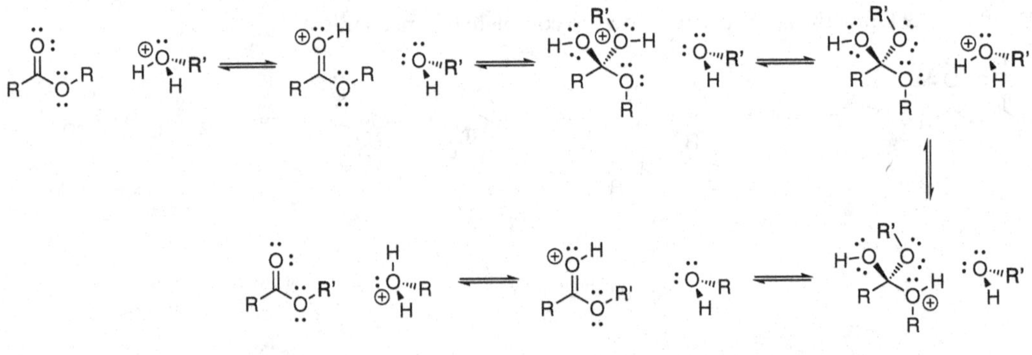

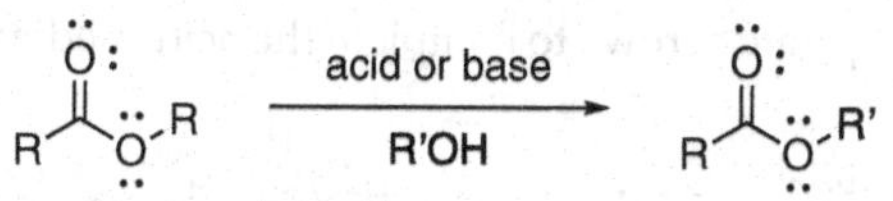

Figure 21.17   Acid- or base-mediated transesterification reaction.

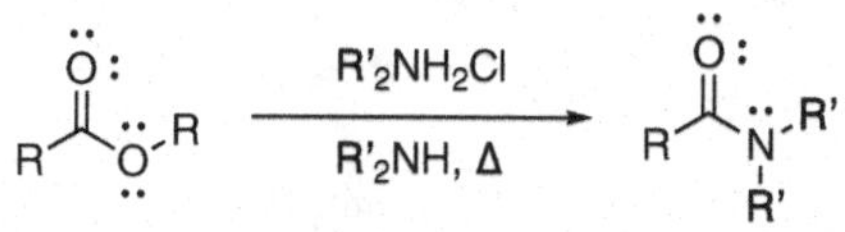

Figure 21.18   Ester–amide exchange reaction.

Problem 21.12 Draw the appropriate arrows to complete the base-mediated mechanism.

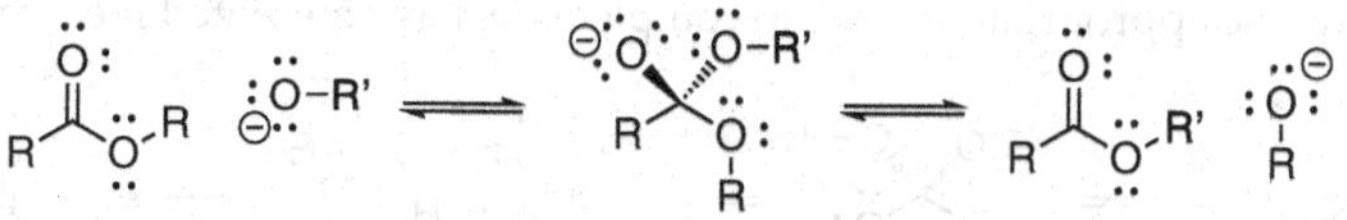

Esters can also be converted into amides through an ester–amide exchange (Figure 21.18).[9] In an ester–amide exchange, an ester is reacted with an amine (which also plays the role of solvent), a small amount of ammonium salt (HCl + the amine), and heat. The ammonium salt helps to speed up the reaction, but it should be noted that it is not a strong enough acid to protonate the carbonyl oxygen. The application of heat (up to 100°C) is what provides the energy to overcome the activation barrier for this reaction.

Problem 21.13. Draw the appropriate arrows to complete the mechanism.

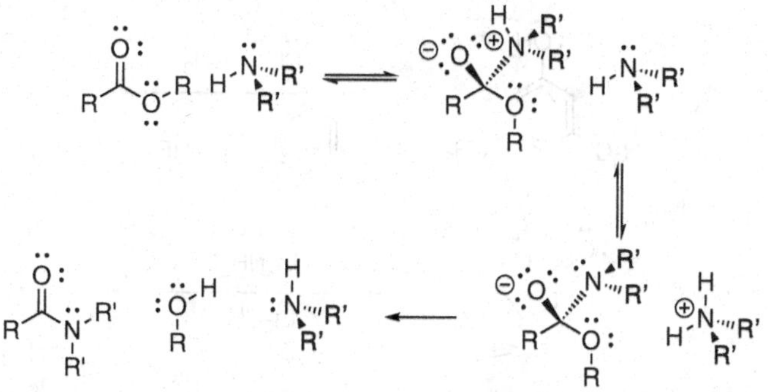

The final reaction to consider for esters is that of hydrolysis (Figure 21.19). As with transesterification, with which it shares similar mechanisms, hydrolysis of an ester requires added acid or base to proceed.[10] Under acidic conditions, this process is the microscopic reverse (every step and intermediate is the same) as Fischer esterification. Under basic conditions, this reaction is called a saponification (soap-making) reaction. The basic hydrolysis of fats produces sodium salts of long-chain, aliphatic carboxylate compounds, which are surfactants, and this is how soap is traditionally made.

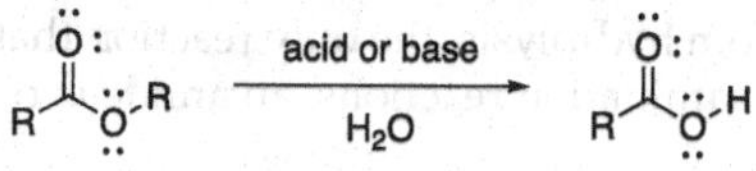

Figure 21.19   Acid- or base-mediated ester hydrolysis.

Problem 21.14 Draw the appropriate arrows to complete the acid-mediated mechanism.

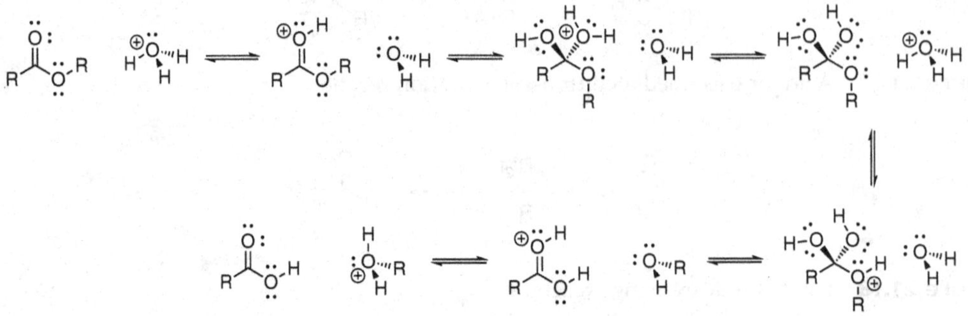

Problem 21.15 Draw the appropriate arrows to complete the base-mediated mechanism.

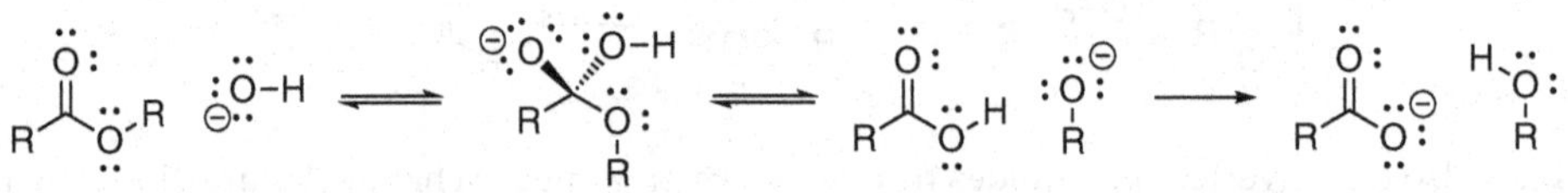

Problem 21.16 Predict the major product(s) for the following reactions.

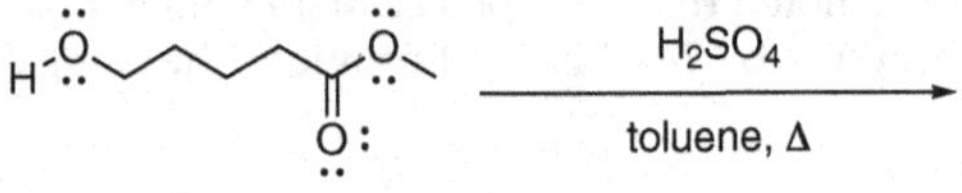

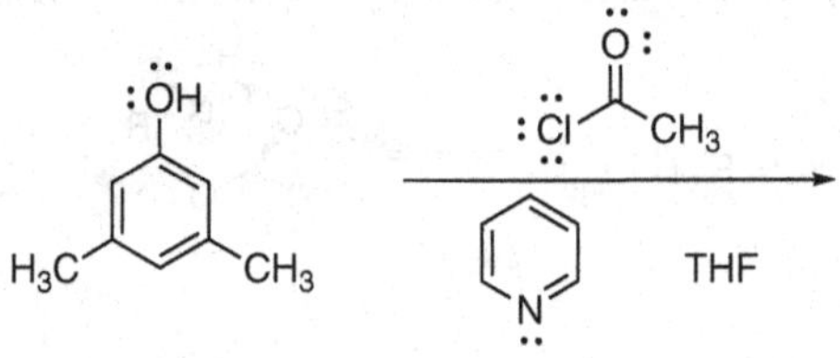

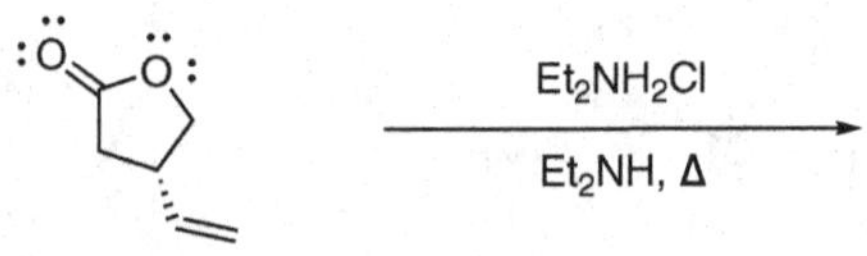

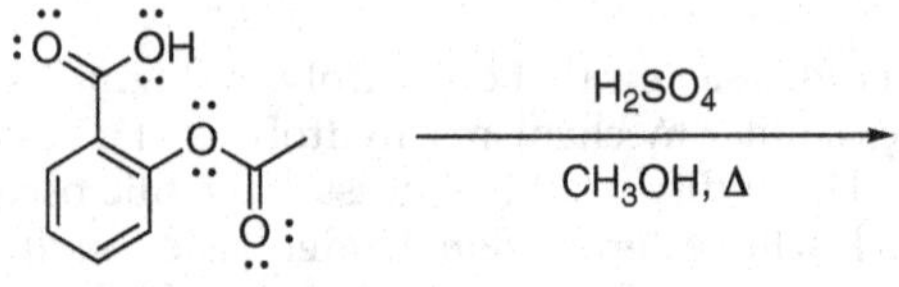

H_2SO_4

toluene, Δ

H_2SO_4

toluene, Δ

Et_2NH_2Cl

Et_2NH, Δ

H_2SO_4

CH_3OH, Δ

AMIDE SUBSTITUTION

As the least reactive carboxylic acid derivatives, amides are the least reactive in terms of nucleophilic acyl substitution. Aside from hydrolysis, the only reaction that we will consider is transamidation (Figure 21.20). In transamidation reactions, an amide is reacted with an amine (which

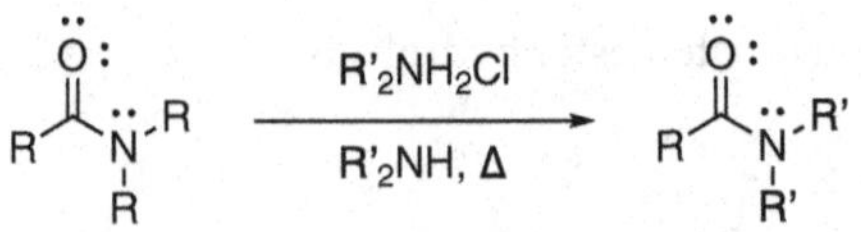

R'_2NH_2Cl

R'_2NH, Δ

Figure 21.20 Transamidation reaction.

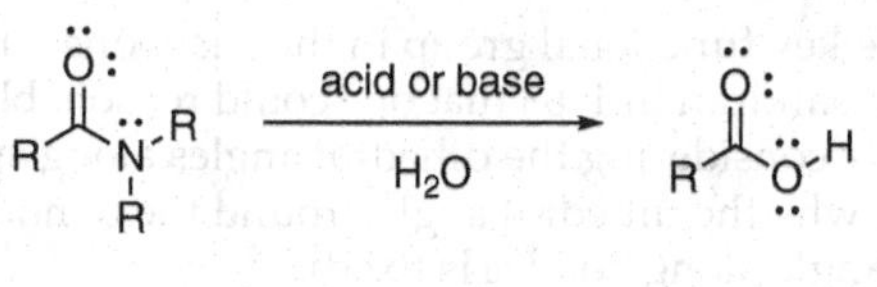

Figure 21.21 Acid- or base-mediated amide hydrolysis.

also plays the role of solvent), a small amount of ammonium salt (HCl + the amine), and heat.[11] The ammonium salt helps to speed up the reaction, but it should be noted that it is not a strong enough acid to protonate the carbonyl oxygen. The application of heat (up to 100 °C) is what provides the energy to overcome the activation barrier for this reaction.

Problem 21.17 Draw the appropriate arrows to complete the mechanism.

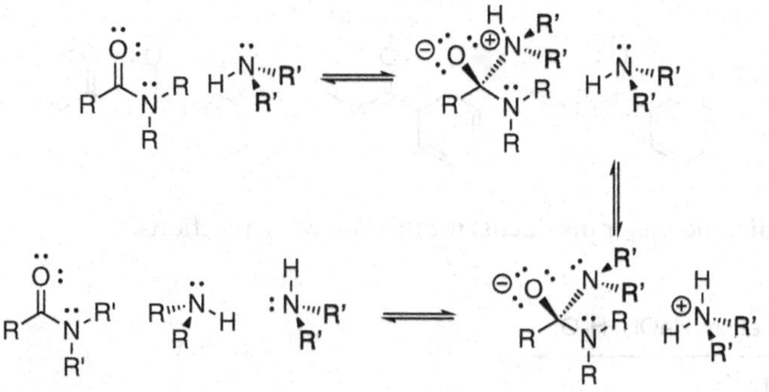

The second, and last, substitution reaction to consider for amides is that of hydrolysis (Figure 21.21). As with hydrolysis of an ester, hydrolysis of an amide requires added acid or base to proceed.[3]

Problem 21.18 Draw the appropriate arrows to complete the acid-mediated mechanism.

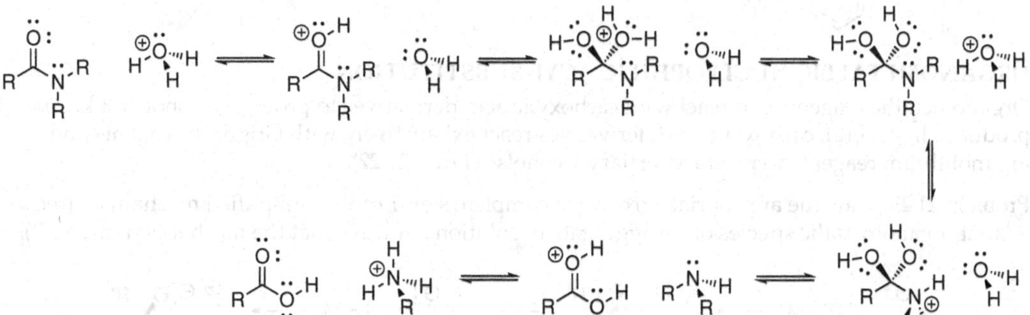

Problem 21.19 Draw the appropriate arrows to complete the base-mediated mechanism.

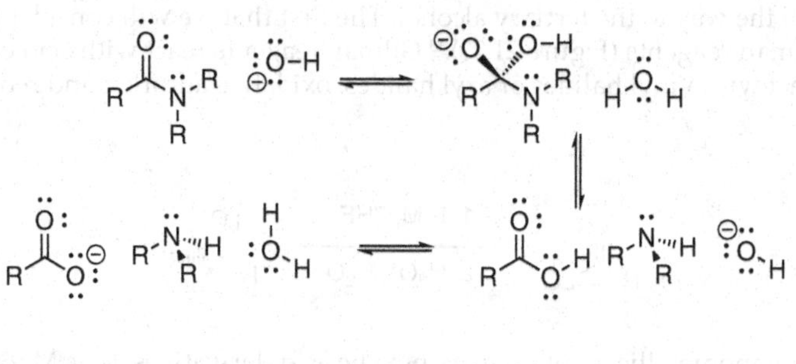

Problem 21.20 Amides are the key functional group in the backbone structure of a protein. It was discovered by G.N. Ramachandran in 1963 that one could reasonably predict likely conformations of a protein simply by considering the dihedral angles along specific bonds, namely Φ and ψ. Provide an explanation why the dihedral angle around the bond labelled ω is not considered (that is why the dihedral angle along bond ω is fixed).

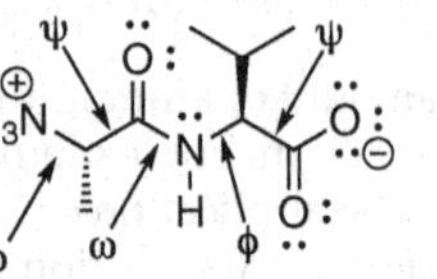

Problem 21.21 Identify the carboxylic acid derivative that is most reactive towards hydrolysis with water.

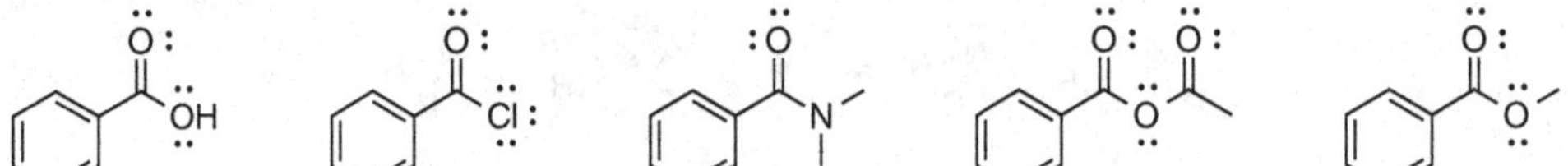

Problem 21.22 Predict the major product(s) for the following reactions.

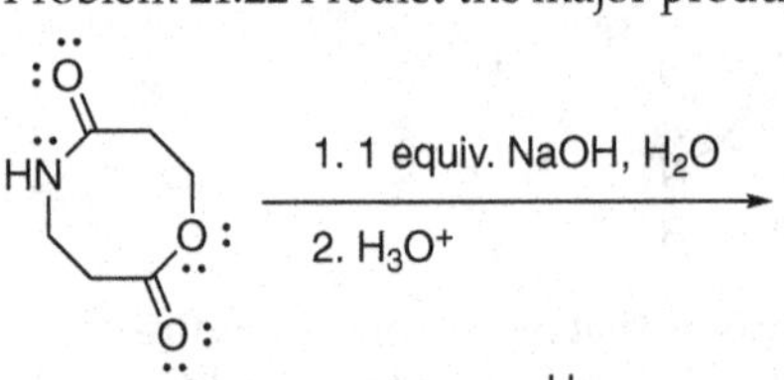

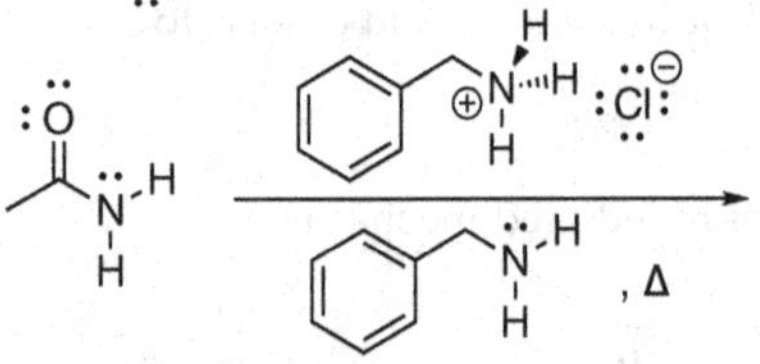

1. 1 equiv. NaOH, H₂O

2. H₃O⁺

ORGANOMETALLIC NUCLEOPHILIC ACYL SUBSTITUTION

Organometallic reagents can react with carboxylic acid derivatives to produce alcohol and ketone products. In general, carboxylic acid derivatives react exhaustively with Grignard reagents and organolithium reagents to produce tertiary alcohols (Figure 21.22).[12]

Problem 21.23 Draw the appropriate arrows to complete step 1 of this simplified mechanism (recall that organometallic species often aggregate in solutions in ways that the mechanism can omit).

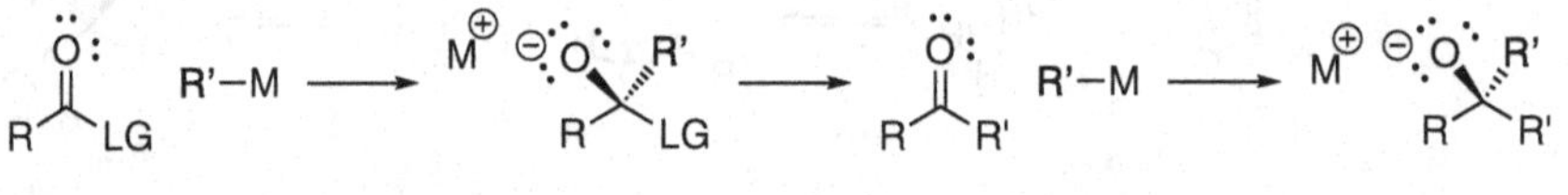

There are two possible methods for producing a ketone product from a carboxylic acid derivative instead of going all the way to the tertiary alcohol. The first that we will consider is using acid chlorides with Gilman reagents (Figure 21.23).[13] Gilman reagents react with acid chlorides the same way they react with vinyl halides or aryl halides: oxidative addition and reductive elimination (Figure 21.24).

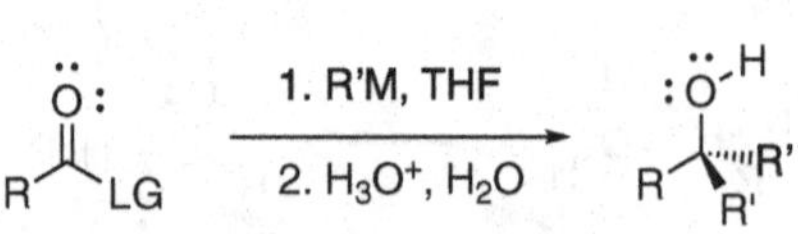

1. R'M, THF

2. H₃O⁺, H₂O

Figure 21.22 Organometallic addition to carboxylic acid derivatives, M is MgBr (Grignard) or Li (organolithium).

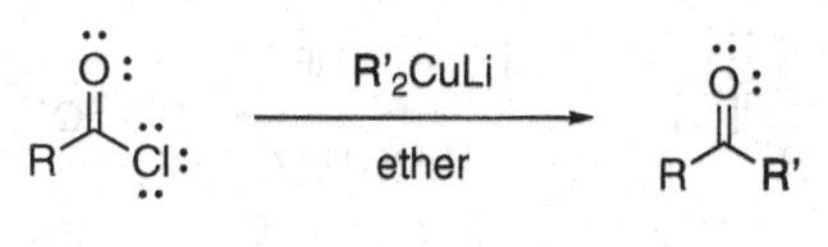

Figure 21.23 Conversion of an acid chloride to a ketone with a Gilman reagent.

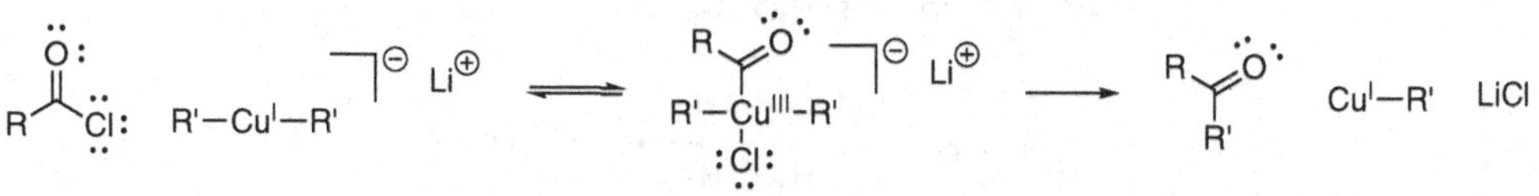

Figure 21.24 Organometallic reaction mechanism (oxidative addition and then reductive elimination) for Gilman reagent substitution reaction.

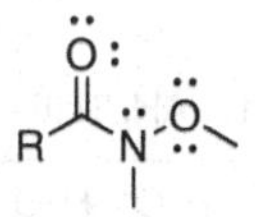

Figure 21.25 A Weinreb amide.

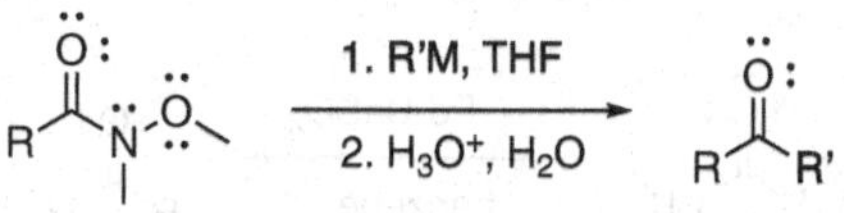

Figure 21.26 General Weinreb amide reaction, M is MgBr (Grignard) or Li (organolithium).

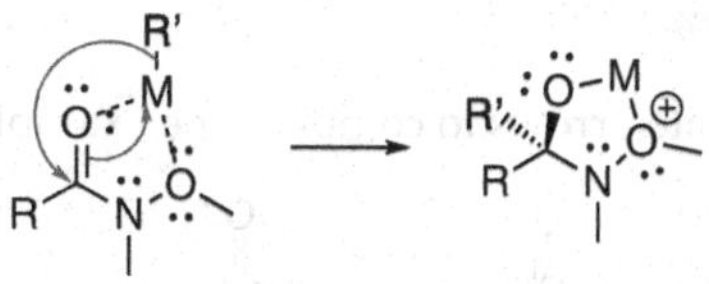

Figure 21.27 Key step of the Weinreb amide reaction showing the chelation of the metal ion.

The second methodology that we will consider producing ketones rather than tertiary alcohols is the Weinreb amide.[14] A Weinreb amide (Figure 21.25) – which this chapter has shown can be produced from acid chlorides, anhydrides, esters, or amides – reacts with Grignard reagents or organolithium reagents to produce, exclusively, ketone products (Figure 21.26).

The reason for the unique outcome for Weinreb amides can be seen in the mechanism (Figure 21.27). Due to the presence of the –OMe group on the nitrogen, once the tetrahedral intermediate forms a metal chelate is produced (a metal atom bound in a ring by two atoms). This chelate prevents the collapse of the tetrahedral intermediate (to give the ketone). As such, no further equivalents of organometallic nucleophiles can be added. Once water is added, the metal chelate is broken and the carbinolamine (hemiaminal) hydrolyses to give the ketone product.

CARBOXYLIC ACID DERIVATIVE REDUCTION

The last reaction type that we will consider is the reduction of carboxylic acid derivatives with lithium aluminium hydride (LAH). LAH will reduce acid chlorides, acid anhydrides, and esters to primary alcohols (Figure 21.28).[15] Note that unlike aldehydes and ketones, lithium aluminium hydride and sodium tetrahydridoborate cannot be used interchangeably. Sodium tetrahydridoborate will reduce acid chlorides to ketones,[16] but sodium tetrahydridoborate cannot reduce acid anhydrides or esters.

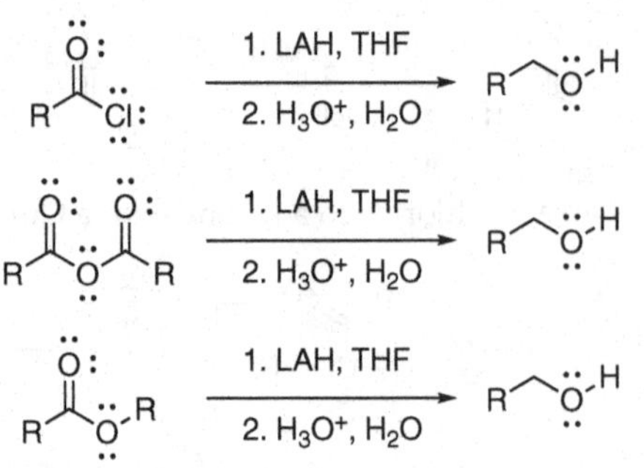

Figure 21.28 LAH reduction of an acid chloride (*top*), anhydride (*middle*), and ester (*bottom*) to produce primary alcohol products.

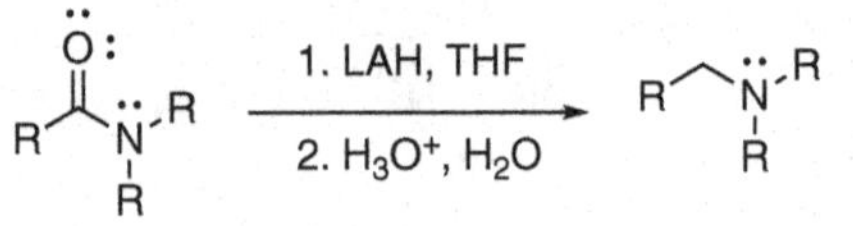

Figure 21.29 LAH reduction of an amide to give a primary amine product.

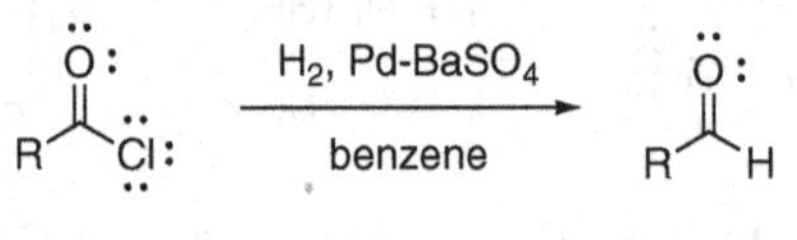

Figure 21.30 Rosenmund reduction of an acid chloride to an aldehyde.

Problem 21.24 Draw the appropriate arrows to complete the example mechanism.

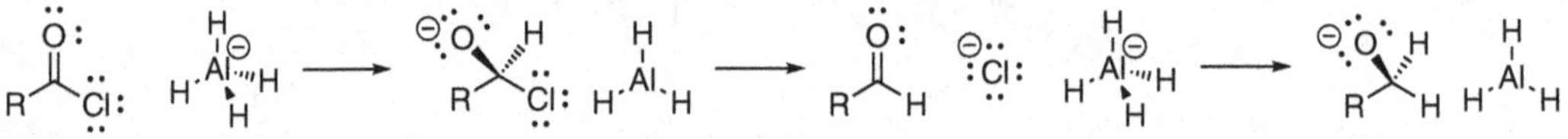

In contrast to the reduction reactions above, amides react with lithium aluminium hydride to produce amines (Figure 21.29).[17] The mechanism of amide reduction is beyond the scope of this chapter.

Like what we saw above with modified organometallic reactions to produce ketones (rather than tertiary alcohols), there are two methodologies to produce aldehydes (rather than primary alcohols) from carboxylic acid derivatives. The first is the Rosenmund reduction (Figure 21.30).[18] Similar to using Lindlar's catalyst to stop reduction of an alkyne at the *cis*-alkene, Pd-BaSO$_4$ is a weak catalyst that can reduce acid chlorides but cannot reduce aldehydes or ketones.

Lastly, we will consider the reagent diisobutylaluminium hydride (DIBALH, Figure 21.31).[19] DIBALH is used to convert esters to aldehydes (Figure 21.32). It is worth noting that reaction does work experimentally, but that it can be finicky and so it can often be simpler to fully reduce a carboxylic acid derivative with LAH and then re-oxidize it to the aldehyde.

Looking at the mechanism (Figure 21.33), we can get a clearer understanding of why the DIBALH reduction of esters produces aldehydes. DIBALH coordinates with the carbonyl oxygen

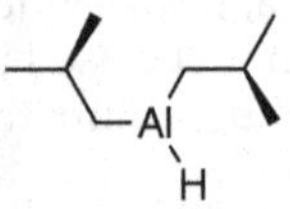

Figure 21.31 Diisobutylaluminium hydride (DIBALH).

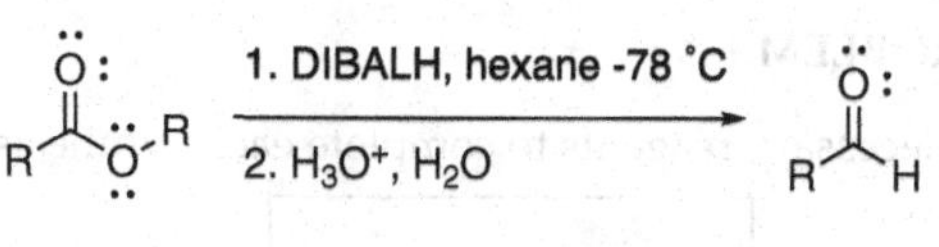
1. DIBALH, hexane -78 °C
2. H3O+, H2O

Figure 21.32 DIBALH reduction of an ester to an aldehyde.

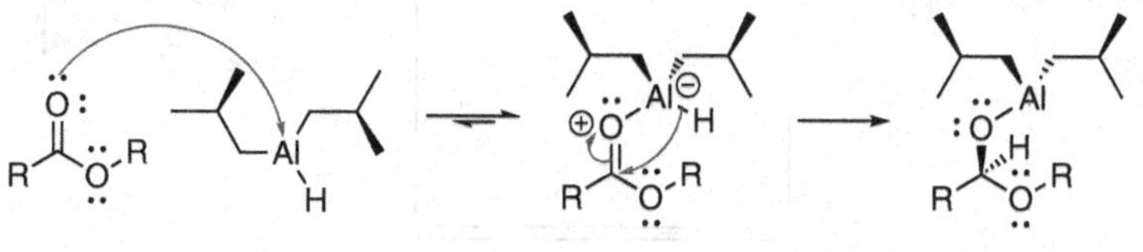

Figure 21.33 First part of the DIBALH reduction mechanism.

and then delivers the hydride to the carbon. At this stage, if the reaction is kept cold (–78°C), the O–Al bond will stay intact, and the tetrahedral intermediate will not collapse to the aldehyde. Only after aqueous, acidic (neutralizing any unreacted DIBALH) and warming up the mixture will the resulting hemiacetal hydrolyze to produce the aldehyde product.

Problem 21.25 Predict the major product(s) for the following reactions.

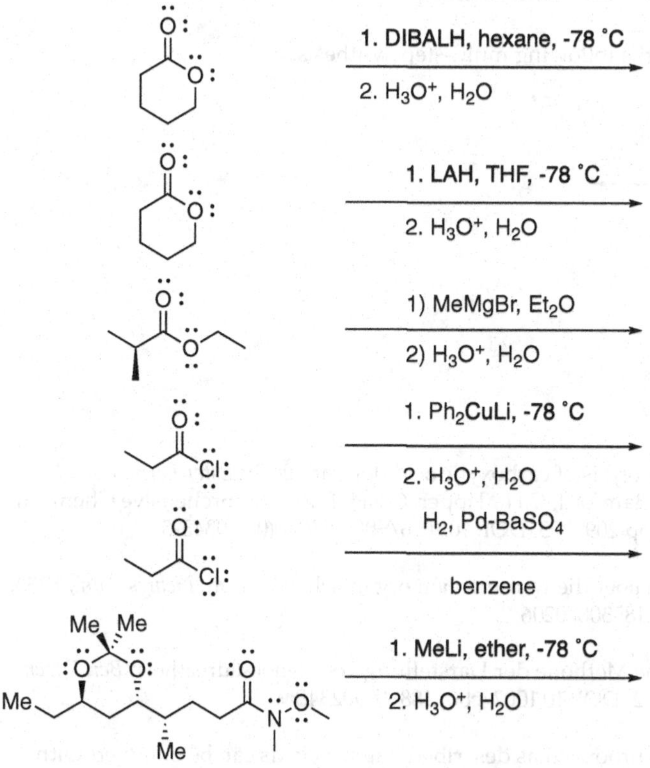
1. DIBALH, hexane, -78 °C
2. H3O+, H2O
1. LAH, THF, -78 °C
2. H3O+, H2O
1) MeMgBr, Et2O
2) H3O+, H2O
1. Ph2CuLi, -78 °C
2. H3O+, H2O
H2, Pd-BaSO4
benzene
1. MeLi, ether, -78 °C
2. H3O+, H2O

GENERAL PRACTICE PROBLEMS

Problem 21.26 Provide the necessary reagents to complete each synthesis.

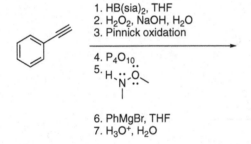

Problem 21.27 Predict the product of the following multi-step synthesis.

1. HB(sia)$_2$, THF
2. H$_2$O$_2$, NaOH, H$_2$O
3. Pinnick oxidation

4. P$_4$O$_{10}$
5. H-N-O

6. PhMgBr, THF
7. H$_3$O$^+$, H$_2$O

NOTES

1. Talbot, R.J.E. Chapter 3 The hydrolysis of carboxylic acid derivatives. In *Ester Formation and Hydrolysis and Related Reactions*; Bamford, C.H., Tipper, C.F.H. Eds.; Comprehensive Chemical Kinetics; Elsevier, 1972; Vol. 10, pp 209–293. DOI: 10.1016/S0069-8040(08)70345-5

2. Gerhardt, C.H. Untersuchungen über die wasserfreien organischen Säuren. *Liebigs Ann.*, **1853**, *87* (2), 149–179. DOI: 10.1002/jlac.18530870206

3. Baumann, E. Ueber eine einfache Methode der Darstellung von Benoësäureäthern *Ber. Dtsch. Chem. Ges.*, **1886**, *19* (2), 3218–3222. DOI: 10.1002/cber.188601902348

4. While esters and amines can be produced as described, higher yields can be obtained with modified conditions. Since these modified conditions come from Schotten's and Baumann's work, these are called Schotten-Baumann reactions or Schotten-Bauman conditions:

   i) Schotten, C. Ueber die Oxydation des Piperidins. *Ber. Dtsch. Chem. Ges.*, **1884**, *17* (2), 2544–2547. DOI: 10.1002/cber.188401702178
   ii) Baumann, E. Ueber eine einfache Methode der Darstellung von Benoësäureäthern *Ber. Dtsch. Chem. Ges.*, **1886**, *19* (2), 3218–3222. DOI: 10.1002/cber.188601902348

5. Moelwyn-Hughes, E.A.; Rolfe, A.C. 32. The kinetics of the esterification of acetic anhydride in ethyl-alcohol solution. *J. Chem. Soc.*, **1932**, 241–246. DOI: 10.1039/JR9320000241

6. Chiozza, L. Vermischte chemischen Mitteilungen. *Liebigs Ann.*, **1854**, *91* (1), 102–113. DOI: 10.1002/jlac.18540910109

7. Orton, K.J.P; Jones, M. CLXXX. – Hydrolysis of acetic anhydride. *J. Chem. Soc., Trans.*, **1912**, *101*, 1708–1720. DOI: 10.1039/CT9120101708

8. i) Fischer, E. Über die Wechselwirkung zwischen Ester- und Alkoholgruppen bei Gegenwart von Katalysatoren. *Ber. Dtsch. Chem. Ges.*, **1920**, *53* (9), 1634–1644. DOI: 10.1002/cber.19200530905
   ii) Reimer, M.; Downes, H.R. The preparation of esters by direct replacement of alkoxyl groups. *J. Am. Chem. Soc.*, **1921**, *43* (4), 945–951. DOI: 10.1021/ja01437a032

9. Glasoe, P.K.; Audrieth, L.F. Acid catalysis in amines. I. The catalytic effect of cyclohexylammonium salts on the reactions between cyclohexylamine and esters. *J. Org. Chem.*, **1939**, *04* (1), 54–59. DOI: 10.1021/jo01213a007

10. Holmberg, B. Verseifung von *l*-Acetyl-äfpelsäure. *Ber. Dtsch. Chem. Ges.*, **1912**, *45* (3), 2997–3008. Polanyi, M.; Szabo, A.L. On the mechanism of hydrolysis. The alkaline saponifications of amyl acetate. *Trans. Faraday Soc.*, **1934**, *30*, 508–512

11. Kelbe, W. Ueber die Einwirkung der Säureamide auf die aromatischen Aminbasen. *Ber. Dtsch. Chem. Ges.*, **1883**, *16* (1), 1199–1201. DOI: 10.1002/cber.188301601268

12. First organometallic reaction (dialkylzinc) with a carboxylic derivative (acid chloride): Freund, A. 1. Ueber die Natur der Ketone. *Liebigs Ann.*, **1861**, *118* (1), 1–21. DOI: 10.1002/jlac.18611180102

13. MacPhee, J.A.; Boussu, M.; Dubois, J.-E. Grignard Reagent–Acid Chloride Condensation in the Presence of Copper(I) Chloride. A Study of Structural Effects by Direct and Competition Methods. *J. Chem. Soc., Perkin Trans. 2*, **1974**, 1525–1530. DOI: 10.1039/P29740001525

14. Nahm, S.; Weinreb, S.M. *N*-Methoxy-*N*-methylamides as effective acylating agents. *Tetrahedron Lett.*, **1981**, *22* (39), 3815–3818. DOI: 10.1016/S0040-4039(01)91316-4

15. Nystrom, R.F. and Brown, W.G. Reduction of organic compounds by lithium aluminium hydride. I. Aldehydes, ketones, esters, acid chlorides and acid anhydrides. *J. Am. Chem. Soc.*, **1947**, *69* (5), 1197–1199. DOI: 10.1021/ja01197a060

16. Chaikin, S.W.; Brown, W.G. Reduction of aldehydes, ketones and acid chlorides by sodium borohydride. *J. Am. Chem. Soc*, **1949**, *71* (1), 122–125. DOI: 10.1021/ja01169a033

17. Nystrom, R.F. and Brown, W.G. Reduction of organic compounds by lithium aluminium hydride. III. Halides, quinones, miscellaneous nitrogen compounds. *J. Am. Chem. Soc.*, **1948**, *70* (11), 3738–3740. DOI: 10.1021/ja01191a057

18. Rosenmund, K.W. Über eine neue Methode zur Darstellung von Aldehyden. 1. Mitteilung. *Ber. Dtsch. Chem. Ges.*, **1918**, *51* (1), 585–593. DOI: 10.1002/cber.19180510170

19. Zakharkin, L.I.; Khorlina, I.M. Reduction of esters of carboxylic acids into aldehydes with diisobutylaluminium hydride. *Tetrahedron Lett.*, **1962**, *3* (14), 619–620. DOI: 10.1016/S0040-4039(00)70918-X

22 Enols, Enolates, and Enamines

In this chapter we will not focus on new reaction types. Instead, we will consider reactions seen in previous chapters but with new nucleophiles: alk**ene** alcoh**ols** (enols), alk**ene** alcoh**olates**[1] (enolates), and alk**ene** **amines** (enamines). Enols (Figure 22.1) are produced from aldehydes and ketones under acidic conditions and are moderate nucleophiles that can be used in reactions with aldehydes and ketones.

Enolates are anionic, strong nucleophiles that can be generated by adding a base to aldehydes and ketones (Figure 22.2), esters (Figure 22.3), or amides (Figure 22.4). We will consider a new base that is commonly used to make enolates below; however, both alkoxide (RO⁻) and amide (R₂N⁻) bases are used to make enolates. Enolates can be employed in S_N2 reactions and in reactions with aldehydes, ketones, and carboxylic acid derivatives.

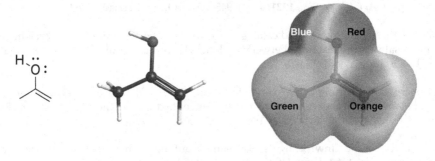

Figure 22.1 Structure (*left*), 3D model (*middle*), and electrostatic potential map (*right*) of an example enol: prop-1-en-2-ol (red corresponds to high electron density and blue corresponds to low electron density).

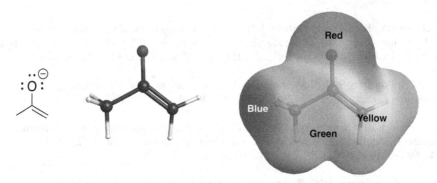

Figure 22.2 Structure (*left*), 3D model (*middle*), and electrostatic potential map (*right*) of an example ketone enolate: prop-1-**en**-2-**olate** (red corresponds to high electron density and blue corresponds to low electron density).

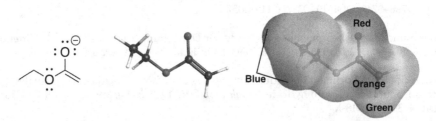

Figure 22.3 Structure (*left*), 3D model (*middle*), and electrostatic potential map (*right*) of an example ester enolate: 1-ethoxyeth**en**-1-**olate** (red corresponds to high electron density and blue corresponds to low electron density).

DOI: 10.1201/9781003479352-22

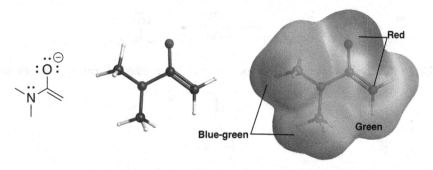

Figure 22.4 Structure (*left*), 3D model (*middle*), and electrostatic potential map (*right*) of an example amide enolate: 1-(dimethylamino)eth**en-1-olate** (red corresponds high electron density and blue corresponds to low electron density).

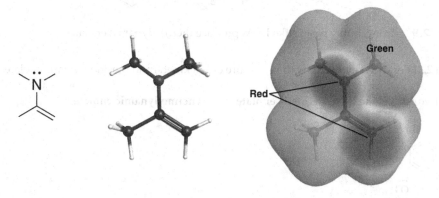

Figure 22.5 Structure (*left*), 3D model (*middle*), and electrostatic potential map (*right*) of an example enamine: *N,N*-dimethylprop-1-**en-2-amine** (red corresponds to high electron density and blue corresponds to low electron density).

Enamines (Figure 22.5) are good, neutral nucleophiles that are produced from secondary amines condensing with aldehydes or ketones (see Chapter 19). Enamines can be used in S_N2 reactions and in reactions with aldehydes, ketones, and carboxylic acid derivatives.

KINETIC AND THERMODYNAMIC ENOLATES

For asymmetric ketones we have two possible enolate isomers that could form (Figure 22.6). In the example shown in Figure 22.6, the key difference between these two isomers is the degree of substitution of the alkene. We term but-1-en-2-olate the kinetic enolate because it results from deprotonation of the more accessible hydrogen atoms (faster to deprotonate). We term *trans*-but-2-en-2-olate the thermodynamic enolate because it is more stable with a more substituted alkene.

In practice, the use of a weaker, hydroxide or alkoxide base will afford the thermodynamic enolate (Figure 22.7) because we have an equilibrium, which favours the more stable enolate.

To create the kinetic enolate, we need a very strong base (to prevent an equilibrium), and it needs to be bulky (steric repulsion favours the least hindered side). Lithium diisopropylamide (LDA) is a very strong, bulk base (Figure 22.8).[2] Using LDA, the kinetic enolate is the dominant species that forms (Figure 22.9).

Figure 22.6 Possible enolate isomers that could form from deprotonation of the asymmetric ketone butan-2-one to give but-1-en-2-olate (*left*) or *trans*-but-2-en-2-olate (*right*).

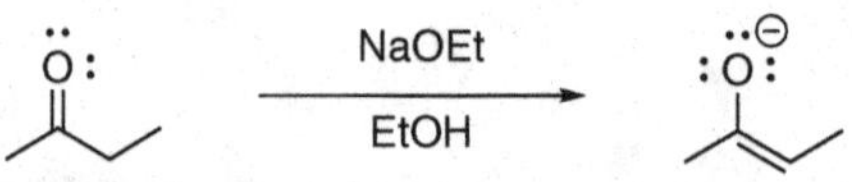

Figure 22.7 Relatively weak bases (like sodium ethoxide) produce thermodynamic enolates.

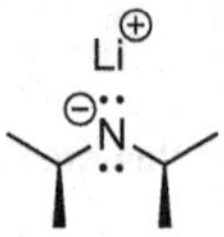

Figure 22.8 Lithium diisopropyl amide (LDA) a strong, bulky base.

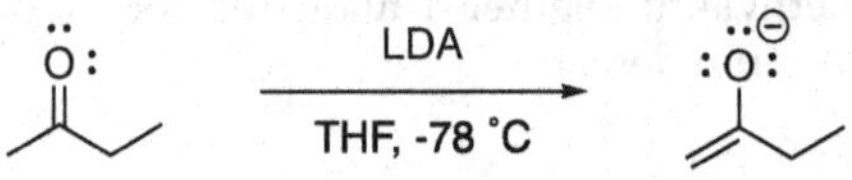

Figure 22.9 Strong, bulky bases (like LDA) produce thermodynamic enolates.

Problem 22.1 For each ketone, draw the structure of the kinetic and thermodynamic enolate.

Aldehyde or ketone	Kinetic enolate	Thermodynamic enolate

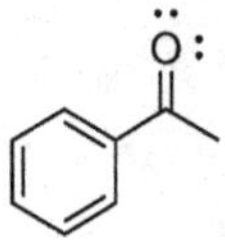

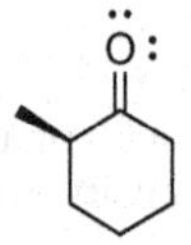

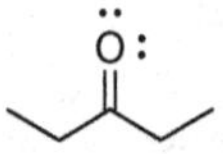

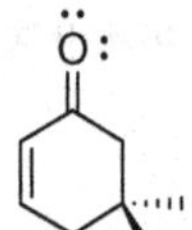

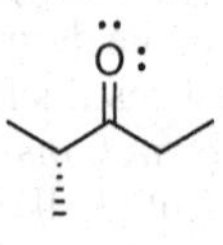

Problem 22.2 For each reaction, draw the correct enolate that will form.

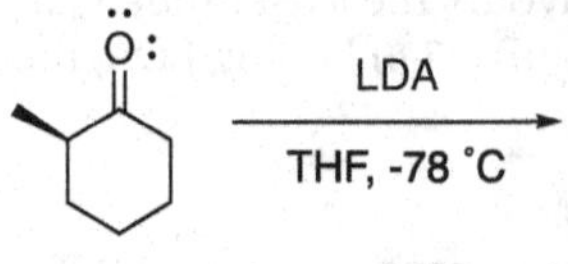

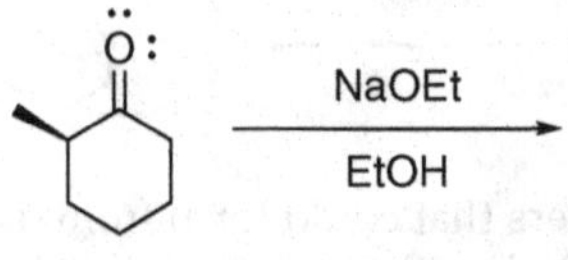

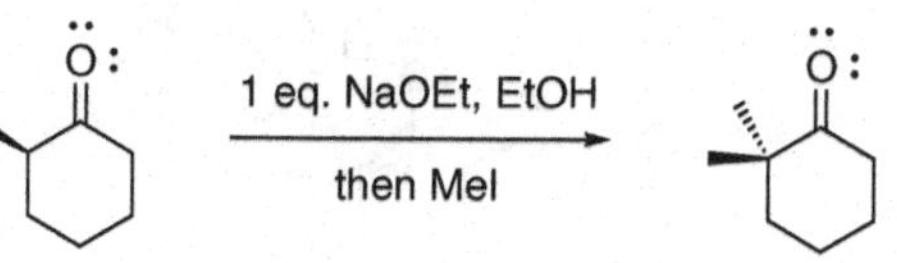

Figure 22.10 Enolate α-alkylation from the thermodynamic enolate.

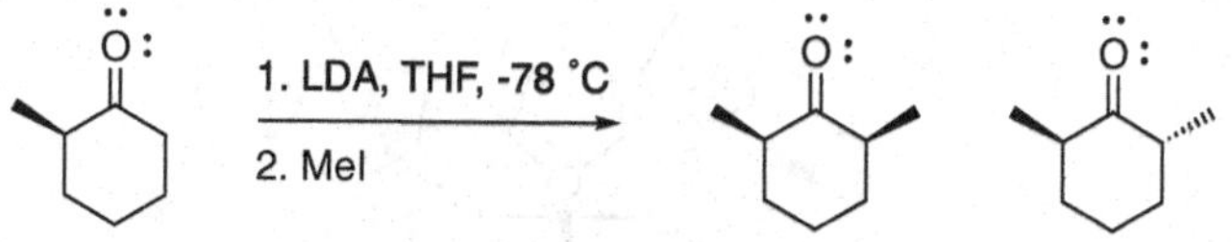

Figure 22.11 Enolate α-alkylation from the kinetic enolate.

ALKYLATION REACTIONS

Now that we have looked at forming enolates in more detail, we will look at some reactions of enols, enolates, and enamines. The first reaction that we will consider is the reaction of enolates as nucleophiles in S_N2 reactions. With asymmetric ketones, we can control the regiochemical outcome to produce substitution on the more substituted side (Figure 22.10).

Problem 22.3 Draw the appropriate arrows to complete the mechanism.

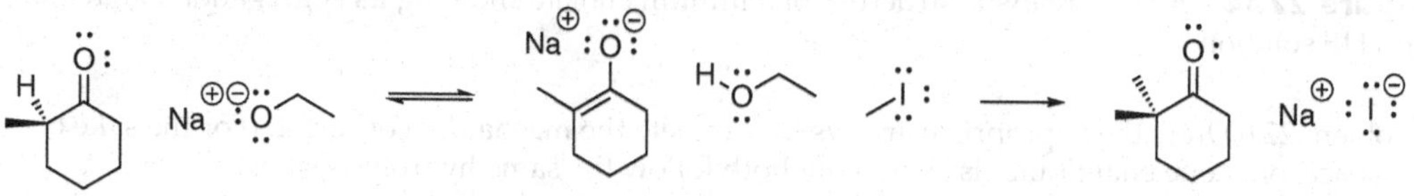

And we can produce substitution on the less substituted side by using our kinetic enolate conditions (Figure 22.11).

Problem 22.4 Draw the appropriate arrows to complete the mechanism.

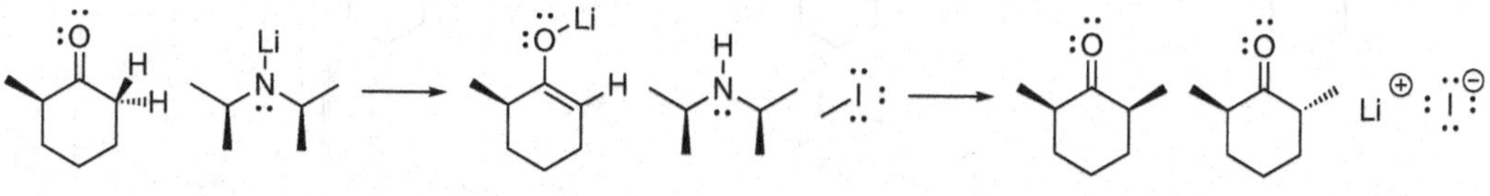

Enamines can also be used as nucleophiles in S_N2 reactions (Figure 22.12).[3] Note that at the end of the reaction, we produce a substituted ketone, not an enamine. Considering the mechanism (Problem 22.5), you can see that the enamine reacts with the electrophile and produces an iminium ion. The iminium ion is then hydrolyzed in the acidic workup to produce the ketone product.

Problem 22.5 Draw the appropriate arrows to complete the mechanism of step 1.

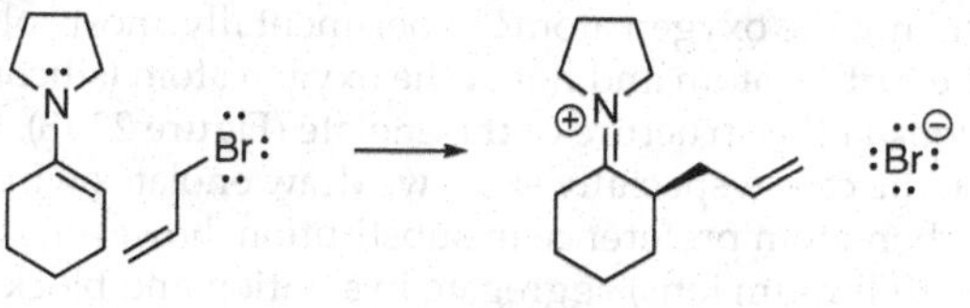

+enantiomer

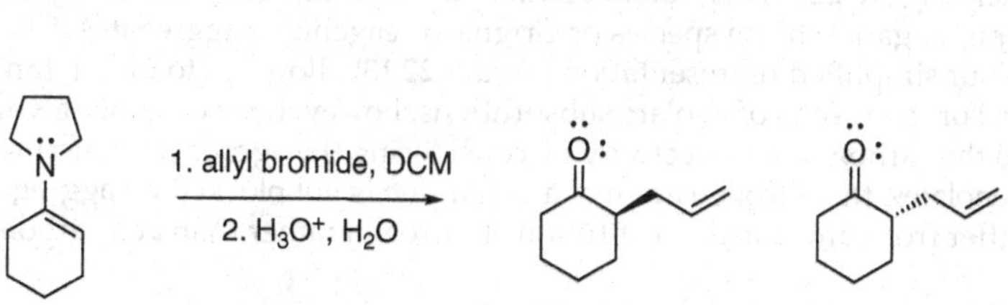

Figure 22.12 Enamine α-alkylation.

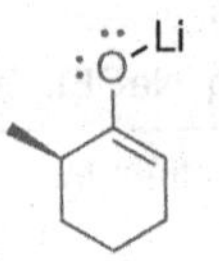

Figure 22.13 Structure of a lithium enolate as typically shown in reaction mechanisms.

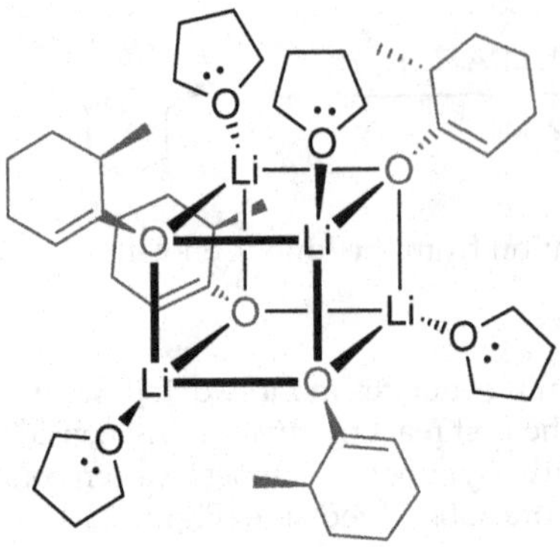

Figure 22.14 A more realistic structure of a lithium enolate showing its aggregation behaviour in THF solution.

Problem 22.6 Draw the appropriate arrows to complete the mechanism of step 2. (For the sake of space, only one enantiomer is shown but both follow the same hydrolysis steps).

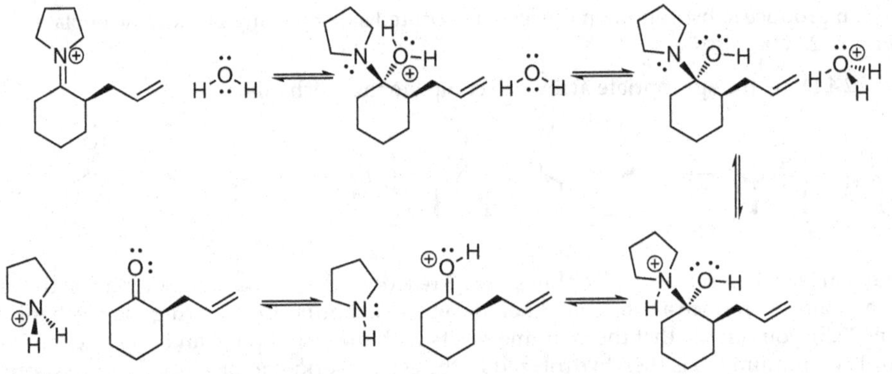

Considering the example reactions above, a question might arise: why does the carbon atom act as the nucleophilic site and not the oxygen atom? Experimentally, most solution phase reactions produce substitution at the carbon atom and not at the oxygen atom (silicon electrophiles are a notable exception). If we look at the structure of the enolate (Figure 22.13), there is not an obvious reason why this should be the case (especially since we draw enolates with the negative charge on the oxygen atom!). The carbon-atom preference in substitution, however, is due to the fact that enolates (particularly those with lithium ions) aggregate in solution and block the oxygen atom from nucleophilic attack (Figure 22.14).[4] We do not tend to draw lithium enolates as aggregates (just as we tend not to draw organolithium species or Grignard reagents as aggregates (Chapter 17)) and most of the time our simplified representation (Figure 22.13) allows us to understand the chemistry we observe. For the results of enolate substitutions, however, we can only give a satisfactory explanation as to the carbon-atom selectivity by considering the aggregate structure (Figure 22.14).

In contrast to enolates, the nitrogen atom of an enamine is not blocked by aggregation. And so enamines can suffer from unwanted substitution at nitrogen rather than at a carbon atom.

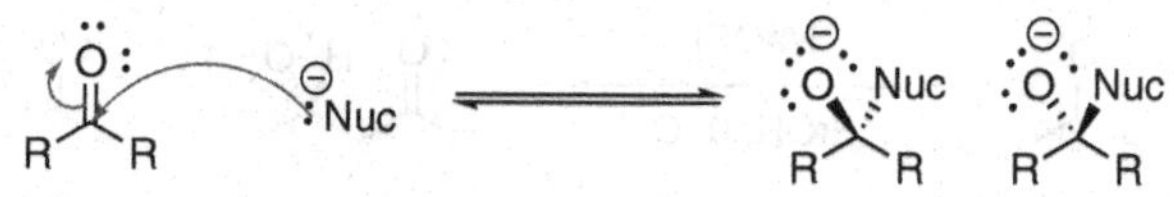
Nuc
R
R
Nuc
R
R
Nuc
R
R

Figure 22.15 Typical aldehyde and ketone reactivity pattern: addition of the nucleophile to the carbonyl carbon atom.

Problem 22.7 Draw the major product(s) for each reaction.

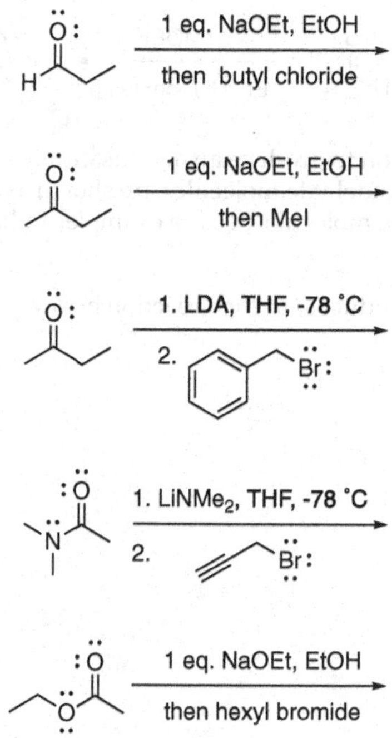
1 eq. NaOEt, EtOH
then butyl chloride
1 eq. NaOEt, EtOH
then MeI
1. LDA, THF, -78 °C
2.
Br
1. LiNMe2, THF, -78 °C
2.
Br
1 eq. NaOEt, EtOH
then hexyl bromide

ALDOL REACTION

The next reaction to consider is that of enols and enamines reacting with aldehyde and ketone electrophiles. Before we dive into this topic, recall that for aldehydes and ketones, a general reactivity pattern is the addition of a nucleophile to the carbonyl carbon (Figure 22.15).

Now if we look at the reaction of an enolate with an aldehyde (Figure 22.16), a ketone will also work here, we see the creation of a new type of product: an **ald**ehyde alcoh**ol** (an **aldol**, even if a ketone is used it is called an aldol reaction and aldol product). The aldol reaction[5,6,7] is an intensely useful reaction because it both makes a new C–C bond and creates complex structures from relatively simple starting materials.

Problem 22.8 Draw the appropriate arrows to complete the mechanism.

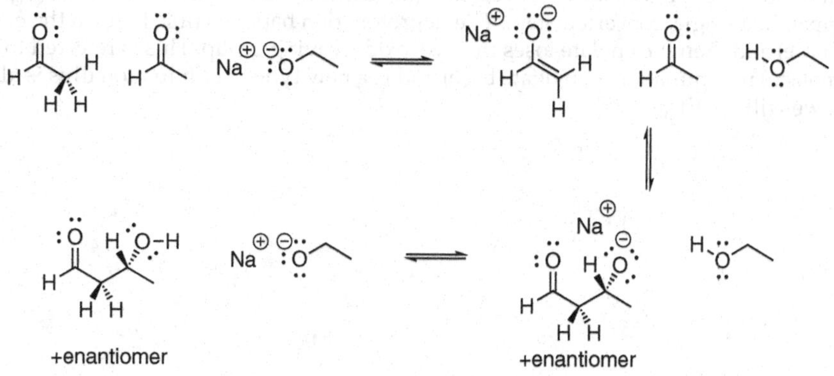
Na
Na
Na
Na
+enantiomer
+enantiomer

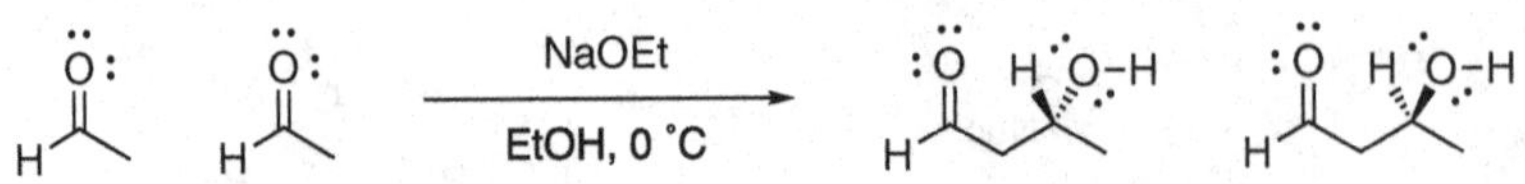

Figure 22.16 Basic addition of an enolate to a carbonyl carbon atom (aldol reaction). Note, for illustrative purposes two acetaldehyde molecules are shown as reactants. Typically, the reactions will only explicitly show one molecule. See the examples in Problem 22.9.

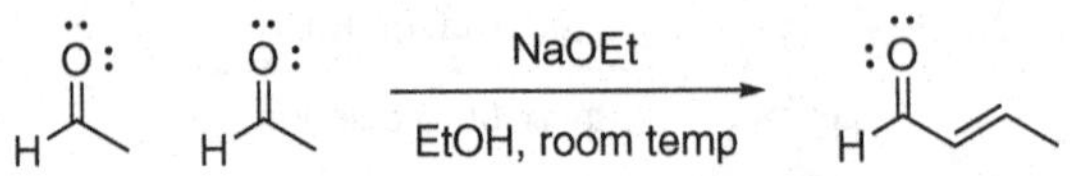

Figure 22.17 Aldol condensation to produce an α,β-unsaturated carbonyl product. Note, again, for illustrative purposes two acetaldehyde molecules are shown as reactants. Typically, the reactions will only explicitly show one molecule. See the examples in Problems 22.10.

Problem 22.9 Predict the major product(s) of each reaction below.

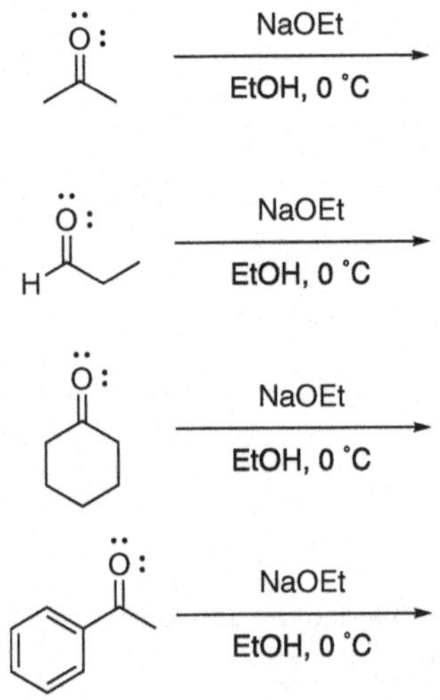

Notice that each of the reactions in Problem 22.9 explicitly lists a cold temperature (0 °C). Under cold, basic conditions like those in Problem 22.9, the aldol reaction will stop at the aldol product. When heated up (room temperature may be enough) the reaction does not stop at the addition reaction but continues to produce the aldol condensation product (an α,β-unsaturated aldehyde or ketone, Figure 22.17).

As you consider the mechanism for this condensation in Problem 22.10, pay particular attention to the elimination step. This is not an E2 reaction (the deprotonation and loss of leaving group do not happen in a single, concerted step). The deprotonation happens first, to form the enolate (a conjugate base) and then the enolate loses the hydroxide leaving group. This is referred to as an $E1_{CB}$ (unimolecular elimination, conjugate base) and is a new type of elimination (this is also the only place we will see $E1_{CB}$).

Problem 22.10 Draw the appropriate arrows to complete the mechanism (notice the first three steps are the same as we saw in Problem 22.8).

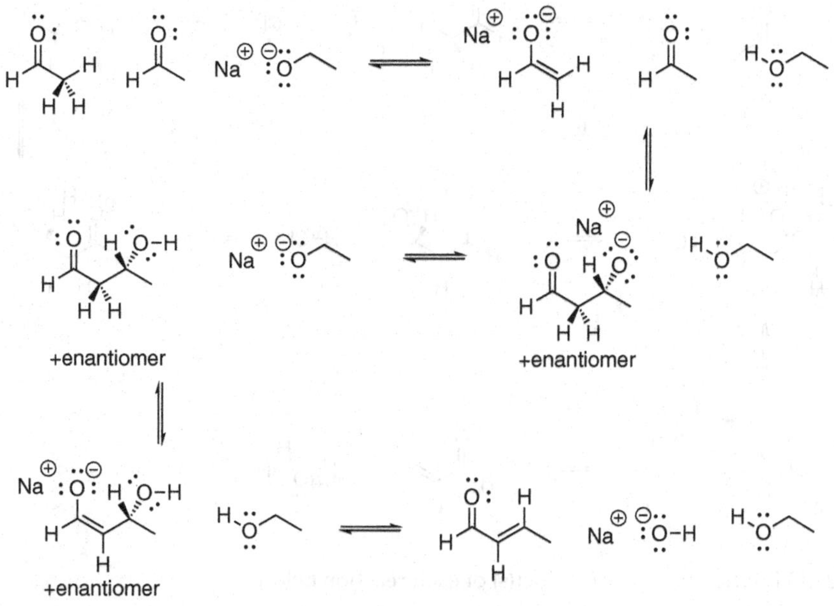

+enantiomer   +enantiomer

+enantiomer

Problem 22.11 Predict the major product(s) of each reaction below.

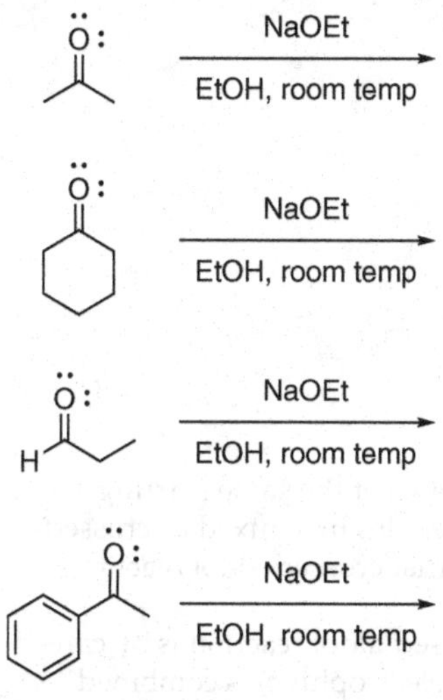

The aldol reaction can also be accomplished under acidic conditions. Under acidic conditions the reaction does not stop at the addition but goes straight through to the condensation product (Figure 22.18).

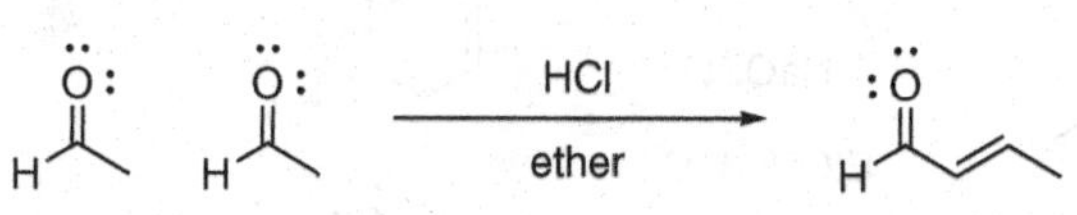

Figure 22.18 Acidic addition of an enol to a carbonyl carbon atom to produce an α,β-unsaturated carbonyl product (aldol condensation). Note, again, for illustrative purposes two acetaldehyde molecules are shown as reactants. Typically, the reactions will only explicitly show one molecule. See the examples in Problems 22.13.

Problem 22.12 Draw the appropriate arrows to complete the mechanism.[8]

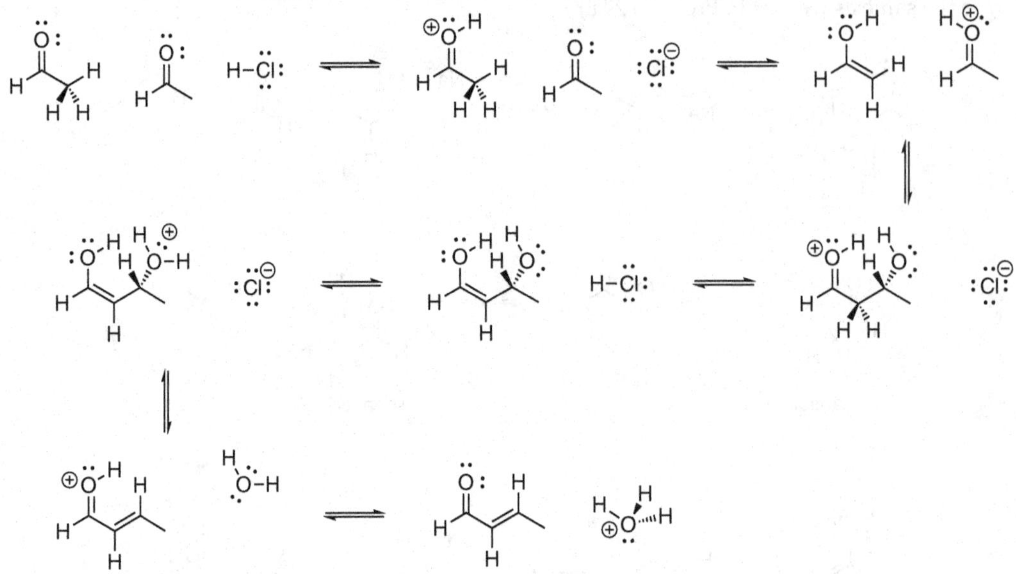

Problem 22.13 Predict the major product(s) of each reaction below.

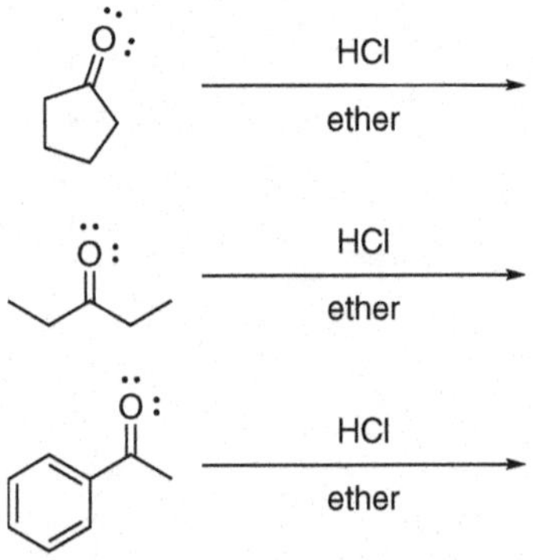

MIXED ALDOL REACTIONS

So far, we have only considered aldol reactions between two equivalents of the same starting material. What if we wanted, however, to combine two different molecules in a mixed or crossed-aldol reaction? The challenge, without careful control and design, is that crossed-aldol reactions can produce a mess of products (Figure 22.19).

The first method of controlling what product is produced in a crossed-aldol reaction is by carefully choosing what reagents are combined. Typically, a ketone (less electrophilic) is combined

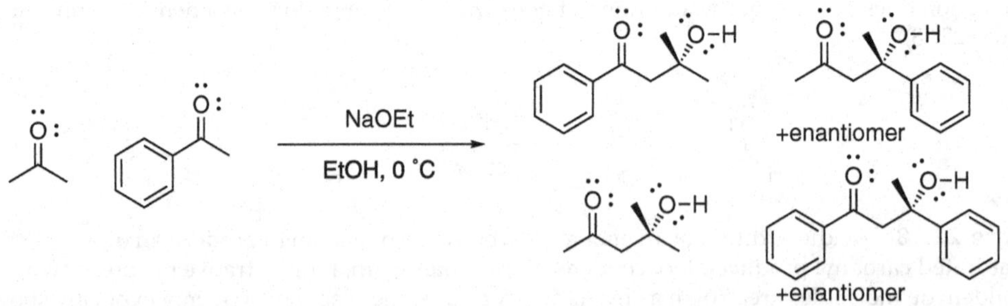

Figure 22.19 Statistical mess of products produced in an uncontrolled crossed-aldol reaction.

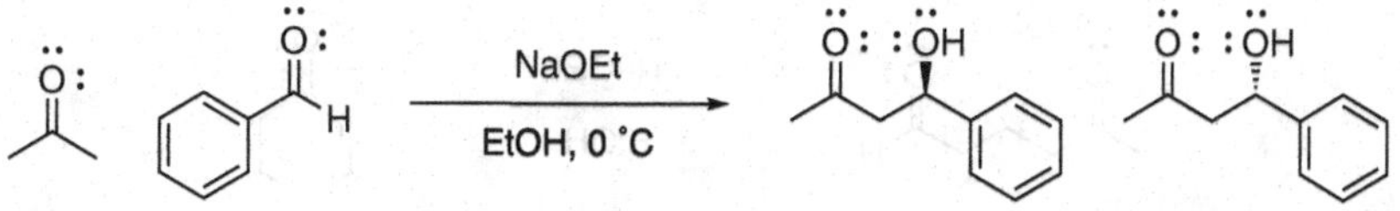

Figure 22.20 A crossed-aldol of a non-enolizable aldehyde (benzaldehyde) and a ketone (acetone).

with a non-enolizable (no hydrogen atoms on the α carbon atom) aldehyde (more electrophilic). The ketone will be converted into an enolate and preferentially attack the aldehyde. Typical aldehydes that are used are benzaldehyde and formaldehyde.

Problem 22.14 Predict the major product(s) of each reaction below.

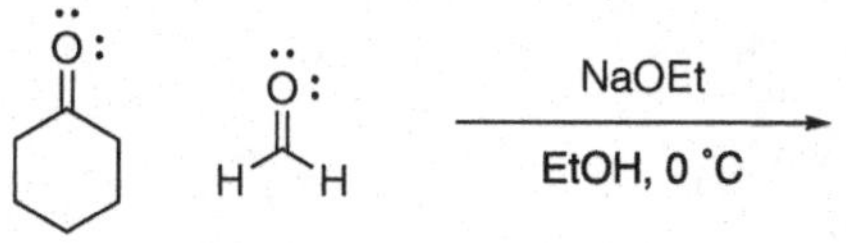

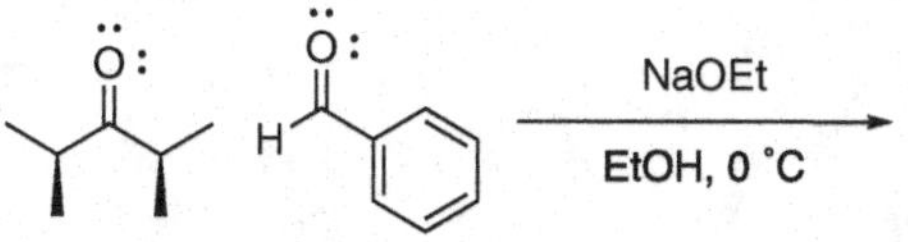

While careful control of the reacting molecules is possible, it is very limiting. The other, general, method of conducting crossed-aldol reactions is to use LDA to form the enolate and then add the aldehyde or ketone electrophile (Figure 22.21).

Problem 22.15 Predict the major product(s) of each reaction below.

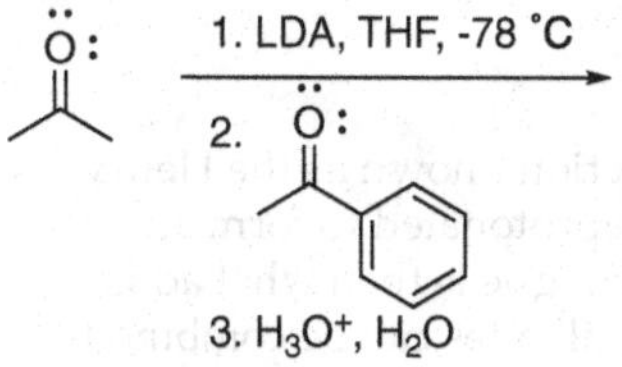

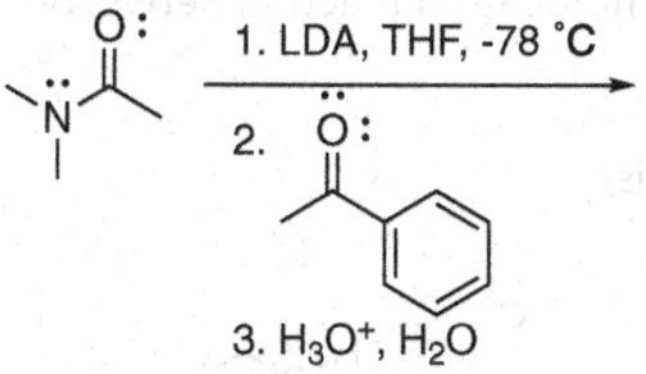

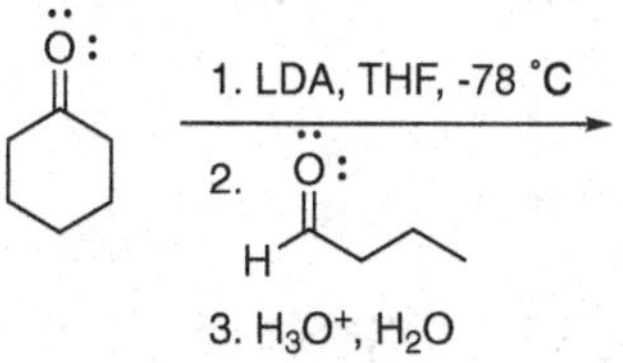

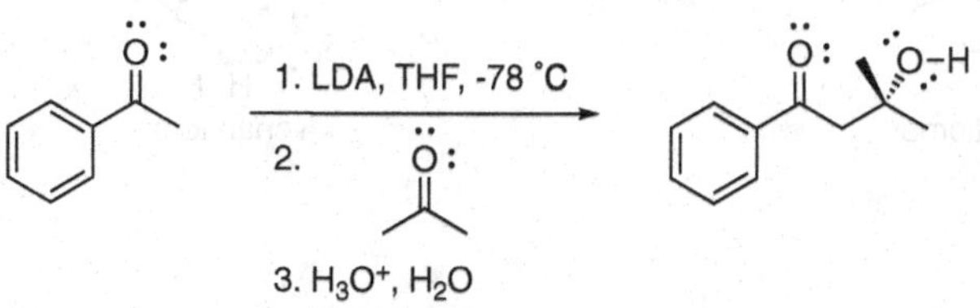

Figure 22.21 A crossed-aldol using LDA to preform the enolate.

263

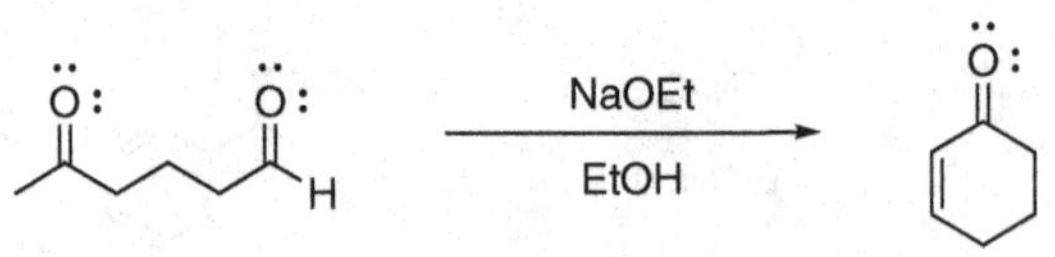

Figure 22.22 An intramolecular aldol reaction.

If a five- or six-membered ring can be produced, an intramolecular aldol takes place (Figure 22.22). The intramolecular aldol reaction almost always goes on to produce the condensation product.

Problem 22.16 Predict the major product(s) of each reaction below.

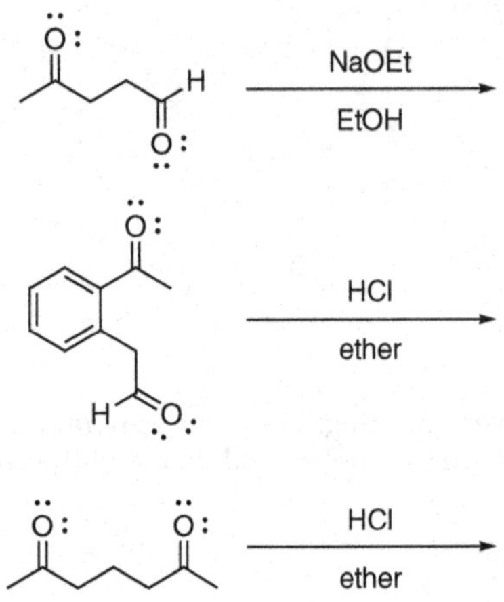

HENRY REACTION

The final variation of the aldol reaction that we will consider is a reaction known as the Henry reaction.[9] In a Henry reaction, a nitroalkane (pK_a values near 10) is deprotonated to form an enolate analogue of the nitroalkane carbanion. This nitro-enolate analogue is then what adds to the aldehyde or ketone. Given their significantly greater acidity, nitroalkanes can be combined in a single reaction vessel with the reacting aldehyde or ketone and the base (Figure 22.23). The nitroalkane will be preferentially deprotonated and there will be no homoaldol reaction between two acetaldehyde molecules.

Problem 22.17 Draw the appropriate arrows to complete the mechanism.

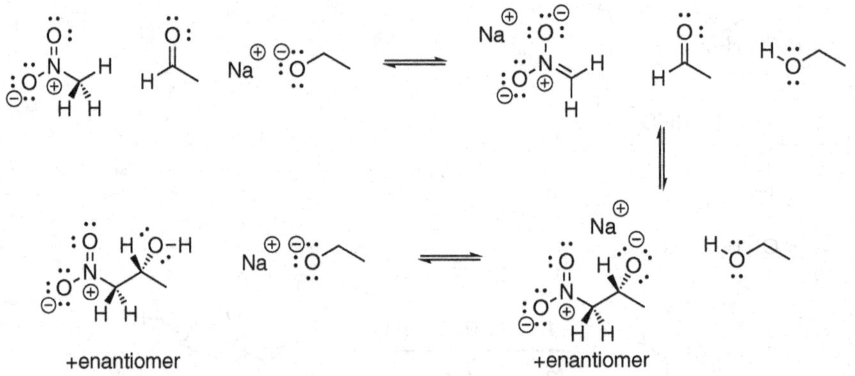

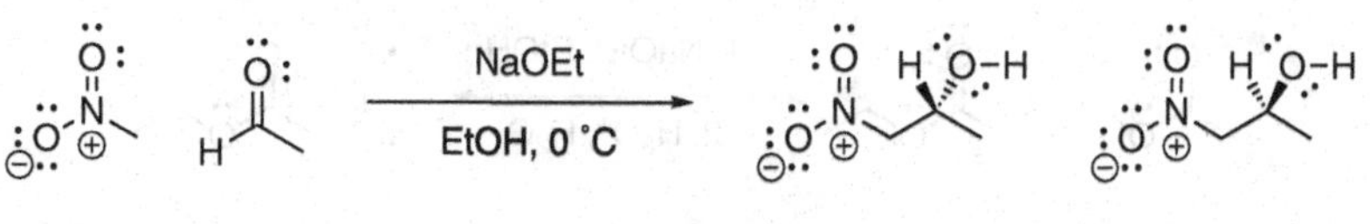

Figure 22.23 Basic addition of a nitroalkane carbanion (enolate analogue) to a carbonyl carbon atom (Henry reaction).

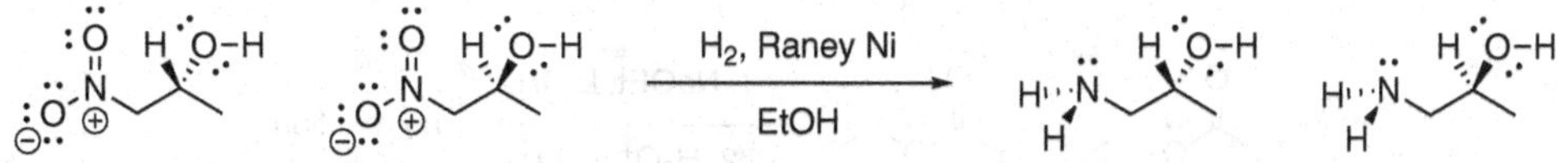

Figure 22.24 Reduction of a Henry reaction product to produce an amino alcohol.

Problem 22.18 Predict the major product(s) of each reaction below.

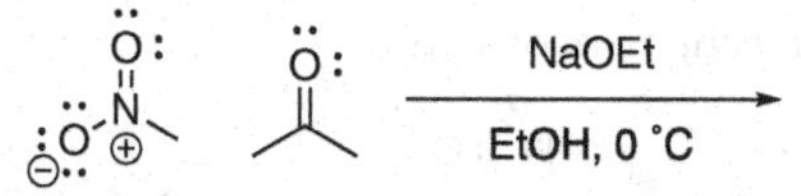

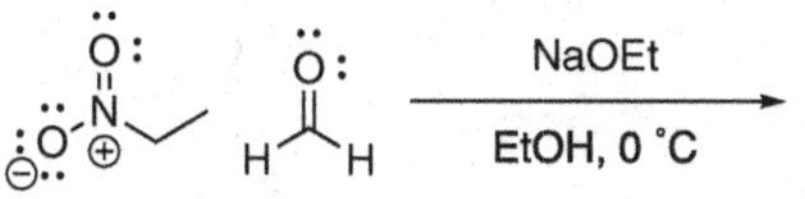

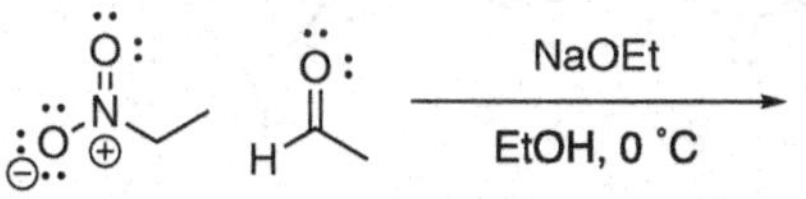

The nitro group can then be reduced to an amine by hydrogenation. The Henry reaction thus provides a route to prepare amino alcohols (Figure 22.24), which have a wide range of applications in inorganic, organic, and bioorganic chemistry. And if the alcohol is a primary alcohol, this is a potential route to synthesize amino acids (after oxidation of the alcohol).

Problem 22.19 Predict the major product(s) of the reaction below.

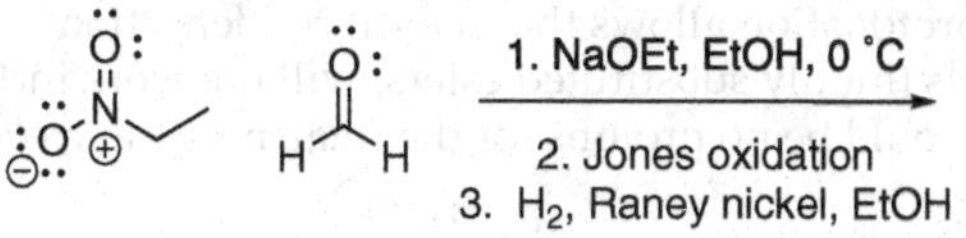

CLAISEN AND DIECKMANN CONDENSATION REACTIONS

As was seen previously, carboxylic acid derivatives (like esters) undergo nucleophilic acyl substitution reactions (Figure 22.25).

The Claisen condensation (Figure 22.26)[10] is the nucleophilic acyl substitution reaction of an ester with an ester enolate nucleophile. The reaction produces a β-ketoester as the final product.

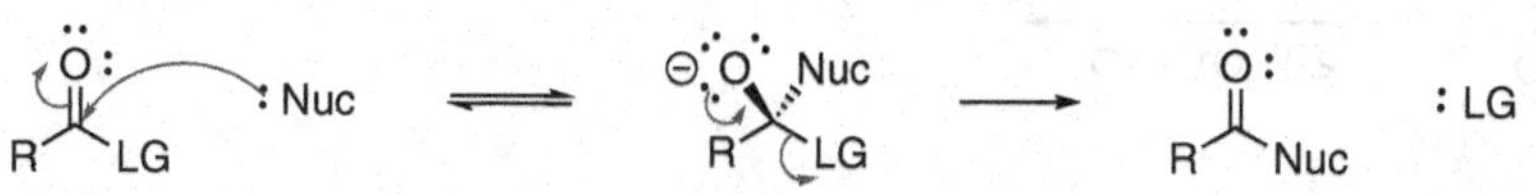

Figure 22.25 General mechanism of nucleophilic acyl substitution reactions: 1) nucleophilic attack to form a tetrahedral intermediate followed by 2) loss of leaving group to reform the carbonyl. LG could be –Cl, –O_2R, –OR, or –NR_2. Claisen, L. Ueber die Einwirkung von Natriumalkylaten auf Benzaldehyd. *Ber. Dtsch. Chem. Ges.*, **1887**, *20* (1), 646–650. DOI: 10.1002/cber.188702001148

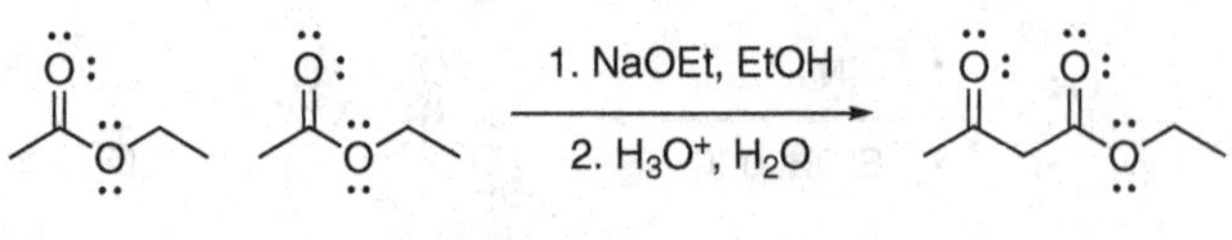
1. NaOEt, EtOH
2. H$_3$O$^+$, H$_2$O

Figure 22.26 Claisen condensation to produce a β-ketoester product. Note, again, for illustrative purposes two ethyl ethanoate molecules are shown as reactants. Typically, the reactions will only explicitly show one molecule. See the examples in Problem 22.22.

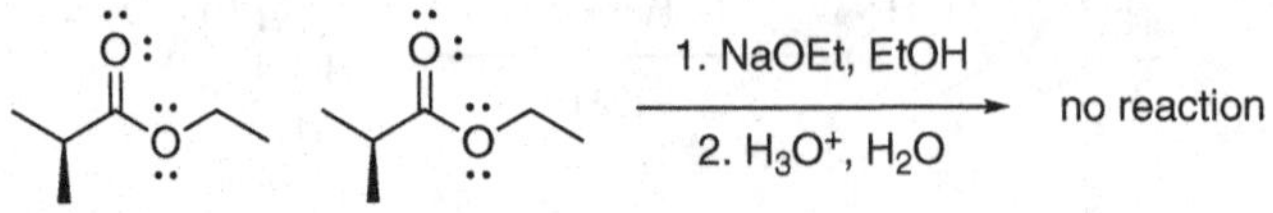
1. NaOEt, EtOH
2. H$_3$O$^+$, H$_2$O
no reaction

Figure 22.27 Claisen condensation does not proceed because no final deprotonation is possible.

Problem 22.20 Draw the appropriate arrows to complete the mechanism of step 1.

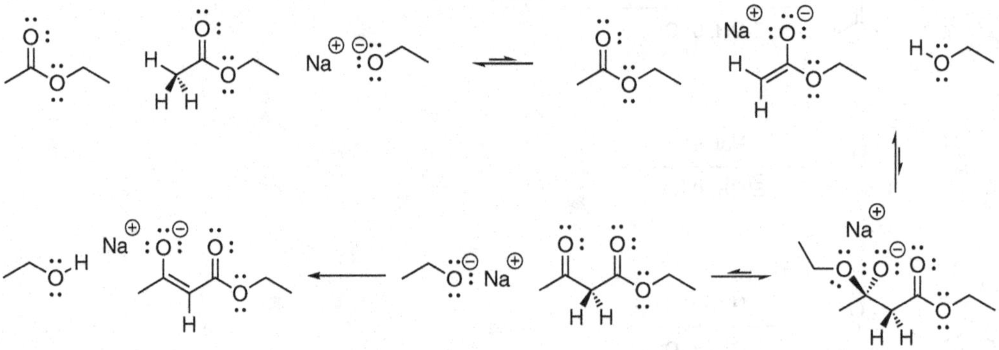

Problem 22.21 Draw the appropriate arrows to complete the mechanism of step 2.

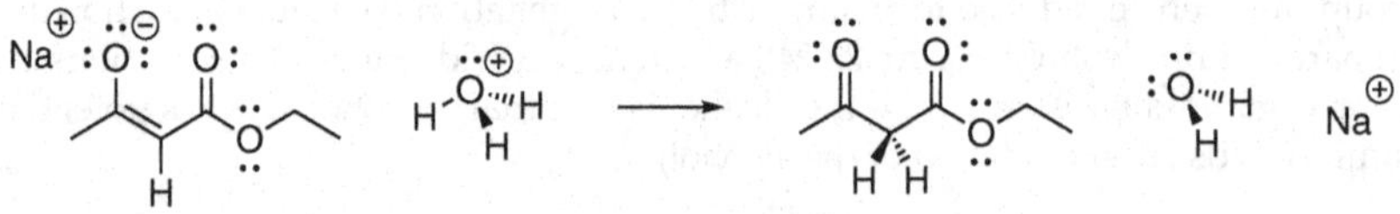

Note in the above mechanism of step 1., that up until the final deprotonation step each step is disfavoured in the forward direction. Only the final deprotonation allows the Claisen condensation to produce a product. As such, some starting materials (highly substituted esters) will not work in a Claisen condensation (Figure 22.27) because there would be no protons for the final, necessary, deprotonation step.

Problem 22.22 Predict the major product(s) of each reaction below.

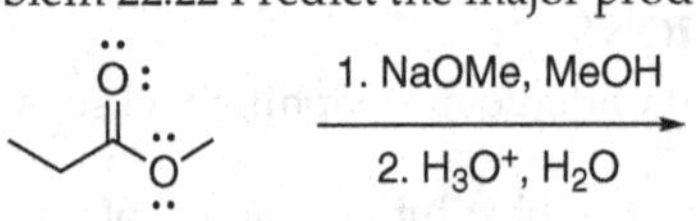
1. NaOMe, MeOH
2. H$_3$O$^+$, H$_2$O

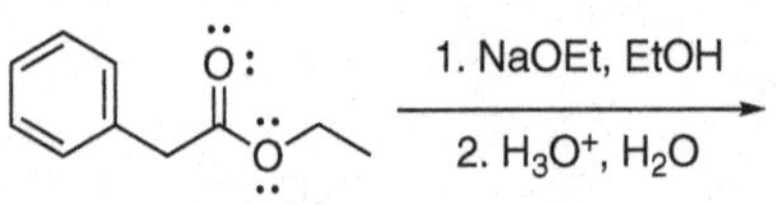
1. NaOEt, EtOH
2. H$_3$O$^+$, H$_2$O

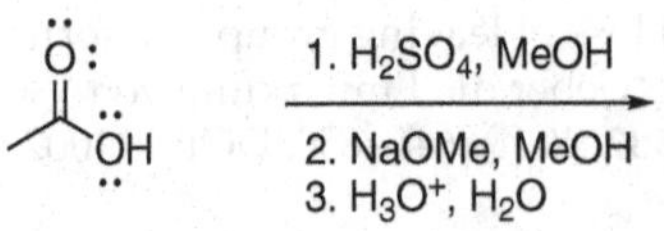
1. H$_2$SO$_4$, MeOH
2. NaOMe, MeOH
3. H$_3$O$^+$, H$_2$O

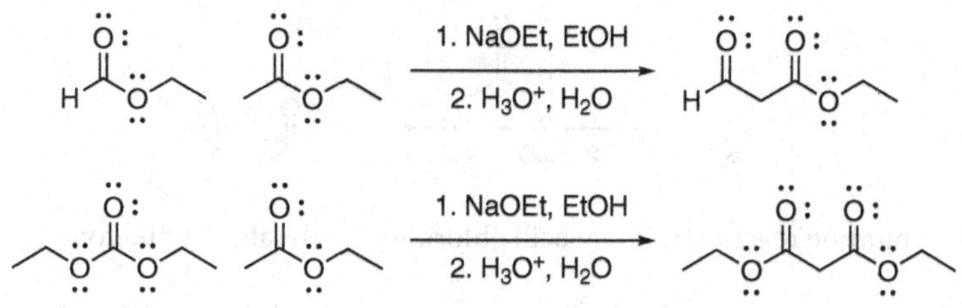

Figure 22.28 Example cross-Claisen condensation with a non-enolizable ester (*top*) and with diethyl carbonate (*bottom*).

All the examples considered so far, have involved the reaction of two equivalents of the same ester condensing together. Crossed Claisen condensation reactions are possible when combining an ester that is enolizable with a non-enolizable ester or diethyl carbonate (Figure 22.28).

Problem 22.23 Predict the major product(s) of each reaction below.

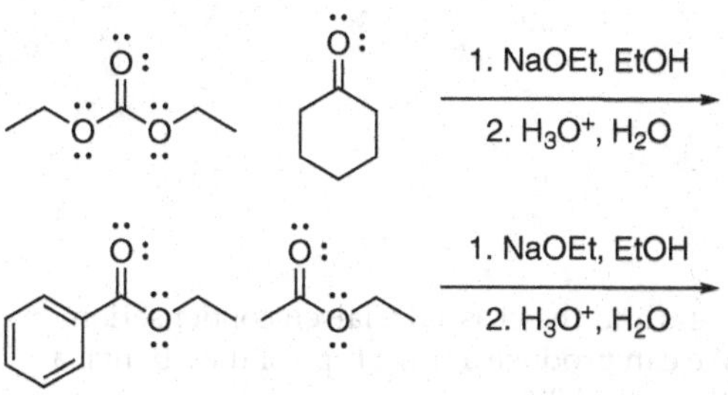

As with the aldol reaction, the Claisen condensation can also occur intramolecularly whenever a five- or six-membered ring is produced. The intramolecular Claisen, however, has its own name: the Dieckmann reaction (Figure 22.29).[11]

Problem 22.24 Draw the appropriate arrows to complete the mechanism of step 1.

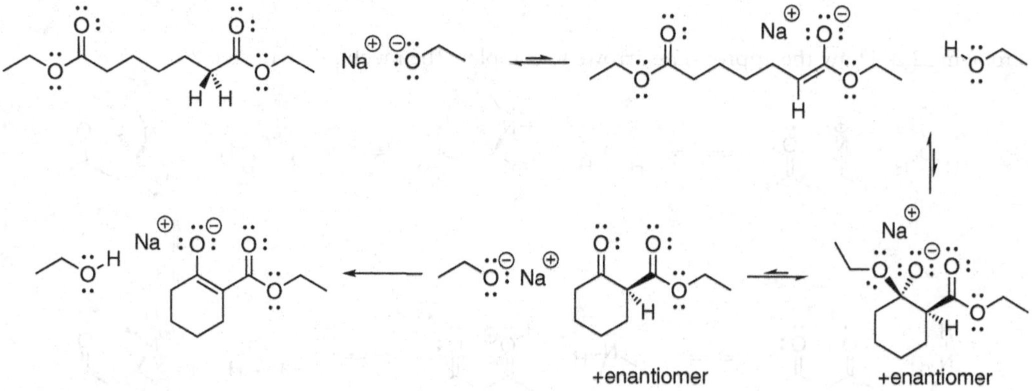

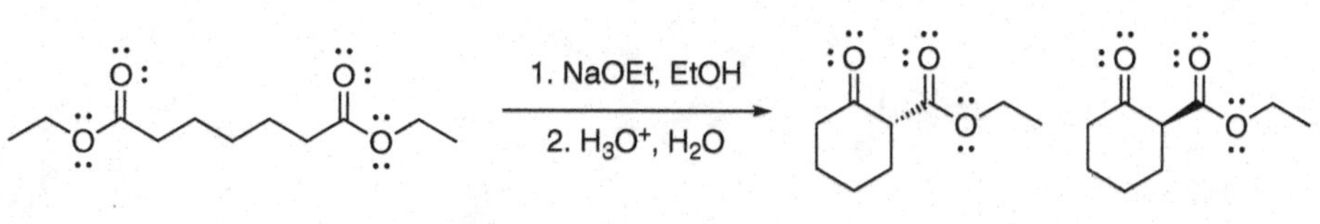

Figure 22.29 Dieckmann reaction.

267

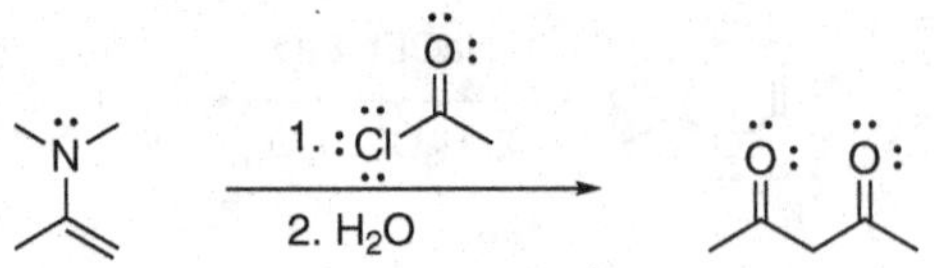

Figure 22.30 Enamine reaction with an acid chloride to generate a β-diketone.

Problem 22.25 Draw the appropriate arrows to complete the mechanism of step 2.

+enantiomer

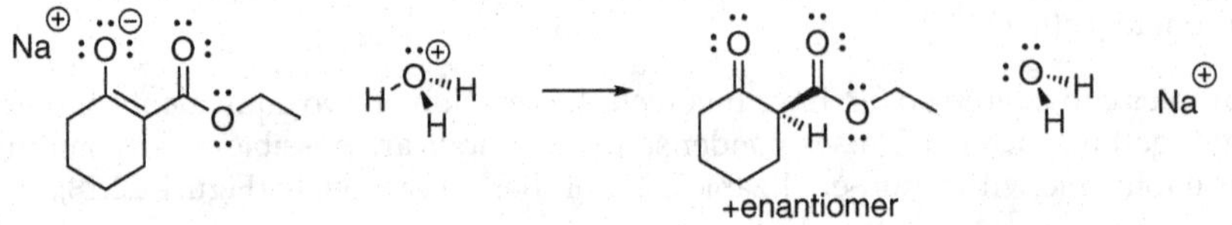

Problem 22.26 Predict the major product(s) of the reaction below.

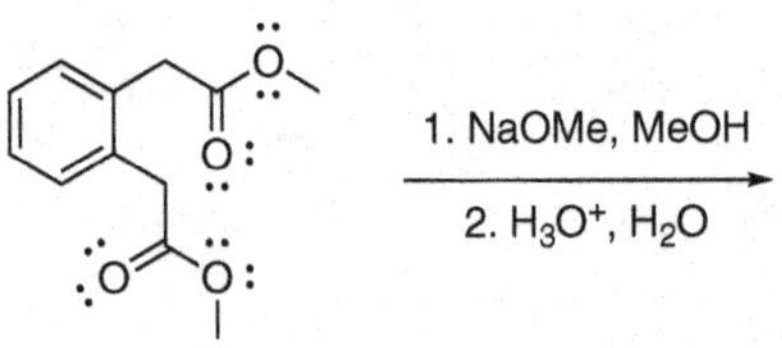

1. NaOMe, MeOH

2. H₃O⁺, H₂O

The Claisen condensation gives us access to β-ketoesters and the crossed Claisen condensation (with diethyl carbonate) gives us access to diesters. We can produce a third type of dicarbonyl, a diketone, by reacting enamines with acid chlorides (Figure 22.30).[4]

Problem 22.27 Draw the appropriate arrows to complete the mechanism of step 1.

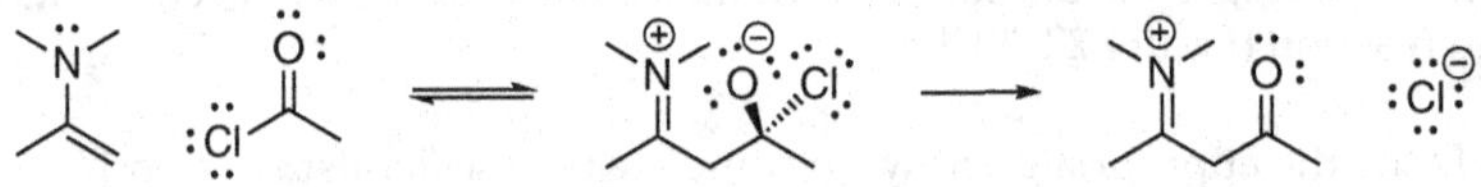

Problem 22.28 Draw the appropriate arrows to complete the mechanism of step 2.

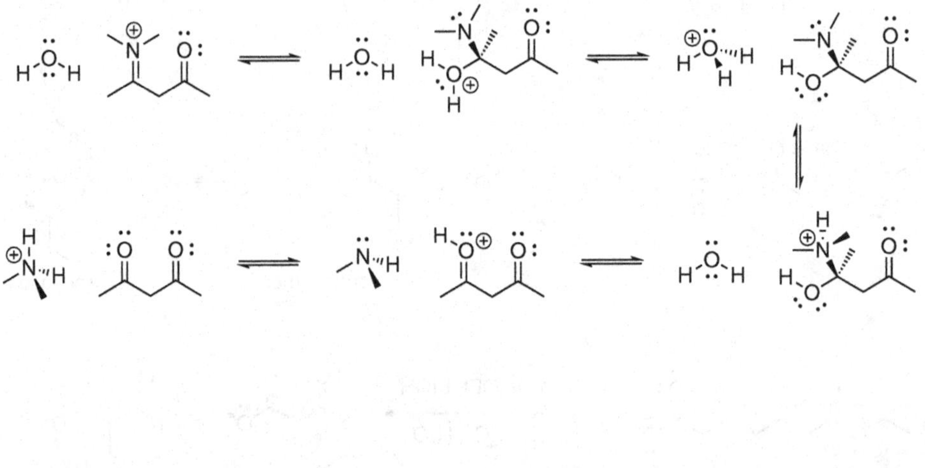

Problem 22.29 Predict the major product(s) of each reaction below.

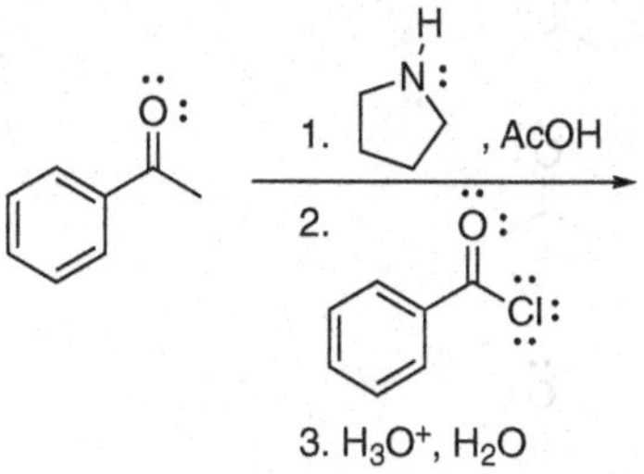

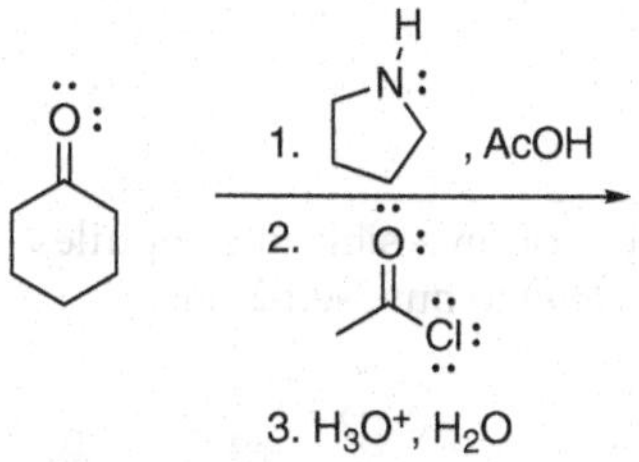

CONJUGATE ADDITION

So far, we have considered only nucleophilic attack at the carbonyl carbon atom. However, α,β-unsaturated aldehydes and ketones are electrophilic at both the carbonyl carbon atom and the b carbon atom (Figure 22.31).

The carbonyl carbon atom is the most electrophilic site on an α,β-unsaturated aldehyde or ketone. And so, nucleophiles always attack the carbonyl carbon atom first (direct addition, Figure 22.15). For nucleophiles like NaBH₄, LiAlH₄, organolithium, and Grignard nucleophiles, this initial attack is irreversible (essentially it is what we would expect based on Chapter 19) and so they only give the direct addition (also called 1,2-addition) products (Figure 22.32).

Less basic nucleophiles (enols, enolates, enamines, cyanide, alkoxides, hydroxide, and thiolates) also attack the carbonyl carbon atom first. However, because they are weaker nucleophiles, this attack is reversible, and so these nucleophiles ultimately produce products from attack at the β-carbon atom (Figure 22.33). These products are referred to as the 1,4-addition, conjugate-addition, or Michael addition products.[12] Note that under acidic conditions, the conjugate addition product is always produced.

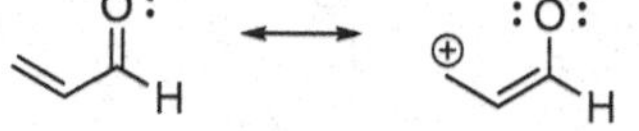

Figure 22.31 Contributing structures of an α,β-unsaturated aldehyde (acrylaldehyde).

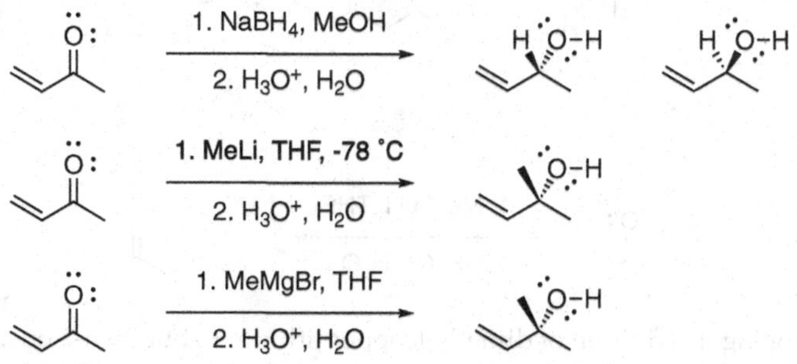

Figure 22.32 1,2-addition (direct addition) of irreversible nucleophiles NaBH₄ (*top*), methyllithium (*middle*), and methylmagnesium bromide (*bottom*) to but-3-en-2-one.

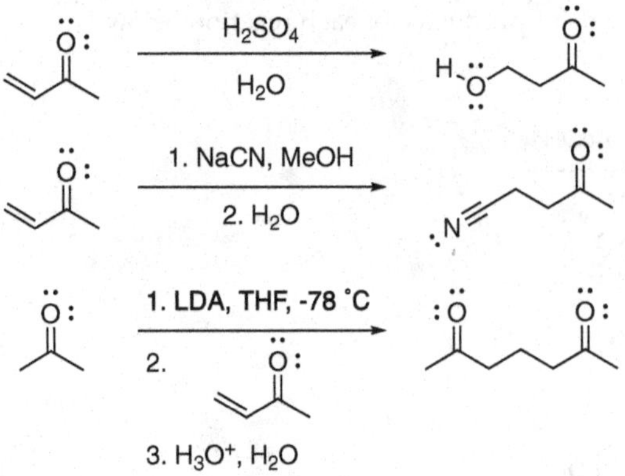

Figure 22.33 1,4-addition (conjugate addition or Michael addition) of reversible nucleophiles water in acid (*top*), sodium cyanide (*middle*), and acetone enolate (*bottom*) to but-3-en-2-one.

Problem 22.30 Draw the appropriate arrows to complete the acidic conjugate addition mechanism.

1. Protonation of the carbonyl

2. Fast, reversible and non-productive attack on the carbonyl carbon atom

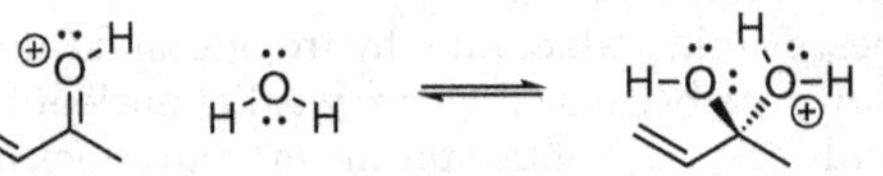

3. Slow, relatively irreversible, productive attack on the β carbon atom.

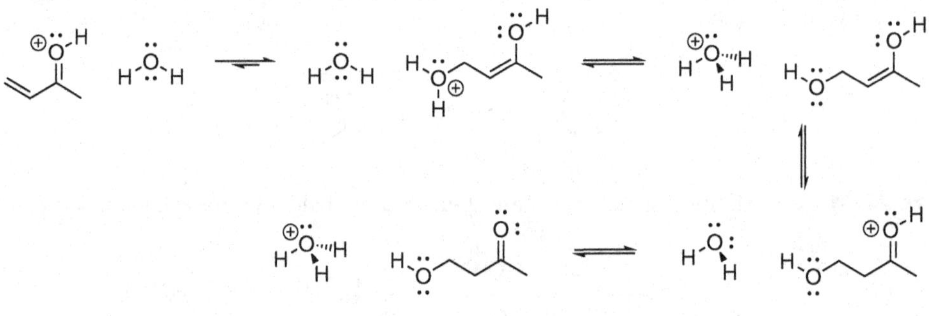

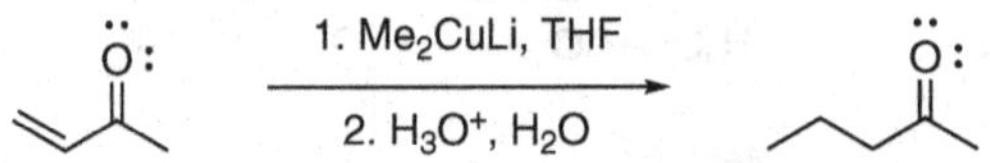

Figure 22.34 Conjugate addition of dimethylcopper lithium to but-3-en-2-one.

Problem 22.31 Draw the appropriate arrows to complete the basic conjugate addition mechanism.

1. Fast, reversible and non-productive attack on the carbonyl carbon atom

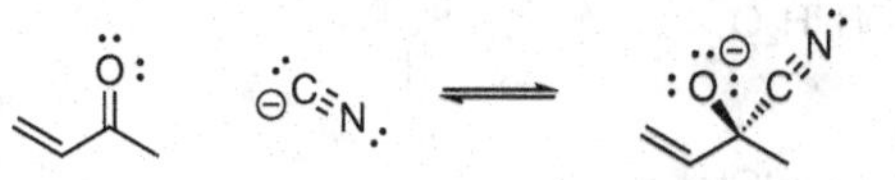

2. Slow, relatively irreversible, productive attack on the β carbon atom.

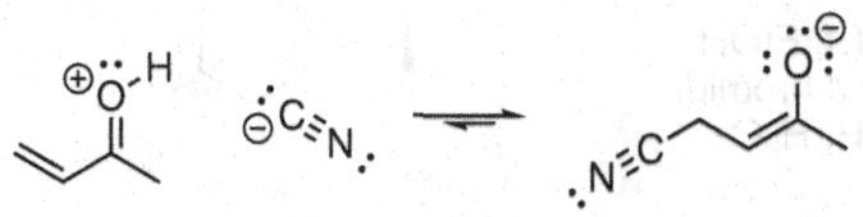

3. Acidic workup to produce the neutral, final product

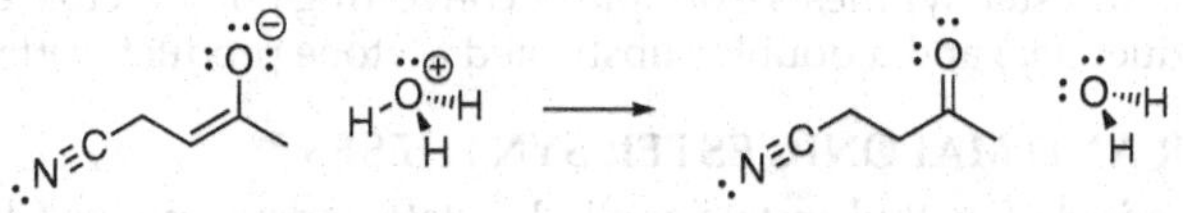

Gilman reagents also produce, preferentially, the conjugate addition product when reacting with α,β-unsaturated aldehydes or ketones (Figure 22.34). The mechanism is beyond the scope of this chapter.

Problem 22.32 Predict the major product(s) of each reaction below.

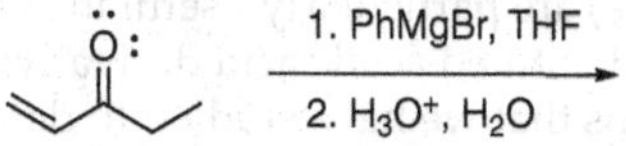

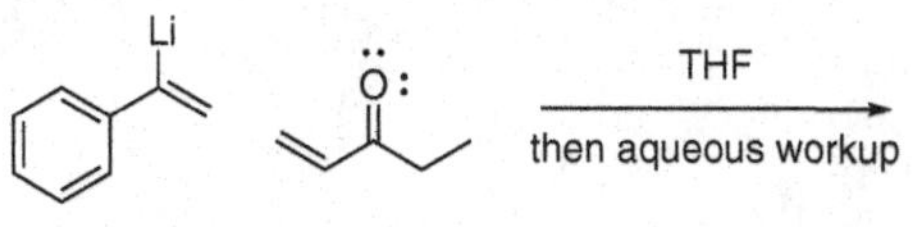

1. PhMgBr, THF

2. H$_3$O$^+$, H$_2$O

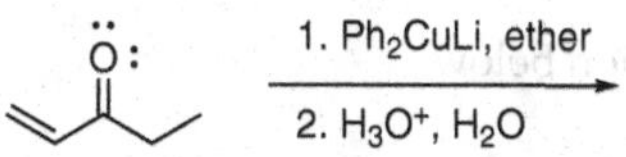

1. Ph$_2$CuLi, ether

2. H$_3$O$^+$, H$_2$O

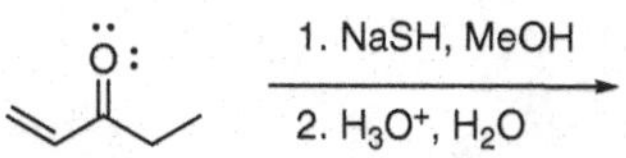

1. NaSH, MeOH

2. H$_3$O$^+$, H$_2$O

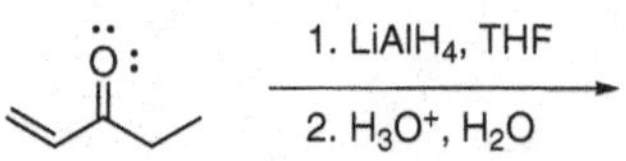

1. LiAlH$_4$, THF

2. H$_3$O$^+$, H$_2$O

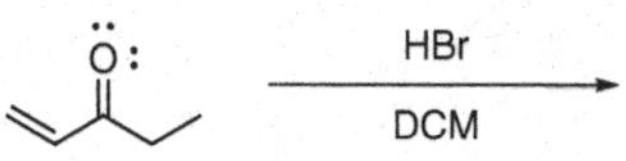

HBr

DCM

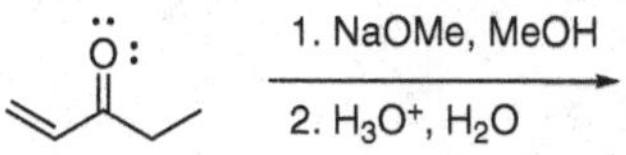

1. NaOMe, MeOH

2. H$_3$O$^+$, H$_2$O

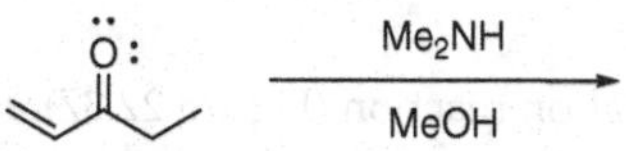

Me$_2$NH

MeOH

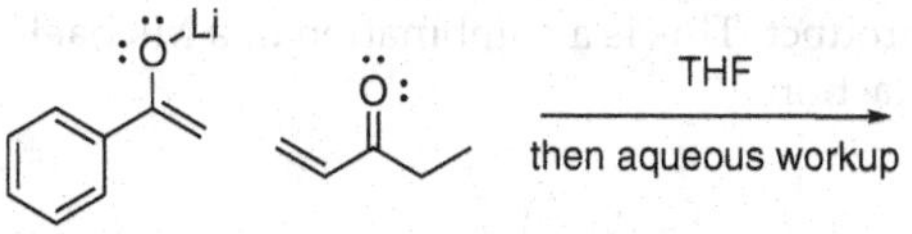

THF

then aqueous workup

THF

then aqueous workup

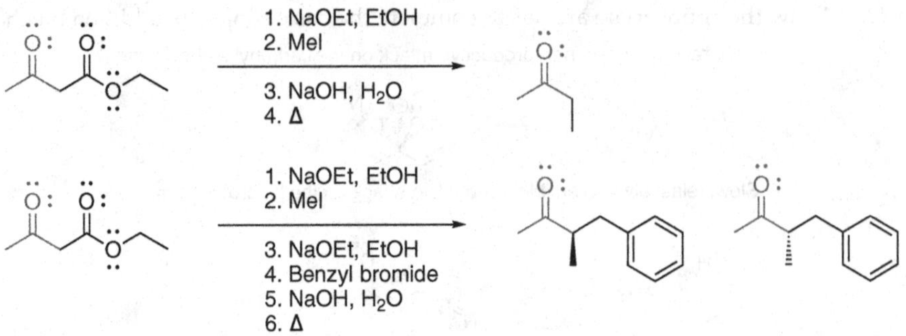

Figure 22.35 Acetoacetic ester synthesis example of converting ethyl acetoacetate into a singly substituted acetone product (*top*) and a doubly substituted acetone product (*bottom*).

ACETOACETIC ESTER AND MALONIC ESTER SYNTHESES

Acetoacetic esters (the product of methyl acetate or ethyl acetate undergoing the Claisen condensation reaction) are particularly useful in synthesis. The acetoacetic ester synthesis is a method to produce highly substituted, asymmetric acetone derivatives (Figure 22.35).[13] This synthetic methodology uses several reactions that we have studied to date: enolate formation, enolate alkylate, ester hydrolysis, and decarboxylation.

Malonic esters (the product of methyl acetate or ethyl acetate undergoing a crossed Claisen condensation reaction with dimethyl or diethyl carbonate) are particularly useful in synthesis. The malonic ester synthesis is a method to produce highly substituted acetic acid derivatives (Figure 22.36).[14] This synthetic methodology uses several reactions that we have studied to date: enolate formation, enolate alkylate, ester hydrolysis, and decarboxylation.

Problem 22.33 Predict the major product(s) of each reaction below.

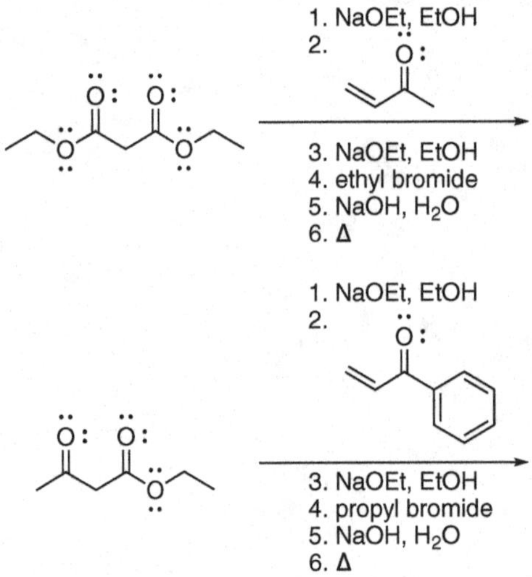

ROBINSON ANNULATION

The last reaction we will consider is the Robinson annulation reaction (Figure 22.37).[15] The Robinson annulation reaction is the reaction of a ketone with an α,β-unsaturated ketone (under acidic or basic conditions) to produce a cyclohexenone product. This is a combination of a Michael addition reaction followed by an intramolecular aldol reaction.

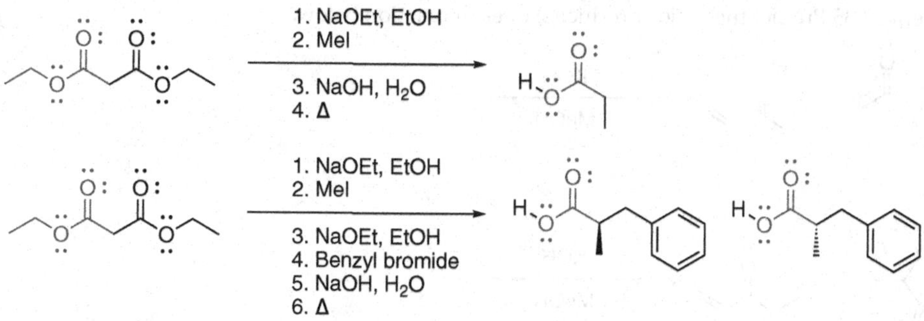

Figure 22.36 Malonic ester synthesis example of converting diethyl malonate into a singly substituted acetic acid product (*top*) and a doubly substituted acetic acid product (*bottom*).

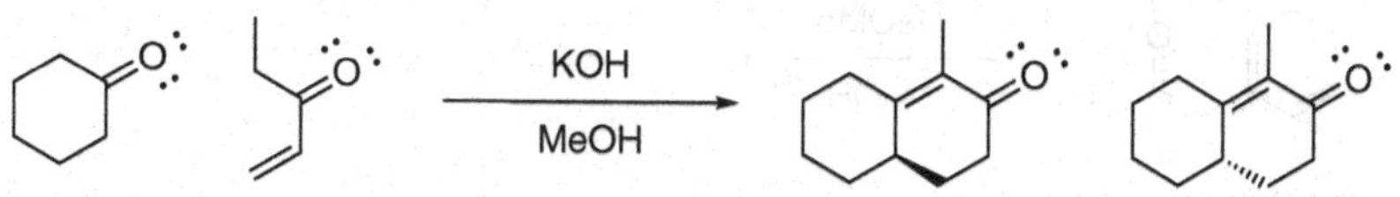

Figure 22.37 Robinson annulation reaction of cyclohexanone with pent-1-en-3-on.

Problem 22.34 Draw the appropriate arrows to complete the mechanism.

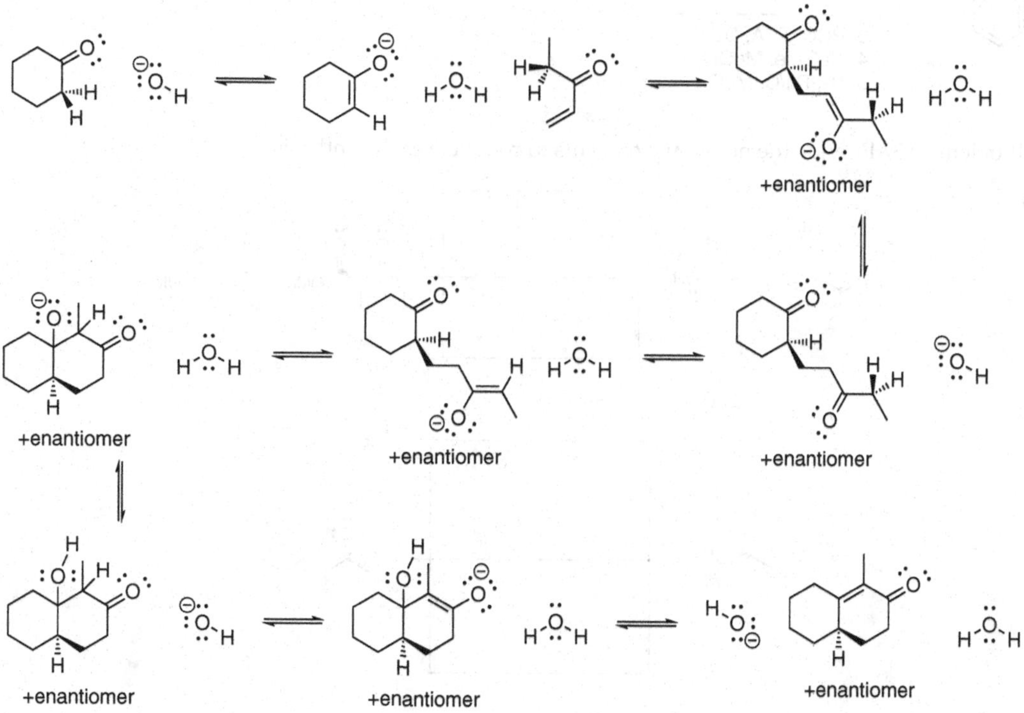

Problem 22.35 Predict the major product(s) of each reaction below.

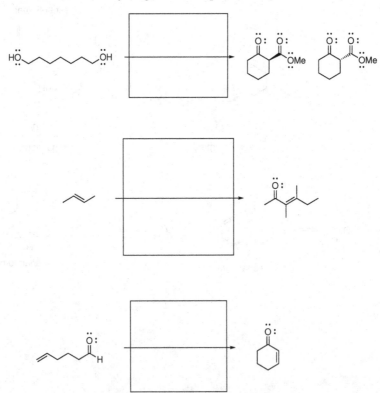

GENERAL PRACTICE PROBLEMS

Problem 22.36 Predict the major product of the following multi-step synthesis.

1. BH₃, THF
2. NaOH, H₂O₂, H₂O

3. PCC, TEA, DCM
4. NaOMe, MeOH, Δ
5. H₂, Pd/C, MeOH

Problem 22.37 Provide the necessary reagents to complete each synthesis.

Problem 22.38 Provide a reasonable mechanism to account for the following transformations.

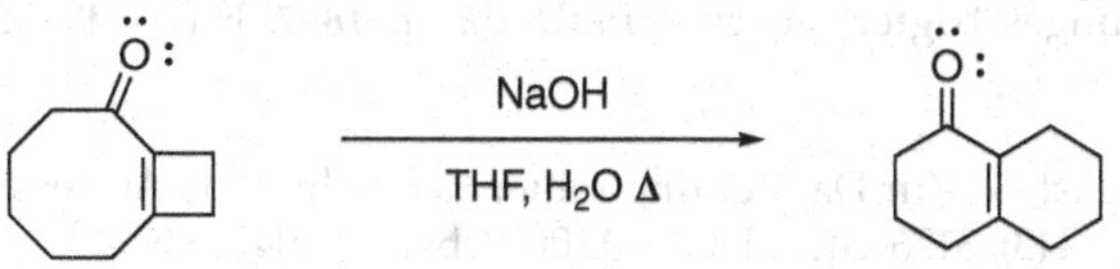

Mechanism:

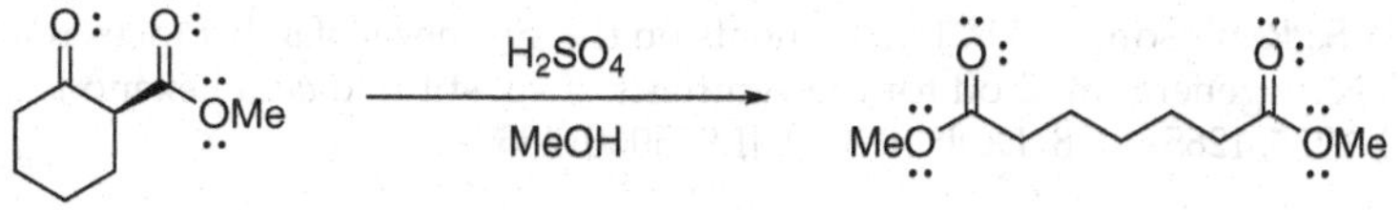

Mechanism:

NOTES

1. Alcoholate is a synonym for alkoxide.

2. i) House, H.O.; Trost, B.M. The chemistry of carbanions. IX. The potassium and lithium enolates derived from cyclic ketones. *J. Org. Chem.*, **1965**, *30* (5), 1341–1348. DOI: 10.1021/jo01016a001

   ii) Stork, G.; Kraus, G.A.; Garcia, G.A. Regiospecific aldol condensations of the kinetic lithium enolates of methyl ketones. *J. Org Chem.*, **1974**, *39* (23), 3459–3460. DOI: 10.1021/jo00937a054

3. Stork, G.; Terrell, R.; Szmuszkovicz, J. A new synthesis of 2-alkyl and 2-acyl ketones. *J. Am. Chem. Soc.*, **1954**, *76* (7), 2029–2039. DOI: 10.1021/ja01636a103

4. Houk, K.N.; Paddon-Row, M.N. Transition structures for C- and O-alkylation of acetaldehyde enolate. Stereoelectronic effects and carbon/oxygen alkylation ratios. *J. Am. Chem. Soc.*, **1986**, *108* (10), 2659–2662. DOI: 10.1021/ja00270a026

5. Kane, R. Ueber den Essiggeist und einige davon abgeleitete Verbindungen. *J. Prakt. Chem.*, **1838**, *15* (1), 129–155. DOI: 10.1002/prac.18380150112

6. Alexander Borodin in: V. von Richter, aus St. Petersburg am 17. October 1869. *Ber. Dtsch. Chem. Ges.*, **1869**, *2* (1), 552–554. DOI: 10.1002/cber.186900201222

7. Wurtz, A. Ueber einen Aldehyd-Alkohol. *J. Prakt. Chem.*, **1872**, *5* (1), 457–464. DOI: 10.1002/prac.18720050148

8. Baigrie, L.M.; Cox, R.A.; Slebocka-Tilk, H.; Tencer, M.; Tidwell, T.T. Acid-catalyzed enolization and aldol condensation of acetaldehyde. *J. Am. Chem. Soc.*, **1985**, *107* (12), 3640–3645. DOI: 10.1021/ja00298a039

9. Henry, L. Formation synthétique d'alcools nitrés. *C. R. Acad. Sci.*, **1895**, *120*, 1265.

10. Claisen, L.; Lowman, O. Ueber eine neue Bildungsweise des Benzoylessigäthers. *Ber. Dtsch. Chem. Ges.*, **1887**, *20* (1), 651–654. DOI: 10.1002/cber.188702001149

11. Dieckmann, W. Zur Kenntniss der Ringbildung aus Kohlenstoffketten. *Ber. Dtsch. Chem. Ges.*, **1894**, *27* (1), 102–103. DOI: 10.1002/cber.18940270126

12. Michael, A. Ueber die Addition von Natriumacetessig- und Natriummalonsäureäthern zu den Aethern ungesättigter Säuren. *J. Prakt. Chem.*, **1887**, *35* (1), 349–356. DOI: 10.1002/prac.18870350136

13. Michael, A.; Wolgast, K. Zur Darstellung reiner Ketone mittels Acetessigester. *Ber. Dtsch. Chem. Ges.*, **1909**, *42* (3), 3176–3177. DOI: 10.1002/cber.19090420340

14. Wislicenus, J. XXXV. Ueber Acetessigestersynthesen. *Liebigs Ann*, **1877**, *186* (2–3), 161–228. DOI: 10.1002/jlac.18771860202

15. Rapson, W.S.; Robinson, R. 307. Experiments on the synthesis of substances related to the sterols. II. New general method for the synthesis of substituted *cyclo*hexenones. *J. Chem. Soc. (Resumed)*, **1935**, 1285–1288. DOI: 10.1039/JR9350001285

23 Amines

Amines are molecules that contain a nitrogen atom. Amines are analogues of ammonia (Figure 23.1), where one or more of the hydrogen atoms on the nitrogen atom are replaced with a carbon substituent. If there is one carbon substituent, the amine is a primary (1°) amine (Figure 23.2). If there are two carbon substituents (Figure 23.3) or three carbon substituents (Figure 23.4), then it is a secondary (2°) or tertiary amine (3°), respectively.

If there are four carbon substituents (Figure 23.5) on the nitrogen atom, then it is a quaternary ammonium cation. Quaternary ammonium cations are quite stable and, depending on the substituents, are generally stable to oxidizing agents, reducing agents, and strong nucleophiles. Quaternary ammonium cations are found in biochemistry as part of the neurotransmitter acetylcholine and the phospholipid phosphatidylcholine. In chemistry, ionic compounds with quaternary ammonium cations are used as phase transfer catalysts – a compound that helps move chemicals between aqueous and organic phases – and as antimicrobials.

Figure 23.1 Structure (*left*), model (*middle*) and electrostatic potential map (*right*) of ammonia (red corresponds to high electron density and blue corresponds to low electron density).

Figure 23.2 Structure (*left*), model (*middle*) and electrostatic potential map (*right*) of methanamine, a primary amine, commonly named methylamine (red corresponds to high electron density and blue corresponds to low electron density).

Figure 23.3 Structure (*left*), model (*middle*) and electrostatic potential map (*right*) of N-methylmethanamine, a secondary amine, commonly named dimethylamine (red corresponds to high electron density and blue corresponds to low electron density).

Figure 23.4 Structure (*left*), model (*middle*) and electrostatic potential map (*right*) of *N,N*-dimethylmethanamine, a tertiary amine, commonly named trimethylamine (red corresponds to high electron density and blue corresponds to low electron density).

Figure 23.5 Structure (*left*) and model (*right*) of tetramethylammonium, a quaternary ammonium cation. No electrostatic potential map is shown as the map for a cation would be uniformly blue.

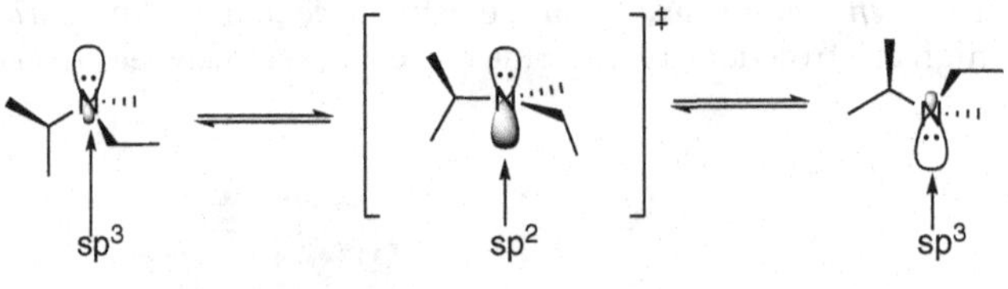

Figure 23.6 Example of nitrogen atom inversion, which prevents neutral amines from being chirality centres.

Primary and secondary amines can react as both electrophiles and nucleophiles. The hydrogen atoms attached to the primary and secondary amine nitrogen atoms are electrophilic (blue in the electrostatic potential map [Figures 23.1–23.4]) and amines are weakly acidic (typical pK_a value about 38). Primary, secondary, and tertiary amine nitrogen atoms have a lone pair and can react as a Brønsted–Lowry and Lewis base (nucleophile).

AMINE STEREOCHEMISTRY

As we saw in Chapter 6, amines cannot be chirality centres. While neutral amines are tetrahedral, and the nitrogen does have four different groups around it (in the example in Figure 23.6, one methyl, one ethyl, one isopropyl and one lone pair), amines rapidly undergo pyramidal inversion at room temperature.[1] This inversion rapidly interconverts and equilibrates the two different configurations at nitrogen (Figure 23.6).

NOMENCLATURE OF AMINES

In terms of priority, amines have greater priority than alkenes and alkynes, but they are lower in priority than alcohol. In IUPAC nomenclature, amines are named with the suffix -amine (Figure 23.7) and an appropriate locant. For secondary and tertiary amines, substituents on nitrogen are identified with the locant N. When an amine is named as a substituent, it is given the name amino.

propan-1-amine
(propylamine)

ethane-1,2-diamine
(ethylenediamine)

N,N-dimethylpropan-1-amine
(dimethylpropylamine)

2-aminoethan-1-ol
(ethanolamine)

Figure 23.7 Example amines showing IUPAC names (and common names).

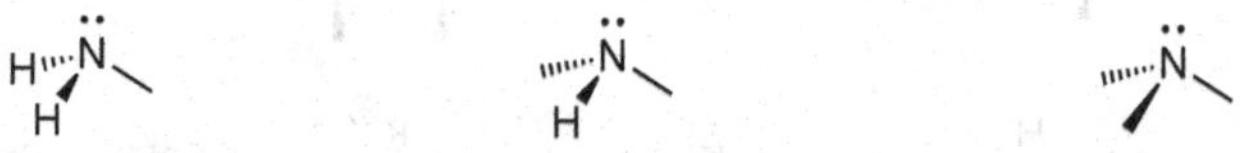

methanamine
dipole moment: 1.47 D
boiling point: -6.6 °C

N-methylmethanamine
dipole moment: 1.03 D
boiling point: 8 °C

N,N-dimethylmethanamine
dipole moment: 0.58 D
boiling point: 5 °C

Figure 23.8 Physical properties of methanamine, N-methylmethanamine, and N,N-dimethylmethanamine.

PHYSICAL PROPERTIES OF AMINES

Primary and secondary amines are relatively polar molecules, due to the polar N–H bonds (Figure 23.8). Tertiary amines are relatively nonpolar molecules because they lack N–H bonds.

AMINES AS ELECTROPHILES

Amines are only weakly electrophilic, and as such they must react with exceptionally basic nucleophiles to act as electrophiles or Brønsted–Lowry acids ($pK_a(NH_3)$ = 38). Butyl lithium is most commonly used to generate azanide conjugate bases (azanides are commonly called amides), which are frequently used in aldol reactions.

Problem 23.1 Draw the appropriate arrows to show the deprotonations below.

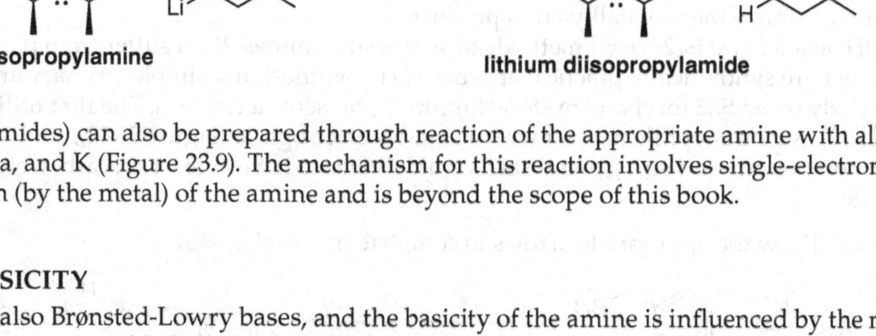

ammonia

lithium azanide
(lithium amide)

diisopropylamine

lithium diisopropylamide

Azanides (amides) can also be prepared through reaction of the appropriate amine with alkali metals Li, Na, and K (Figure 23.9). The mechanism for this reaction involves single-electron transfer reduction (by the metal) of the amine and is beyond the scope of this book.

AMINE BASICITY

Amines are also Brønsted-Lowry bases, and the basicity of the amine is influenced by the number of alkyl substituents. Added alkyl substituents increase the energy of the lone pair (Figure 23.10), which correlates with increased basicity.[2] Note that for N,N-dimethylmethanamine the pK_a is lower than would be expected based on the lone pair energy because the structure reduces the ability of water to hydrate and stabilize the ammonium ion.[3]

SYNTHESIS OF AMINES

Amines play an important role in organic chemistry. Amines are common nucleophiles in synthesis with carbonyls and carboxylic acid derivatives, and amine products are common components of dyes and pharmaceuticals. In biochemistry, amines are present in amino acids and proteins,

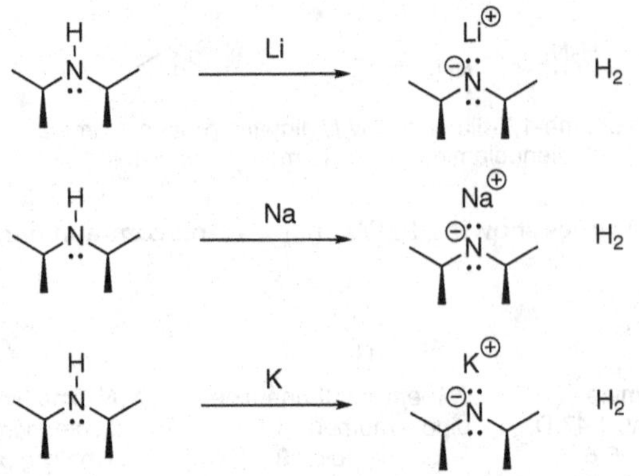

Figure 23.9 Examples of diisopropylamides formation using different alkali metals.

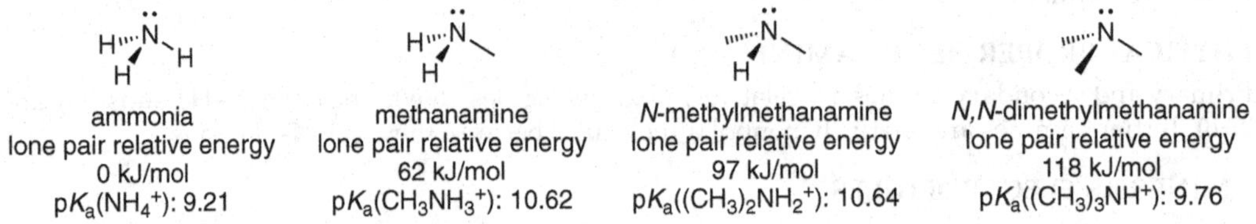

ammonia	methanamine	N-methylmethanamine	N,N-dimethylmethanamine
lone pair relative energy 0 kJ/mol	lone pair relative energy 62 kJ/mol	lone pair relative energy 97 kJ/mol	lone pair relative energy 118 kJ/mol
$pK_a(NH_4^+)$: 9.21	$pK_a(CH_3NH_3^+)$: 10.62	$pK_a((CH_3)_2NH_2^+)$: 10.64	$pK_a((CH_3)_3NH^+)$: 9.76

Figure 23.10 Effect on lone pair energy (relative to ammonia) and pK_a (of conjugate base) of added alky substituents.

neurotransmitters, and hormones. Given the importance of amines, a great deal of chemistry has been developed to synthesize amines.

Substitution

In part, the effort to develop amine syntheses has been driven by the need to circumvent a key challenge: amines cannot be cleanly synthesized by S_N2 substitution. The issue arises due to the increasing basicity, and nucleophilicity, of amines after alkylation. For example, attempts to monomethylate phenylmethanamine (benzylamine) with iodomethane will produce a mixture of secondary, tertiary, and quaternary ammonium products and leave behind a significant amount of unreacted starting material (Figure 23.11). The reaction could be forced to make 100% of the quaternary ammonium cation product, if excess iodomethane were used, but there is no way in an S_N2 reaction to maximize the monoalkylation product.

We will consider two S_N2-based methods to synthesize amines. Both suffer from poor atom economy, but are synthetically practical approaches to synthesizing simple, primary amines – since they rely on an S_N2 mechanism steric hindrance presents a concern. The first utilizes azide (Figure 23.12) as a nucleophile to react with a suitable electrophile (Figure 23.13), azides can be explosive compounds and so a general safety guideline is to have at least six carbon atoms per every azide.[4]

Problem 23.2 Draw the appropriate arrows to complete the mechanism.

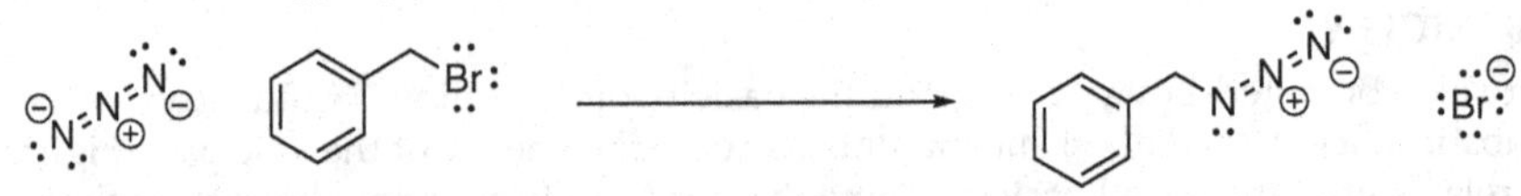

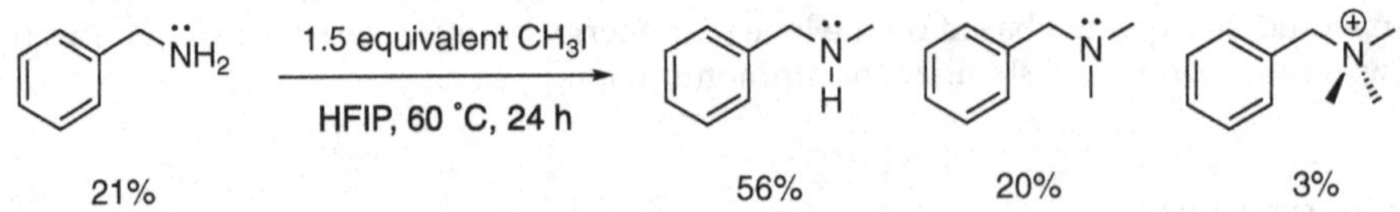

21%		56%	20%	3%

Figure 23.11 Percentage of each species after the reaction of phenylmethanamine with iodomethane. Lebleu, T.; Ma, X.; Maddaluno, J.; Legros, J. Selective monomethylation of primary amines with simple electrophiles†. *Chem. Commun.*, **2014**, *50*, 1836–1838. DOI: 10.1039/c3cc48997c

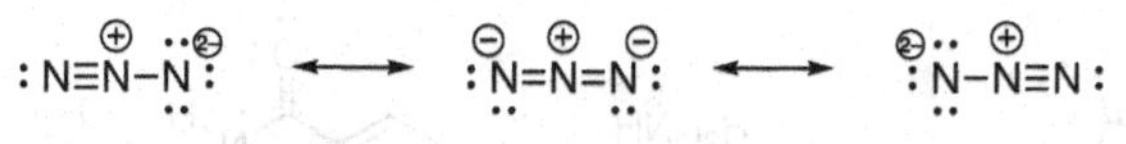

Figure 23.12 Contributing structures of azide.

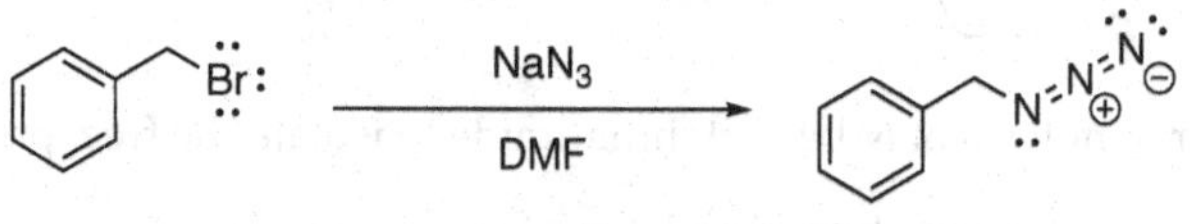

Figure 23.13 Synthesis of (azidomethyl)benzene by S_N2 reaction.

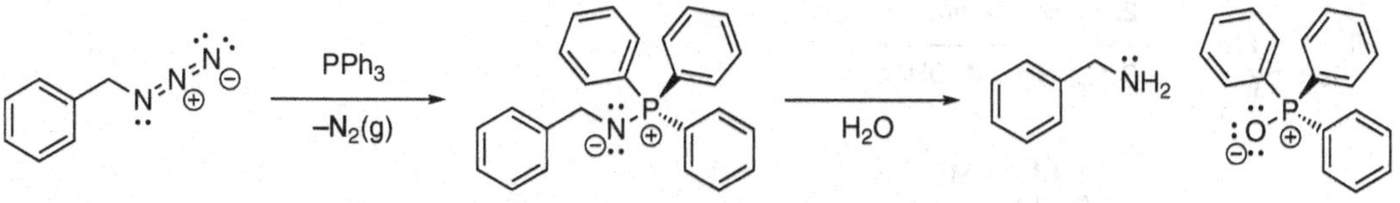

Figure 23.14 Staudinger reduction of (azidomethyl)benzene to phenylmethanamine.

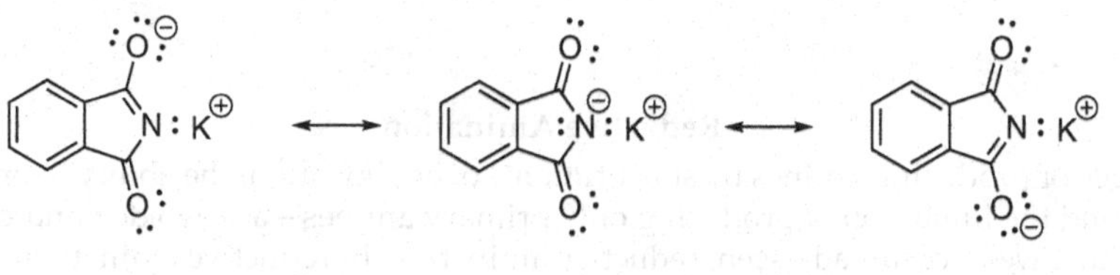

Figure 23.15 Selected contributing structures of potassium phthalimide showing delocalization of nitrogen atom lone pair electrons.

Azides can then be reduced to amines through a Staudinger reduction.[5] First triphenylphosphine converts the azide into an iminophosphorane, which is then hydrolyzed into triphenylphosphine oxide and the amine (Figure 23.14). The mechanism of this reaction is beyond the scope of this book.

Primary amines can also be synthesized through the Gabriel synthesis,[6] which uses potassium phthalimide (Figure 23.15) as the nitrogen-atom source. Potassium phthalimide can be generated from phthalimide with a base like potassium carbonate (K_2CO_3).

Potassium phthalimide will react only with primary haloalkanes, not secondary haloalkanes, via an S_N2 mechanism to generate an N-alkylphthalimide.

Problem 23.3 Provide an explanation of why potassium phthalimide will not react with secondary haloalkanes.

Problem 23.4 Draw the appropriate arrows to complete the mechanism.

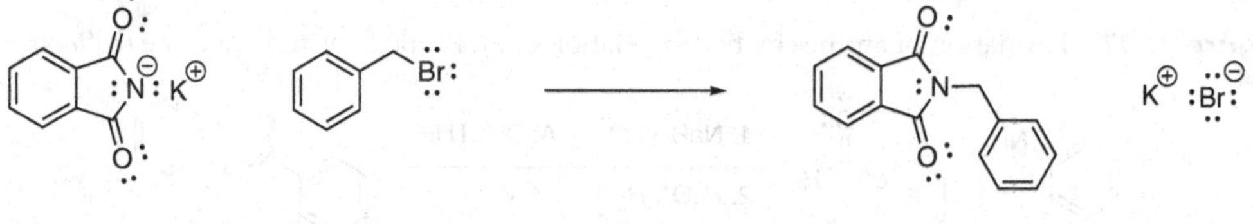

To isolate the free amine, the imide functional group can be hydrolyzed with strongly acidic or strongly basic conditions but hydrazinoloysis is typically the preferred method (Figure 23.16).[7]

Problem 23.5 Provide a mechanism for the hydrazinolysis of N-benzylphthalimide. The reaction is a double transamidation (Figure 21.20).

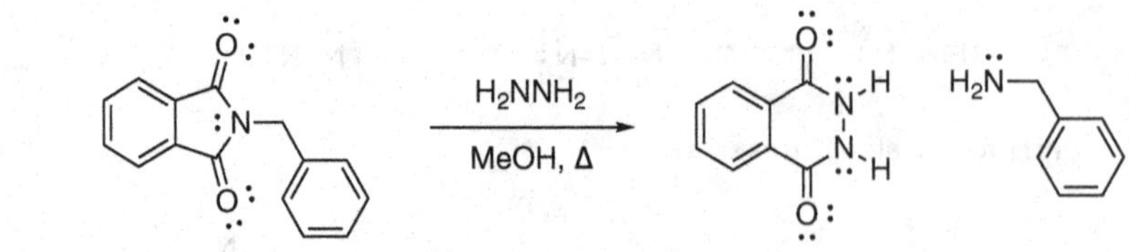
H2NNH2
MeOH, Δ

Figure 23.16 Hydrazinolysis of *N*-benzylphthalimide to isolate the free, primary amine.

Problem 23.6 Predict the major product of each reaction.

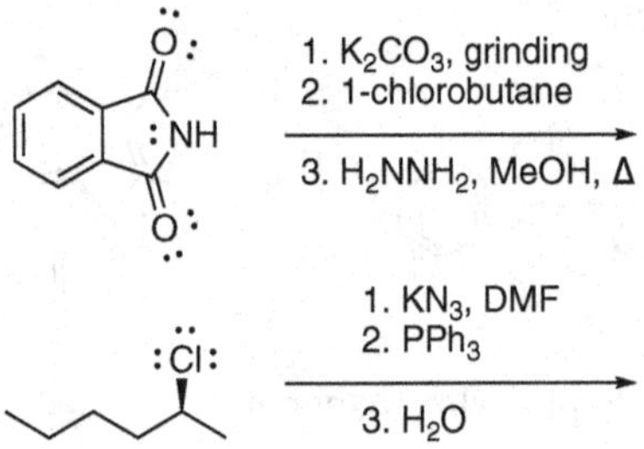
1. K2CO3, grinding
2. 1-chlorobutane
3. H2NNH2, MeOH, Δ
1. KN3, DMF
2. PPh3
3. H2O

Reductive Amination

The challenges of producing amines by substitution – over alkylation, the ability to only use certain amines and the limitation of producing only primary amines – are avoided and circumvented by a process that we have already seen, reductive amination. In reductive amination, an imine, enamine, or iminium ion is generated by the reaction of an amine with an aldehyde or ketone, which can then be reduced by hydrogenation (Figure 23.17) and with hydride-reducing agents

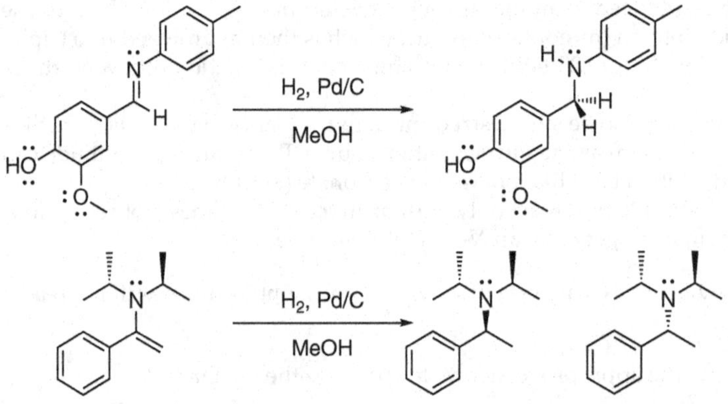
H2, Pd/C
MeOH
H2, Pd/C
MeOH

Figure 23.17 Formation of amines by hydrogenation of an imine (*top*) and enamine (*bottom*).

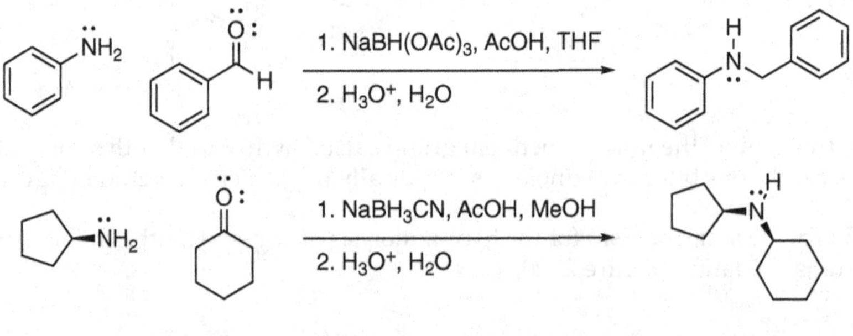
1. NaBH(OAc)3, AcOH, THF
2. H3O+, H2O
1. NaBH3CN, AcOH, MeOH
2. H3O+, H2O

Figure 23.18 Example formation of amines by reductive amination with sodium triacetoxyhydroborate (*top*) and sodium cyanotrihydridoborate (*bottom*).

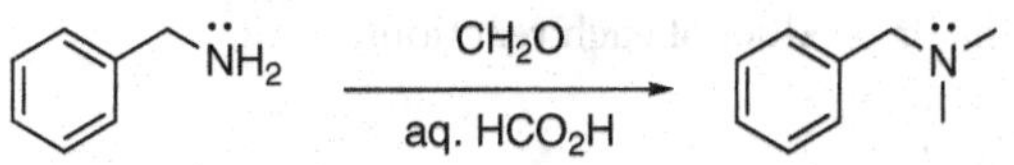

Figure 23.19 Eschweiler–Clarke reaction of phenylmethana
mine.

(Figure 23.18). In reductive amination reactions, the imine and enamine can be preformed (Figure 23.17) or made *in situ* (Figure 23.18).

A new reductive amination reaction to add to our synthetic toolbox is the Eschweiler–Clarke reaction,[8] which uses formaldehyde and formic acid (Figure 23.19) to add methyl groups to amine nitrogen atoms. This process produces only tertiary amines. In this reaction, formic acid serves as the hydride source.

Problem 23.7 Draw appropriate arrows to complete the mechanism.

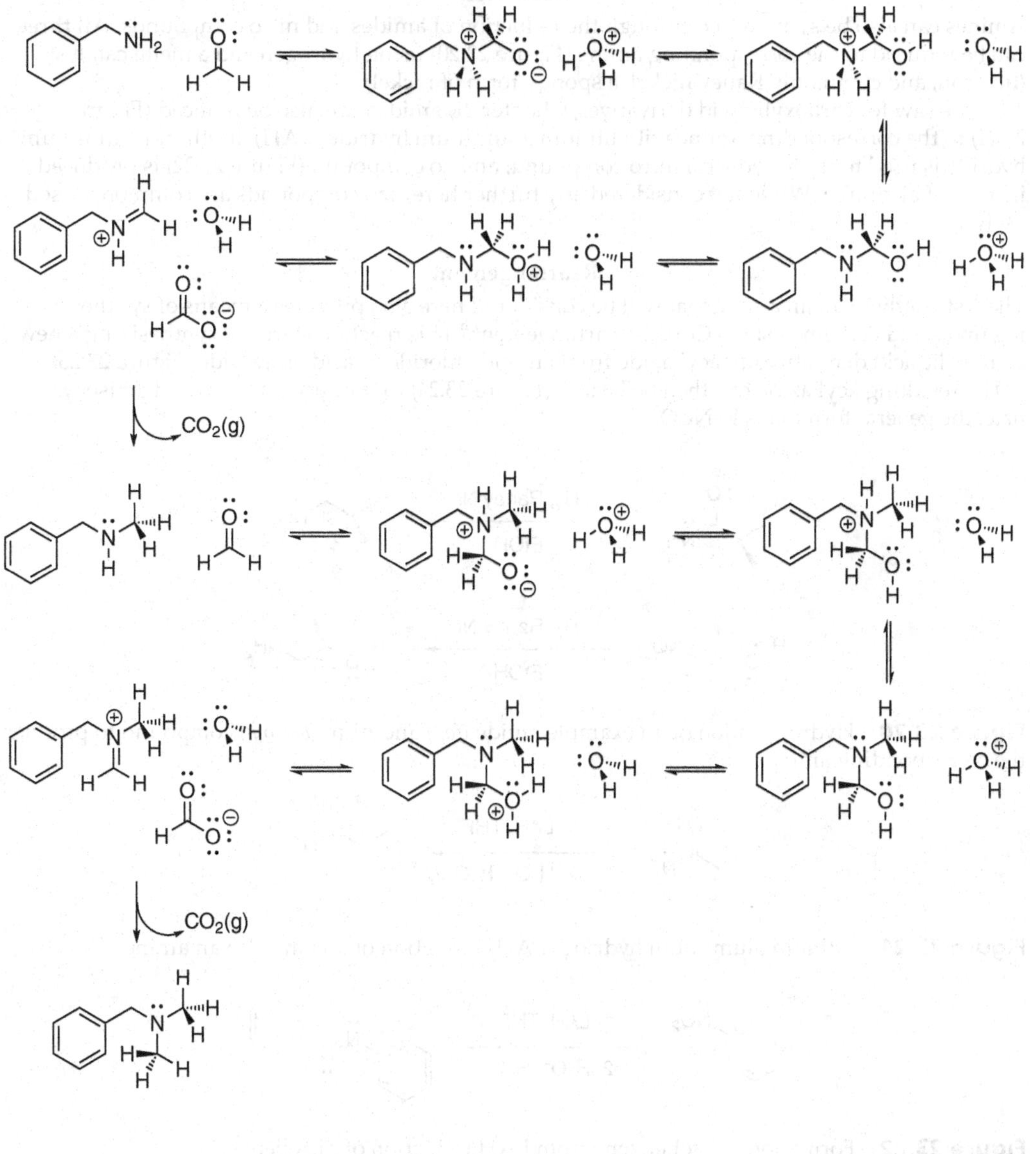

Problem 23.8 Predict the major product of each reaction.

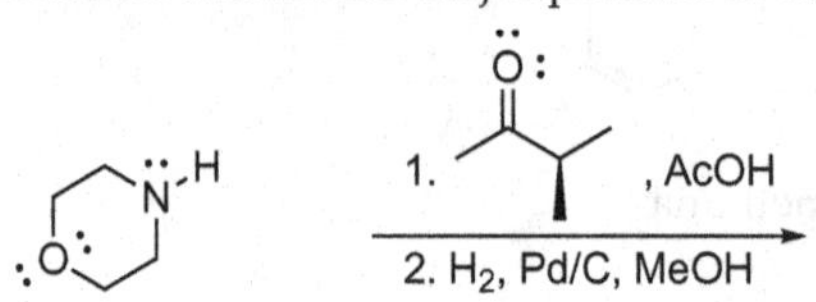

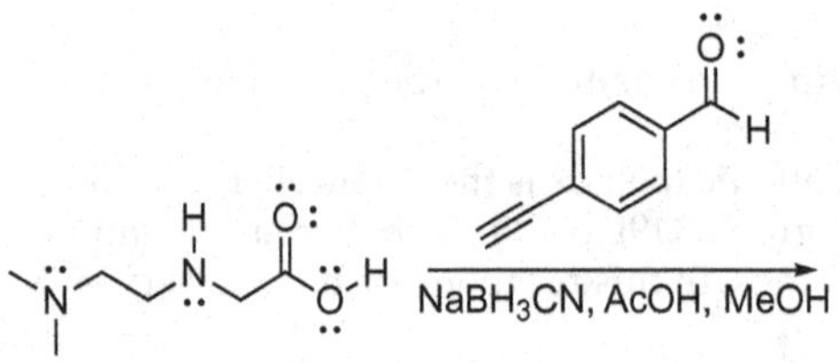

Reduction

Amines can also be synthesized through the reduction of amides and nitro compounds. All three can be reduced to the corresponding amine (Figure 23.20) using hydrogen and a metal catalyst (tin, iron, and commonly Raney nickel, a spongy form of nickel).

As we saw for carboxylic acid derivatives, Chapter 21, amides can also be reduced (Figure 23.21) to the corresponding amine with lithium aluminium hydride (LAH). If lithium aluminium hydride is used to try to reduce a nitro compound, an azo compound (Figure 23.22) is produced instead of an amine. While not considered any further here, azo compounds are commonly used as dyes.

Rearrangement

The last synthesis of an amine that will be considered here also provides a means of synthesizing ureas and carbamates: the Curtius rearrangement.[9] This reaction starts by synthesizing a new carboxylic acid derivative, an acyl azide from an acid chloride or acid anhydride (Figure 23.23).

The resulting acyl azide can then be heated (Figure 23.24) to produce nitrogen and an isocyanate, the general formula is R–NCO.

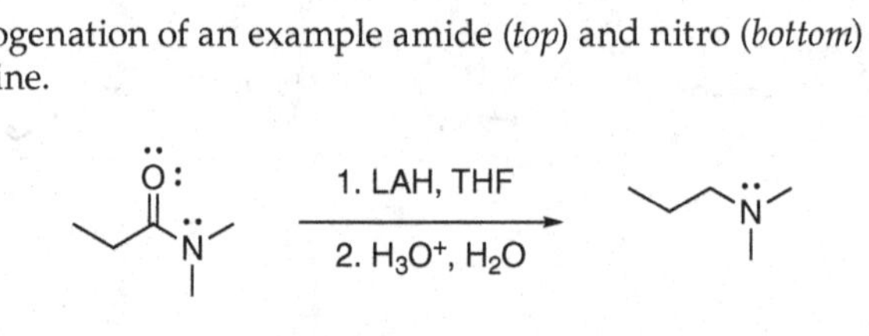

Figure 23.20 Hydrogenation of an example amide (*top*) and nitro (*bottom*) compound to produce the corresponding amine.

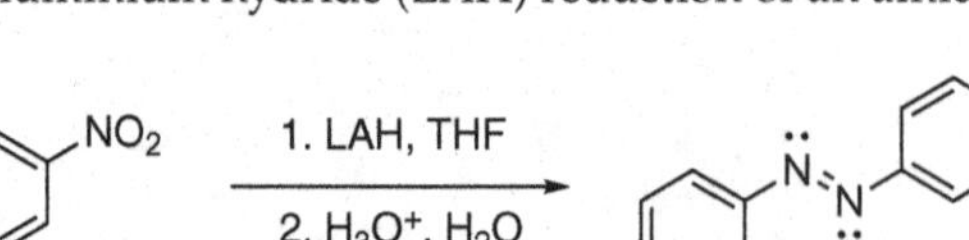

Figure 23.21 Lithium aluminium hydride (LAH) reduction of an amide to an amine.

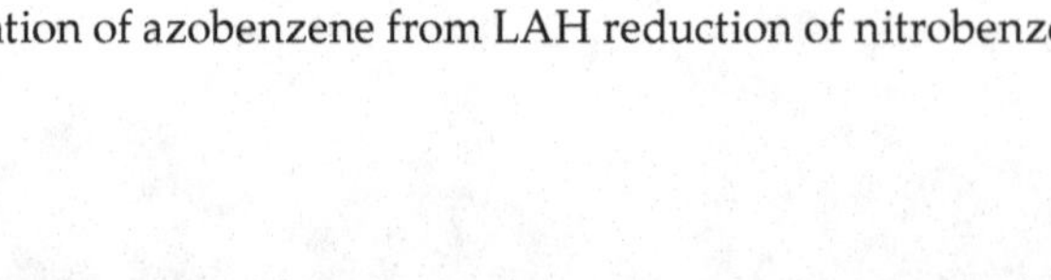

Figure 23.22 Formation of azobenzene from LAH reduction of nitrobenzene.

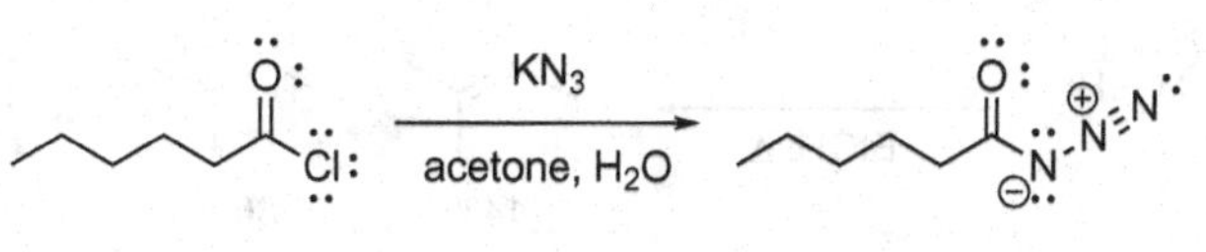

Figure 23.23 Synthesis of hexanoyl azide, an acyl azide, from hexanoyl chloride, an acid chloride.

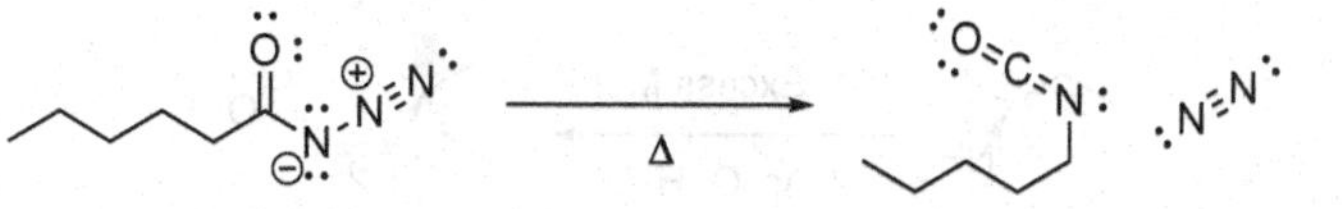

Figure 23.24 Formation of pentyl isocyanate from hexanoyl azide.

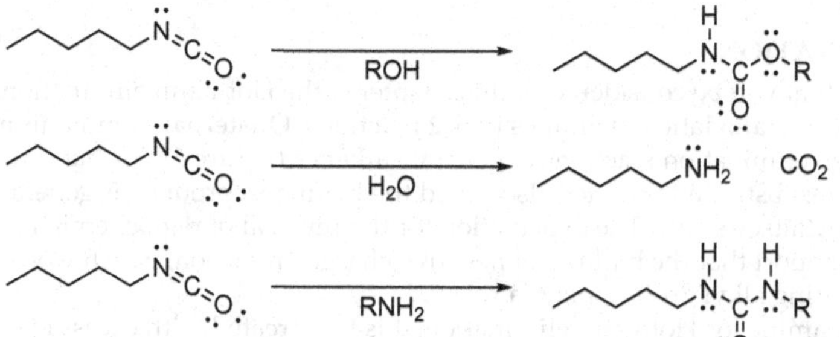

Figure 23.25 Reaction of pentyl isocyanate with an alcohol to produce a carbamate (*top*), with water to produce an amine (*middle*), and an amine to produce a urea (*bottom*).

Problem 23.9 Annotate Figure 23.24 with appropriate arrows to show the mechanism (this reaction happens in a single concerted step).[10]

Isocyanates are electrophilic at the carbon atom and can react with alcohols, water, and amines (Figure 23.25). The reaction with alcohols produces carbamates and with amines ureas are produced, both have a carbon atom sandwiched between heteroatoms. With water, a similar structure is originally produced, carbamic acid, but this decomposes into an amine and carbon dioxide. It is important to note that when producing amines with the Curtius rearrangement, the resulting carbon chain is one atom shorter than it was originally.

Problem 23.10 Draw the appropriate arrows to show the mechanism for pentyl isocyanate reacting with water.[11] Note the mechanism for an alcohol or an amine reacting with an isocyante is the same up to the formation of carbamic acid.

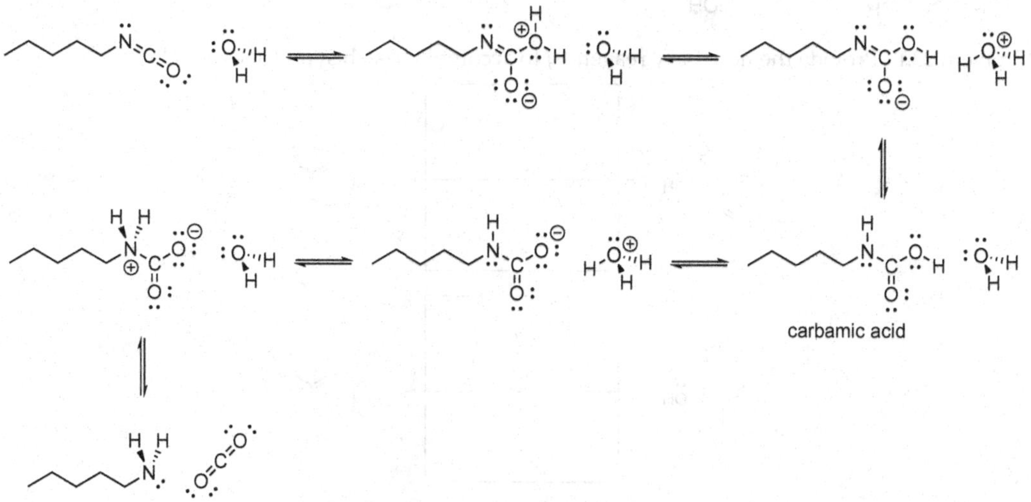

carbamic acid

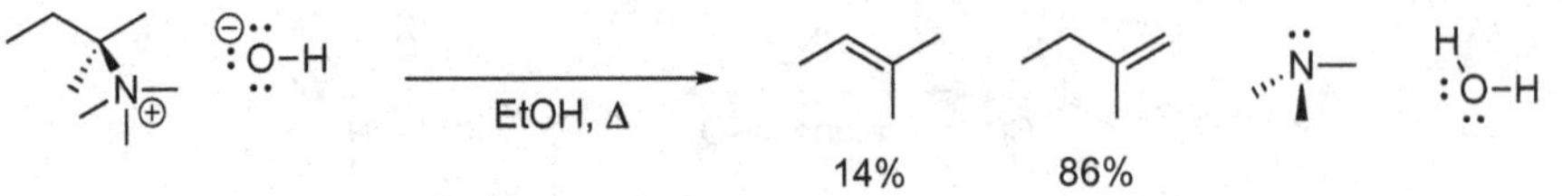

Figure 23.26 Hofmann elimination showing a preference for the less-substituted alkene, the Hofmann product.

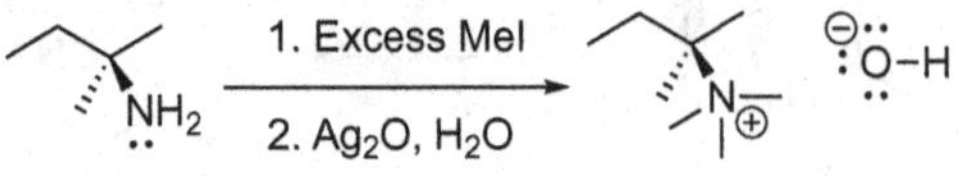

Figure 23.27 Preparation of an amine for Hofmann elimination.

AMINE ELIMINATION

A final reaction that will be considered in this chapter is the Hofmann elimination,[12] which takes advantage of the overalkylation of amines in S_N2 reactions. Quaternary ammonium hydroxide salts can undergo elimination reactions to generate alkenes (Figure 23.26), but unlike typical E2 reactions the less-substituted product, also called the Hofmann product, is generated, which is in contravention to Zaitsev's rule. The explanation for the reversal of regiochemistry is unsettled, but evidence does support that the buildup of negative charge on carbon, which would be destabilized by electron-donating alkyl groups, plays a role.[13]

To prepare an amine for Hofmann elimination, it is first treated with excess methyl iodide, which generates exclusively the quaternary ammonium salt. Then the iodide anion is replaced with hydroxide by using silver oxide and water (Figure 23.27).

The Hofmann elimination can be used to generate alkenes with a reversal of regiochemistry; it is more commonly a concern for materials that contain quaternary ammonium ions: phase transfer catalysts, polymers, and antimicrobials.

GENERAL PRACTICE PROBLEMS

Problem 23.11 Predict the major product(s) of each reaction.

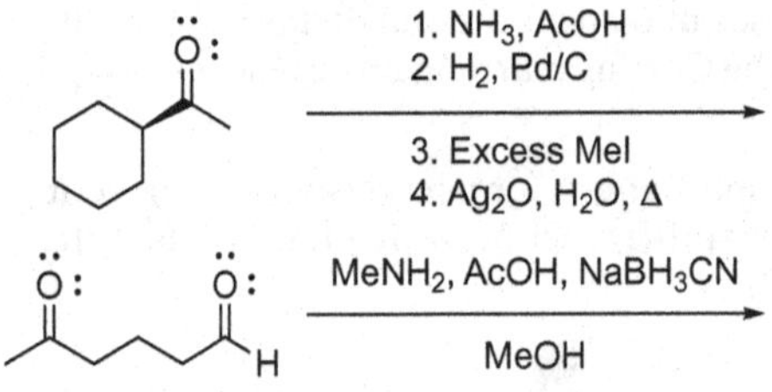

Problem 23.12 Provide the necessary reagent(s) to accomplish each synthesis.

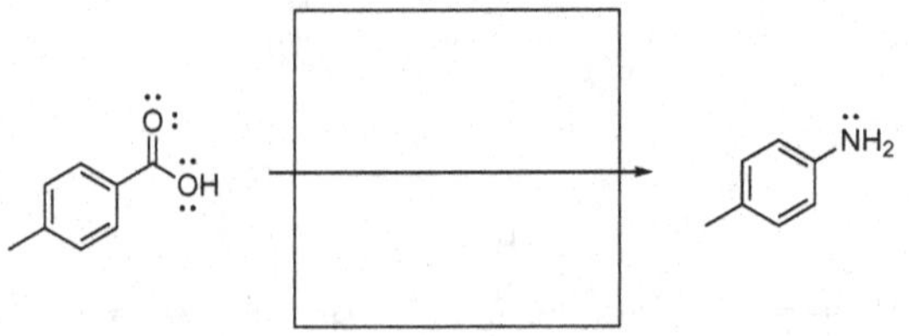

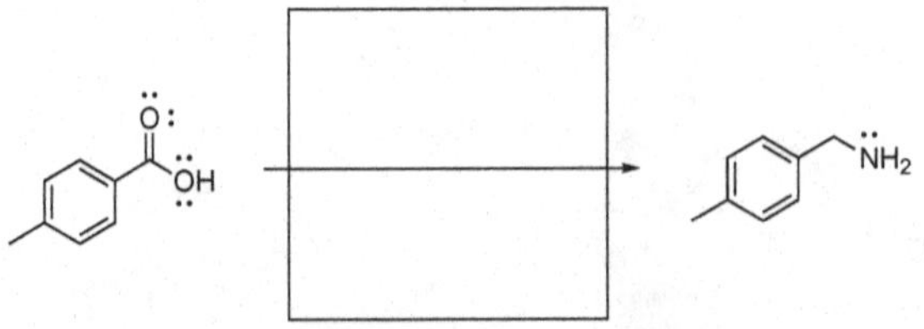

NOTES

1. Dennison, D.M. The infra-red spectra of polyatomic molecules. Part II. *Rev. Mod. Phys.*, **1940**, *12* (3), 175–214. DOI: 10.1103/RevModPhys.12.175

2. Brauman, J.I.; Riveros, J.M.; Blair, L.K. Gas-phase basicities of amines. *J. Am. Chem. Soc.*, **1971**, *93* (16), 3914–1916. DOI: 10.1021/ja00745a016

3. Nagy, P. Theoretical calculations on the basicity of amines: Part 2. The influence of hydration. *J. Mol. Struct.*, **1989**, *201*, 271–286. DOI: 10.1016/0166-1280(89)87081-2

4. Kolb, H.C.; Finn, M.G.; Sharpless, K.B. Click Chemistry: Diverse chemical function from a few good reactions. *Angew. Chem. Int. Ed.*, **2001**, *40* (11), 2004–2021. DOI: 10.1002/1521-3773(20010601)40:11<2004::AID-ANIE2004>3.0.CO;2-5

5. Staudinger, H.; Meyer, J. Über neue organische Phosphorverbindungen III. Phosphinmethylenderivate und Phosphinimine. *Helv. Chim Acta*, **1919**, *2* (1), 635–646. DOI: 10.1002/hlca.19190020164

6. Gabriel, S. Ueber eine Darstellungsweise primärer Amine aus den enstprechenden Halogenverbindungen. *Ber. Dtsch. Chem. Ges.*, **1887**, *20* (2), 2224–2236. DOI: 10.1002/cber.18870200227

7. Gibson, M.S.; Bradshaw, R.W. The Gabriel synthesis of primary amines. *Angew. Chem. Int. Ed.*, **1968**, *7* (12), 919–930. DOI: 10.1002/anie.196809191.

8. i) Eschweiler, W. Ersatz von an Stickstoff gebundenen Wasserstoffatomen durch die Methylgruppe mit Hülfe von Formaldehyd. *Ber. Dtsch. Chem. Ges.*, **1905**, *38* (1), 880–882. DOI: 10.1002/cber.190503801154.
   ii) Clarke, H.T.; Gillespie, H.B.; Weisshaus, S.Z. The Action of Formaldehyde on Amines and Amino Acids[1], **1933**, *55* (11), 4571–4587. DOI: 10.1021/ja01338a041

9. i) Curtius, Th. Ueber Stickstoffwasserstoffsäure (Azoimid) N_3H. *Ber. Dtsch. Chem. Ges.*, **1890**, *23* (2), 3023–3033. DOI: 10.1002/cber.189002302232
   ii) Curtius, Th. 20. Hydrazide und Azide organischer Säuren I. Abhandlung. *J. Prakt. Chem.*, **1894**, *50* (1), 274–294. DOI: 10.1002/prac.18940500125

10. i) L'Abbé, G. Decomposition and addition reactions of organic azides. *Chem. Rev.*, **1969**, *69* (3), 345–363. DOI: 10.1021/cr60259a004
    ii) Rauk, A. and Alewood, P.F. A theoretical study of the Curtius rearrangement. The electronic structures and interconversions of the CHNO species. *Can. J. Chem.*, **1977**, *55* (9), 1498–1510. DOI: 10.1139/v77-209

11. Johnson, S.L.; Morrison, D.L. Kinetics and mechanism of decarboxylation of *n*-arylcarbamates. Evidence for kinetically important zwitterionic carbamic acid species of short lifetime. *J. Am. Chem. Soc.*, **1972**, *94* (4), 1323–1334. DOI: 10.1021/ja00759a045

12. i) von Hofmann, A.W. XIV. Researches into the molecular constitution of the organic bases. *Phil. Trans. R. Soc.*, **1851**, *141*, 357–398. DOI: 10.1098/rstl.1851.0017
    ii) von Hofmann, A.W. Beiträge zur Kenntniss der flüchtigen organischen Basen. *Liebigs Ann.*, **1851**, *78* (3), 253–286. DOI: 10.1002/jlac.18510780302

13. Cockerill, A.F. Chapter 3 Elimination Reactions. In *Comprehensive Chemical Kinetics*. Volume 9.; Elsevier, 1973. DOI: 10.1016/S0069-8040(08)70194-8

24 Conjugated Diene Chemistry

In Chapter 12, we considered the chemistry of alkenes. In that chapter, we focused almost exclusively on molecules that had only a single double bond. In this chapter, we will consider molecules that have two double bonds (dienes) or more (polyenes). For a molecule that has two double bonds, those double bonds can be one right after the other (cumulated dienes, Figure 24.1). Cumulenes will not be a significant focus of this chapter, but they have some unique properties both chemically and structurally. The double bonds in a molecule can be on adjacent pairs of atoms (Figure 24.2). We call this configuration a conjugated diene. We have discussed conjugation previously in Chapter 2. The important point to remember is that conjugation implies that the electrons are not localized, and we show the delocalization of electrons with contributing structures. The final configuration for double bonds in a diene is non-conjugated double bonds (Figure 24.3). This is where the double bonds are separated by one or more sp^3-hybridized atoms. Non-conjugated dienes behave as we would expect any alkene to behave from our previous work with alkenes and their chemistry will not be considered further in this chapter.

ALKENE STABILITY

There are several methods that we might consider for evaluating the behaviour of the different types of diene molecules mentioned above. One of the most straightforward, experimental data we might consider are the enthalpy of reaction ($\Delta_r H°$) associated with hydrogenation of these molecules to alkanes (Figure 24.4).[1] Notice that $\Delta_r H° < 0$ for each of these reactions, which indicates an exothermic reaction. Alkanes (with all single bonds) are more stable than the analogous alkene because σ bonds are lower in energy than π bonds.

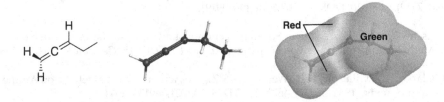

Figure 24.1 Structure (*left*), 3D model (*middle*), and electrostatic potential map (*right*) of an example cumulated diene (cumulene): penta-1,2-diene (red corresponds high electron density and blue corresponds to low electron density).

Figure 24.2 Structure (*left*), 3D model (*middle*), and electrostatic potential map (*right*) of an example conjugated diene: penta-1,3-diene (red corresponds to high electron density and blue corresponds to low electron density).

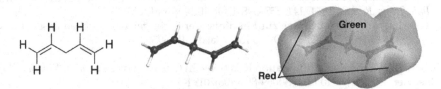

Figure 24.3 Structure (*left*), 3D model (*middle*), and electrostatic potential map (*right*) of an example non-conjugated diene: penta-1,4-diene (red corresponds to high electron density and blue corresponds to low electron density).

DOI: 10.1201/9781003479352-24

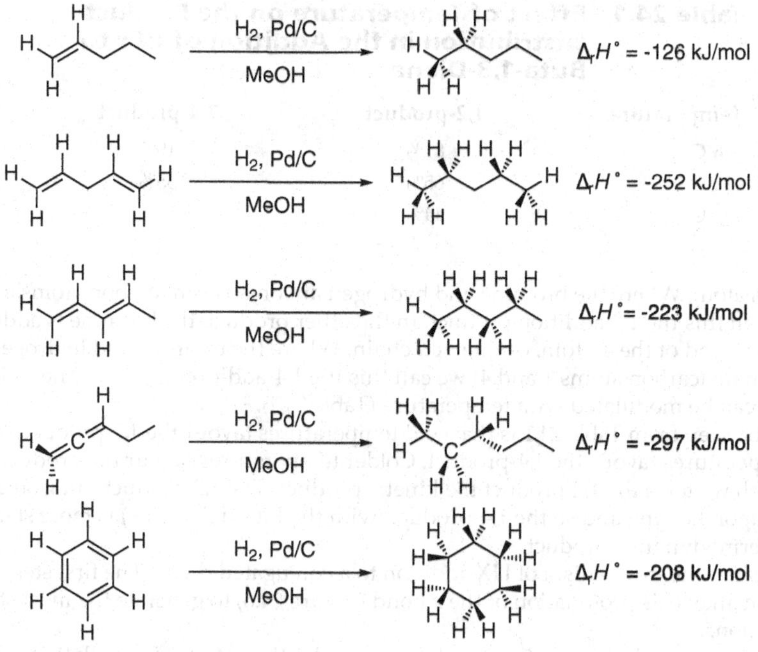

Figure 24.4 Enthalpy of hydrogenation (in kJ/mol) of selected alkenes (*in order from top to bottom*): penta-1-ene, penta-1,4-diene, penta-1,3-diene, penta-1,2-diene, and benzene.

Looking at Figure 24.4, the enthalpy of reaction for pent-1-ene, the hydrogenation of a mono-substituted monoalkene, is –126 kJ/mol. Looking now at penta-1,4-diene, a non-conjugated diene, the enthalpy of reaction is –252 kJ/mol, which is exactly twice the enthalpy of reaction of pent-1-ene. This contrasts with penta-1,3-diene, where the enthalpy of reaction is –223 kJ/mol. The enthalpy of reaction for penta-1,3-diene is 29 kJ/mol less than we would expect or put another way the conjugated diene is 29 kJ/mol more stable than two, non-conjugated π bonds. As we continue our analysis, we can see that location is incredibly important. Consider penta-1,2-diene, where the enthalpy of reaction is –297 kJ/mol, which is 45 kJ/mol greater than we would expect. So, while conjugated π bonds are more stable, cumulated π bonds are significantly less stable.

We will consider the significance of this in Chapter 25 but let us wrap up this enthalpy of hydrogenation analysis by considering benzene. The enthalpy of reaction is –208 kJ/mol for a molecule with 3, putative, π bonds. We would expect, given three π bonds, that the enthalpy of reaction should be –378 kJ/mol. We can infer, then, that benzene is 170 kJ/mol more stable than three, nonconjugated π bonds. This exceptional stability is a hallmark of benzene, and we will consider this stability in Chapter 25.

CONJUGATED DIENE ADDITION REACTIONS

Conjugation increases the stability of dienes, and conjugation impacts the type of chemistry that dienes undergo. Let us consider two reactions that we are familiar with: HX addition and X_2 addition. For HX addition, if we add one equivalent of HX (HCl, HBr, or HI) to a conjugated diene then we get out different possible products (Figure 24.5).

The two products are differentiated by the relative position of the bromine atom. In the first product, the bromine was added to the carbon atom immediately adjacent to that which grabbed

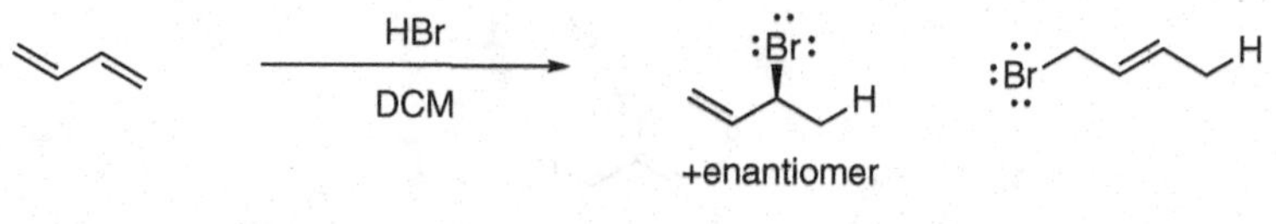

Figure 24.5 Possible products from the addition of HBr to buta-1,3-diene to give the 1,2-addition product (*left*) and 1,4-addition product (*right*).

Table 24.1 **Effect of Temperature on the Product Distribution in the Addition of HBr to Buta-1,3-Diene**

Temperature	1,2-product	1,4-product
–78°C	81%	19%
–12°C	65%	35%
–25°C	44%	56%

the hydrogen atom. Where the bromine and hydrogen are on adjacent carbon atoms (carbon atoms 1 and 2) we call this the 1,2-addition product. In the other product, the bromine is added to the carbon atom at the end of the 4-atom, conjugated chain. Where the bromine and hydrogen are on the ends of the chain (carbon atoms 1 and 4) we call this the 1,4-addition product. The ratio of 1,2- and 1,4-products can be modulated with temperature (Table 24.1). [2]

What can be seen from Table 24.1 is that cold temperatures favour the 1,2-product and that warmer temperatures favour the 1,4-product. Colder temperatures favour the product that forms fastest, as such we term the 1,2-product the kinetic product. The 1,2-product can isomerize to the 1,4-product upon heating and so the 1,4-product (with the internal alkene) is more stable and is called the thermodynamic product.

Let us consider the mechanism of HX addition to a conjugated diene. The first step, as with HX addition to an alkene, is protonation of the π bond (Figure 24.6) to generate the most stable, possible carbocation.

Before moving on with this mechanism, let us consider the nature of the allylic cation intermediate a little further. There are two good contributing structures that we can draw for this allylic cation (Figure 24.7). The first contributing structure is the one that we would expect from the electron-pushing arrows shown in Figure 24.6. This contributing structure shows a terminal alkene and a secondary carbocation. The other contributing structure (Figure 24.7) has an internal alkene and a primary carbocation.

From the consideration of the carbocation, the contributing structure with the more substituted carbocation should be a more significant contributing structure (that is the actual structure of the molecule is more like that contributing structure than the other). And so, the actual structure of the molecule (if we take a weighted average of the two) would be something akin to the hybrid structure in Figure 24.8. That is there is a partial double bond across all three carbon atoms and the two terminal atoms have some positive charge (the δ symbol indicates partial charge). What is important to note is that the distribution of positive charge is higher on the internal carbon and lower on the terminal carbon, which makes the internal carbon more electrophilic and the site most likely to be attacked.

Given this understanding of the allylic cation, let us consider the rest of the mechanism for HX addition. The first product to be produced is the 1,2-addition product, which results from

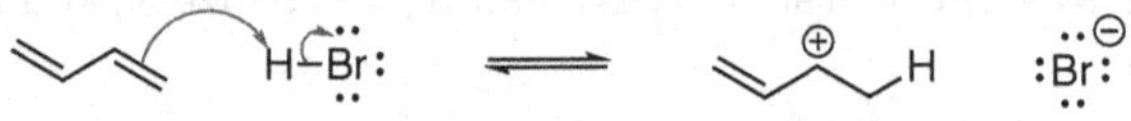

Figure 24.6 First step of HX addition: protonation of the diene to generate an allylic cation.

more significant   less significant

Figure 24.7 Two contributing structures for the allylic cation produced from buta-1,3-diene protonation.

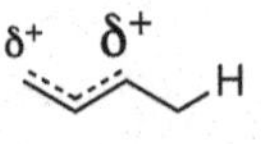

Figure 24.8 Hybrid structure for the allylic cation produced from buta-1,3-diene protonation, which shows the unequal distribution of positive charge.

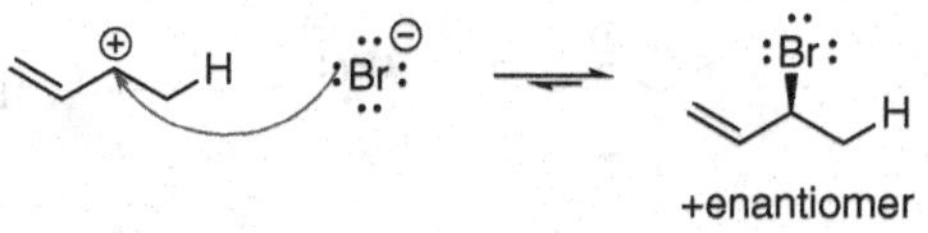

+enantiomer

Figure 24.9 Second step of HX addition: formation of the 1,2-addition product by nucleophilic attack on the more substituted, more electrophilic carbon atom.

nucleophilic attack by the bromide on the more substituted, more electrophilic carbon atom (Figure 24.9). Under cold conditions, this is the major pathway for the reaction, which is why 1,2-addition products constitute most of the product produced.

For HX addition, the 1,4-addition product requires warmer temperatures because it requires the dissociation and recombination of the allylic bromide. That is formation of the 1,4-addition product is the result of isomerization of the 1,2-addition product to the more stable (more highly substituted alkene) 1,4-addition product (Figure 24.10).

For HX addition to a diene, then, we need to consider three different pieces of information: 1) when the allylic cation is formed, what are the two electrophilic sites and which site is more electrophilic? 2) which attachment site will give the most stable alkene product? 3) what are the conditions (cold or hot), and which product will predominate? The following questions will help to build up this approach to understanding the reaction.

Problem 24.1 Draw the carbocation intermediate that would result from protonation of each diene. Draw the two contributing structures for each and identify the more electrophilic carbon atom.

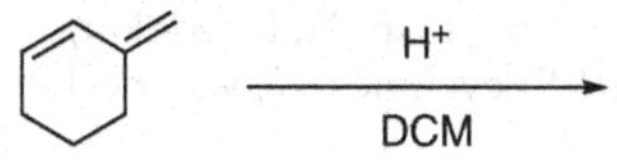

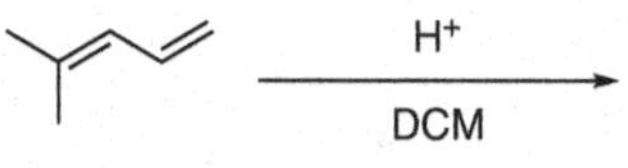

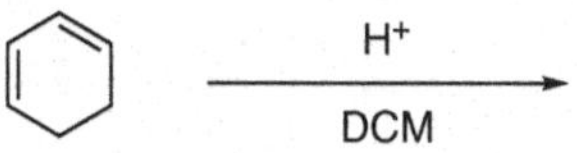

Problem 24.2 Draw the two possible products that would be produced from each reaction. Which product would form fastest? Which would be more stable?

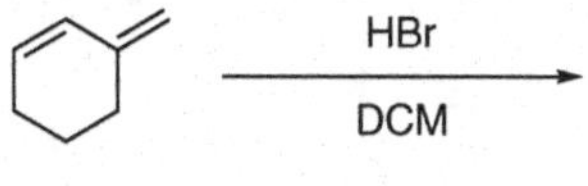

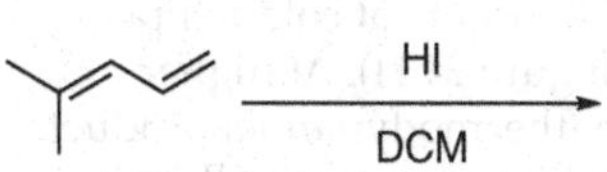

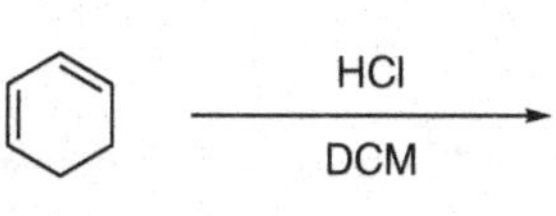

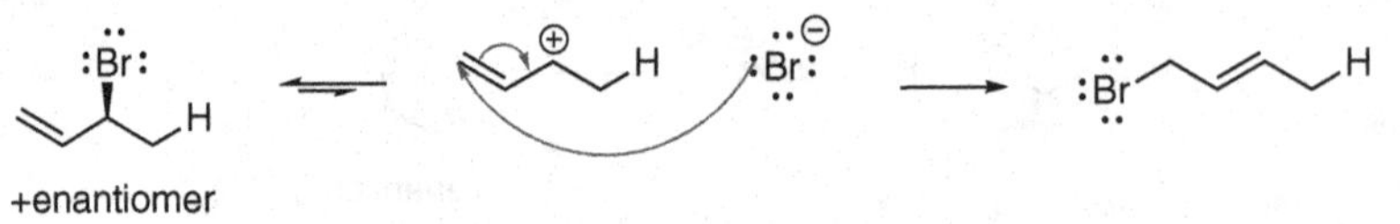

Figure 24.10 Isomerization of the 1,2-addition product to the 1,4-addition product.

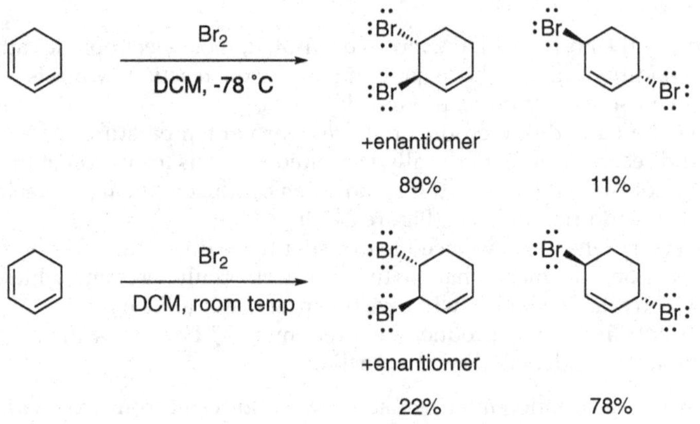

Figure 24.11 Addition of bromine to cyclohexa-1,3-diene at cold temperatures (*top*) and warm temperatures (*bottom*) and the effect of temperature on product distribution. Han, X.; Khedekar, R.N.; Masnovi, J.; Baker, R.J. Bromination of cyclohexa-1,3-diene and (*R,S*)-cyclohexa-3,5-diene-1,2-diol. *J. Org. Chem.*, **1999**, *64* (14), 5245–5250. DOI: 10.1021/jo990473s

Problem 24.3 With the conditions provided, predict the major product(s).

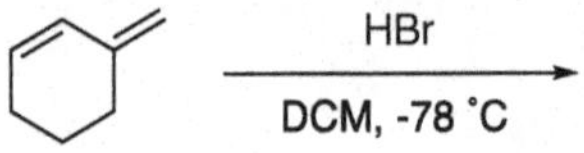

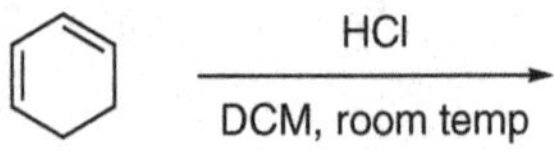

The addition of one equivalent of X_2 (Cl_2 or Br_2) to a diene is a very similar story to that seen for HX addition to an alkene.[3] As with the addition of HX, the addition of X_2 occurs at cold temperatures (–78 °C) to give preferentially the 1,2-addition (kinetic) product (Figure 24.11). At higher temperatures (room temperature) the major product is the 1,4-addition (thermodynamic) product.

The mechanism of the addition of X_2 to a diene to produce the 1,2-addition product will look very familiar from Chapter 12 (Figure 24.12).

The thermodynamic product of X_2 addition to a diene is produced through isomerization of the kinetic product (Figure 24.13).

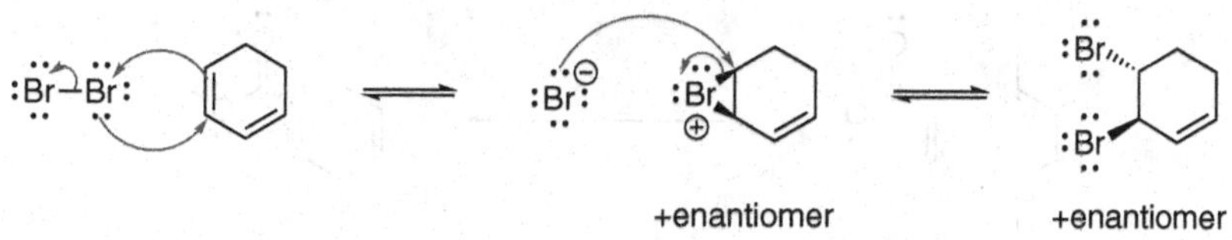

Figure 24.12 Mechanism for the addition of bromine to cyclohexa-1,3-diene to produce the kinetic (1,2-addition) product.

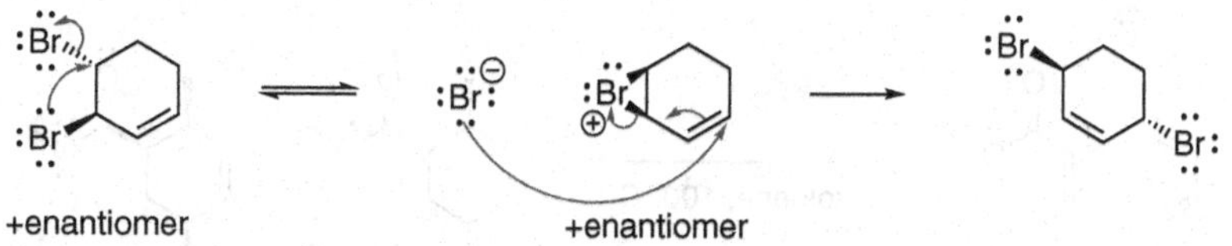

Figure 24.13 Mechanism for isomerization of kinetic (1,2-addition) product to the thermodynamic product in Br_2 to cyclohexa-1,3-diene.

Problem 24.4 Predict the major product(s) of each of the following reactions.

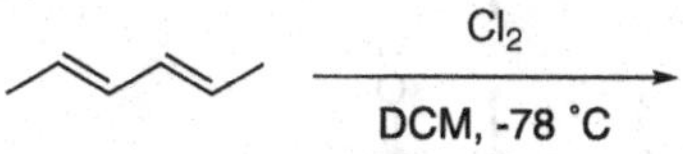

PERICYCLIC REACTIONS – DIELS–ALDER

In this second half of the chapter, we will consider reaction types that are entirely without precedent in our previous work. These reactions are pericyclic reactions, that is reactions that involve the concerted, cyclical flow of electrons. These reactions are exceptionally green reactions and can frequently be run without added solvent in addition to having 100% atom economy. We will first look at the Diels-Alder reaction. The Diels–Alder reaction (Figure 24.14) is a cycloaddition (that is a

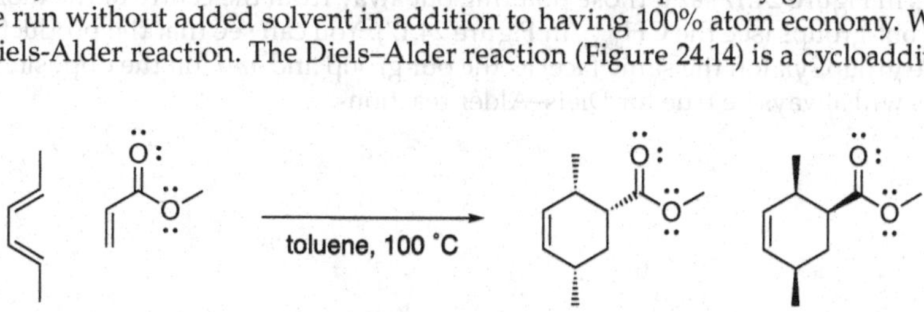

Figure 24.14 Example Diels–Alder reaction of (2E,4E)-hexa-2,4-diene and methyl acrylate.

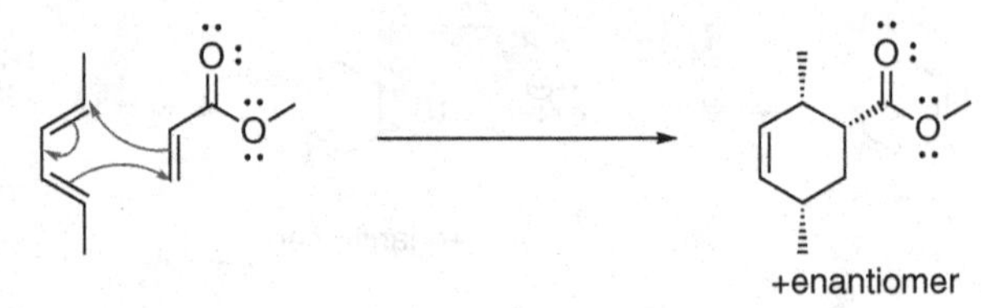

+enantiomer

Figure 24.15 Diels–Alder reaction mechanism.

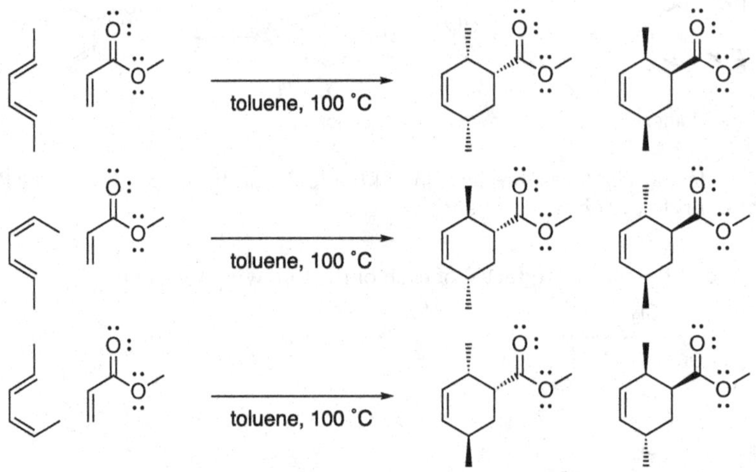

Figure 24.16 Impact of diene geometry on Diels–Alder product stereochemistry.

reaction that adds two molecules (or portions of a molecule) together to form a ring).[4] The reaction involves a conjugated diene being heated together with a dienophile (an alkene with an electron-withdrawing group attached).

The mechanism for the Diels–Alder reaction involves a single step showing the concerted, cyclical flow of electrons (Figure 24.15).

The Diels–Alder reaction is incredibly useful in synthesis because it takes two different molecules (or two different pieces of the same molecule) and builds a cyclohexene ring. Moreover, this is accomplished stereoselectively. Notice in Figure 24.13 that the methyl groups and ester substituent are all syn (on the same face) of the cyclohexene ring. The syn relationship between the methyl groups stems from their initial relative position (both alkene geometries of the diene were *E*), and in the single step of the mechanism there is no opportunity for the relative geometry to change. Consider Figure 24.16 to see how the initial geometry of the methyl groups impacts their stereochemistry in the product.

In Figure 24.16, we can see that the stereochemical relationship between groups on that had been on the end of the diene is determined by the relative geometry of those groups. Let us consider the last example more closely (Figure 24.17). There are two different positions ascribed to diene groups. Those that point towards the middle of the molecule are referred to as in-groups (see the $CH_{3(in)}$ in Figure 24.17) and those pointing out away from the centre of the molecule are referred to as out-groups (see the $CH_{3(out)}$ in Figure 24.17). You can see that the product always has the dienophile group *syn* (on the same face) as the out-group and *anti* (on the opposite face) as the in-group. This will always be true for Diels–Alder reactions.

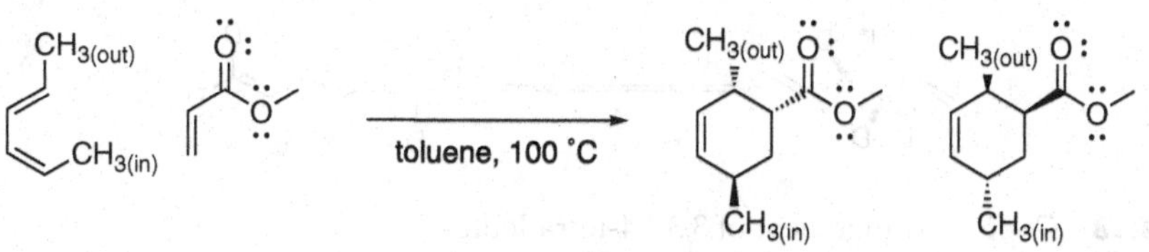

Figure 24.17 Diels–Alder reaction with in and out designations explicit.

Problem 24.5 Predict the major product(s) of each of the following reactions.

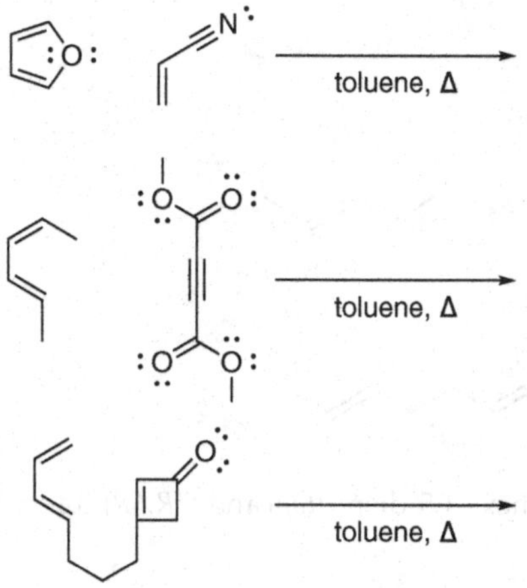

PERICYCLIC REACTIONS – SIGMATROPIC SHIFT

The final type of reaction that we will consider is a sigmatropic shift. A sigmatropic shift involves a non-conjugated hexa-1,5-diene (or related structures), in which the location of σ-bonds and π-bonds shifts. The first sigmatropic reaction to consider is the Cope rearrangement (Figure **24.18**) involving hexa-1,5-dienes.[5]

In the Cope rearrangement, the molecule must fold up into a cyclohexane-like structure. Understanding this folding arrangement is key to understanding the stereochemistry of this reaction. The mechanism, like the Diels–Alder, involves the cyclic flow of electrons (Figure 24.19).

Consider the following two Cope rearrangements (Figure 24.20). These reactions are product-favoured because they produce more substituted (more stable) alkenes. Let us consider how to understand the double-bond geometries that result.

To elucidate the structures of these rearrangement products, we need to think of the reactants folding up like the lowest energy cyclohexane-ring-like conformation (Figure 24.21). It is from this perspective that we can see we (3R,4R)-3,4-dimethylhexa-1,5-diene ends up with both alkenes E and (3R,4S)-3,4-dimethylhexa-1,5-diene ends up with one alkene E and one alkene Z.

Problem 24.6 Predict the major product(s) of each of the following reactions.

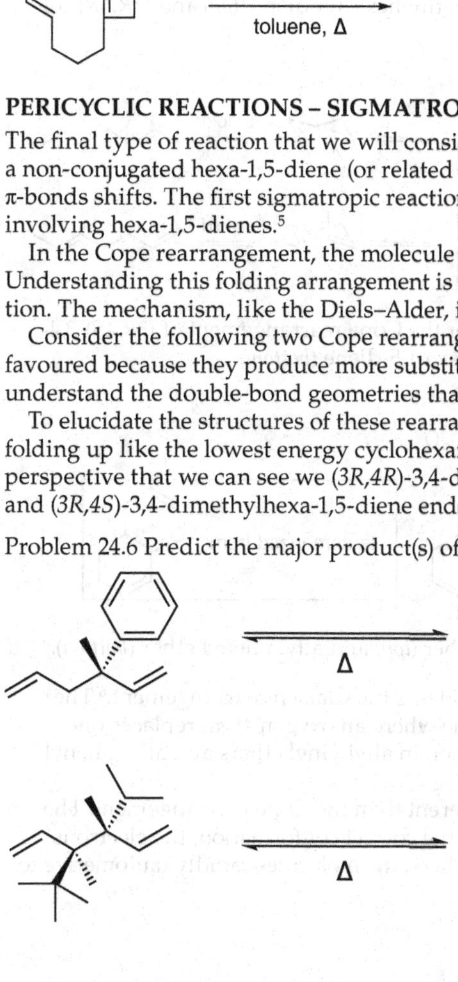

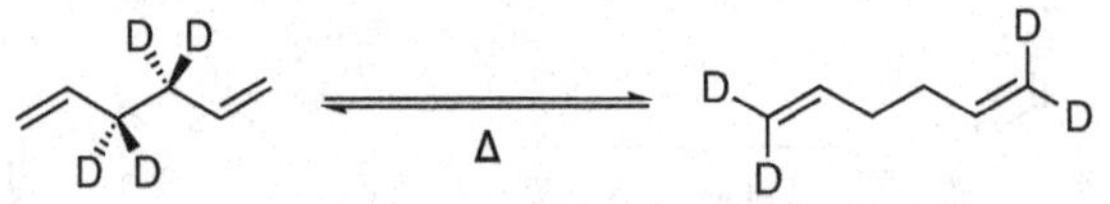

Figure 24.18 Cope rearrangement of 3,3,4,4-tetradeute rohexa-1,5-diene.

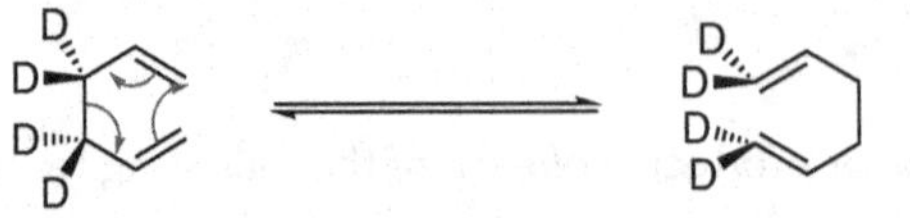

Figure 24.19 Cope rearrangement mechanism.

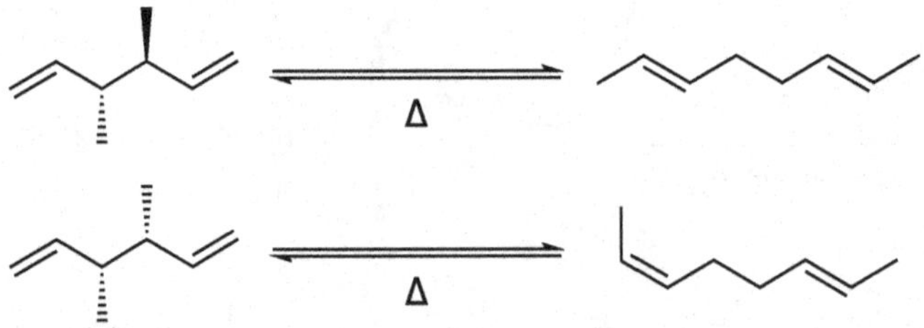

Figure 24.20 Cope rearrangement of (3*R*,4*S*)-3,4-dimethylhexa-1,5-diene (*top*) and (3*R*,4*R*)-3,4-dimethylhexa-1,5-diene (*bottom*).

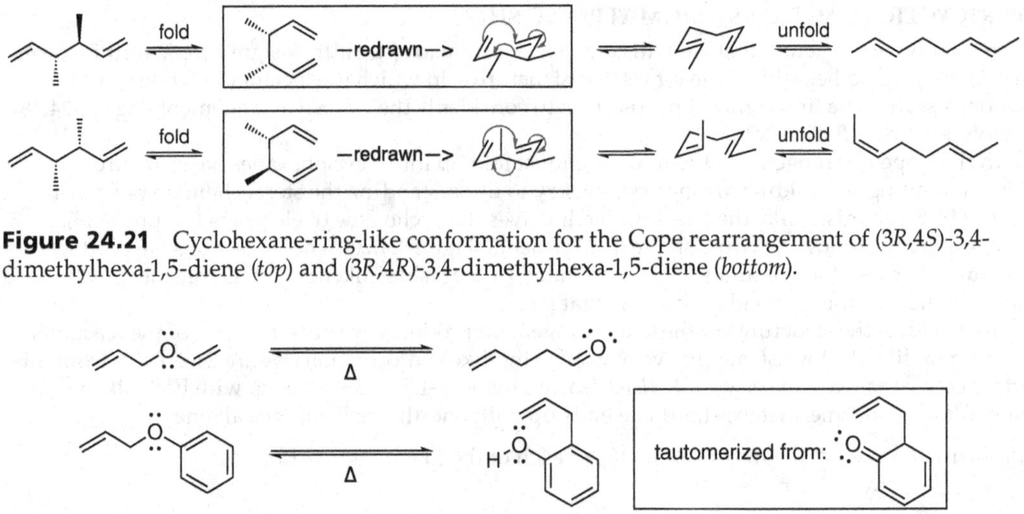

Figure 24.21 Cyclohexane-ring-like conformation for the Cope rearrangement of (3*R*,4*S*)-3,4-dimethylhexa-1,5-diene (*top*) and (3*R*,4*R*)-3,4-dimethylhexa-1,5-diene (*bottom*).

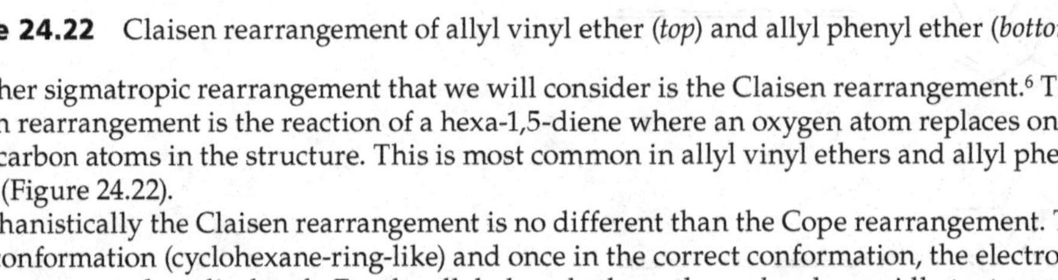

Figure 24.22 Claisen rearrangement of allyl vinyl ether (*top*) and allyl phenyl ether (*bottom*).

The other sigmatropic rearrangement that we will consider is the Claisen rearrangement.[6] The Claisen rearrangement is the reaction of a hexa-1,5-diene where an oxygen atom replaces one of the carbon atoms in the structure. This is most common in allyl vinyl ethers and allyl phenyl ethers (Figure 24.22).

Mechanistically the Claisen rearrangement is no different than the Cope rearrangement. The same conformation (cyclohexane-ring-like) and once in the correct conformation, the electrons flow in a concerted, cyclical path. For the allyl phenyl ethers, the molecules rapidly tautomerize to re-establish the benzene ring structure.

Problem 24.7 Predict the major product(s) of each of the following reactions.

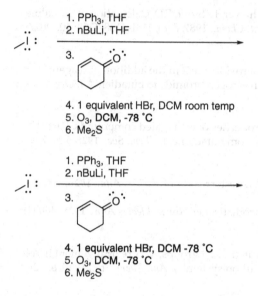

GENERAL PRACTICE PROBLEMS

Problem 24.8 Predict the major product of the following multi-step synthesis.

1. PPh$_3$, THF
2. nBuLi, THF

3.

4. 1 equivalent HBr, DCM room temp
5. O$_3$, DCM, -78 °C
6. Me$_2$S

1. PPh$_3$, THF
2. nBuLi, THF

3.

4. 1 equivalent HBr, DCM -78 °C
5. O$_3$, DCM, -78 °C
6. Me$_2$S

Problem 24.9 Provide the necessary reagents to complete the synthesis.

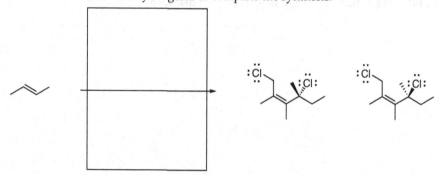

Problem 24.10 Conjugation allows for reaction pathways that are otherwise unavailable to molecules. For example, in Chapter 16 we looked at the epoxidation of alkenes using *m*CPBA (an electrophilic epoxidation reaction). For cyclohexenone, the reaction with *m*CPBA does not produce any products. Provide a possible explanation for why this reaction fails.

$$\text{cyclohexenone} \xrightarrow[\text{benzene}]{m\text{CPBA}} \text{no reaction}$$

Problem 24.11 To epoxidize α,β-unsaturated aldehydes or ketones, nucleophilic epoxidation conditions must be employed. Provide a reasonable electron-pushing mechanism to account for this transformation.

NOTES

1. i) Kistiakowsky, G.B.; Ruhoff, J.R.; Smith, H.A.; Vaughan, W.E. Heats of organic reactions. IV. Hydrogenation of some dienes and of benzene. *J. Am. Chem. Soc.*, **1936**, *58* (1), 146–153. DOI: 10.1021/ja01292a043

   ii) Ibrahim, M.R.; Fataftah, Z.A.; von Ragué Schleyer, P.; Stout, P.D. Calculation of enthalpies of hydrogenation of hydrocarbons. *J. Comput. Chem.*, **1987**, *8* (8), 1131–1138. DOI: 10.1002/jcc.540080810

2. Kharasch, M.S.; Margolis, E.T.; Mayo, F.R. The peroxide effect in the addition of reagents to unsaturated compounds. XIII. The addition of hydrogen bromide to butadiene*. *J. Org. Chem.*, **1936**, *01* (4), 393–404. DOI: 10.1021/jo01233a008

3. Farmer, E.H.; Lawrence, C.D.; Thorpe, J.F. C.–Properties of conjugated compounds. Part IV. The formation of isomeric additive dibromides from butadiene. *J. Chem. Soc.*, **1928**, 729–739. DOI: 10.1039/JR9280000729

4. i) Diels, O.; Alder, K. Über die Ursachen der "Azoesterreaktion". *Liebigs Ann.*, **1926**, *450* (1), 237–254. DOI: 10.1002/jlac.19264500119

   ii) Diels, O.; Alder, K. Synthesen in der hydroaromatischen Reihe. *Liebigs Ann.*, **1928**, *460* (1), 98–122. DOI: 10.1002/jlac.19284600106

5. Cope, A.C.; Hardy, E.M. Introduction of substituted vinyl groups. V. A rearrangement involving the migration of an allyl group in a three-carbon system[1]. *J. Am. Chem. Soc.*, **1940**, *62* (2), 441–444. DOI: 10.1021/ja01859a055

6. Claisen, L. Über Umlagerung von Phenol-allyläthern in C-Allyl-phenole. *Ber. Dtsch. Chem. Ges.*, **1912**, *45* (3), 3157–3166. DOI: 10.1002/cber.19120450348

25 Aromaticity

The uniqueness of benzene (Figure 25.1) has been alluded to at several points throughout this book. Benzene has been known since 1825, and both its empirical formula (C_1H_1) and unique properties were identified during its discovery.[1] After its discovery, there were many attempts to correctly deduce a structure for benzene. The correct structure for benzene is generally attributed to August Kekulé,[2] though our modern understanding of benzene's structure and the reasons for its stability would not be determined until 60 years later by Hückel.

BENZENE STRUCTURE

Looking at the electrostatic potential map (Figure 25.1), we can see a continuous ring of high electron density around the middle. This is a result of the delocalization of π bond electrons around the ring, which also means that the benzene ring has completely equal C–C bond lengths (139.5 pm) and C–H bond lengths (108 pm). To show this delocalization we can show two contributing structures for benzene (Figure 25.2), and a common shorthand[3] is to show the hexagonal ring shape with a circle drawn in the middle (Figure 25.3) with the circle representing the aromatic sextet (six delocalized π bond electrons).[4] Throughout this chapter we will draw one or the other contributing structure and not the aromatic sextet depiction; however, the aromatic sextet depiction is often encountered in a variety of contexts.

BENZENE REACTIVITY

As has been alluded to, benzene's chemistry is unique. Consider the reaction of bromine with unsaturated hydrocarbons (Figure 25.4). In every example except benzene, bromine adds to the C=C double bond. As shall be seen in Chapter 26, bromine does not undergo any sort of addition reaction, rather it undergoes substitution reactions.

AROMATICITY

Also, as was seen in Chapter 24 Conjugated dienes, the hydrogenation of benzene is 170 kJ/mol less exothermic than expected for a triene. Or put in positive terms, benzene is 170 kJ/mol more

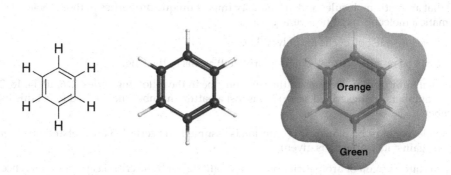

Figure 25.1 Structure (*left*), 3D model (*middle*), and electrostatic potential map (*right*) of benzene (red corresponds to high electron density and blue corresponds to low electron density).

Figure 25.2 Contributing structures of benzene.

Figure 25.3 Aromatic sextet depiction of benzene.

DOI: 10.1201/9781003479352-25

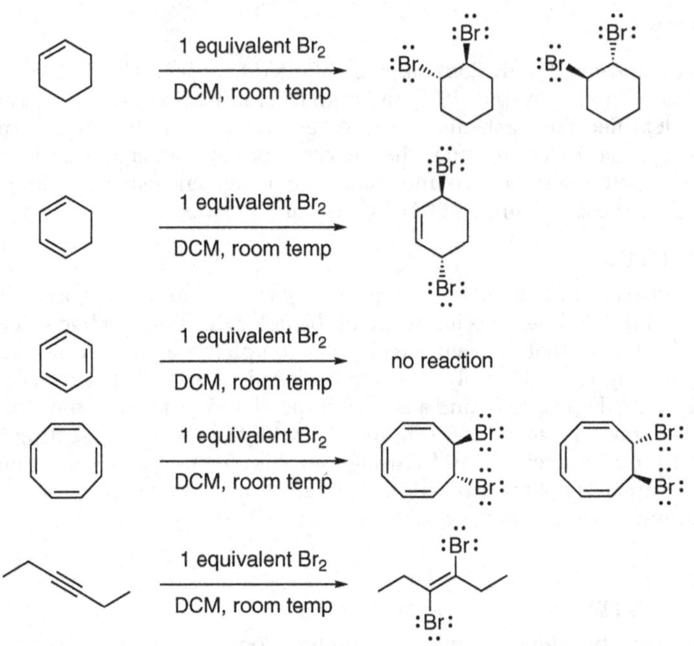

Figure 25.4 Reaction of bromine with unsaturated hydrocarbons.

stable than we would expect for a molecule with three double bonds. Given all of this, how do we explain this behaviour of benzene? Benzene is an example of an aromatic molecule – the term aromaticity was first used to differentiate aliphatic carboxylic acids from those with a benzene core.[5] Aromatic molecules show significant stabilization from the cyclic delocalization of certain numbers of electrons.[6] The foundational reasons for this stability are not the focus of this chapter. Instead, we will focus our attention on 1) being able to identify aromatic molecules and 2) recognizing that aromatic molecules and aromaticity impart unique properties to those molecules. To be aromatic a molecule must fit several criteria:

1. The molecule must be cyclic (or polycyclic).

2. There can be no fully saturated (sp³-hybridized) atoms in the ring.

3. The total number of π electrons in the ring must be in the following series: 2, 6, 10, 14, 18, 22, etc. These numbers are described as $4n+2$ (n is just a natural number that is used to get this list of numbers).[7]

4. The molecule must be planar (this criterion is assumed if criteria 1–3 are met and no other indication against its planarity is given).

Benzene is our paragon of aromaticity but let us walk through the criteria above in reference to the structure of benzene (Figure 25.5). As we can see in Figure 25.5, benzene meets all the required criteria to be aromatic.

While benzene is an aromatic molecule, it is not the only aromatic molecule. There are other hydrocarbons that are aromatic (Figure 25.6). Review the examples paying particular attention to the π electron counts.

Heterocycles (molecules that have other atoms besides carbon in the ring) can also be aromatic. Figure 25.7 walks through some examples of aromatic heterocycles. It is important to note that an atom with a lone pair that is making a π bond cannot have both its π bond and its lone pair in conjugation (the orbital requirements make the π bond and lone pair perpendicular to one another).

Figure 25.5 Benzene and checking the aromatic criteria.

1. Cyclic? Yes, toluene is a 6-membered ring.
2. No sp^3 atoms in the ring? Yes, all atoms are sp^2 hybridized (the sp^3 methyl group is not part of the ring)
3. 4n+2 π electrons? Yes, there are 3 π bonds, which totals 6 π electrons
4. Planar? Yes, all atoms are sp^2-hybridized and trigonal planar

1. Cyclic? Yes, naphthalene is a fused ring system of 10 atoms
2. No sp^3 atoms in the ring? Yes, all atoms are sp^2 hybridized.
3. 4n+2 π electrons? Yes, there are 5 π bonds, which totals 10 π electrons
4. Planar? Yes, all atoms are sp^2-hybridized and trigonal planar

1. Cyclic? Yes, cyclopentadienyl anion is a 5-membered ring.
2. No sp^3 atoms in the ring? Yes, all atoms are sp^2 hybridized.
3. 4n+2 π electrons? Yes, there are 2 π bonds and 1 lone pair, which totals 6 π electrons
4. Planar? Yes, all atoms are sp^2-hybridized and trigonal planar

Figure 25.6 Toluene (*top*), napthalene (*middle*), and cyclopentadienyl anion (*bottom*) with their aromaticity criteria.

1. Cyclic? Yes, pyridine is a 6-membered ring.
2. No sp^3 atoms in the ring? Yes, all atoms are sp^2 hybridized
3. 4n+2 π electrons? Yes, there are 3 π bonds, which totals 6 π electrons (the nitrogen lone pair electrons cannot count because nitrogen is already making a π bond)
4. Planar? Yes, all atoms are sp^2-hybridized and trigonal planar

1. Cyclic? Yes, pyrrole is a 5-membered ring
2. No sp^3 atoms in the ring? Yes, all atoms are sp^2 hybridized.
3. 4n+2 π electrons? Yes, there are 2 π bonds and 1 lone pair, which totals 6 π electrons (the nitrogen lone pair does count because the nitrogen is not already making a π bond)
4. Planar? Yes, all atoms are sp^2-hybridized and trigonal planar

1. Cyclic? Yes, furan is a 5-membered ring.
2. No sp^3 atoms in the ring? Yes, all atoms are sp^2 hybridized.
3. 4n+2 π electrons? Yes, there are 2 π bonds and 1 lone pair, which totals 6 π electrons (only one oxygen lone pair can count towards the ring electron total because lone pair cannot occupy the same volume of space. As such, if one is a p-orbital lone pair that counts, the other must be, regardless of hybridization, perpendicular)
4. Planar? Yes, all atoms are sp^2-hybridized and trigonal planar

Figure 25.7 Pyridine (*top*), pyrrole (*middle*), furan (*bottom*) with their aromaticity criteria.

As such, the lone pair on an atom with a π bond will never count towards the π electron total. Similarly, if an atom has two lone pairs, only one can be in conjugation with the ring, and so one lone pair will count towards the π electron count, and one cannot and will not.

NONAROMATIC MOLECULES

We have now seen a wide variety of molecules that are aromatic. It is worth remembering that most molecules we have seen so far in previous chapters (and a supermajority overall) are non-aromatic. We have seen lots of examples of molecules that are nonaromatic. Figure 25.8 provides some examples of molecules that are nonaromatic and highlights how they fail one (or more) of the aromatic criteria.

ANTIAROMATICITY

There is a special type of molecule that is nonaromatic. These are molecules that otherwise fit all the aromatic criteria, but where the total number of π electrons are in the following series: 4, 8, 12, 16, 20, etc. These numbers are described as 4n (n is just a natural number that is used to get the above list). If these molecules were to stay fully conjugated, they would be antiaromatic (that is especially unstable).[8] The antiaromatic state is a transition state (a high-energy state that exists for one molecular vibration and is represented by the brackets and ‡ notation in Figure 25.9) and these molecules exist in some distorted shape to disrupt conjugation. Because the antiaromatic state is

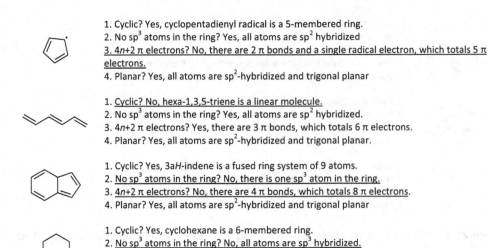

1. Cyclic? Yes, cyclopentadienyl radical is a 5-membered ring.
2. No sp³ atoms in the ring? Yes, all atoms are sp² hybridized
3. 4n+2 π electrons? No, there are 2 π bonds and a single radical electron, which totals 5 π electrons.
4. Planar? Yes, all atoms are sp²-hybridized and trigonal planar

1. Cyclic? No, hexa-1,3,5-triene is a linear molecule.
2. No sp³ atoms in the ring? Yes, all atoms are sp² hybridized.
3. 4n+2 π electrons? Yes, there are 3 π bonds, which totals 6 π electrons.
4. Planar? Yes, all atoms are sp²-hybridized and trigonal planar.

1. Cyclic? Yes, 3aH-indene is a fused ring system of 9 atoms.
2. No sp³ atoms in the ring? No, there is one sp³ atom in the ring.
3. 4n+2 π electrons? No, there are 4 π bonds, which totals 8 π electrons.
4. Planar? Yes, all atoms are sp²-hybridized and trigonal planar

1. Cyclic? Yes, cyclohexane is a 6-membered ring.
2. No sp³ atoms in the ring? No, all atoms are sp³ hybridized.
3. 4n+2 π electrons? No there are 0 π bonds and 0 lone pair.
4. Planar? No, all carbon atoms are tetrahedral geometry.

Figure 25.8 Nonaromatic examples and the aromatic criterion/criteria they fail have been underlined.

always avoided by molecular distortions any molecule that otherwise fits the aromatic criteria but has 4n π electrons should be identified as nonaromatic (Figure 25.9).

Problem 25.1 For each of the following molecules, identify the number of π electrons and determine whether the species is aromatic or is nonaromatic. If nonaromatic, state why it is nonaromatic.

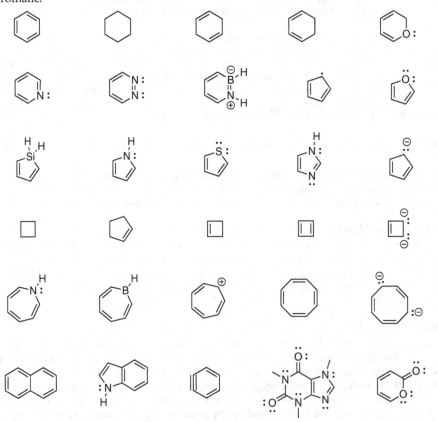

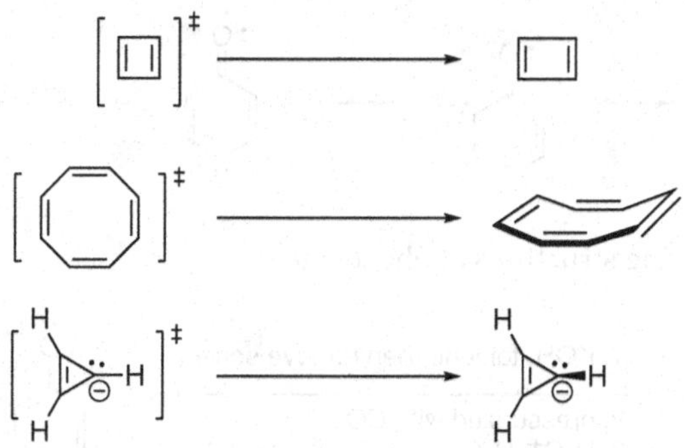

Figure 25.9 Molecules that would be antiaromatic if fully conjugated (a transition state indicated by brackets and ‡). These molecules distort into nonaromatic conformations (after the arrow).

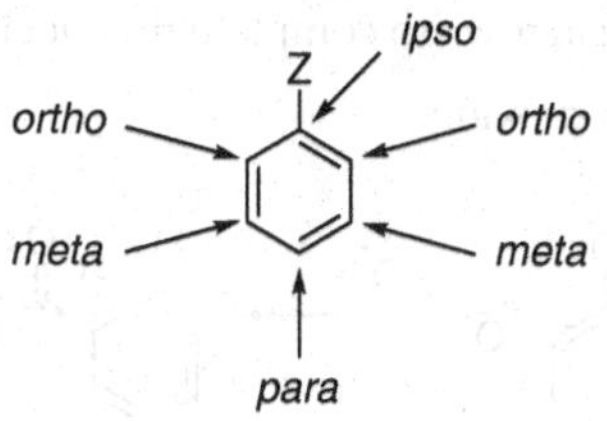

Figure 25.10 Benzene nomenclature for position relative to a substituent (Z).

BENZENE NOMENCLATURE

In Chapter 26, we will consider reactions involving benzene. It is worth noting that benzene has its own nomenclature system for referring to positions on the ring (Figure 25.10), relative to a given substituent (Z). The carbon with the substituent (Z, position one) is referred to as the *ipso* position (Latin "itself"). The positions next to the substituent (positions two and six) are called *ortho* (Greek "right"). The sites on the ring that are one removed from Z (position three and five) are called *meta* (Greek "after"). Finally, the spot exactly opposite Z on the ring (position four) is called *para* (Greek "at one side").

PHENOL

We will end this chapter by giving special attention to phenol (hydroxybenzene, Figure 25.11). Phenols are common in chemistry and biochemistry, and they are a useful bridge between some chemistry that we have seen in previous units and the chemistry of aromatic substitution that we will see in Chapter 26.

Compared to a typical alcohol (pK_a values near 16) a phenol is much more acidic (pK_a equals 9.95) due to the delocalization stabilization of the benzene ring (Figure 25.12). Phenols can be readily deprotonated (to form the phenoxide ion) and used as nucleophiles in several reactions including in S_N2 reactions (Williamson ether syntheses).

Looking more closely at phenol and phenoxide, they should look like an enol and an enolate, respectively. In this way we can understand how, and why, phenols and phenoxides are nucleophilic. We will consider only one reaction of phenols, the Kolbe–Schmitt carboxylation.[9] The Kolbe–Schmitt carboxylation of phenol itself produces salicylic acid (an acne medication and a precursor to aspirin) from phenoxide and carbon dioxide (Figure 25.13).

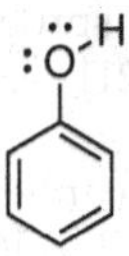

Figure 25.11 Phenol.

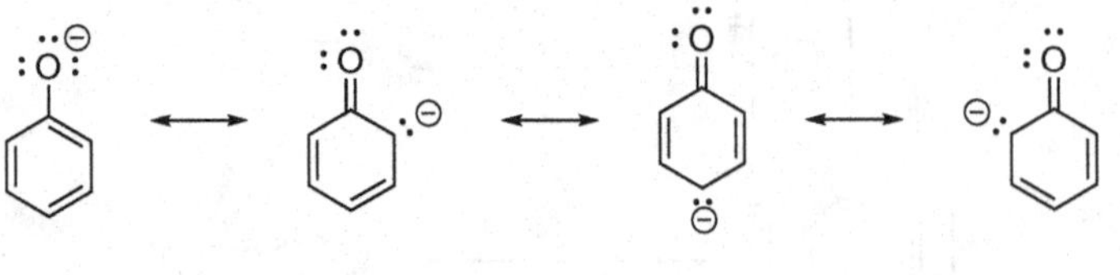

Figure 25.12 Contributing structures of phenoxide.

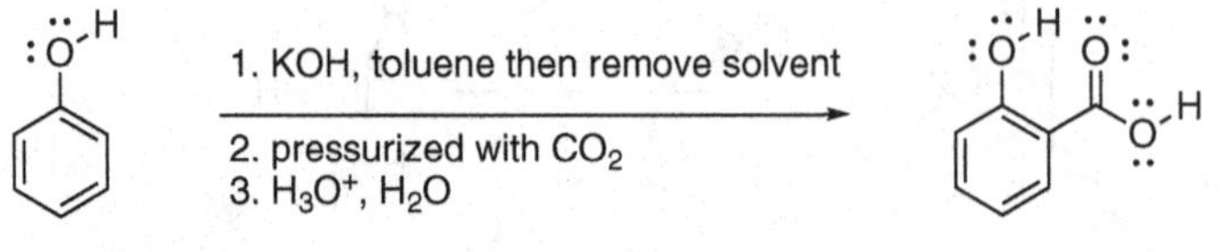

Figure 25.13 Kolbe–Schmitt carboxylation of phenol.

Problem 25.2 Draw the appropriate arrows to complete the mechanism of step 2.

First, nucleophilic attack on carbon dioxide

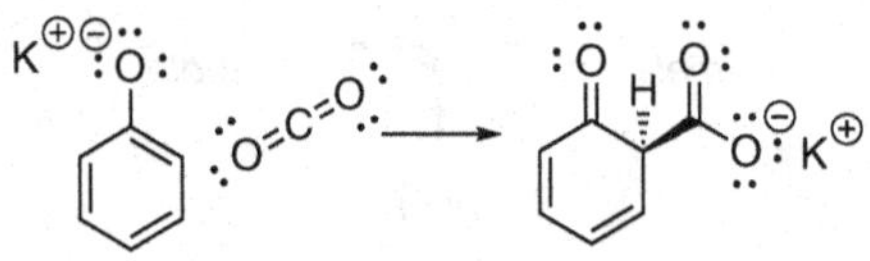

+enantiomer

Second, self-tautomerization (only one enantiomer shows, both follow the same steps).

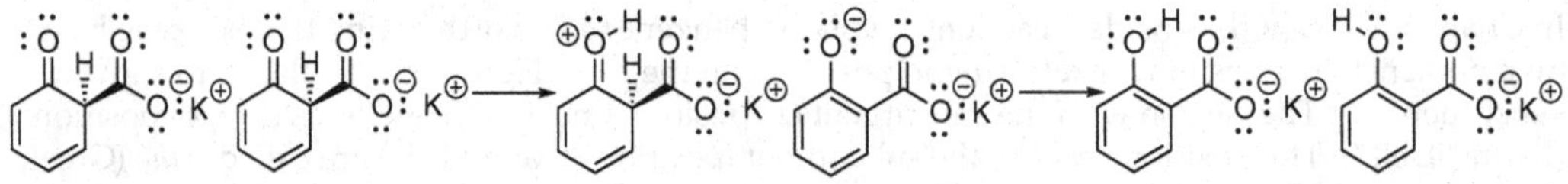

The Kolbe–Schmitt carboxylation is a rather specialized reaction; however, it is worth identifying some key points that will be important as we move forward. The reaction of phenoxide with carbon dioxide looks like the reaction of an enolate, but it is more important to think of it as a nucleophilic attack by a benzene ring (here made more nucleophilic by the electron-donating oxide). Also, unlike an enolate, the molecule quickly reestablishes the benzene ring structure. This will be a frequent trope in Chapter 26: whenever benzene acts as a nucleophile and aromaticity is disrupted, aromaticity will be re-established in a subsequent step.

NOTES

1. Faraday, M. XX. On new compounds of carbon and hydrogen, and on certain other products obtained during the decomposition of oil by heat. *Phil. Trans. R. Soc.*, **1825**, *115*, 440–466. DOI: 10.1098/rstl.1825.0022

2. i) Kekulé, A. Ueber einige Condensationsproducte des Aldehyds†. *Liebigs Ann.*, **1872**, *162* (1), 77–124. DOI: 10.1002/jlac.18721620110

   ii) Kekulé, A. Ueber einige Condensationsproducte des Aldehyds. *Liebigs Ann.*, **1872**, *162* (2–3), 309–320. DOI: 10.1002/jlac.18721620211

3. Armit, J.W.; Robinson, R. CCXI.–Polynuclear heterocylic aromatic types. Part II. Some anhydronium bases. *J. Chem. Soc., Trans.*, **1925**, *127*, 1604–1618. DOI: 10.1039/CT9252701604

4. Crocker, E.C. Application of the octet theory to single-ring aromatic compounds. *J. Am. Chem. Soc.*, **1922**, *44* (8), 1618–1630. DOI: 10.1021/ja01429a002

5. von Hofmann, A.W. I. On insolinic acid. *Proc. R. Soc. Lond.*, **1857**, *8*, 1–3. DOI: 10.1098/rspl.1856.0002

6. i) Hückel, E. Quantentheoretische Beiträge zum Benzolproblem I. Die Elektronenkonfiguration des Benzols und verwandter Verbindungen. *Z. Physik*, **1931**, *70*, 204–286. DOI: 10.1007/BF01339530

   ii) Hückel, E. Quantentheoretische Beiträge zum Benzolproblem II. Quantentheorie der induzierte Polaritäten. *Z. Physik*, **1931**, *72*, 310–337. DOI: 10.1007/BF01341953

   iii) Hückel, E. Quantentheoretische Beiträge zum Problem der aromatischen und ungesttigten Verbindungen. III. *Z. Physik*, **1932**, *76*, 628–648. DOI: 10.1007/BF01341936

7. Doering, W. von E.; Detert, F.L. Cycloheptatrienylium oxide. *J. Am. Chem. Soc.*, **1951**, *73* (2), 876–877. DOI: 10.1021/ja01146a537

8. Breslow, R.; Brown, J.; Gajewski, J.J. Antiaromaticity of cyclopropenyl anions. *J. Am. Chem. Soc.*, **1967**, *89* (17), 4383–4390. DOI: 10.1021/ja00993a023

9. i) Kolbe, H. Ueber Synthese der Salicylsäure. *Liebigs Ann.*, **1860**, *113* (1), 125–127. DOI: 10.1002/jlac.18601130120

   ii) Schmitt, R. Beitrag zur Kenntniss der Kolbe'schen Salicylsäure Synthese. *J. Prakt. Chem.*, **1885**, *31* (1), 397–411. DOI: 10.1002/prac.18850310130

26 Aromatic Substitution

As established in the previous chapter, benzene possesses unique stability. However, benzene, and other aromatic molecules, are not unreactive. There is indeed a broad and rich chemistry associated with aromatic substitution reactions. In this chapter, we will look at electrophilic aromatic substitution (EAS), nucleophilic aromatic substitution, and radical aromatic substitution of benzene and its derivatives.

ELECTROPHILIC AROMATIC SUBSTITUTION REACTIONS

Most of this chapter will focus on electrophilic aromatic substitution (EAS) reactions. In general, this process occurs through the addition of a strongly electrophilic reagent (E[+]), which substitutes one of the hydrogen atoms on the ring (Figure 26.1).[1] As shall be seen, the mechanism for substitution on the benzene ring is similar from one reaction to the next. The reactions differ in the formation of E[+].

The key intermediate generated in this process (Figure 26.2) is generally called an arenium ion (or in the case where it is unsubstituted benzene the benzenium ion).[2] This has also, historically, been known as a Pfeiffer–Wizinger complex,[3] a Wheland intermediate,[4] a Brown σ complex,[5] and a benzenonium ion.[6] Understanding the stability of this intermediate will be central to understanding the regiochemical outcome of multiple EAS reactions.

HALOGENATION

The first EAS reactions that we will consider are chlorination and bromination. The chlorination of benzene will no proceed with chlorine by itself, and it requires the addition of a Lewis acid (AlCl$_3$ or FeCl$_3$) to proceed (Figure 26.3).[7] It should be noted that for many EAS reactions of benzene or simple derivatives the reactant is also the solvent (and no solvent will be indicated below the arrow). Otherwise, nitromethane and DCM are common solvents in EAS reactions.

Problem 26.1 Draw the appropriate arrows to complete the mechanism.

Generation of the electrophile (E[+])

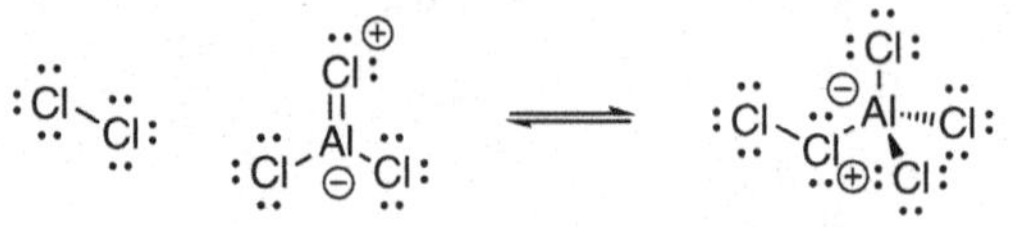

Electrophilic aromatic substitution

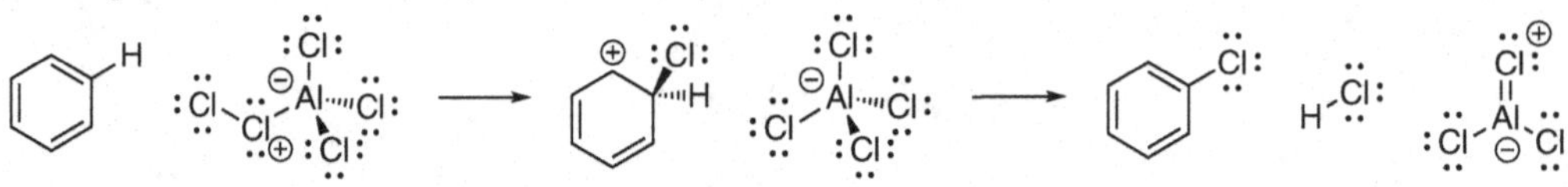

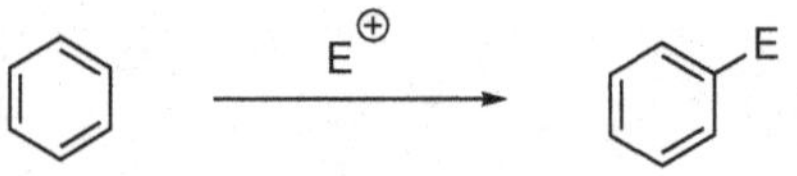

Figure 26.1 General reaction of electrophilic aromatic substitution (EAS) where E[+] is a strongly electrophilic reagent (usually generated in the reaction mixture).

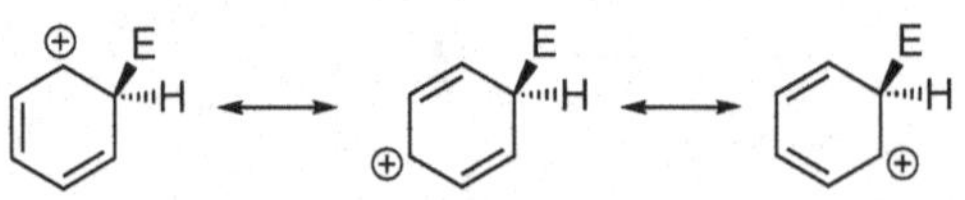

Figure 26.2 Contributing structures of the key arenium ion intermediate formed in EAS reactions.

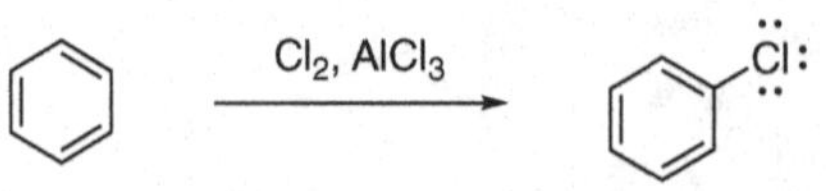

Figure 26.3 Chlorination of benzene.

DOI: 10.1201/9781003479352-26

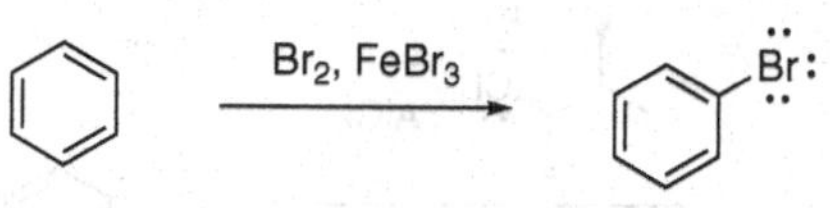

Figure 26.4 Bromination of benzene.

Bromination of benzene, like chlorination, will not proceed without an added Lewis acid catalyst. The most common Lewis acid for bromination is $FeBr_3$ (Figure 26.4), $AlCl_3$ and $FeCl_3$ can also be used.[8] As you review the mechanism in Problem 26.2, it should look very similar to the chlorination mechanism in Problem 26.1.

Problem 26.2 Draw the appropriate arrows to complete the mechanism.

Generation of the electrophile (E^+)

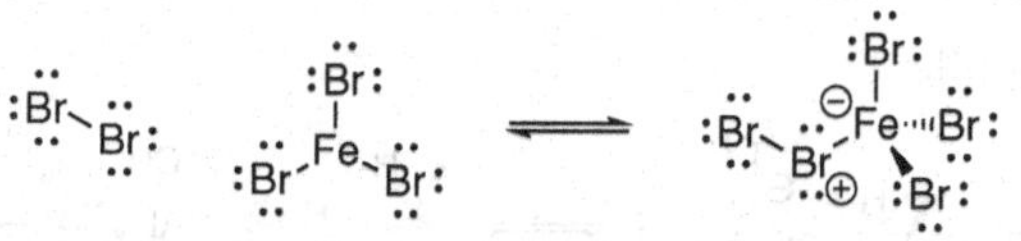

Electrophilic aromatic substitution

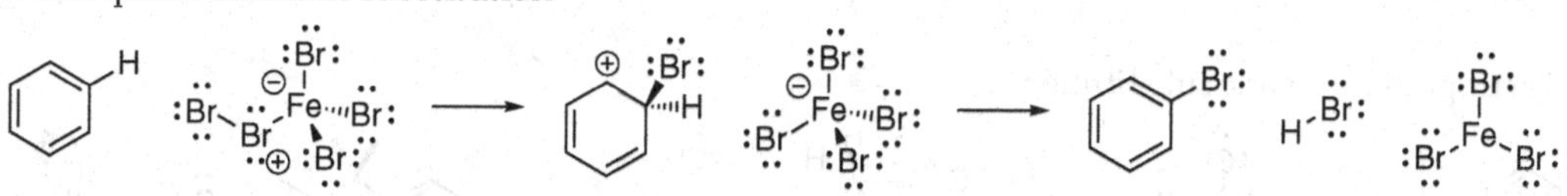

NITRATION

Nitration of benzene (Figure 26.5) is one of the oldest reactions of benzene, which was first reported in 1834, only nine years after benzene's discovery.[9] The reaction involves the combination of nitric acid and sulfuric acid to generate the strongly electrophilic nitronium ion (NO_2^+). As we saw with the Henry reaction, the nitro group can be reduced to give an amine (aminobenzene is called aniline).

Problem 26.3 Draw the appropriate arrows to complete the mechanism.

Generation of the electrophile (E^+)

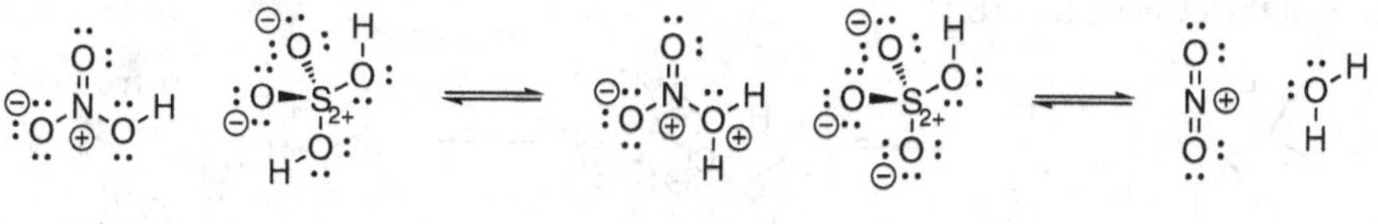

Electrophilic aromatic substitution

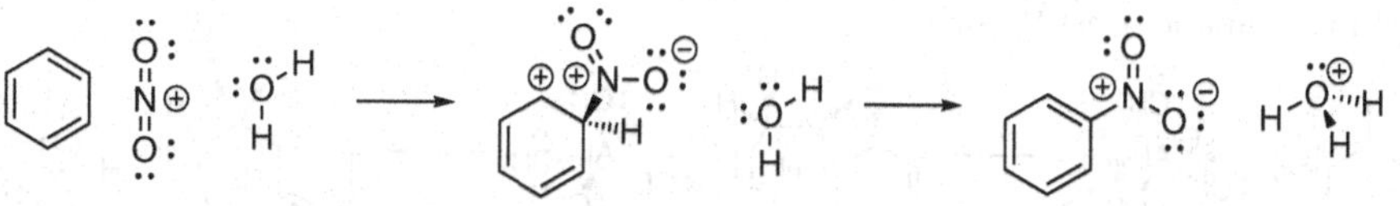

FRIEDEL–CRAFTS ALKYLATION

There are two common methods for substituting a benzene ring with carbon chains. The first that we will consider is the Friedel–Crafts alkylation (Figure 26.6).[10] This reaction uses a strong Lewis acid ($AlCl_3$ or $FeCl_3$) with an alkyl chloride to generate a carbocation electrophile. Carbocation intermediates can also be generated by adding an alkene with a strong Brønsted-Lowry acid (H_2SO_4). Because a carbocation intermediate is generated, carbocation rearrangements can and

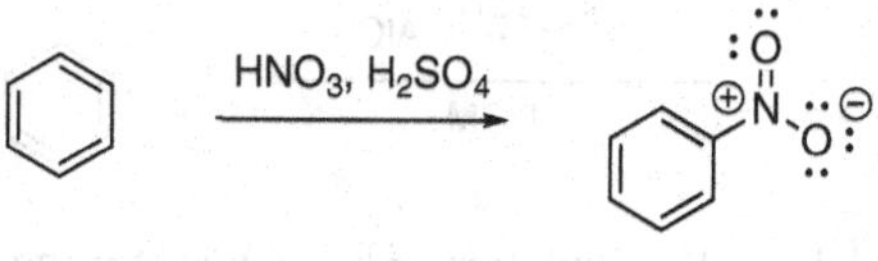

Figure 26.5 Nitration of benzene.

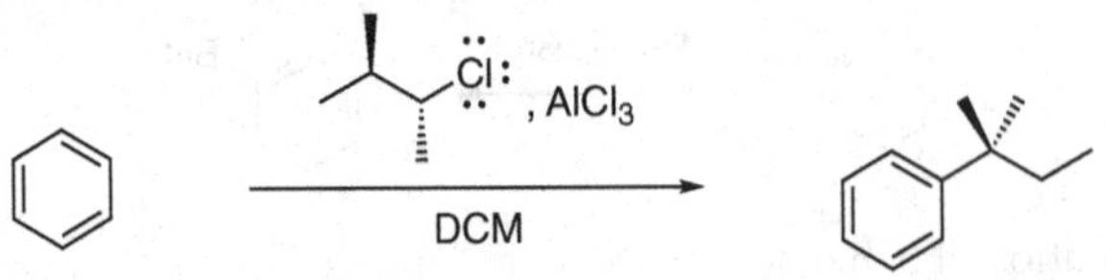

Figure 26.6 Example Friedel–Crafts alkylation of benzene showing carbon skeleton rearrangement.

will occur. In addition, this reaction has the limitation that it does not work with benzene rings that have electron-withdrawing groups – carbonyls, sulfonates, nitriles, trifluoromethyl, or trichloromethyl – attached.

Problem 26.4 Draw the appropriate arrows to complete the mechanism.

Generation of the electrophile (E^+)

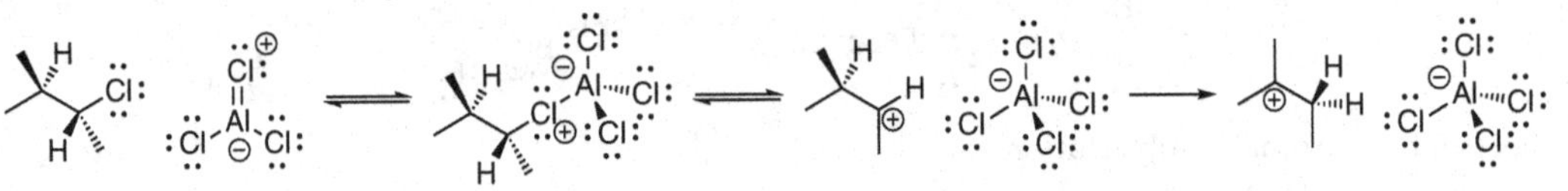

Electrophilic aromatic substitution

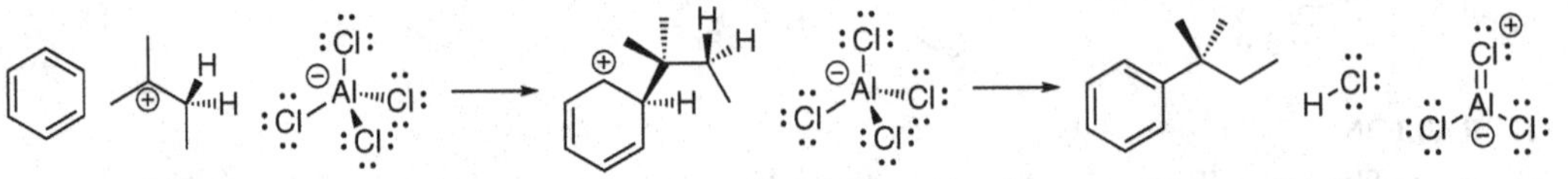

A significant disadvantage of the Friedel–Crafts alkylation is that it cannot be used to attach a primary carbon atom to the benzene ring. Primary alkyl chlorides will undergo skeletal rearrangements to give products where a secondary or tertiary carbon atom is the site of attachment (Figure 26.7). Considering the mechanism in Problem 26.5, you can see that this is the result of concomitant loss of the $AlCl_4$ leaving group and skeletal rearrangement.

Problem 26.5 Draw the appropriate arrows to complete the mechanism.

Generation of the electrophile (E^+)

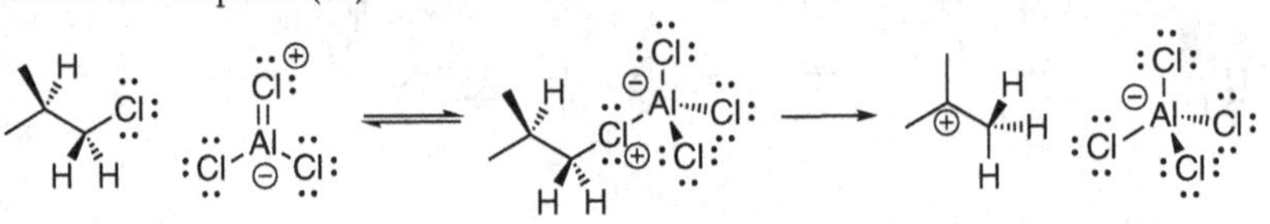

Electrophilic aromatic substitution

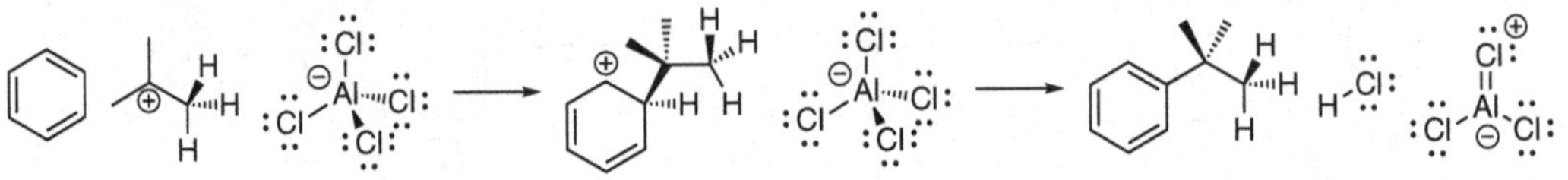

FRIEDEL–CRAFTS ACYLATION

The second method of substituting benzene rings with carbon chains is the Friedel–Crafts acylation reaction (Figure 26.8).[11] Here an acid chloride (or acid anhydride) is used to generate the strongly electrophilic acylium ion ($R-C\equiv O^+$). The acylium ion does not rearrange, unlike a

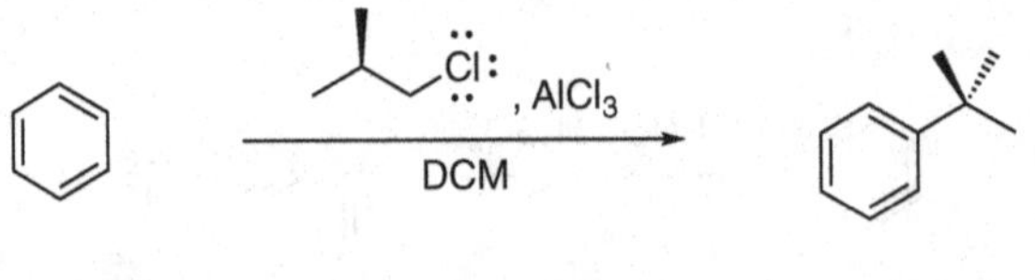

Figure 26.7 Example Friedel–Crafts alkylation of benzene showing carbon skeleton rearrangement of a primary alkyl chloride.

carbocation in the Friedel–Crafts alkylation. The reaction also requires a water workup step to remove the aluminium.

Problem 26.6 Draw the appropriate arrows to complete the mechanism.

Generation of the electrophile (E⁺)

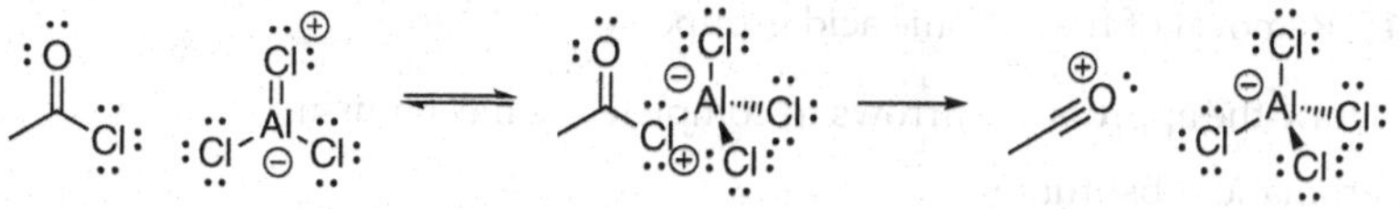

Electrophilic aromatic substitution

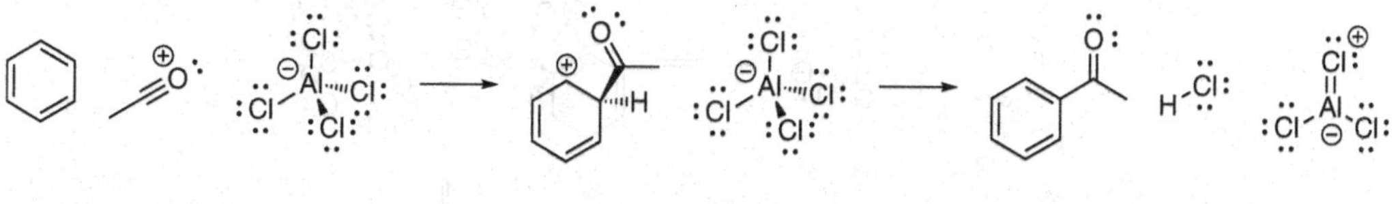

The Friedel–Crafts acylation reaction also offers a means of overcoming a limitation of the Friedel–Crafts alkylation: attaching a primary carbon to the benzene ring. First, a Friedel–Crafts acylation adds a carbonyl to the benzene ring and then Wolff-Kishner reduction reduces the carbonyl to an alkane (Figure 26.9). Contrast the results of the reactions in Figure 26.7 and in Figure 26.9. It is not uncommon to see the approach used in EAS problems that otherwise would be unsolvable.

SULFONATION

The final EAS reaction that we will consider is the sulfonation reaction (Figure 26.10).[10] Sulfonation is another very old reaction (reported in 1834). Sulfonation of benzene proceeds by mixing benzene with a combination of sulfur trioxide (the strongly electrophilic species) and sulfuric acid. This combination is also referred to as fuming sulfuric acid and, in older literature, as oleum.

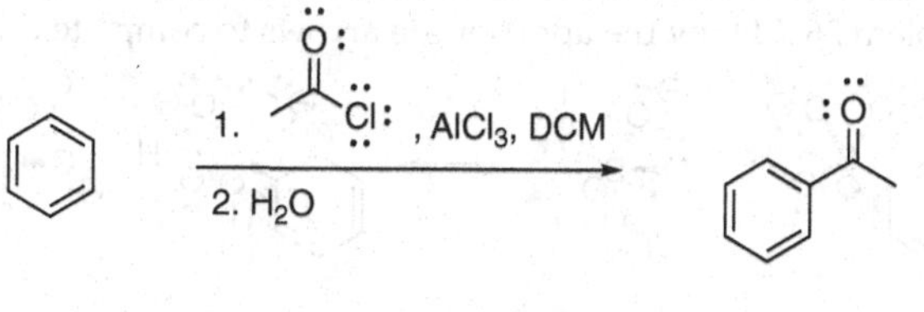

Figure 26.8 Friedel–Crafts acylation of benzene with acetyl chloride.

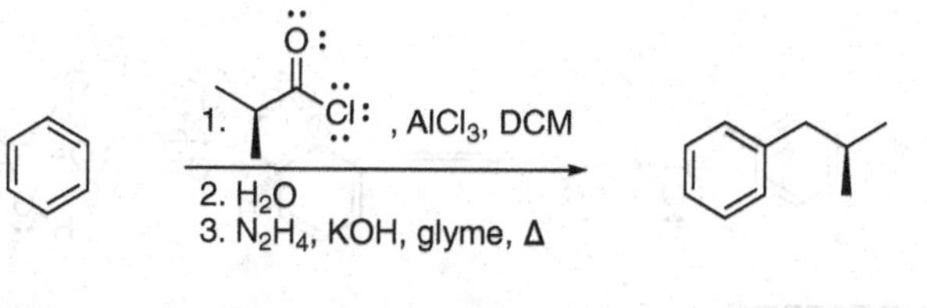

Figure 26.9 Friedel–Crafts acylation of benzene and Wolff-Kishner reduction, formally the attachment of a primary carbon atom to the ring.

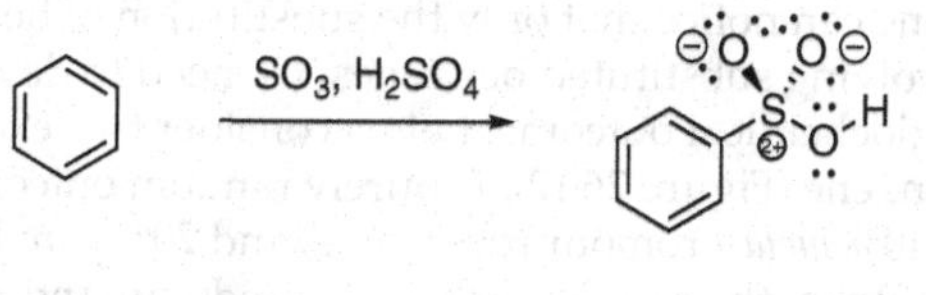

Figure 26.10 Sulfonation of benzene.

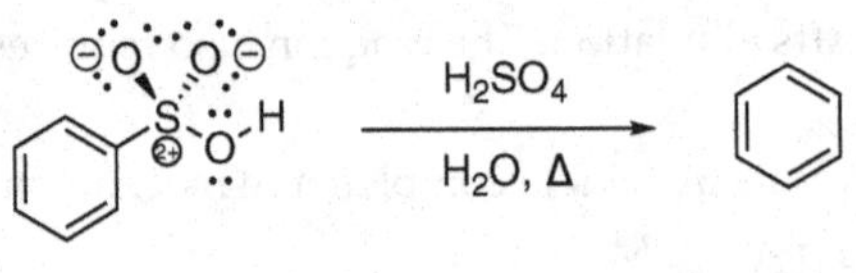

Figure 26.11 Removal of the sulfonic acid group.

Problem 26.7 Draw the appropriate arrows to complete the mechanism.

Electrophilic aromatic substitution

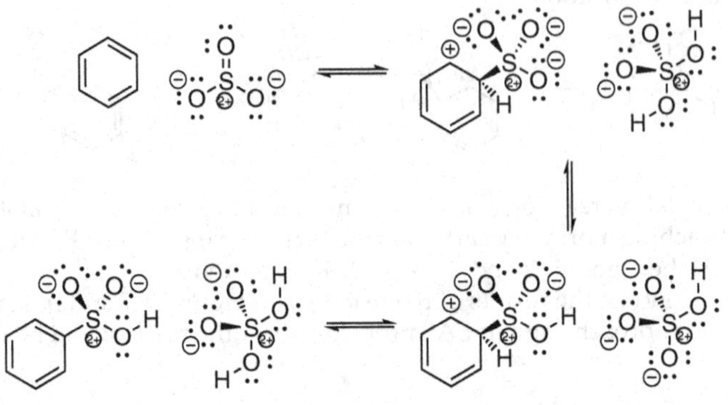

Sulfonation of benzene is unique among all the EAS reactions that we have considered because it is reversible. Upon heating benzene sulfonic acids in a mixture of sulfuric acid and water, the sulfonic acid group can be removed (Figure 26.11). This reversibility of sulfonation will be useful as a blocking/protecting group in synthesis.

Problem 26.8 The mechanism for the removal of the sulfonic acid group is the exact reverse of the mechanism in Problem 26.7. Draw the appropriate arrows to complete the mechanism.

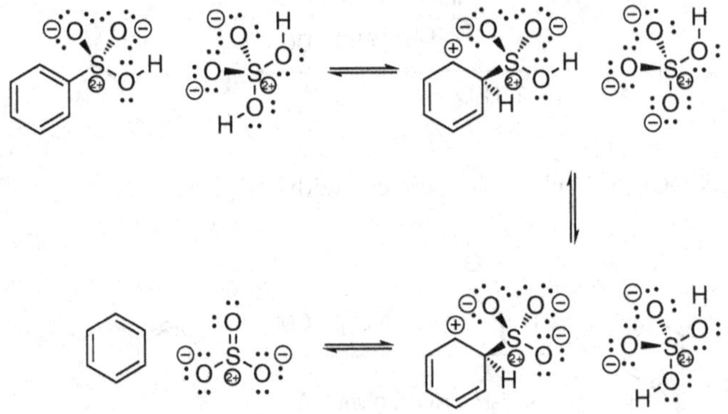

DIRECTING GROUP EFFECTS

Above are all the EAS reactions that we will consider in this chapter. There are other EAS reactions, but the reactions presented in this chapter are some of the most common. Reviewing the reaction examples above, one can notice that only the substitution of benzene itself was considered. For EAS reactions involving substituted benzenes, we need to develop an understanding of what influences their regiochemical outcome. Let us consider two examples. The first example is the nitration of bromobenzene (Figure 26.12). A purely random outcome would predict a 40% *ortho*-bromonitrobenzene, 40% *meta*-bromonitrobenzene, and 20% *para*-bromonitrobenzene. Instead, we see (Figure 26.12) that the reaction produces predominantly *ortho*-bromonitrobenzene and *para*-bromonitrobenzene.

The second example to consider is the bromination of nitrobenzene (Figure 26.13), the inverse of the reaction in Figure 26.12. Again, a purely random outcome would predict a 40% *ortho*-bromonitrobenzene, 40% *meta*-bromonitrobenzene, and 20% *para*-bromonitrobenzene. Instead, we see (Figure 26.13) that the reaction produces predominantly *meta*-bromonitrobenzene.

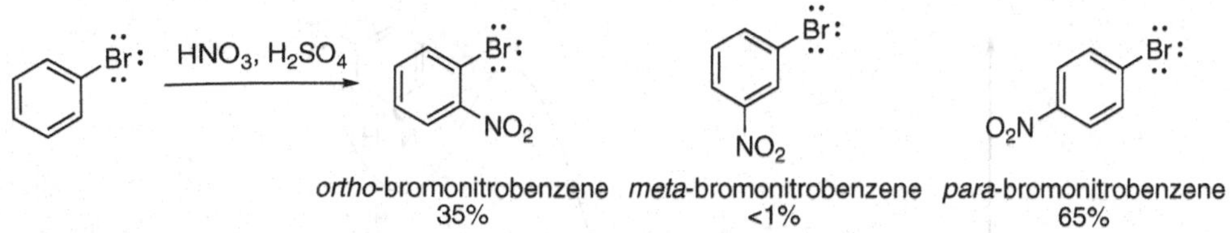

ortho-bromonitrobenzene   meta-bromonitrobenzene   para-bromonitrobenzene
35%   <1%   65%

Figure 26.12   Nitration of bromobenzene with product distribution. Coste, J.H.; Parry, E.J. Ueber die Nitrirung von Brombenzol. *Ber. Dtsch. Chem. Ges.*, **1896**, *29* (1), 788–792. DOI: 10.1002/cber.189602901142

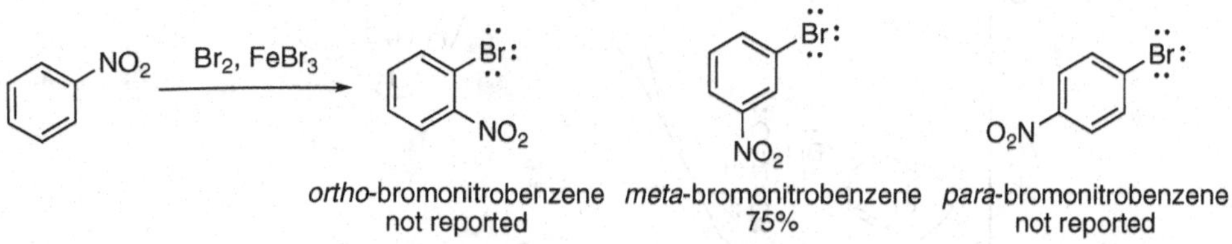

ortho-bromonitrobenzene   meta-bromonitrobenzene   para-bromonitrobenzene
not reported   75%   not reported

Figure 26.13   Bromination of nitrobenzene with an amount of *meta*-bromonitrobenzene product (the amount of *ortho*- and -*para* product was not reported). Scheufelen, A. Ueber Eisenverbindungen als Bromüberträger. *Liebigs Ann.*, **1885**, *231* (2), 152–195. DOI: 10.1002/jlac.18852310204

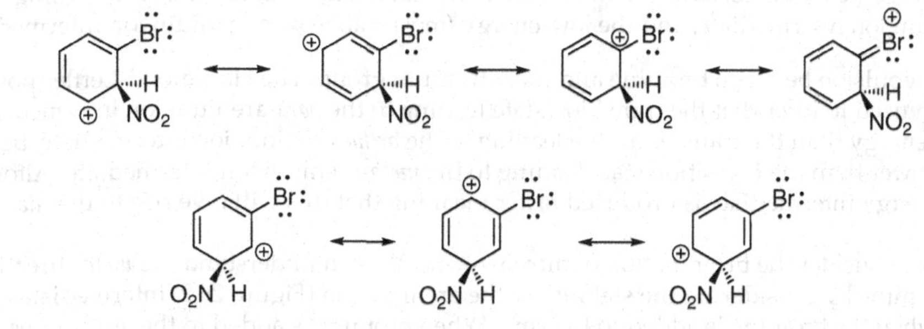

Figure 26.14   Arenium ions formed from the addition of nitronium to the *ortho*- (*top*), meta- (*middle*), and *para*-positions of bromobenzene.

The two example reactions (Figure 26.12 and Figure 26.13) produce the same product (bromonitrobenzene) but different regiochemical outcomes: nitration of bromobenzene gives predominantly *ortho*- and *para*-products while bromination of nitrobenzene gives predominantly the *meta*-product. Therefore, we can say that the group already on the ring determines where the incoming electrophile ends up (the incoming electrophile has no role in determining its location).

Let us first consider the nitration of bromobenzene. We can understand the *ortho*-/*para*- directing effect of bromine by considering the stability of the arenium ion (Figure 26.2) intermediates that form when the nitronium group adds to the ring. When the nitronium is added to the *ortho*- or *para*-positions, the bromine can help to stabilize the positive charge through lone-pair donation (Figure 26.14). In contrast, when the nitronium adds to the *meta*-position the bromine cannot help to stabilize the arenium ion.

Now if a reaction coordinate diagram is considered (Figure 26.15), the arenium ions where the nitronium has added to the *ortho*- and *para*-position lie energetically below the arenium ion where the nitronium has added to the *meta*-position. The arenium ion where the nitronium added to the *para*-position is lower in energy than the *ortho*-intermediate because the *para*-intermediate does not

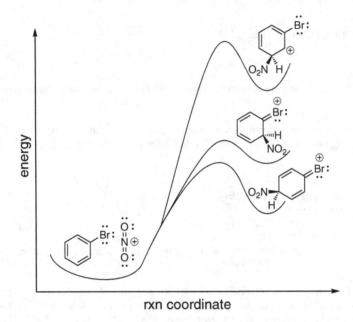

Figure 26.15 A reaction coordinate diagram showing the different energy pathways leading to the high-energy (least stable) *meta*-arenium ion intermediate, the middle-energy (middling stable) *ortho*-arenium ion intermediate, and the low-energy (most stable) *para*-arenium ion intermediate.

have steric repulsion between bromine and the nitronium group. The Hammond-Leffler postulate[11] is then used to infer that the transition state leading to the *para*-arenium ion intermediate is lower in energy than the transition state leading to the *ortho*-arenium ion intermediate, both of which are lower than the transition state leading to the *meta*-arenium ion intermediate. Altogether, the lower energy intermediate is produced faster meaning that this will give rise to the major product(s).

Let us now consider the bromination of nitrobenzene. We can understand the *meta*-directing effect of bromine by considering the stability of the arenium ion (Figure 26.2) intermediates that form when the bromine is added to the ring. When bromine is added to the *ortho*- or *para*-positions, the arenium ion intermediate is destabilized by the positive charge being adjacent to the electron-withdrawing nitro group (Figure 26.16). In contrast, when bromine adds to the

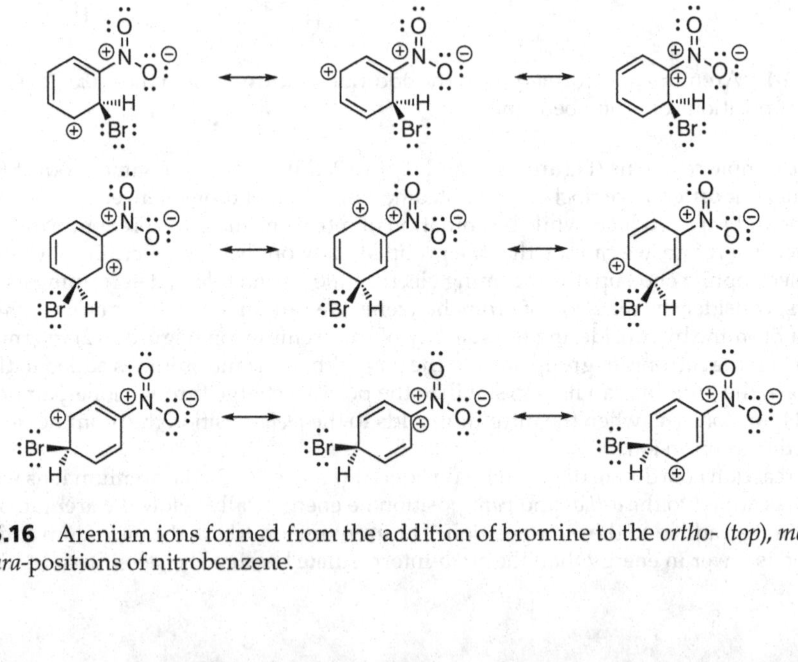

Figure 26.16 Arenium ions formed from the addition of bromine to the *ortho*- (*top*), *meta*- (*middle*), and *para*-positions of nitrobenzene.

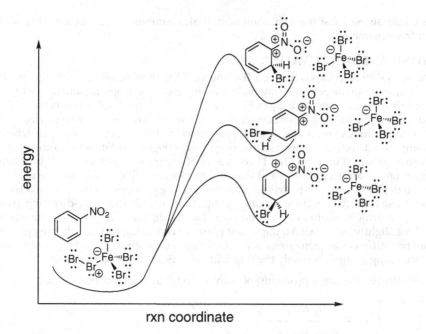

Figure 26.17 A reaction coordinate diagram showing the different energy pathways leading to the high-energy (least stable) *ortho*-arenium ion intermediate, the middle-energy (middling stable) *para*-arenium ion intermediate, and the low-energy (most stable) *meta*-arenium ion intermediate.

meta-position the positive charge is not immediately adjacent to the electron-withdrawing nitro group and so it is not significantly destabilized.

Now if a reaction coordinate diagram is considered (Figure 26.17), the arenium ions where bromine has been added to the *ortho*- and *para*-position lie energetically above the arenium ion where bromine has been added to the *meta*-position. The arenium ion where bromine is added to the *para*-position is lower in energy than the *ortho*-intermediate because the *para*-intermediate does not have steric repulsion between bromine and the nitronium group. The Hammond-Leffler postulate[14] is then used to infer that the transition state leading to the *meta*-arenium ion intermediate is lower in energy than the transition state leading to the *para*-arenium ion intermediate, both of which are lower than the transition state leading to the *ortho*-arenium ion intermediate. Altogether, the lower energy intermediate is produced faster meaning that this will give rise to the major product(s).

Different substituents on benzene are either *ortho*- and *para*-directing groups (like bromine) or they are *meta*-directing groups (like the nitro group).[12] In general, *ortho*- and *para*-directing groups (Figure 26.18) have a lone pair on the atom immediately adjacent to the benzene ring (halogen atoms, nitrogen atoms, oxygen atoms, and sulfur atoms) or they donate electron density through hyperconjugation (alkyl substituents). *Meta*-directing groups (Figure 26.19) either have a conjugated withdrawing group (a nitro group, nitrile group, or carbonyl) or they have a strong,

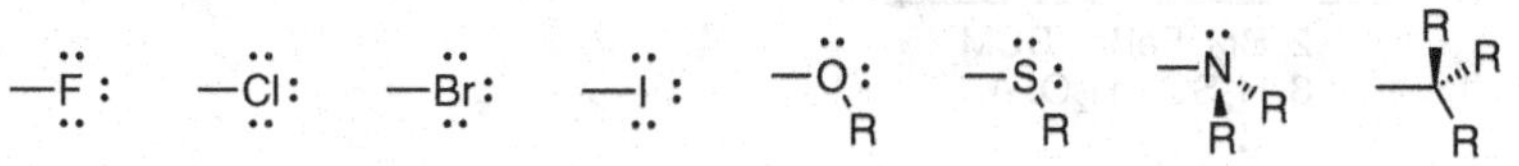

Figure 26.18 *Ortho*- and *para*-directing groups.

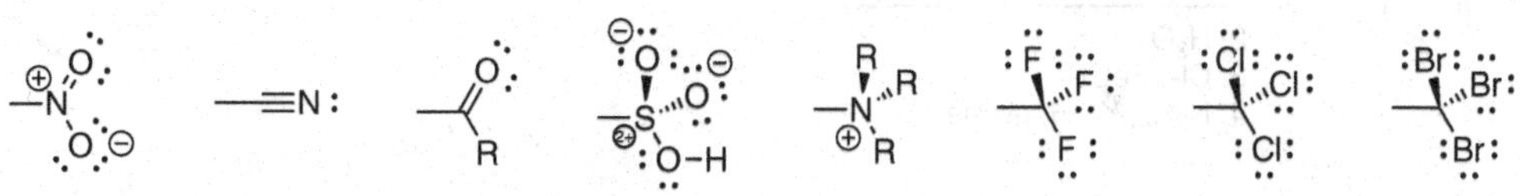

Figure 26.19 *Meta*-directing groups.

inductive withdrawing substituent (sulfonic acid, trialkylammonium, trifluoromethyl, trichloro-methyl, or tribromomethyl).

ACTIVATING AND DEACTIVATING GROUPS

The last topic in EAS reactions is that of activating and deactivating groups. Substituents that add electron density to the benzene ring make the ring more nucleophilic, and this activates the benzene ring to react more quickly in EAS reactions. In contrast, substituents that withdraw electron density make the ring less electrophilic, and this deactivates the benzene ring, which reacts more slowly in EAS reactions. The division between the two groups nearly matches up with the *ortho*- and *para*-directing groups as activating groups (Figure 26.20) and *meta*-directing groups as deactivating groups (Figure 26.21).[13] There is a set of groups which crosses over between the two designations: chlorine, bromine, and iodine. Chlorine, bromine, and iodine are poor electron donors (though good enough to be *ortho*- and *para*-directing groups) and are quite electronegative. Altogether, these three (Cl, Br, I) are the only groups that are *ortho*- and *para*-directing groups and deactivating. Fluorine is a better electron donor than Cl, Br, and I and so it is as nucleophilic as benzene if not slightly more so.[14] An important point for activating and deactivating groups is that if there are two different benzene rings in one molecule (one with an activating group and one with a deactivating group), then only the ring with an activating group will react.

Problem 26.9 Predict the major product(s) of each of the following reactions.

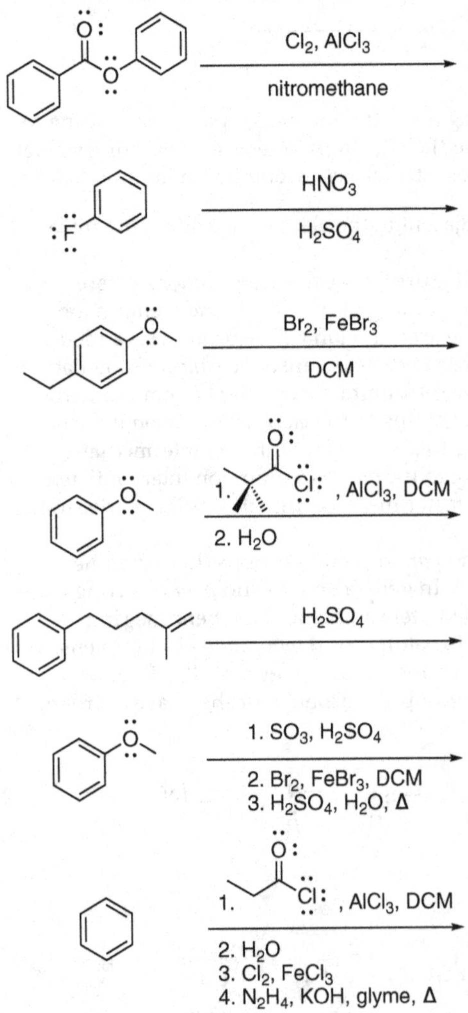

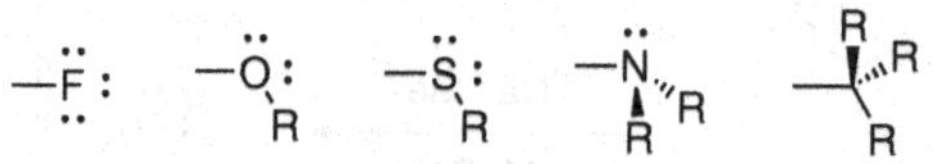

Figure 26.20 Activating groups.

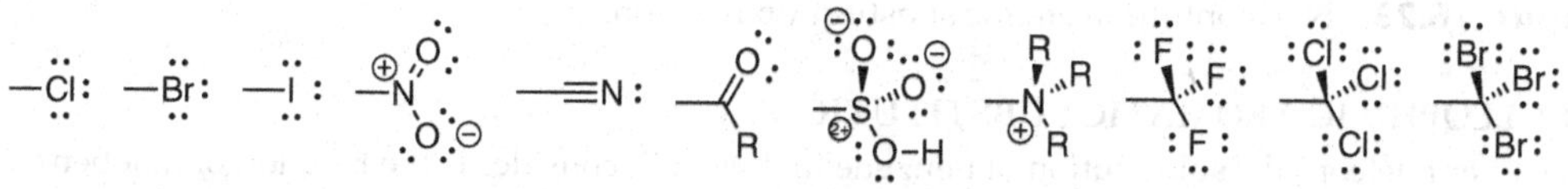

Figure 26.21 Deactivating groups.

BENZYNE CHEMISTRY

We will now turn our attention to nucleophilic substitutions of benzene. The first nucleophilic substitution reaction that we will consider is the reaction of halobenzenes – without electron-withdrawing groups – with strongly basic nucleophiles like potassium hydroxide (KOH) or sodium amide (NaNH$_2$). Consider the substitution reactions in (Figure 26.22) of 4-methylbromide. Note to proceed with hydroxide as the base the two reagents are melted together (fuse) at a very high temperature.

These substitution reactions produce phenols (hydroxy-substituted benzene) and anilines (amino-substituted benzene), which we cannot produce, directly, with the EAS reactions we saw above. The mechanism combines aspects that we have seen but are unique in combination: E2 elimination produces a benzyne intermediate and then the base adds to the strained alkyne-like intermediate to produce the substitution (and mix of regiochemical products) observed.[15]

Problem 26.10 Draw the appropriate arrows to complete the mechanism.

Generation of benzyne

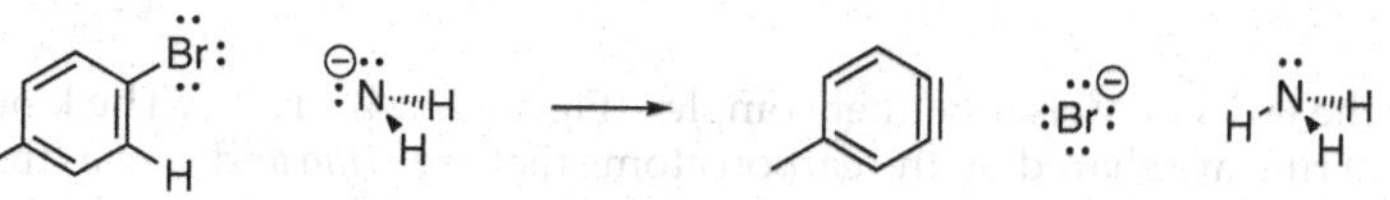

Production of one regiochemical outcome

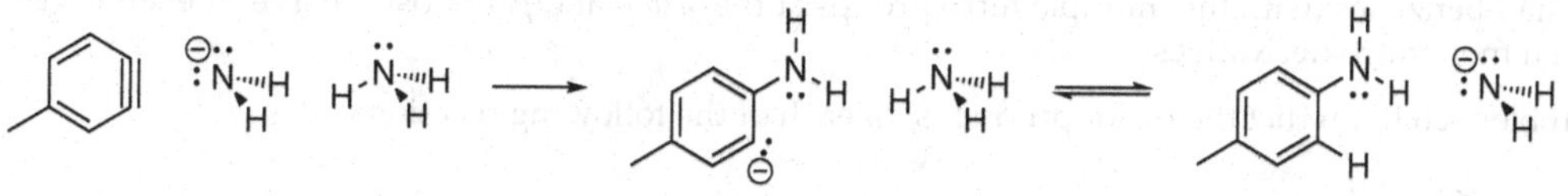

Production of the other regiochemical outcome

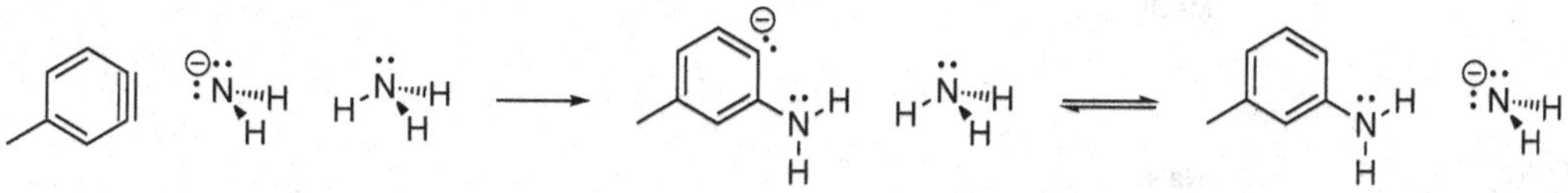

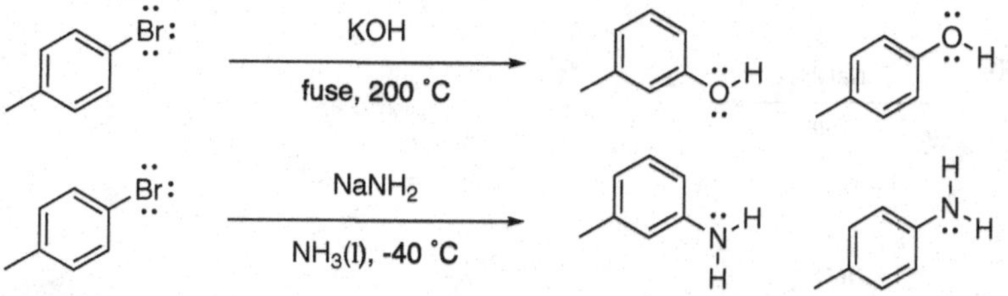

Figure 26.22 Benzyne substitution of halobenzene with potassium hydroxide (*top*) sodium amide (*bottom*).

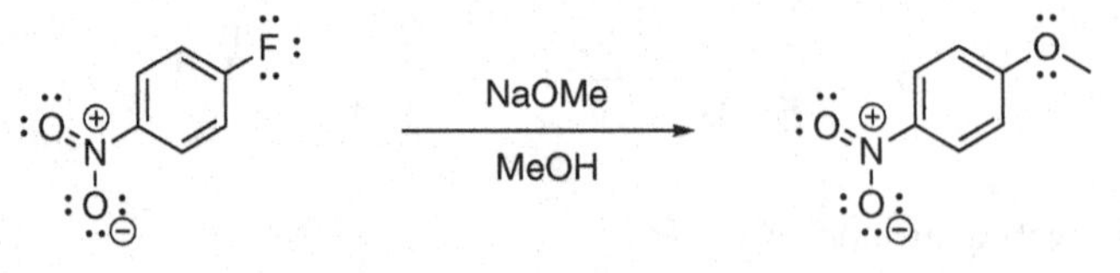

Figure 26.23 Nucleophilic aromatic substitution reaction.

NUCLEOPHILIC AROMATIC SUBSTITUTION

The other nucleophilic substitution of benzene that we will consider is the reaction of halobenzenes with electron-withdrawing groups (Figure 26.23). Nitro groups are the most common electron-withdrawing group in these reactions, though any other deactivating group could be used. This is called a nucleophilic aromatic substitution reaction (S_NAr).[16] Unlike the benzyne reaction (Figure 26.22), this reaction is regiospecific and only substitutes the existing halogen atom and no other positions on the ring.

As you consider the mechanism of the S_NAr reaction in Problem 26.11, you should notice the similarities between this reaction mechanism and the mechanism for nucleophilic acyl substitution from Chapter 21. First, a nucleophile attacks a planar, sp^2-hybridized carbon atom to generate a tetrahedral intermediate – here called the Jackson-Meisenheimer complex[17] – and then loss of the leaving group reforms the stable functional group (a carbonyl in nucleophilic acyl substitution and a benzene ring here).

Problem 26.11 Draw the appropriate arrows to complete the mechanism.

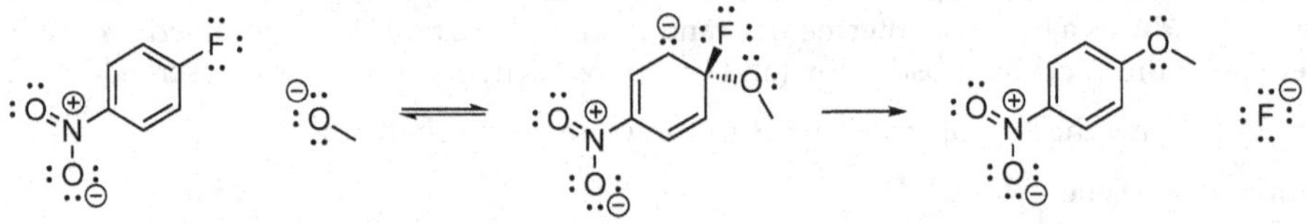

Let us consider the Jackson-Meisenheimer complex (Figure 26.24) further. The lone pair and negative charge in the ring are shared by the carbon atoms that are *ortho* and *para* to the tetrahedral carbon. As such, it makes sense that nitro-group substitution at the *ortho*- and *para*-positions helps to stabilize the negative charge thus making the reaction faster. Dinitro and trinitro derivatives of halobenzenes (with the multiple nitro groups at the *ortho*- and *para*-positions) react even faster than mononitro derivatives.

Problem 26.12 Predict the major product(s) of each of the following reactions.

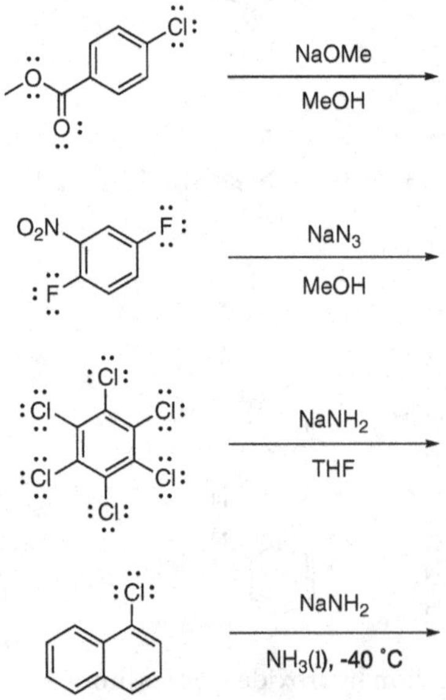

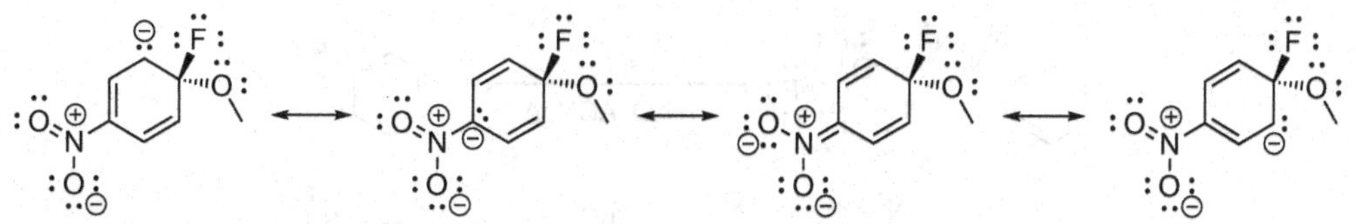

Figure 26.24 Jackson-Meisenheimer complex contributing structures.

AROMATIC RADICAL SUBSTITUTION REACTIONS

Phenyldiazonium ions (Figure 26.25) are used as the reagents in radical substitution reactions.

Phenyl diazonium ions are synthesized from anilines (aminobenzenes) using a combination of sodium nitrite ($NaNO_2$) and acid (Figure 26.26).[18] These reactions usually use water as the solvent (or in combination with water-miscible solvents) to avoid decomposition of the unstable diazonium salt. The diazotization mechanism, while long, should look familiar, in pieces, to lots of different mechanisms we have seen through this book.

Problem 26.13 Draw the appropriate arrows to complete the mechanism.

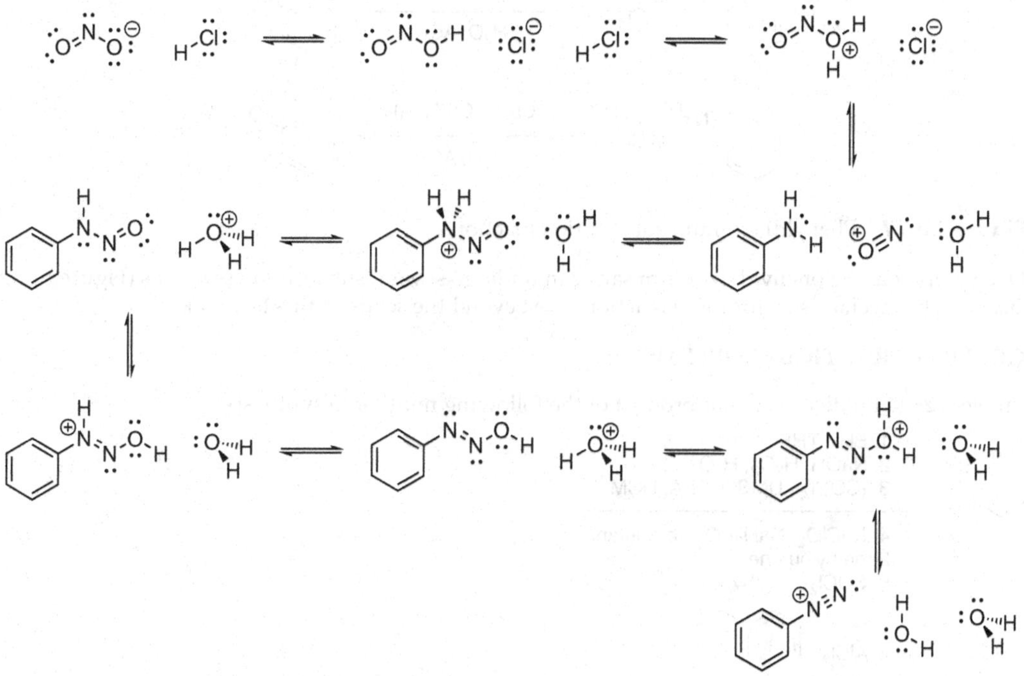

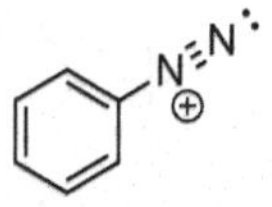

Figure 26.25 Phenyldiazonium.

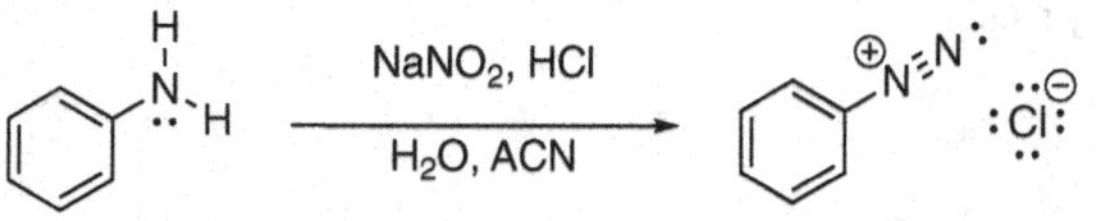

Figure 26.26 Synthesis of phenyldiazonium chloride (the anion counterion identity is determined by the acid used).

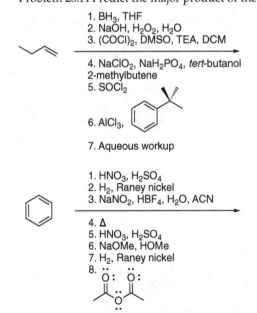

Figure 26.27 Phenyldiazonium substitution reactions.

Once prepared, the phenyldiazonium salts can undergo several substitution reactions (Figure 26.27).[19] The mechanisms for these reactions are beyond the scope of this textbook.

GENERAL PRACTICE PROBLEMS

Problem 26.14 Predict the major product of the following multi-step synthesis.

1. BH_3, THF
2. NaOH, H_2O_2, H_2O
3. $(COCl)_2$, DMSO, TEA, DCM

4. $NaClO_2$, NaH_2PO_4, *tert*-butanol
2-methylbutene
5. $SOCl_2$

6. $AlCl_3$,

7. Aqueous workup

1. HNO_3, H_2SO_4
2. H_2, Raney nickel
3. $NaNO_2$, HBF_4, H_2O, ACN

4. Δ
5. HNO_3, H_2SO_4
6. NaOMe, HOMe
7. H_2, Raney nickel
8.

Problem 26.15 Provide the necessary reagents to complete the synthesis.

Problem 26.16 Draw a reaction mechanism that accounts for the following transformation.

NOTES

1. Olah, G.A. Aromatic Substitution. XXVIII. Mechanism of electrophilic aromatic substitutions. *Acc. Chem. Res.*, **1971**, *4* (7), 240–248. DOI: 10.1021/ar50043a002

2. Olah, G.A. Stable carbocations. CXVIII. General concept and structure of carbocations based on differentiation of trivalent (classical) carbenium ions from three-center bound penta- of tetracoordinated (nonclassical) carbonium ions. Role of carbocations in electrophilic reactions. *J. Am. Chem. Soc.*, **1972**, *94* (3), 808–820. DOI: 10.1021/ja00758a020

3. Pfeiffer, P.; Wizinger, R. Zur Theorie der Halogensubstitution. *Liebigs Ann.*, **1928**, *461* (1), 132–154. DOI: 10.1002/jlac.19284610108

4. Wheland, G.W. A Quantum mechanical investigation of the orientation of substituents in aromatic molecules. *J. Am. Chem. Soc.*, **1942**, *64* (4), 900–908. DOI: 10.1021/ja01256a047

5. Brown, H.C.; Brady, J.D. Solubility of hydrogen chloride at low temperatures. A measure of the basic properties of aromatic nuclei; p- and s-complexes and their role in aromatic substitution. *J. Am. Chem. Soc.*, **1952**, *74* (14), 3570–3582. DOI: 10.1021/ja01134a032

6. Doering, W.v.E.; Saunders, M.; Boynton, H.G.; Earhart, H.W.; Wadley, E.F.; Edwards, W.R.; Laber, G. The 1,1,2,3,4,5,6-heptamethylbenzenonium ion. *Tetrahedron*, **1958**, *4* (1–2), 178–185. DOI: 10.1016/0040-4020(58)88016-3

7. Mouneyrat, A.; Pouret, C. Chloruraion de la benzine en presence du chlorure d'aluminium. *C. R. Acad. Sci*, **1898**, *127*, 1025–1028.

8. Couper, A. Ueber einige Derivate des Benzols. *Liebigs Ann.*, **1857**, *104* (2), 225–227. DOI: 10.1002/jlac.18571040217

9. Mitscherlich, E. Ueber das Benzin und die Verbindungen desselben (1834) in *Ostwald's Klassiker Der Exakten Wissenschaften Nr. 98*. Wislicenus, J., Ed.; Wilhelm Engelmann: Leipzig. 1898. 1–39.

10. Friedel, C.; Crafts, J.-M. Sur une méthode Générale nouvelle de synthèse d'hydrocarbures, d'acétones, etc. *C. R. Acad. Sci.*, **1877**, *84*, 1450–1454.

11. i)   Leffler, J.E. Parameters for the description of transition states. *Science*, **1953**, *117*, 340–341. DOI: 10.1126/science.117.3039.340

ii) Hammond, G.S. A correlation of reaction rates. *J. Am. Chem. Soc.*, **1955**, *77* (2), 334–338. DOI: 10.1021/ja01607a027

12. Brown, A.C.; Gibson, J. XXX.–A rule for determining whether a given benzene mono-derivative shall give a meta-di-derivative or a mixture of ortho- and para-di-derivatives. *J. Chem. Soc.*, **1892**, *61*, 367–369. DOI: 10.1039/CT8926100367

13. Knowles, J.R.; Norman, R.O.C.; Radda, G.K. 948. A quantitative treatment of electrophilic aromatic substitution. *J. Chem. Soc.*, **1960**, 4885–4896. DOI: 10.1039/JR9600004885

14. Rosenthal, J.; Schuster, D.I. The anomalous reactivity of fluorobenzene in electrophilic aromatic substitution and related phenomena. *J. Chem. Educ.*, **2003**, *80* (6), 679–690. DOI: 10.1021/ed080p679

15. Roberts, J.D.; Simmons, H.E.; Carlsmith, L.A.; Vaughan, C.W. Rearrangement in the reaction of chlorobenzene-1-c^{14} with potassium amide. *J. Am. Chem. Soc.*, **1953**, *75* (13), 3290–3291. DOI: 10.1021/ja01109a523

16. Janovsky, J.V.; Erb, L. Zur Kenntniss der directen Brom- und nitrosubstitutionsproducte der Azokörper. *Ber. Dtsch. Chem. Ges.*, **1886**, *19* (2), 2155–2158. DOI: 10.1002/cber.188601902113

17. i) Jackson, C.L.; Gazzolo, F.H. On certain colored substances derived from nitro compounds. Third paper. *Proceedings of the American Academy of Arts and Sciences*, **1900**, *35* (14), 263–281. DOI: 10.2307/25129930
    ii) Meisenheimer, J. Ueber Reactionen aromatischer Nitrokörper. *Liebigs Ann.*, **1902**, *323* (2), 205–246. DOI: 10.1002/jlac.19023230205

18. Griess, P. Vorläufige Notiz über die Einwirkung von salpetriger Säure auf Amidinitro- und Aminitrophenylsäure. *Liebigs Ann.*, **1858**, *106* (1), 123–125. DOI: 10.1002/jlac.18581060114

19. i) Sandmeyer, T. Ueber die Ersetzung der Amid-gruppe durch Chlor, Brom und Cyan in den aromatischen Substanzen. *Ber. Dtsch. Chem. Ges.*, **1884**, *17* (2), 2650–2653. DOI: 10.1002/cber.188401702202

    ii) Balz, G.; Schiemann, G. Über aromatische Fluorverbindungen, I.: Ein neues Verfahren zu ihrer Darstellung. *Ber. Dtsch. Chem. Ges.*, **1927**, *60* (5), 1186–1190. DOI: 10.1002/cber.19270600539
    iii) Cohen, T.; Dietz Jr., A.G.; Miser, J.R. A simple preparation of phenols from diazonium ions via the generation and oxidation of aryl radicals by copper salts. *J. Org. Chem.*, **1977**, *42* (12), 2053–2058. DOI: 10.1021/jo00432a003

Common Organic Functional Groups and Nomenclature

The following is an example of common organic functional groups, their structures, and their IUPAC names. The part(s) of each name that identifies the functional group (whether prefix or suffix) is bold. Common names are also included in parentheses.

Name	General Structure	Shorthand	Example	IUPAC nomenclature (trivial)
alkane		R–H		ethane
alkene		n/a		ethene (ethylene)
alkyne		n/a		ethyne (acetylene)
arene		Ar		benzene
haloalkane		R–X		bromoethane (ethyl bromide)
alcohol		R–OH		ethanol (ethyl alcohol)
phenol		PhOH		phenol
thiol		R-SH		ethanethiol (ethyl mercaptan)
ether		R–O–R		ethoxyethane (diethyl ether)
sulfide		R–S–R		(ethylsulfanyl)ethane (ethyl sulfide)
epoxide		n/a		oxirane (ethylene oxide)
aldehyde		RCHO		ethanal (acetaldehyde)
ketone		RC(O)R		propan-2-one (acetone)

Name	General Structure	Shorthand	Example	IUPAC nomenclature (trivial)
carboxylic acid		RCO_2H		**ethanoic acid** (acetic acid)
acid anhydride		$RC(O)OC(O)R$		**ethanoic anhydride** (acetic anhydride)
acid chloride		$RC(O)Cl$		**ethanoyl chloride** (acetyl chloride)
ester		RCO_2R		**methyl** **ethanoate** (methyl acetate)
amide		$RC(O)NR_2$		eth**anamide** (acetamide)
nitrile		RCN		eth**anenitrile** (acetonitrile)
amine		RNH_2 (1°) R_2NH (2°) R_3N (3°)		*N*-**methyl**meth**anamine** (dimethylamine)
imine		n/a		eth**animine**
enamine		n/a		*N,N*-**dimethyl**eth**enamine**
nitro		RNO_2		**nitro**ethane
sulfoxide		$RS(O)R$		**methanesulfinyl**methane (dimethyl sulfoxide)
sulfone		RSO_2R		**methanesulfonyl**methane (dimethyl sulfone)

**Note: in any of the above examples, R is being used in its broadest form to include all hydrocarbons and benzene. That is each functional group could be on a benzene ring **

Index

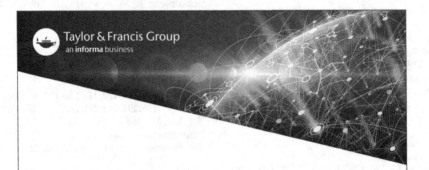

Printed in the United States
by Baker & Taylor Publisher Services